Fundamentals of
Radiation Biology

Fundamentals of Radiation Biology

Susan B Klein
Indiana University, USA

Marc S Mendonca
Indiana University, USA & Purdue University, USA

World Scientific

NEW JERSEY • LONDON • SINGAPORE • BEIJING • SHANGHAI • HONG KONG • TAIPEI • CHENNAI • TOKYO

Published by

World Scientific Publishing Co. Pte. Ltd.
5 Toh Tuck Link, Singapore 596224
USA office: 27 Warren Street, Suite 401-402, Hackensack, NJ 07601
UK office: 57 Shelton Street, Covent Garden, London WC2H 9HE

British Library Cataloguing-in-Publication Data
A catalogue record for this book is available from the British Library.

FUNDAMENTALS OF RADIATION BIOLOGY

ISBN 978-981-125-765-0 (hardcover)
ISBN 978-981-125-891-6 (paperback)
ISBN 978-981-125-766-7 (ebook for institutions)
ISBN 978-981-125-767-4 (ebook for individuals)

For any available supplementary material, please visit
https://www.worldscientific.com/worldscibooks/10.1142/12884#t=suppl

Typeset by Stallion Press
Email: enquiries@stallionpress.com

Dedication

To our teachers and our students

Contents

Preface

We, the authors, each began our journeys as instructors of radiation biology in 1983 at the University of California, Berkeley campus. We were concurrent graduate students, teaching assistants in the Biophysics Department, both learning to lecture under the extraordinary tutelage of Professor Ed Alpen. Over the years, we have traveled separate paths lecturing in many areas of physics, biology, biophysics, and medicine, serendipitously meeting up again at the Indiana University Cyclotron Facility in 1997. Marc had settled in at the Indiana University School of Medicine, Radiation Oncology Department in Indianapolis and Susan was transitioning from the Biology Department to the Physics Department on the Bloomington campus. We each continued to lecture in several areas for the next 23 years teaching one course in common on separate campuses — Radiation Biology. In 2019, we decided to collaborate on this book and bequeath our legacy on to you. The book is laid out very much like our lectures; each chapter builds on the preceding chapters as the fundamentals are patiently revealed. We hope you enjoy the journey of discovery as much as we have enjoyed our combined eight decades of discovering and sharing the complexities of biological responses to radiation.

We have some advice for the reader to help you grasp the information in this text. First, learn to read formulas as if they were statements: sentences or paragraphs. Equations are mathematical shorthand. They not only allow you to find a solution to a problem, such as, "find the potential energy equivalent of an electron at rest" ($E = m_e c^2 = 0.511$ MeV), equations also tell you something about the universe. In this case, the formula tells you that there is a relationship between energy and mass such that mass is proportional to energy, and the constant of proportionality is the speed of light, squared. Therefore, it is possible to convert mass to energy . . . and so on. Volumes have been written about what this formula tells us. As you read this text, we encourage you to spend some time examining the formulas and their development. Think about what they are telling you. Ask yourself questions such as, "What happens when this term goes to zero? What would a graph of this look like? Which term dominates the outcome?" As you can see from Einstein's simple equation for the relationship between energy and mass, a lot of phenomenology can be packed into the pithiest of expressions.

Second, we have provided supplementary information as text enclosed in gray boxes. This gray-framed text offers "a deeper dive" into the topic being discussed with additional examples and explanations. The boxes are intended to answer questions such as, "what would that look like" or "how do they do that?" The reader can skip over these remarks with impunity — but the interested reader may find them beneficial. To answer the inevitable student inquiry, the exam will not hold you responsible for information included within gray boxes.

Pertinent to Chapter 9, this is not a cancer biology text. Our review of cancer biology is intended to be sufficient to enable the reader to understand how radiation induces tumors and influences cancer pathology. The topic of cancer encompasses an enormous body of ever evolving literature representing extensive research, clinical experiences, computational models, and therapeutic endeavors. What we have selected to present herein barely represents the tip of the iceberg and we apologize if our overview misrepresents any aspect of this complex disease. We have also determined that we will not discuss biological responses to cancer therapy in this text, although Chapter 10 examines radiation therapy delivery techniques. We present neither radiation control of tumors nor chemotherapy, immunotherapy, targeted therapy, hormonal therapy, or surgery. Although fascinating and relevant, these topics are not fundamental, they are rather applications of the fundamentals that we hope are provided in this text.

Finally, the references at the end of each chapter have been intentionally limited in hopes of engaging the reader. The list is so far from exhaustive that it resides in a different universe. This is done with purpose. The number of references has been held to an extremely small sampling to encourage students to seek out and peruse these publications; it is a manageable compilation. By-and-large, the selected publications are either reviews or example publications that are both enlightening and enjoyable. It is not possible to include comprehensive information in even the most voluminous of tombs, so we hope that you will avail yourself of these suggested publications to enhance your understanding of topics that piqued your curiosity while reading our text. These publications will lead you, through the references therein, to the sum total of spectacular works that innumerable researchers have provided to enrich our understanding radiation biology. We apologize to all our colleagues that we have disrespected by not citing their excellent contributions but console ourselves in the truth that we have been evenhanded in our exclusions.

A few sample problems are provided at the end of each chapter. These inquiries are intended to provide exercises that will assist students in testing their conceptual understanding of the material and their ability to apply concepts to novel circumstances. Often, you will be required to seek outside resources for information to develop your answer, and you are encouraged to use additional texts, the Internet, and classmate assistance. We have intentionally not provided true/false, multiple choice, and short answer questions intended to test the students' retention. Instructors are encouraged to develop these independently. We have also neglected to provide mathematical examples in the text, although we recognize the utility of such inclusions. Having been classically trained in biophysics, we strongly believe that working through examples is the stuff of great lectures, and so we leave that task to the instructor as well.

We hope you enjoy this book.

Chapter 1

The Origins of Radiation

1.1. Introduction

We begin our investigation of radiation biology by examining the physics of radiation production and absorption. Understanding fundamental physical principles is critical to understanding radiation absorption by biological systems at atomic, molecular, cellular, tissue, and organism levels, and the resultant response to radiation exposure. In general, this text will provide sufficient biological background to allow the reader to understand the radiobiological concept being presented but will refer the reader to other resources for more comprehensive and in-depth reviews of biology, biochemistry, anatomy, and physiology. Thus, the bulk of this text will interrogate the biophysical responses to radiation insult with relevant fundamental biology reviewed on a case by case basis as needed. Let us begin.

Radiation originates one of two ways: radioactive isotopes or machine-produced radiation. The former, radioisotopes, are created by three mechanisms. Primordial radionuclides result from stellar events such as nucleosynthesis during stellar fusion reactions or supernova explosions. Pertinent to our concerns, these isotopes along with their stable element counterparts materialize from planetary dust that collides over millennia and eventually compresses into planets such as Earth, becoming components of our natural environment. The second mechanism derives from the first, as these radionuclides decay to create secondary radionuclides. Uranium-238 represents a primordial radionuclide that decays into secondary radionuclides (Figure 1-1). Cosmogenic radionuclides also are continuously produced by the third mechanism: cosmic rays — high-energy interstellar radiation — pass through our atmosphere inducing showers of secondary particles. The continuous fixed rate production of atmospheric ^{14}C produced by this third mechanism facilitates carbon dating.

1.2. Radiation Produced by Radionuclides

1.2.1. *Nuclear Structure*

In 1932, James Chadwick published his discovery of the neutron (Chadwick, 1932), and for this work, he received the Nobel Prize in Physics in 1935. Based on the description of a massive, uncharged nuclear particle, Werner Heisenberg proposed the proton–neutron model of the atomic nucleus that same year. According to the model, all atomic nuclei are composed of protons (mass = 1.007276 u and charge = +1) and neutrons (mass = 1.008665 u and charge = 0). These particles are compressed together into a roughly spherical space with a diameter of less than 15×10^{-15} m, such that it becomes difficult to explain why the charge repulsion between protons does not cause the nucleus to blow apart. The eventual explanation disclosed a "nuclear force" or "strong force" that acts between two protons, a proton and a neutron, and two neutrons; it counteracts the repulsive Coulombic forces. Examination of the strong force is an active area of nuclear physics research and we will not trouble ourselves with the quantum mechanical details here. For our purposes, a general description of the nucleus and the forces that bind its components suffice for the study of radiation biology. For smaller nuclei, when the proton number equals the neutron

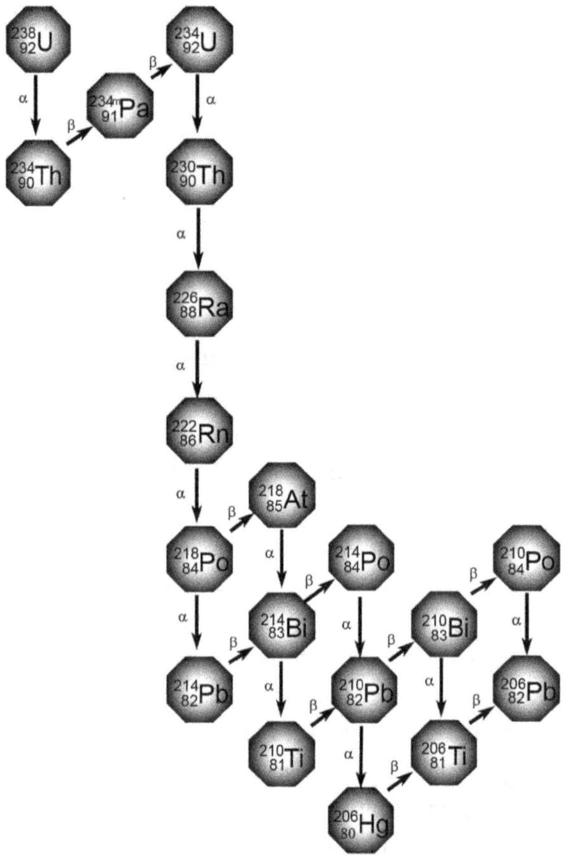

Figure 1-1 Uranium (or radium) series decay chain of the primordial radionuclide ^{235}U. Each of the radioactive products of decay represents a secondary radionuclide. Arrows indicate transitions from one nuclide to the decay product. Reproduced and modified with permission through CC, from https://commons.wikimedia.org/wiki/ File:Decay_chain(4n%2B2,_Uranium_series).svg.

number the nucleus is stable, but as nuclei become larger, the share of neutrons must increase to maintain stability. In addition, nuclei are stable when both neutrons and protons are paired (even numbers), and even more stable when nuclei are composed of sets of two protons plus two neutrons. The remainder of nuclei — the unstable nuclei — are "radioactive"; that is, they spontaneously eject particles as a mechanism to reduce excess energy. All nuclei with more than 83 protons (larger than Bismuth) are radioactive.

The strong force applied over the nuclear distance is indicative of the work expended to hold the nucleus together and therefore is an expenditure of energy. The nuclear energy resides in two forms: the energy required to disassemble a nucleus into its component parts and the energy derived from converting the mass of the nuclear particles into energy (Einstein's famous relationship $E = mc^2$). Because the transfer of energy to biological matter causes damage, a core tenant of radiation biology requires us to follow how energy is dissipated or deposited in biological systems.

1.2.2. *Binding Energy*

Consider a nucleus composed of Z protons and $(A - Z)$ neutrons, where A is the "atomic mass" or total number of protons plus neutrons. Because the masses of both protons and neutrons are close to 1.0 u, the actual mass of the atom very nearly equals the number of protons plus the number of neutrons. The formation of a nucleus from its component parts can be represented as

$$Z \text{ protons} + (A - Z) \text{ neutrons} \rightarrow \left({}_{Z}^{A}X \right)^{+Z} + BE \tag{1.1}$$

where X is the chemical representation of the atom containing the nucleus of interest. For example, the periodic table reveals that calcium has 20 protons. If you assemble a nucleus containing 20 protons, the chemical symbol "Ca" replaces "X," the atomic mass of calcium "A," equals 40, the number of protons "Z" equals 20, and the number of neutrons ($A - Z$, or 40 – 20) equals 20. To compress these protons and neutrons together to create the nucleus, work must be performed, and energy must be expended. How much energy must be expended? To answer that question, we will examine a highly imprecise but logically simplistic description that ignores the subtleties of nuclear quantum physics. The resistance to compression comes from the Coulombic repulsive force of the charge, q, on the protons. The attractive strong (nuclear) force must overcome the repulsive Coulombic force at some point in order to hold the nucleus together. Imagine a person scrambling desperately up a steep, gravel bank to avoid falling into a pit of quicksand. The closer he or she comes to the pit, the more desperately he or she scrambles to get away. But once the pit is encountered, the victim is quickly sucked into the morass. Such is the fate of the proton. Figure 1-2A is a nonquantitative representation of this phenomenon. As a proton approaches a nucleus of any size greater than the single proton of a hydrogen nucleus to the largest theoretical mass of protons and neutrons, the repulsive force increases as the distance decreases ($1/r^2$). Because the x-axis of Figure 1-2 represents distance from the center of the nucleus, the proton approaches from the right. If the force pushing the proton toward the nucleus is removed at any position (r), the proton will move away from the nucleus with an energy ($\frac{1}{2}mv^2$) equivalent to the work expended to hold it at "r." Therefore, just prior to release, the proton at "r" has a potential energy (the y-axis of Figure 1-2) equal to the instantaneous kinetic energy once the force is removed. More and more energy must be expended to push the proton toward the nucleus as r decreases until it suddenly encounters the strong force, and immediately, at that distance, energy must be applied to remove the proton from the nucleus. Once that threshold is crossed, regardless of the distance from the center of the nucleus, the same amount of energy must be

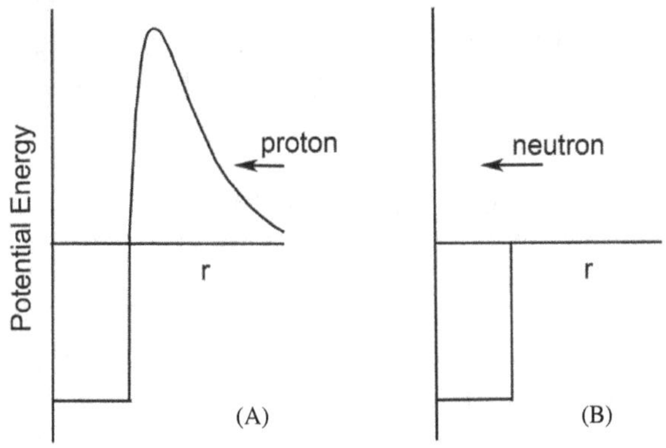

Figure 1-2 Potential energies of a (**A**) proton and (**B**) neutron as each approaches the nucleus. Because the proton carries a positive charge, electrostatic repulsion increases potential energy as the proton approaches the positively charged nucleus. When the proton intercepts the strong force, it is drawn into the nucleus by that force and the potential energy is minimized (**A**). Because the neutron carries no charge, it is not repelled by the positively charged nucleus and freely approaches the nucleus until captured by the strong force (**B**).

expended to remove the proton. Because neutrons carry no charge and are not repelled by either the protons or the neutrons within the nucleus, a neutron may effortlessly approach a nucleus until it reaches the range of the strong force whereupon it requires the same energy as a proton to be removed (Figure 1-2B). The energy required to remove a nuclear particle, be it proton or neutron, is called the binding energy (BE). An examination of Equation 1.1 discloses that the resultant nucleus is represented by the number of protons and neutrons $({}^{A}_{Z}X)$ plus the BE, or potential energy, that could be released from the nucleus if it were disassembled. The energy is as much of a component of the nucleus as the particles.

1.2.3. *Radioactive Decay*

Energy can be added to nuclei through fusion reactions, or particle bombardment. Many isotopes used in medicine are created artificially in cyclotrons or neutron reactors through processes that add particles to the nuclei of stable nuclides. Isotopes are forms of an element that have the same number of protons, Z (and therefore the same chemical description, X) but different numbers of neutrons (so different A). When we refer to isotopes in our context, we implicitly mean radioactive isotopes. Unstable or "radioactive" nuclei spontaneously eject particles to reduce excess energy within the nucleus. More generally, the term radionuclide refers to any unstable nucleus inclusive of radioactive isotopes or radioisotopes. The terms may be used interchangeably. Manufactured radioisotopes, primordial radionuclides, secondary radionuclides, and cosmogenic radionuclides release nuclear particles to attain stable "magic numbers" that represent a balance between the strong force and the Coulombic force.

Radionuclides release energy through a process known as radioactive "decay." To reach a stable nuclear structure, radioactive decay ejects nuclear particles: protons, neutrons, gamma rays (high-energy "photons"), or alpha particles (two protons and two neutrons) through predictable, easily understood mechanisms. Let us call the nuclide under consideration the "parent" and the nuclide following decay the "daughter." The parent is radioactive, and the daughter may be radioactive or may be stable. When the parent nuclide decays and transforms into the daughter nuclide, we represent this as:

$$\mathrm{P} \rightarrow \mathrm{D}. \tag{1.2}$$

If there are originally five parent nuclides, and one decays into a daughter nuclide, there are (5 – 1) parent nuclides and one daughter nuclide (Figure 1-3). Thus, the number of remaining parent nuclides will be equal to

$$N(t) = N_{\mathrm{P}}(0) - N_{\mathrm{D}}(t), \tag{1.3}$$

where $N(0)$ indicates the original number at $t = 0$, and $N(t)$ indicates the number at any time "t".

There is no way to predict whether a given nucleus will decay and thereby decrease the number of parent nuclides and increase the number of daughter nuclides because the process is random and

Figure 1-3 One of five parent nuclei (red) decays, transforming into a daughter nucleus (blue). $N(t) = N_{\mathrm{P}}(0) - N_{\mathrm{D}}(t) = 5 - 1 = 4$.

stochastic. If a large number of parent nuclides exist, a decay constant, λ, can describe the probability that any one of the parent nuclides will decay. The decay constant for each known isotope has been determined and can be found easily in a chart of the nuclides provided on electronic databases. We can now describe the decrease in the number of parent nuclides (dN) thus:

$$-dN = \lambda N(t)dt \tag{1.4}$$

where the decrease in the number of parent nuclides will depend on the probability that a nucleus will decay into a daughter nuclide times the number of parent nuclides that exist at a given time ($N(t)$). That differential equation is readily solved:

$$N(t) = N(0)e^{-\lambda t}. \tag{1.5}$$

A graph of this exponential function (Figure 1-4) illustrates that for any original number of parent nuclei, the remainder of parent nuclides will be half of the original number at some time "$T_{1/2}$." This amount of time is called the "half-life" of the radionuclide. This extremely useful relationship provides a more meaningful measure of the stability of a given nuclide, so much so that the half-life rather than the decay constant is listed in most resources.

$$\frac{N(t)}{N(0)} = e^{-\lambda t} \longrightarrow \frac{1}{2} = e^{-\lambda T_{1/2}} \tag{1.6}$$

Taking the natural log of both sides of the equation and because the negative natural log of ½ is equal to the positive natural log of 2, we find:

$$\ln 2 = \lambda T_{1/2} \tag{1.7}$$

$$0.693 = \lambda T_{1/2} \tag{1.8}$$

$$T_{1/2} = \frac{0.693}{\lambda}. \tag{1.9}$$

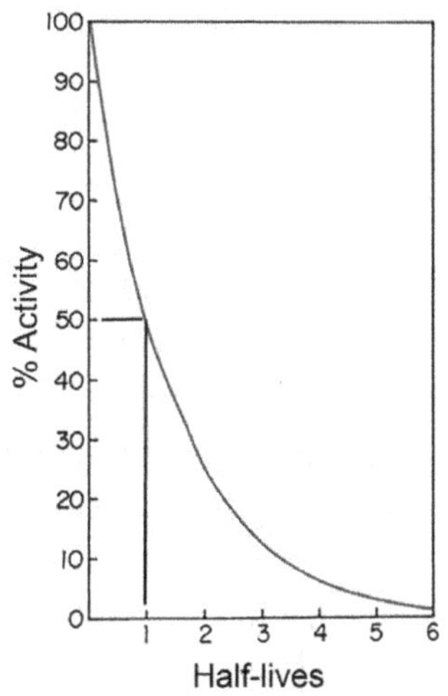

Figure 1-4 Exponential decay of an imaginary isotope. One "half-life" is the time elapsed until ½ of the original number (or ½ of the original activity as $A(0) = \lambda N(0)$) of nuclei remain. After two half-lives, ¼ of the original number of nuclides remain, and so on.

How then do we determine the number of daughter nuclei? We subtract the number of parent nuclei at time t from the original number of parent nuclei, thus

$$N_D(t) = N_P(0) - N_P(t), \tag{1.10}$$

$$N_D(t) = N_P(0) - N_P(0)e^{-\lambda_P t}, \tag{1.11}$$

$$N_D(t) = N_P(0)(1 - e^{-\lambda_P t}). \tag{1.12}$$

The mean life or average lifetime of a radioactive nucleus is represented by the Greek letter Tau (τ) and can be determined by summing over all the individual lifetimes for all the individual nuclei and dividing by the original number of nuclei, just as your average travel time to Chicago can be found by adding all your travel times to Chicago and dividing by the number of trips. Therefore, the mean life, t, is calculated with Equation 1.13, where

$$\tau = \frac{1}{N_0}\int_{N_0}^{0} t\,dN. \tag{1.13}$$

Substituting for dN and $N(0)$ using Equations 1.4 and 1.5 yields

$$\tau = \frac{T_{1/2}}{0.693} = 1.44T_{1/2}. \tag{1.14}$$

This relationship becomes useful, for example, when a person ingests or inhales a given quantity of a radioactive substance that completely decays while inside the subject's body.

The actual number of nuclei under investigation is in general a very large number and for our purposes it is unnecessary to know precisely. A much more useful quantity is the "activity." Imagine you are listening to popcorn kernels bursting open in your microwave as they heat. The activity of the conversion from corn kernel to popcorn decreases with time. Similarly, the conversion of the parent nuclei to daughter nuclei exhibit an activity. The activity depends on the number of parent nuclei and the probability that any given nucleus will decay (λ):

$$A = \lambda N(t) \tag{1.15}$$

and

$$A(t) = A(0)e^{-\lambda t}. \tag{1.16}$$

Activity is properly measured in decays per second — or becquerel (Bq) — although the arcane term, Curie (Ci), is still in common use. One Ci is equal to 3.700×10^{10} Bq.

One last concern: sometimes, a parent decays to a radioactive daughter that subsequently decays to a granddaughter, resulting in "chain decay"; in other words, the nucleus of the daughter product continues to possess excess energy and will attempt to become stable through further decay. An example of chain decay is presented in Figure 1-1. In this case of ^{238}U, radon gas (specifically the isotope ^{222}Ra) presents a serious health risk because it is readily inhaled, and the decay emits alpha particles that are particularly biologically harmful. A radiation biologist needs to predict the activity of ^{222}Ra. To begin, examine the simplest case of chain decay: a three-component chain. The decay of strontium-90 illustrates such a case.

Figure 1-5 The flow of nuclides into and out of the daughter isotope pool. The number of daughter nuclei at any given time depends on the rate of decay of parent nuclides into the pool as well as the decay of daughter nuclei that removes nuclides from the pool.

$$ {}^{90}_{38}\mathrm{Sr} \xrightarrow{T_{1/2}=29.1\text{ y}} {}^{90}_{39}\mathrm{Y} \xrightarrow{T_{1/2}=64\text{h}} {}^{90}_{40}\mathrm{Zn}(\text{stable}) \tag{1.17} $$

Both parent and daughter in this case decay through β^- emission. For simplicity, let's denote the parent as nuclide 1 and the daughter as nuclide 2, which will allow us to extend the progression through granddaughters and great-grand daughters, and so on. At the initial timepoint,

$$ \frac{dN_1(t)}{dt} = -\lambda N_1(t), \tag{1.18} $$

or the rate of decrease in the number of parent nuclei is proportional to the probability that a nucleus will decay. The rate of change in the number of daughter nuclei at any given time, t, is determined by the number of parent nuclei transforming into daughter nuclei less the number of daughter nuclei transforming into granddaughter nuclei (Figure 1-5),

$$ \frac{dN_2(t)}{dt} = \lambda_1 N_1(t) - \lambda_2 N_2(t). \tag{1.19} $$

And the accumulation of granddaughter nuclei depends only on the rate at which the daughter nuclei transform into granddaughter nuclei, because N_3 is stable.

$$ \frac{dN_3(t)}{dt} = \lambda_2 N_2(t) \tag{1.20} $$

Determining $N_2(t)$ in terms of $N_2(0)$ — the second term in differential Equation 1.19 — is simply Equation 1.5. Determining $N_2(t)$ in terms of $N_1(t)$ — the first term in differential Equation 1.19 — requires integrating over all the decays of the parent during the time interval of interest:

$$ \int_0^t dt'[\lambda_1 N_1(t')e^{-\lambda(t-t')}]. \tag{1.21} $$

This integral must be evaluated analytically. Recalling that $N_1(t) = N_1(0)e^{-\lambda t}$,

$$ N_2(t) = N_2(0)e^{-\lambda_2 t} + \frac{\lambda_1}{\lambda_2 - \lambda_1} N_1(0)[e^{-\lambda_1 t} - e^{-\lambda_2 t}]. \tag{1.22} $$

The first term on the right side of the equation, $N_2(0)e^{-\lambda_2 t}$, represents the number of existing daughter nuclei that are decaying into granddaughter nuclei. Similarly, $N_1(0)e^{-\lambda_1 t}$ represents the number of parent nuclei decaying into daughter nuclei. And $-N_1(0)e^{-\lambda_2 t}$ represents the number of newly transformed daughter nuclei decaying into granddaughter nuclei. The ratio of the decay constant for the parent to the difference between the decay constants is indicative of the flow from parent to granddaughter due to the different probabilities of decay. The number of granddaughter nuclei can be found by integrating Equation 1.20. The solution is presented here; the derivation is left to the student.

$$N_3(t) = N_3(0) + N_2(0)[1 - e^{-\lambda_2 t}] + \frac{1}{\lambda_2 - \lambda_1} N_1(0)[\lambda_2(1 - e^{-\lambda_1 t}) - \lambda_1(1 - e^{-\lambda_2 t})] \tag{1.23}$$

The first term on the right side of the equation is simply the number of existing granddaughter nuclei. If there are none at $t = 0$, this term drops out. When multiplied through, the second term represents the number of daughter nuclei at $t = 0$ (which may also be 0) minus those daughter nuclei that have decayed at the time of interest. The ratio of the decay constant to the difference between the decay constant for the transformation of the daughter and the decay constant for the transformation of the parent must now be separated into two measures of flow: one for the decay of the daughter (λ_2) and one for the decay of the parent (λ_1), and so those constants are separated and associated with their appropriate transformation of parent or daughter. For clarity, multiply through by the fraction. And when multiplied through by $N_1(0)$, the third term reflects the original number of parent nuclei minus the number of decays at the time of interest and subtracts the original number of parent nuclei minus the number of decays of the transformed parent from daughter to granddaughter.

To find the activities of the parent (Equation 1.5) and daughter (Equation 1.22), multiply through by λ_1 and λ_2 respectively, because $A = \lambda N$. The granddaughter (Equation 1.23) is stable and therefore the activity equals 0. The mathematical details of three chain decay are interesting because they insinuate the complexity of the relationships in long chains such as that of ^{238}U (Figure 1-1). Fortunately, when the half-life of the parent is much larger than the half-life of the daughter, the ability to decay is limited by the parent. Let's do a Gedanken-Experiment (German for "thought experiment""). Suppose, on average, a nuclide of an imaginary isotope decays every minute, and the daughter decays every 5 seconds. Imagine that there are no daughter nuclei initially. Suppose a parent nucleus decays, i.e., transforms into a daughter nucleus — pop! Five seconds later, the daughter decays — pop! Now, there are no daughter nuclei, again. Another parent nucleus decays 55 seconds later, and 5 seconds after that the daughter decays — pop! And so on. Despite the 5-second half-life of the daughter isotope, the daughter can decay only once a minute because that it is the rate of flow (transformation) of the parent state into the daughter state. The two isotopes are said to be in *secular equilibrium*. This Gedanken-Experiment can be confirmed mathematically using the expression for the activity of the daughter isotope,

$$A_2(t) = A_2(0)e^{-\lambda_2 t} + \frac{\lambda_2}{\lambda_2 - \lambda_1} A_1(0)e^{-\lambda_1 t}[1 - e^{-(\lambda_2 - \lambda_1)t}]. \tag{1.24}$$

The first term on the right side of the equation $(A_2(0)e^{-\lambda_2 t})$ equals zero because we stated that there were no daughter nuclei initially. In any case, it is negligible and can be ignored because if the half-life is small, λ_2 is large ($T_{1/2} = 0.693/\lambda$) and because $e^{-x} = 1/e^x$. Because λ_1 is very small compared with λ_2 (in cases of secular equilibrium, the parent has a very long half-life often on the order of thousands of years), λ_1 can be ignored in the fraction $(\frac{\lambda_2}{\lambda_2 - \lambda_1})$ and it resolves to 1. For the same reason, $e^{-(\lambda_2 - \lambda_1)t}$ is approximately $e^{-\lambda_2 t}$, and just as for the first term on the right side of Equation 1.24 the exponential becomes diminishingly small, leaving

$$A_2(t) \cong A_1(0)e^{-\lambda_1 t} = A_1(t). \tag{1.25}$$

The parent isotope, and the daughter product, exhibit the same activity.

What if the parent half-life is longer than the daughter half-life, but the parent half-life is not extremely long or the difference between them is not so great? In that case, the exponents behave in much the same way but the ratio of the decay constant to the difference between the decay constants does not go to 1:

$$A_2(t) \cong \frac{\lambda_2}{\lambda_2 - \lambda_1} A_1(0)e^{-\lambda_1 t} = \frac{\lambda_2}{\lambda_2 - \lambda_1} A_1(t). \tag{1.26}$$

In this case, the parent and daughter decay at the same rate. This situation is called *transient equilibrium*. Secular and transient equilibrium simplify the assessment of radioactivity in the case of chain decay and become useful to the radiation biologist when establishing risk due to exposure (Figure 1-6).

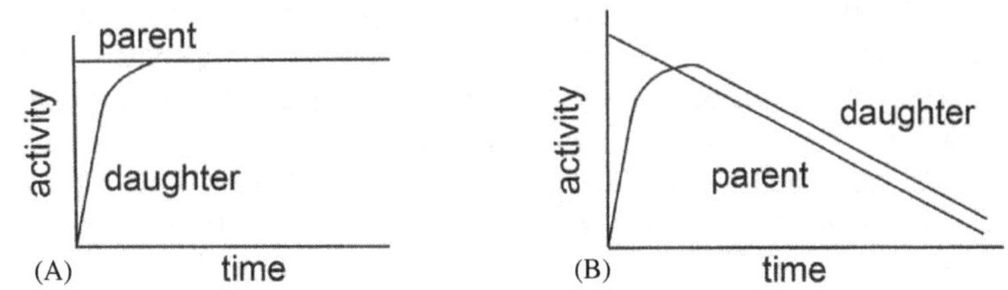

Figure 1-6 Secular and transient equilibrium. In the case of **(A)** secular equilibrium, the half-life is long, and the activity of the parent may not appear to change during the period of observation. The daughter activity increases until it approximately equals that of the parent. In **(B)** transient equilibrium, the activity of the daughter product increases until it approximates the activity of the parent and then the activity of both isotopes decrease at the same rate. The activities are not the same, they differ by $\lambda_1/(\lambda_2 - \lambda_1)$; they are proportional.

1.2.4. *Decay Energy, Q*

Happily, for students of radiation biology, there are only a few ways that unstable isotopes decay to release energy because the nucleus favors specific configurations (magic numbers; pairing of nuclear particles). Table 1-1 lists the primary decay modes of radioactive decay and presents the symbolic reactions for each mode.

To determine the energy released when an isotope decays and the percentage of that energy transferred to the biological target, we begin by examining each of the mechanisms through which unstable nuclei release energy. A gamma photon or "gamma ray" is electromagnetic radiation that has energy greater than 0.01 MeV. A photon is a quantized particle of light. Light exhibits both particle and wave characteristics and so may be referred to as either a photon or a wave depending on

Table 1-1 Spontaneous decay reactions.

Decay Type	Reaction
Gamma (γ)	${}^{A}_{Z}P^{*} \rightarrow {}^{A}_{Z}P + \gamma$
Alpha (α)	${}^{A}_{Z}P \rightarrow {}^{A-4}_{Z-2}D + \alpha$
Negatron (β–)	${}^{A}_{Z}P \rightarrow {}^{A}_{Z+1}D + \beta^{-} + \bar{\nu}$
Positron (β+)	${}^{A}_{Z}P \rightarrow {}^{A}_{Z-1}D + \beta^{+} + \nu$
Electron capture	${}^{A}_{Z}P + e^{-} \rightarrow {}^{A}_{Z-1}D^{*} + \nu$

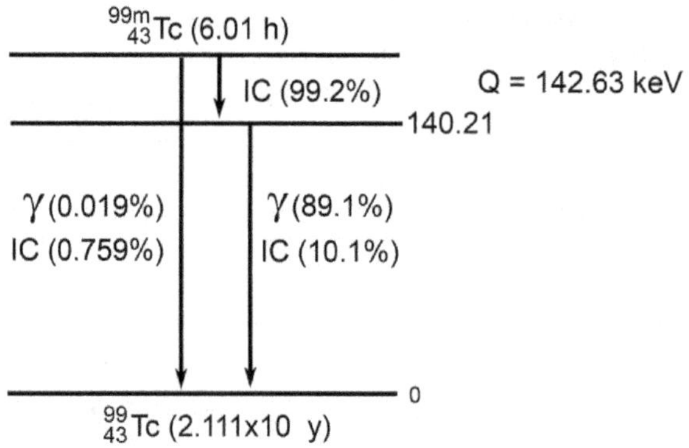

Figure 1-7 Decay scheme for TC-99m. The format for decay schemes represents energy on the y-axis and change in proton number on the x-axis, however both axes are implied only. Because emission of a gamma does not alter Z, the decay is represented as a vertical drop in nuclear energy. The reaction energy (Q) is presented here to the right of the initial state. Intermediate energies are presented to the right of the intermediate states. The half-life of the parent is listed in parentheses, and because the daughter is not stable, a half-life is also listed for the product. The branching ratio, or percent of nuclei that decay by a particular mode, is represented in parentheses to the right of each mode. The isomeric transition is designated "γ"; IC is an internal conversion reaction.

the behavioral physics at the time of observation. The γ-ray differs from an x-ray in that it is released from a nucleus during decay, whereas an x-ray is an atomic emission generated within the electron orbitals. Release of a γ photon is often called an isomeric transition because the atomic number of the parent nuclide does not change, and therefore the only difference between parent and daughter is a more stable, less energetic nucleus. Table 1-1 shows the reaction for an isomeric transition denoting that neither A nor Z is affected by the decay. The metastable isotope technicium-99m, commonly used for nuclear medicine imaging studies, provides a typical example of an isomeric transition. We can determine the amount of energy released during isomeric transition as follows (Figure 1-7):

$$E({}^{A}_{Z}\mathrm{P}^*) \equiv E({}^{A}_{Z}\mathrm{P}) + E^*. \tag{1.27}$$

The energy of the unstable parent (* implies excited) is equal by definition to the energy of the stable daughter (retaining the same A and Z), plus the excess energy. Because $E = mc^2$, and energy must be conserved in the reaction, this can be rewritten

$$M({}^{A}_{Z}\mathrm{P}^*)c^2 \equiv M({}^{A}_{Z}\mathrm{P})c^2 + E^* = M({}^{A}_{Z}\mathrm{P})c^2 + \left(E_{\mathrm{P}} + E_{\gamma}\right), \tag{1.28}$$

where E_{P} is the energy expended by the recoil of the nucleus (Newton's third law) and $E\gamma$ is the energy of the escaping gamma photon. The energy of the reaction, i.e., the Q-value, is therefore

$$Q\gamma = E_{\mathrm{P}} + E_{\gamma}. \tag{1.29}$$

Momentum must also be conserved, so if the emitted gamma has a momentum equivalent to the momentum for a massless particle

$$p_{\gamma} = E_{\gamma}/c \tag{1.30}$$

and the recoil nucleus has momentum equal to the momentum for a massive particle,

$$p_P = M_P v_P = \sqrt{2M_P E_P}. \tag{1.31}$$

given $E_P = ½\, mv^2$, then the two momenta must equal each other:

$$2M_P E_P = E_\gamma^2 c^2. \tag{1.32}$$

We can substitute into Equation 1.29 and rearrange to get

$$E_\gamma = Q_\gamma \left[1 + \frac{E_\gamma}{2M_P c^2}\right]^{-1}. \tag{1.33}$$

Because the energy equivalent of the mass of the nucleus is orders of magnitude greater than the energy of the escaping gamma, the fraction contributes a negligible amount to the quantity in brackets. As logic would tell you, the energy of the gamma is approximately equal to the reaction energy, Q.

Release of energy in the form of gamma emission, an isomeric transition, is unique because the high-energy photon is massless. The remainder of nuclear transformations take place through the release of massive particles. Therefore, we can define the reaction energy, Q, as the difference between the rest mass of the parent (plus any particles or energy added to the parent to create an unstable parent in the case of manufactured nuclides) and the daughter plus escaping nuclear particles; in the terms of chemistry, the reactants minus the products. But notice that we have just said that an *energy* (Q) is equal to the difference in *masses* (M). No worries because $E = mc^2$. Thus, in the case of alpha particle emission (two protons bound together with two neutrons e.g., Figure 1-8):

$$Q/c^2 = M({}^{A}_{Z}\mathrm{P}) - \{[M({}^{A-4}_{Z-2}D)^{2-}] + m({}^{4}_{2}\alpha)\}. \tag{1.34}$$

Why does the daughter product carry a double negative charge? Because the charge neutral parent possessed a number of electrons equal to the number of protons. When two protons were ejected from the nucleus, the electrons remained associated with the daughter atom. In other words, n nuclear protons

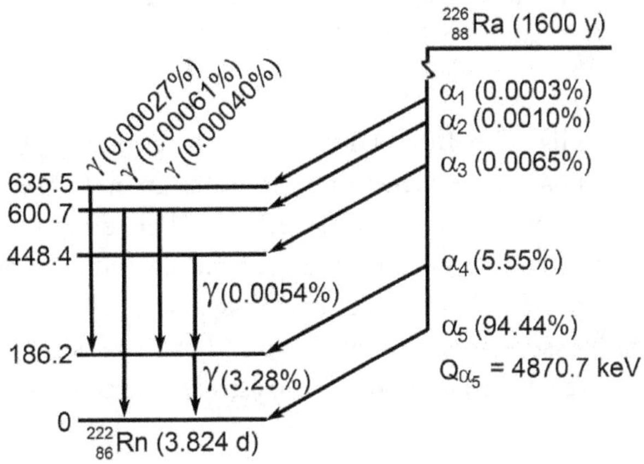

Figure 1-8 Decay scheme for radon-226. Radon decays by alpha emission through several modes. The lower energy decay modes release the remaining energy from the nucleus by gamma emission. Because alpha emission results in the loss of two protons, the daughter products are shown shifted (2 × x) to the left. The gamma emissions do not result in a change in Z and therefore are represented as vertical drops in energy.

plus n orbital electrons became $n - 2$ nuclear protons plus n orbital electrons. The mass of the reaction products must account for the masses of these two extra electrons that will not be included in the mass of the *uncharged* daughter: the mass listed in the periodic table, for example. The reaction is:

$$Q/_{c^2} = M(^{A}_{Z}\text{P}) - \{[M(^{A-4}_{Z-2}D)] + 2m_e + m(^{4}_{2}\alpha)\}. \tag{1.35}$$

For the calculation, the electrons can be imagined associated with the alpha particle to form a helium atom

$$Q/_{c^2} = M(^{A}_{Z}\text{P}) - \{[M(^{A-4}_{Z-2}D)] + M(^{4}_{2}\text{He})\}, \tag{1.36}$$

providing three masses that easily can be obtained from a look up table or periodic chart.

The Difference in Mass: Protons vs. Neutrons

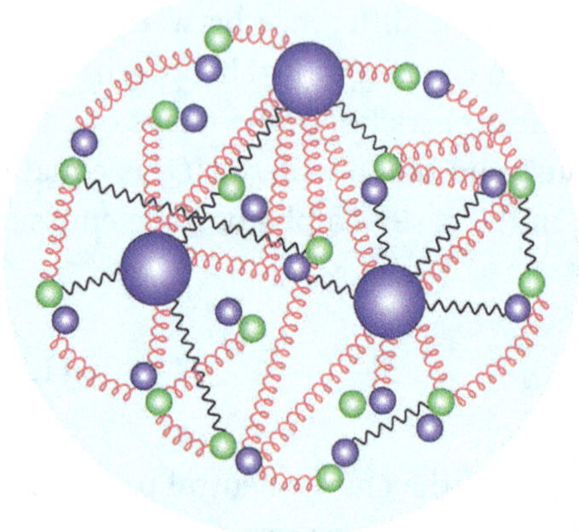

Reprinted with permission from Physics Today 68, 6, 17 (2015).

The mass difference between the neutron and proton is 0.14%. Quantum chromodynamics (QCD), the physics theory elucidating the strong force, proposes that neutrons and protons have three valence quarks (pictured here as large purple balls), many gluons (red springs), and quark (purple)–antiquark (green) pairs. The quarks and antiquarks interact by exchanging photons (gray wavy lines). Proton valence quarks are [up, up, down] and neutron quarks are [up, down, down]. It turns out that the "up" quark is slightly less massive than the "down" quark. Gluons facilitate the strong force between quarks similar to photon interactions with the electromagnetic force between electrons, except that gluons also carry "color charge" and therefore interact with each other. The quarks and antiquarks interact electromagnetically and kaon decay, which pairs strange quarks or antiquarks with up or down antiquarks or quarks, implies a symmetry-breaking process called charge–parity violation. Precision calculation of mass requires re-evaluating light quark masses and the application of electromagnetic corrections. (Chang, 2015)

The next two decay reactions listed in Table 1-1 represent the emission of a particle with a mass equal to an electron (5.4858×10^{-4} u) possessing either a negative charge (negatron decay) or a positive charge (positron decay). In general, negatron emission is referred to as "beta emission" and the ejected β particle is indistinguishable from an orbital electron. The charge of the ejected particle depends on the necessity of the nucleus to rid itself of a specific charge. In negatron decay, a neutron is transposed into a proton. The logic is this: the neutron possesses equal negative and positive charges and so it is neutral. If the neutron ejects the negative charge, it now becomes a positively charged proton. The reaction looks like this:

$$^{A}_{Z}\text{P} \rightarrow [^{\;\;A}_{Z+1}D]^{+} + {}^{\;\;0}_{-1}e + \overline{\nu}, \tag{1.37}$$

where the reaction products include the daughter nuclide $[^{\;\;A}_{Z+1}D]$, the ejected beta particle (${}^{\;\;0}_{-1}e$, electron) and an antineutrino ($\overline{\nu}$). Conservation of momentum during the decay process requires ejection of an

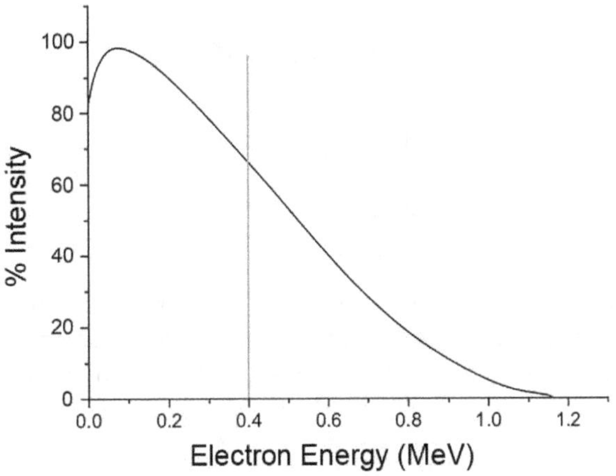

Figure 1-9 Beta emission energy spectrum. During negatron or positron emission, the reaction energy, Q, is divided between the β-particle and the neutrino or antineutrino through an unpredictable allocation. This spectrum represents the relative possible energies allotted to a particle resulting from a 1.15 MeV decay. The average β energy for this decay is 0.4 MeV; the most probable energy for an escaping β-particle is 0.08 MeV. The maximum β energy for this decay is 1.15 MeV.

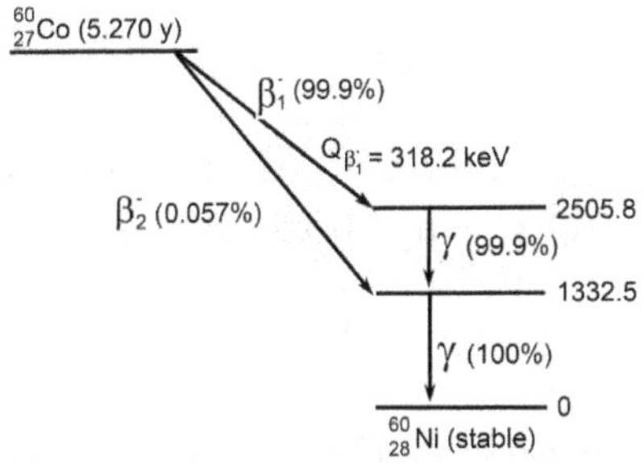

Figure 1-10 Decay scheme for cobalt-60. The negatron decay of ^{60}Co adds a proton to the nucleus through two modes. The daughter products are represented shifted to the right to indicate an increase in Z. The excess nuclear energy is emitted in the form of gammas, which are represented as vertical drops in energy to indicate no change in Z.

antineutrino. This third product was predicted by Pauli (Geiger & Scheel, 1933) in 1933 and formalized by Fermi (Fermi, 1934) in 1934. The existence of neutrinos and antineutrinos was confirmed in 1970 at Argon National Laboratory. These virtually massless particles pass through humans without depositing significant energy and therefore are of minimal interest to the radiation biologists. Nonetheless, the energy released during decay is divided between the electron and the antineutrino, and there is no way to predict how the energy will be allocated (Figure 1-9). For this reason, only the maximum possible beta energy can be determined as follows (as shown in Figure 1-10):

$$Q/_{c^2} = M({}^{A}_{Z}\mathrm{P}) - \{M([{}_{Z+1}^{\ A}D]^{+}) + m_{\beta-} + (m_{\bar{\nu}})\}. \tag{1.38}$$

We must deal with the excess positive charge in the nucleus — because again the number of orbital electrons did not change when the transmutation occurred, and we now have one more proton than the original number of electrons in the parent atom. So, when calculating the reaction energy, we borrow a free electron from the universe at large and add it to the daughter atom to create a neutral daughter product,

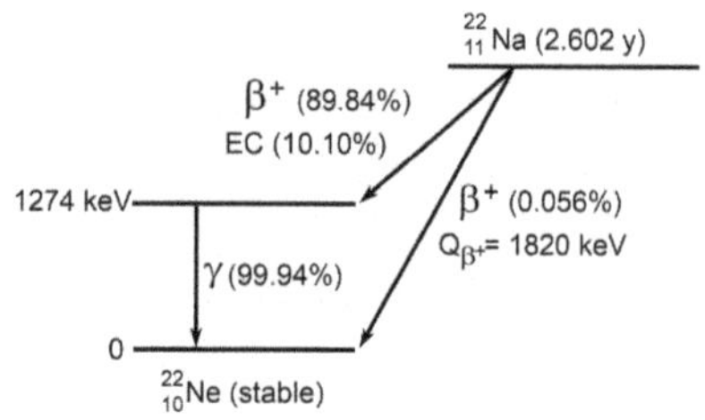

Figure 1-11 Decay scheme for sodium-22. Because positron decay results in the loss of one proton, the daughter products are shifted to the left. Nearly 90% of the time ^{22}Na decays through ejection of a positron with a maximum energy of 546 keV. Ten percent of the time ^{22}Na will release energy through electron capture. Positron emission and electron capture are competing decay reactions.

$$Q/_{c^2} = M({}^{A}_{Z}\mathrm{P}) + m_{e-} - \{[M({}_{Z+1}^{\;\;\;A}D)] + m_{\beta-} + m_{\bar{\nu}}\}. \tag{1.39}$$

The masses of the added electron and the ejected β^- cancel and the mass of the antineutrino is negligible yielding,

$$Q/_{c^2} = M({}^{A}_{Z}\mathrm{P}) - [M({}_{Z+1}^{\;\;\;A}D)]. \tag{1.40}$$

The reaction energy for positron emission (Figure 1-11) is calculated in an analogous manner except that a proton is transmuted into a neutron leaving a negative charge on the atom and a neutrino is expelled.

$${}^{A}_{Z}\mathrm{P} \rightarrow [{}_{Z-1}^{\;\;\;A}D]\; + {}^{\;\,0}_{+1}e + \tag{1.41}$$

The calculation for the reaction energy generated by positron emission is therefore:

$$Q/_{c^2} = M({}^{A}_{Z}\mathrm{P}) - \{[M({}_{Z-1}^{\;\;\;A}\mathrm{D})] + m_{\beta+} + m_{e-}\}. \tag{1.42}$$

It is important to remember that Q represents the maximum possible energy attributed to the beta particle (positron or negatron) because the energy imparted to the neutrino or antineutrino is unknowable for any specific emission. In general, the mean energy of all beta particles emitted from a given isotope equals about one-third of the maximum Q.

A nucleus alternatively can increase the n/p ratio and thus gain stability by electron capture. This process competes with positron emission and represents the only mechanism available to nuclei with less than 1.022 MeV excess energy. We'll examine that energy boundary a little later. During electron capture, an orbital electron falls into the nucleus and combines with a proton to create a neutron. The number of electrons in the atom thereby decreases by one; the number of protons in the nucleus decreases by one, and the number of neutrons in the nucleus increases by one:

$${}^{A}_{Z}\mathrm{P} + {}^{\;\,0}_{-1}e \rightarrow {}_{Z-1}^{\;\;\;A}D + \nu\,. \tag{1.43}$$

The reaction produces only a monoenergetic neutrino of no consequence to radiation biology. However, the missing orbital electron will instigate a cascade of electrons falling to lower energy levels. These will emit characteristic x-rays that will deposit energy in biological material and so are deserving of some scrutiny. We'll discuss characteristic x-rays a little later.

Sometimes an escaping gamma from an isomeric transition collides with and deposits its energy in an electron orbiting the same atom. The energized electron, an "Auger electron," escapes its orbit. This event is termed an internal conversion and the empty orbital results in an electron cascade analogous to electron capture. The energy is thus divided among the escaping electron and the characteristic x-rays that arise.

Rarely, neutron-rich nuclei may eject a neutron and proton rich-nuclei may eject a proton in order to reach stability. These isotopes are not common and the energetic neutrons and protons that concern radiation biologists are more commonly produced through nuclear activation reactions or machine-produced radiation. Derivation of the decay reactions and reaction energies for these rare isotopes is trivial.

1.2.5. *Particle Kinetic Energy, T*

Radiation biology is concerned with the energy transferred to biological matter. The reader might suspect that the total reaction energy Q is not necessarily available for transfer. Some, we learned, is absorbed in the recoil of the nucleus following particle emission. The expelled particles possess a kinetic energy determined by velocity (½ mv^2) and a potential energy determined by mass ($E = mc^2$). So, our first task is to determine the kinetic energy (T) of the escaping particle. In the case of isomeric transition, the mass-less gamma carries all the reaction energy:

$$T_\gamma = Q, \tag{1.44}$$

as $m = 0$. In the case of internal conversion, the Auger electron carries away the gamma energy (=Q) minus the energy expended in breaking the electron out of orbit:

$$T_{\text{Auger}} = Q - \text{BE}. \tag{1.45}$$

For massive particles, the kinetic energy of the escaping particle depends on the mass of the particle compared with the mass of the nucleus that recoils opposite the escape vector. For β emission, the mass of the ejected electron or positron is four orders of magnitude less than a single nuclear particle. Recoil is insignificant and

$$T_{\beta^{+/-}} = Q\frac{M_{\text{P}}}{M_{\text{P}} + m_\beta} \cong Q. \tag{1.46}$$

However, for alpha emission the ejected α-particle consists of a significant portion of the nucleus: two protons and two neutrons. The kinetic energy is less than the reaction energy by a factor that reflects the portion of the nucleus that was ejected.

$$T_\alpha = Q\frac{M_{\text{A}-4,\text{Z}-2}}{M_{\text{A}-4,\text{Z}-2} + \text{M}_{4,2}} \cong Q\frac{A_{\text{P}} - 4}{A_{\text{P}}} \tag{1.47}$$

Two equivalents are enormously useful when calculating Q and T: the rest mass equivalent energy for an electron is 0.511 MeV and the energy equivalent for 1 atomic mass unit (amu or u) is 931 MeV. Thus, multiplying the generalized equation for Q through by c yields

$$c[Q/c] = \{\text{mass of the reactants} - \text{mass of the products}\}c \tag{1.48}$$

or, in MeV:

$$Q\ [\text{MeV}] = \{\text{mass of the reactants } [u] - \text{mass of the products } [u]\}\ (931\ \text{MeV}/u). \quad (1.49)$$

1.3. Machine-Produced Radiation

1.3.1. *History*

Both isotope-produced radiation and machine-produced radiation were identified at the fin de siècle, just short of 1900. Marie Skłodowska Curie discovered two naturally occurring radioisotopes in 1898: radium and polonium, coined the term "radioactivity," and shared the Nobel Prize for her work in Physics in 1903. Wilhelm Röntgen described x-ray radiation in 1895, earning him the Nobel Prize in Physics in 1901 — the first prize in Physics ever awarded. By the time Marie and Wilhelm had won their Nobel Prizes, their discoveries had been applied to imaging internal anatomy and to treating several human maladies, and the fields of medical physics and health physics (or radiation protection) were firmly established. We discussed isotopes and nuclear radiation in Section 1.2. Here, we will examine machine-produced radiation generated through the interaction of energetic charged particles with metals, or "high Z" materials. Although several alternatives exist for the generation of radiation to which humans may be exposed intentionally or accidentally (e.g., Van de Graaff generators), the health industry dominates machine radiation exposure of humans. Thus, we present the examples of clinical radiation machines here and encourage the reader to explore the broader subject of machine-produced radiation. We will examine the physics of charged particles and high Z materials in Chapter 9; for now, we will limit the discussion of radiation production to engineering.

Contemporary mythology reports that Röntgen was operating a Crookes tube placed on a lab bench that included a drawer. In that drawer was a coin sitting on a photographic plate. When Röntgen retrieved the plate, he noticed that it was fogged everywhere except where the coin had been resting and he reasoned that "invisible light" had been emitted from a source above the plate, that it had penetrated the lab bench, but it had not penetrated the coin. He surmised that the source must have been the Crookes tube. True story or not, Röntgen reasoned that the Crookes tube emitted unique electromagnetic radiation which he named "x-rays" (Röntgen, 1895). In Röntgen's Crookes tube, electrons were pulled off gas molecules using a potential difference of up to a few hundred kilovolts between a relatively negative cathode and a relatively positive anode (Figure 1-12). The small initial number of energetic electrons collided with gas molecules, knocking off more electrons. The positive ions created when electrons were removed were strongly attracted to the cathode, striking it and knocking electrons out of the cathode, which in

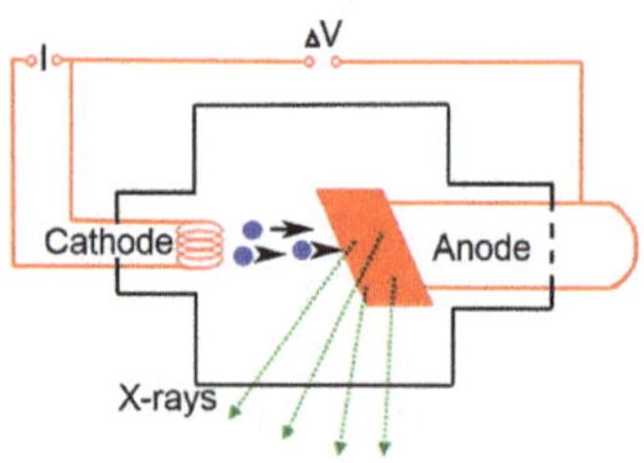

Figure 1-12 Schematic of a modern cathode ray tube, the precursor of the modern diagnostic x-ray tube. Current (I) is passed through the cathode to provide free electrons (blue) that are propelled through the vacuum by the voltage difference (ΔV) between cathode and anode. Energetic electrons collide with the atoms of the anode, releasing x-rays (green) that pass through the housing (black).

turn were repelled by the cathode and attracted to the anode. These electrons were termed "cathode rays." The cathode rays (energetic electrons) interacted with the anode and glass housing to create x-rays. At the time, the x-rays were an unintended and unknown complication of cathode ray production. Cathode ray tubes (CRT) dominated our culture for half a century in the form of televisions and computer monitors.

1.3.2. *Diagnostic X-ray Machines*

Modern x-ray tubes used (Figure 1-13) in diagnostic imaging improve on Crookes tube technology in only a few ways but represent incredible leaps in technology. The tube is evacuated to extremely low pressures: 0.1 to 0.005 Pa, to prevent electron collisions with the gas. Current passes through the cathode filament heating the filament and boiling off electrons; the greater the current, the hotter the filament and the greater the number of electrons liberated into the vacuum tube. The voltage difference between the relatively negative cathode and the relatively positive anode accelerates the electrons through the vacuum to the anode. The voltage difference determines the energy of the x-rays. The number of electrons released from the cathode determines the number of x-rays, or the fluence. The anode is angled to reflect the emitted x-rays toward a beryllium window through which they exit the tube. During the time it takes to generate sufficient x-rays to take a single radiograph, the spot on the anode where the electrons strike can heat to more than 2,500°C, necessitating two more design elements. First, a heat-tolerant tungsten–rhenium target is mounted on a molybdenum heat sink core; and the anode is flattened into a large disk that rotates on a spindle like a vinyl phonograph record. The electrons contact the anode as the needle contacts the vinyl, tracing out a path around the anode such that the same number of electrons per second are now dispersed over the circumference traced out around the surface. This design both reduces the heat deposition and provides time for the anode to cool. Liquid gallium lubricated fluid dynamic bearings facilitate extremely low friction between the mechanical parts while being resistant to diffusion into the vacuum. Lastly, because of the extreme heat generation, the tube housing is blown from high-borate borosilicate thermal tolerant glass that matches the expansion of the anode material.

1.3.3. *Therapeutic X-ray Machines*

Therapy for cancer requires x-ray energies several orders of magnitude greater than diagnostic imaging. Imaging contrast is enhanced by x-ray interactions that occur in the range of 1–50 keV. Therapy requires that x-rays penetrate deeply and deposit large amounts of energy in the target, demanding energies in the range of 5–30 MeV. The x-ray source continues to be a metal target struck by high-energy electrons, but the electrons must be accelerated to considerably greater energy. This requires a particle accelerator.

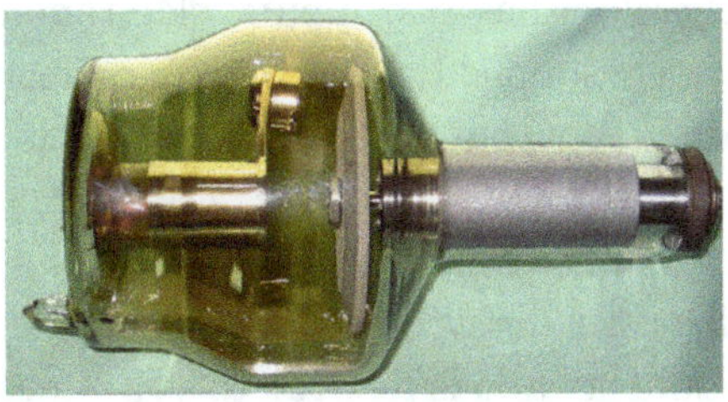

Figure 1-13 Modern rotating anode x-ray tube. Electrons are boiled off the cathode and accelerated toward the rotating anode through a voltage differential of tens to hundreds of kilovolts. Original image by Daniel W. Rickey, reproduced with permission through CC BY-SA 3.0.

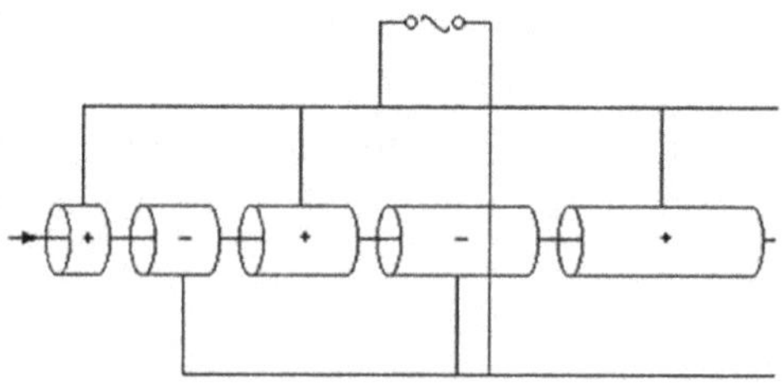

Figure 1-14 Schematic for a Louis Alverez linear accelerator based on alternating current. Electrons are generated at the left (arrow) and accelerated across the gaps between tubes. The length of each tube increases along the path such that the transit time within the tube remains constant. The current alternates so that as electrons exit each tube, the voltage becomes relatively positive at the entrance to the next tube.

Louis Alverez designed the first drift tube linear accelerator at Lawrence Berkeley Laboratory in 1955. Imagine a parallel plate capacitor with a voltage difference between the two plates. Now imagine rolling the plates into cylinders and rotating them so that their central axes align (Figure 1-14). Within the cylinders the electric field is neutral, but between the cylinders a potential difference exists that is equal to the difference in potential that existed between the parallel plates. An electron placed in the gap will be accelerated into the more positive cylinder where it will drift along according to the momentum imparted within the gap. Alvarez connected several of these cylinders and provided alternating current so that when the drifting electron reached the next gap, the polarity reversed, the electron resides at the edge of a negative cylinder with a positive cylinder at the other side of the gap. The electron accelerates across the gap and drifts through the positive cylinder until the polarity switches again. Subsequent accelerations increase the velocity of the electron, so the length of each cylinder in the sequence increases over the last such that the electron drifts through each tube for the same time. Now, imagine a bolus of many electrons bunched together replacing the single electron. Modern electron accelerators use traveling or, more often, standing electromagnetic waves in place of the alternating current. Just as a perpendicular electric field is generated by a voltage difference within a gap between oppositely charged plates, an electromagnetic wave provides an electric field without requiring alternating current and this field becomes the accelerating force. In the modern electron accelerator, the electromagnetic wave is tuned by adjusting resonance cavities and the drift spaces are eliminated from the path producing a compact constant acceleration. The physics and engineering of this design are fascinating but beyond the scope of this book and the reader is referred to the references at the end of the chapter for a more detailed description.

In modern therapy machines, a klystron or magnetron supplies the electromagnetic wave. Though the designs are different, both klystrons and magnetrons pump electrons up and down to create electromagnetic waves that are transferred into the accelerator through wave guides (not unlike fiber optic cable). The electron accelerator therefore has two inputs: electrons emitted from an electron gun (a cathode) and an electromagnetic wave supplied by a klystron or magnetron. A modulator (electronic circuitry) synchronizes the electron gun output with the electromagnetic wave input to the accelerator. The electron gun and accelerator reside in the neck of the radiation machine, whereas the power supply, modulator, and klystron/magnetron are in a cabinet behind the rotating portion of the machine (Figure 1-15). In an act of vainglorious obscurity, the radiation machine was named for the linear electron accelerator at the heart of its functionality and is referred to as a "linear accelerator" or simply a "linac." Thus, when one refers to the linear accelerator, no one is ever certain whether he or she is referring to the electron accelerator or the entire machine. Beyond the electron accelerator, in the head of the machine, the electron beam is redirected from horizontal to vertical projecting the output x-rays toward the patient. Charged particles with a velocity greater than zero deflect in the presence of a magnetic field

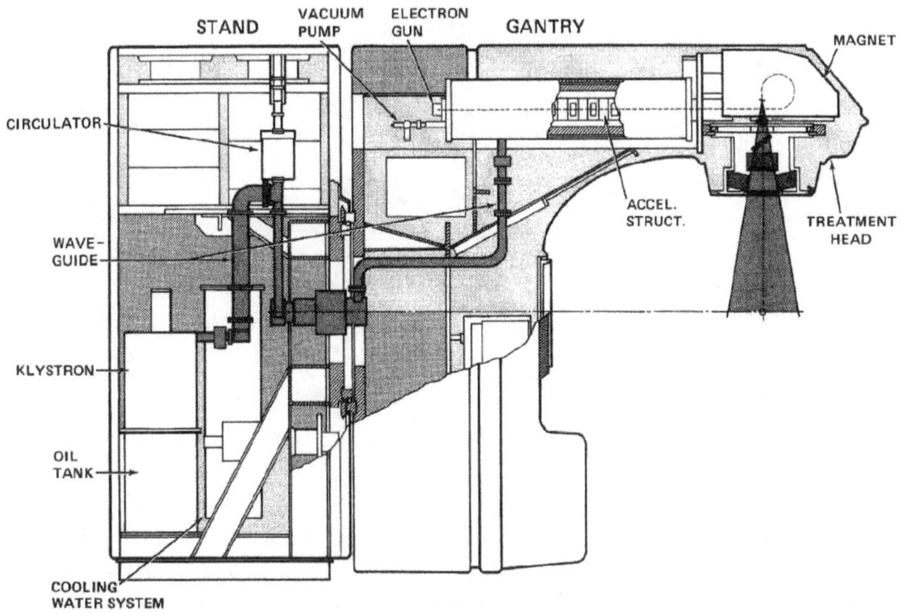

Figure 1-15 Schematic of a typical linear accelerator or linac. The stand is stationary and houses a cabinet containing the electronics, power supply, and klystron. The gantry is free to rotate 360 degrees around a patient located at the isocenter of the emitted beam. The electron linear accelerator is located in the horizontal neck of the machine. At the extreme end of the gantry the electron beam is directed at a thin metal target to produce the desired x-ray output. Reproduced with permission of Medical Physics Publishing from Karzmark and Morton, 1998.

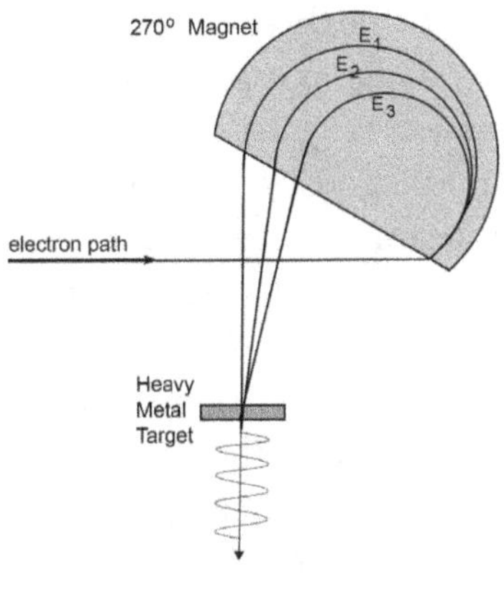

Figure 1-16 Three electron paths represented within the magnetic field (perpendicular to the page) of a 270° bending magnet. The most energetic electrons will circumscribe a larger radius (E_1) than the less energetic electrons (E_2 and E_3). The result is that all paths will converge at a triple focus point. If a target is placed at that focal point, x-rays are produced. If no target is present, a narrow electron beam will exit the treatment head.

perpendicular to the direction of travel. Employing a 270-degree bending magnet not only changes the direction of the electron beam by 90 degrees, but it also improves beam optics. Electrons traveling slower than the mean velocity are deflected more by the magnetic field and define a smaller arc, whereas electrons traveling faster than the mean velocity define a larger arc through the magnetic field (Figure 1-16). The radius of the arc is determined by the simple equation

$$r = {}^{mv}\!/_{q\mathbf{B}} \tag{1.50}$$

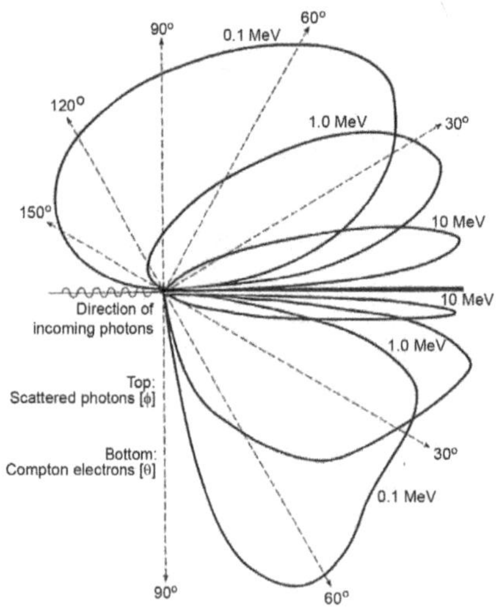

Figure 1-17 Radiation scatter. An x-ray enters from the left and impacts an orbital electron at the polar origin. Depending on the energy of the incoming photon, it will deflect according to the probability densities outlined by the ellipsoids labeled 0.1 MeV, 1.0 MeV, and 10 MeV (top, above 0 degree angle). Higher energy collisions lead to more forward scattered x-rays. The impacted electron will scatter according to the probability densities outlined by the ellipsoids labeled 0.1 MeV, 1.0 MeV, and 10 MeV (bottom, below 0 degree angle). Higher energy collisions lead to more forward scattered electrons. The angles are depicted above and below the horizontal plane for clarity. In reality, both x-rays and electrons scatter throughout 360 degrees. Reproduced with permission of Wiley, from Hendee and Ritenour (2001).

where r is the radius of curvature, m is the mass of the charged particle, $\mathbf{v}$ is the velocity vector, q is the charge of the particle, and $\mathbf{B}$ is the magnetic field vector. Thus the 270-degree magnet releases electrons of a bandwidth of energies ($E = \frac{1}{2} mv^2$) along exit paths that will intersect at some point beyond the magnetic field edge. If a metal transmission target is placed at this point in space, the smallest possible focal spot impinges on the target. As in a diagnostic x-ray tube, the energetic electrons interact with the metal atoms to produce x-rays. X-ray beam modifying devices may be positioned beyond the metal target. If the target is removed from the beam path, the electron beam will continue out the treatment head and can be used for electron beam therapy. We learned in Section 1.3.2 that diagnostic x-ray machines employ a reflective anode that emits x-rays back toward the cathode off the surface of the anode (Figure 1-12), while in this section we saw that linacs exploit a transmission target that emits x-rays forward from the front edge. Both targets are composed of high Z absorbing material, generally a tungsten alloy, effective for converting energetic electrons into x-rays. Why are these designs different? They are different because the physics of particle collisions depend on the energy of the particle. Figure 1-17 illustrates the scatter of x-rays of three energies: 0.1 MeV, 1.0 MeV, and 10 MeV. The graph displays, in polar coordinates, high-energy radiation entering from the left, colliding with orbital electrons at the origin. From this representation it is easily seen that at diagnostic energies, around 100 keV, x-rays will backscatter significantly at 120–150 degrees and relatively few will forward scatter. Thus, a reflective target simplifies the x-ray tube engineering. On the other hand, X-rays at therapeutic energies, on the order of 10 MeV, scatter strongly forward, and so a thin target struck from the back will emit x-rays in the forward direction.

1.3.4. *Heavy Particle Therapy Machines*

Proton radiation, carbon ion radiation, and neutron radiation are produced by machines not unlike the x-ray linac described earlier. Analogous to an electron treatment beam configuration, heavy charged particles leave the treatment head without impinging on a target. Analogous to an x-ray treatment beam configuration, accelerated deuterium impinges on a beryllium target to produce neutrons:

$$^{2}_{1}\mathrm{H}+^{9}_{4}\mathrm{Be}\rightarrow ^{10}_{5}\mathrm{B}+^{1}_{0}n. \tag{1.51}$$

The difference between manipulating electrons and manipulating heavy charged particles is that protons are 2×10^3 times more massive than electrons, and carbon ions are $12 \times 2 \times 10^3$ times more massive. Because $F = ma$, three orders of magnitude more force must be applied to accelerate, turn, and focus heavy charged particles. Heavy particle therapy machines and all the components must be much larger. Two formats of clinical heavy charged particle accelerators are commercially available: the multiple room facility and the single room facility. Multiroom facilities have a single accelerator that pumps energy into charged particles by applying a radiofrequency pulse at specific intervals as the particles circle the accelerator achieving energies of 70–350 MeV as in Figure 1-18. The accelerator is circular rather than linear to conserve space and can be either a cyclotron or a synchrotron. The first cyclotron was constructed by Ernest Lawrence in 1930; it fit in the palm of his hand. Facilities constructed in the 1970s in Indiana, Upsala, and Cape Town housed cyclotrons weighing as much as World War II (WWII) battleships. Modern cyclotrons have diameters that range from 1.5 to 4.4 m. In the case of cyclotrons, charged particles spiral outward (r increases) according to Equation 1.50 as they gain energy ($\frac{1}{2}mv^2$) from radiofrequency (rf) pulses. Synchrotrons accelerate a bolus of charged particles around an evacuated ring that includes several bending magnets. These bending magnets are ramped up, i.e., the field strength (**B**) is increased, each rotation of the bolus such that the proton path radius (r) remains constant as the energy ($\frac{1}{2}mv^2$) of the particles increases. Understanding this engineering is unnecessary for our investigation of radiation biology and so is left to the interested student to explore. The protons pass out of the accelerator into an evacuated beam line supporting switching magnets at the entrance to each treatment room. When activated, these magnets send the charged particle beam into the associated treatment room. The gantry and beam modifying devices reside in the treatment room at the end of the

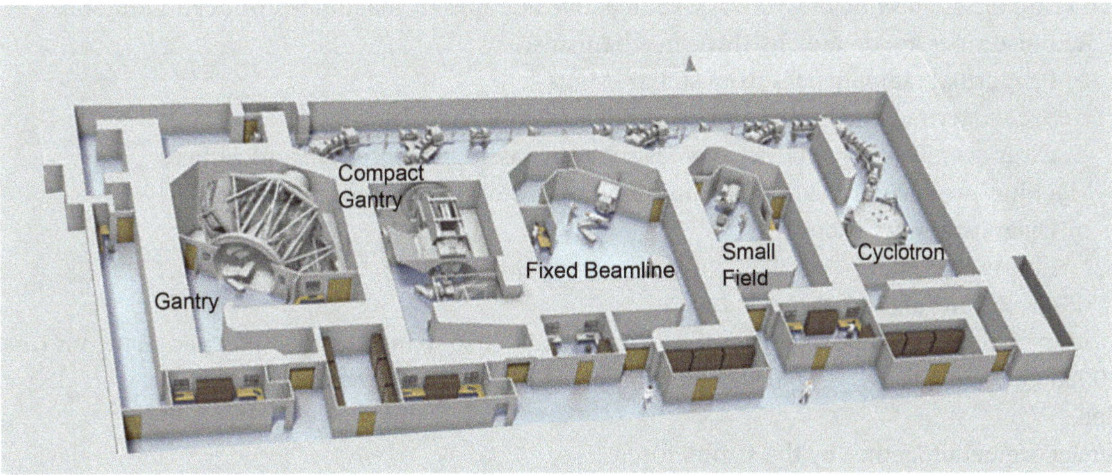

Figure 1-18 Multiroom IBA proton therapy facility in Seattle Washington. High-energy protons exit the cyclotron at far right and are directed through the beamline at the back (top) of the facility. Switching magnets steer the beam into a given room when activated by the operator. Two different IBA designs for a 360-degree rotating gantry are incorporated in this facility: the standard gantry at far left and the compact design gantry, center left. Two fixed beam lines are included: one patient beamline, center right, and one small field beamline closest to the cyclotron. Reproduced with the permission of Ion Beam Applications, Belgium.

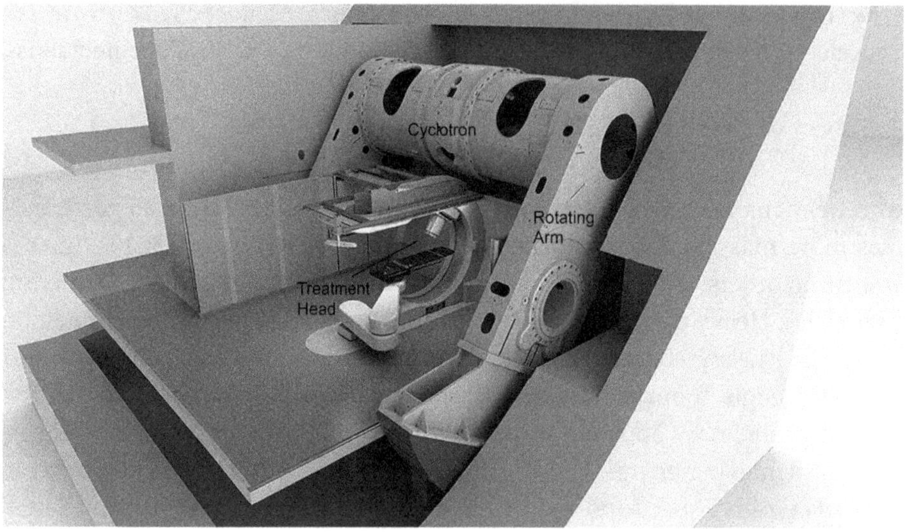

Figure 1-19 Single room proton therapy facility designed by Mevion. The cyclotron that supplies high-energy protons is located between two rotating arms that translate the treatment head 180 degrees. Patient orientation using the treatment couch enables 360-degree beam delivery. Image courtesy of Mevion Medical Systems, Inc. All rights reserved.

beamline. The single room solution takes the advantage of Henry Blosser's design for a superconducting cyclotron (Figure 1-19). Using superconducting technology enables sufficient accelerator size reduction to mount the accelerator directly on top of the gantry. The beamline and components can be similarly compacted. Although no clinical facilities use linear particle accelerators, several research facilities continue to make use of linear particle accelerators.

1.4 Summary

- Radiation originates one of two ways: radioactive isotopes or machine-produced radiation.
  - Radioisotopes are created by three mechanisms:
    - Primordial radionuclides from stellar events
    - Secondary radionuclides from primordial radionuclide decay
    - Cosmogenic radionuclides from cosmic rays passing through the atmosphere.
  - Machine-produced radiation is dominated by radiology and therapy units:
    - Diagnostic x-ray tubes
    - Linear accelerators
    - Cyclotrons and synchrotrons.
- Atomic nuclei are composed of protons (mass = 1.007276 u and charge = +1) and neutrons (mass = 1.008665 u and charge = 0) plus the BE that could be released from the nucleus if it were disassembled.
- Nuclei are held together by the strong force.
- Unstable nuclei release energy through radioactive decay.
  - The rate of decay is described by the "activity."
  - The half-life is defined by the length of time that is required to reduce the original number of unstable nuclei to half.

- If an unstable nucleus decays to unstable descendants, the resulting chain decay can be described by identifying transient equilibrium or secular equilibrium instances within the chain.
  - — Transient equilibrium occurs when the half-life of the parent is much longer than the half-life of the daughter; the activity of the daughter is proportional to the activity of the parent.
  - — Secular equilibrium occurs when the half-life of the parent is much, much longer than the half-life of the daughter; the activity of the daughter equals the activity of the parent.
- There are five primary modes of decay:
  - — Gamma
  - — Alpha
  - — Beta minus
  - — Beta plus
  - — Electron capture.
- Nuclei become more stable by releasing energy. The energy released during decay is designated "Q." It can be calculated as [(the mass of the reactants – the mass of the products) × c^2].
- The energy available to be transferred to biological material depends on the kinetic energy imparted by Q to the escaping particle(s). It is, in general, less than Q.
- Diagnostic x-ray machines produce radiation by accelerating electrons across a voltage drop within a vacuum. The electrons (cathode ray) impact a heavy metal target to produce x-rays.
- Therapy radiation machines use linear accelerators to produce high-energy electrons that are either used directly or collided with a heavy metal target to produce x-rays.
- Heavy ions are accelerated by cyclotrons or synchrotrons.

1.5. Problems

1. Several examples were provided to illustrate the method for determining the reaction energy released during decay (e.g., Equation 1.34). Provide the equation for electron capture: $(Q/c^2)_{EC}$.
2. Rarely, a heavy isotope will release energy by expelling a neutron or proton from the nucleus. Provide the reactions (e.g., Equation 1.42) for neutron and proton decay.
3. How much energy is required to remove a neutron from sodium?
4. Neutrons are generated intentionally for neutron therapy and generated unintentionally during proton therapy when beryllium is bombarded with protons. Provide the balanced equation (e.g., Equation 1.42) for this reaction. Determine the energy released (Q) per reaction.
5. ^{235}U decays to ^{231}Th by alpha decay. Find the reaction energy (Q) and the kinetic energy of the escaping alpha particle.
6. Derive Equation 1.23 by integrating Equation 1.20.
7. The atomic mass, m, of ^{64}Cu is 63.929757 amu. It undergoes positron decay with a half-life of 12.9 h. The product of the decay is ^{64}Ni. The mass, m, of this product is 63.927956 amu. Show the balanced reaction equation. What is the total energy of the positron plus the neutrino resulting from the decay? Is the product liable to be stable or be radioactive? Why?
8. The activity of a radioisotope decreases by 35% in 5 days. What is the half-life and mean life of this isotope?
9. A source of ^{99m}Tc arrives at the laboratory for use at 10:00 am on Monday morning, at which time the daughter product is eluted for diagnostic use. The parent, ^{99}Mo, had a decay constant (λ) of 0.01039 h^{-1}. If, after the separation of the daughter, the parent was found to have an

activity of 5.0×10^9 Bq, what is the activity of the parent and the daughter the following Thursday at 10:00 am?

10. The nuclide ^{131}I has a half-life of 8.06 days. Find the decay constant (λ) for this radionuclide. A source of this isotope has an activity of 2.500 mCi. Find the activity remaining after the elapse of 12 days from the measurement just given. Express your result in millicuries and bequerels.
11. How many nanograms of ^{137}Cs are there in a 2.5 mCi source (a pure source containing only ^{137}Cs)?
12. A radioactive source initially contains only ^{90}Sr which decays over time to ^{90}Y. How much time must pass before the activity of the daughter reaches 5% of the activity of the parent?
13. For the following hypothetical decay scheme, find the maximum kinetic energy of the β^+ particle and the energy of the γ_2 photon (in keV) if the energy of the γ_1 photon = 250 keV and the energy of the γ_3 photon = 400 keV. Provide the absolute branching ratio (probability per decay) for γ_3 and γ_1.

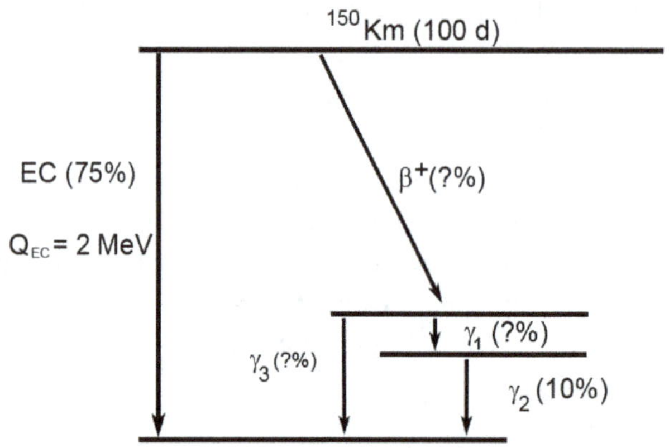

14. The fluence 50 cm away from a point isotropic radiation source is 10^8 particles per cm^2. What is the particle fluence in a human cell located 100 cm away from the source?
15. A 100 mm cell culture dish is irradiated using a ^{60}Co source. Five years ago, the source had an activity of 30 Ci. The source is located 1 m above the culture dish. How long must the culture dish be irradiated to cause each cell to be intercepted by an average of 20 photons? To simplify calculations, assume that at 1 m, the fluence is composed of parallel photon rays, that there is no absorption by the petri dish lid and that a cell has a diameter of 10 μm.

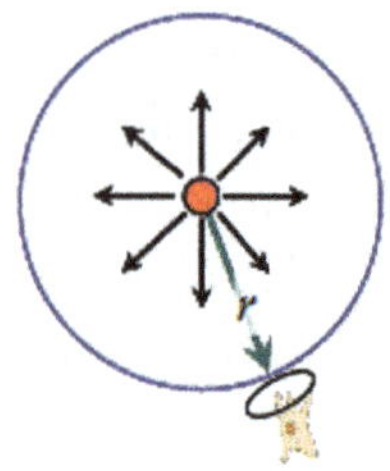

1.6. Bibliography

Bernal, S., 2018. *A Practical Introduction to Beam Physics and Particle Accelerators.* 2nd ed. Bristol: IOP Publishing.

Blosser, H. G., 1989. Beam interactions with materials and atoms. *Nuclear Instruments and Methods in Physics,* pp. 1326–1330.

Borsanyi, S. *et al.*, 2015. Ab initio calculation of the neutron-proton mass difference. *Science,* pp. 1452–1455.

Chadwick, J., 1932. Possible existence of a neutron. *Nature,* 129, p. 312.

Chadwick, J., 1932. The existence of a neutron. *Proceedings of the Royal Society,* pp. 692–708.

Chang, S., 2015. The neutron and proton weigh in, theoretically. *Physics Today,* p. 17.

Cowan, C. L. *et al.*, 1956. Detection of the free neutrino: a confirmation. *Science,* pp. 103–104.

DeLaney, T. F. & Kooy, H. M., 2008. *Proton and Charged Particle Radiotherapy.* Philadelphia: Lippincott Williams and Wilson.

Fermi, E., 1934. Versuch einer Theorie der beta-Strahlen. *Zeitschrift für Physik,* p. 161 (in German).

Geiger, H. & Scheel, K., 1933. *Handbuch der Physik,* Volume 24, Part 1. Berlin: Springer.

Hendee, W. R. & Ritenour, E. R., 2001. *Medical Imaging Physics.* 4th ed. New York: Wiley.

Mladjenovic, M., 1992. *History of Nuclear Physics: 1896–1931 — Radioactivity and Its Radiations,* volume 1. Singapore: World Scientific Publishing.

Morton, C. J. & Karzmark, R. J., 2018. *A Primer on Theory and Operation of Linear Accelerators in Radiation Therapy.* 3rd ed. Madison: Medical Physics Publishing.

Röntgen, W., 1895. Ueber eine neue Art von Strahlen. Vorläufige Mitteilung. *Aus den Sitzungsberichten der Würzburger Physik.-medic,* pp. 137–147.

Shultis, J. K. & Faw, R. E., 2007. *Fundamentals of Nuclear Science and Engineering.* 3rd ed. Boca Raton: CRC Press.

Chapter 2

Interactions Between Radiation and Absorbing Materials

2.1. Introduction

In Chapter 1, we examined the sources of radiation. In this chapter, we will investigate what happens when radiation encounters an absorbing material. For the purposes of exploration into fundamental radiation biology, the Bohr model of the atom is sufficient. The Bohr model is based on Rutherford's theory developed after he, Geiger and Marsden demonstrated that a thin gold foil scattered α-particles off a dense, positively charged nucleus encircled by negatively charged electrons (Geiger & Marsden, 1909).

2.2. The Structure of Matter

In our working model, atoms are composed of a nucleus containing protons and neutrons, with electrons circling the nucleus in orbits. The atomic diameter is about 10^{-10} m. According to classical dynamics, static negative particles placed near a positive nucleus should be drawn into the nucleus according to the relationship for Coulombic attractive forces:

$$F = \frac{Ze^2}{4\pi\varepsilon_0 r^2}, \tag{2.1}$$

F being force, Z the number of protons, e the charge of an electron, ε_0 the permittivity of free space, and r the radius which is the distance between the electron and the nucleus. But, if the charged particles encircle the nucleus, the force due to outward acceleration of the rotational bodies may oppose the Coulombic force:

$$F = \frac{Ze^2}{4\pi\varepsilon_0 r^2} = \frac{mv^2}{r}, \tag{2.2}$$

m being the mass of the rotational body, v the velocity. Imagine a mass at the end of a rope swung around by a spinning person like a shotput prior to release. The tension of the rope keeps the mass from flying off: the inward centripetal force. For the electron/nucleus system, the centripetal force is the attractive Coulombic force (Equation 2.1). The equal and opposite outward force ($F = \boldsymbol{ma} = mv^2/r$; Equation 2.2) is provided by the acceleration ($\boldsymbol{a} = v^2/r$).

When energy is added to an atom, it is distributed among the electrons. According to classical dynamics, energy added to an orbiting charge at constant velocity will accelerate the charged particle outward defining a larger radius. As the energy is released, the charge will spiral back to its original orbit. However, excited atoms do not emit a continuous energy spectrum as electrons collapse into their nonexcited states but rather emit discrete quantities of energy, in the form of wavelengths of

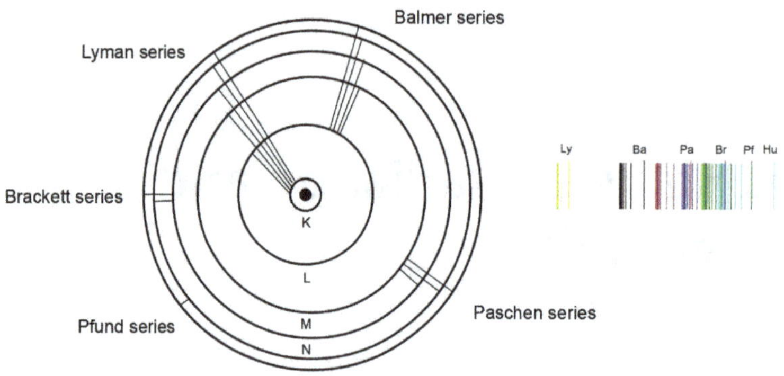

Figure 2-1 The Bohr model of the hydrogen atom. The concentric circles represent allowed orbitals. The lines indicate transitions between orbitals; electrons are elevated to higher orbitals when excited and fall to lower orbitals when energy is released. The emitted energy spectrum is identified by the lowest energy orbital in the series. The orbitals are denoted K, L, M, N, and so on.

electromagnetic radiation (inset Figure 2-1). Bohr explained this phenomenon by quantizing the electron orbits based on the following postulates:

1. An electron moves in a circular orbit about the nucleus obeying the laws of classical mechanics.
2. Instead of an infinity of orbits, only those orbits whose angular momentum (L) is an integral multiple of $h/2\pi$ are allowed.
3. Electrons radiate energy only when moving from one allowed orbit to another.

Figure 2-1 illustrates that the emitted radiation results from the difference between the binding energies of the quantized states, called orbitals or shells (in three dimension [3D]). The binding energy is the amount of energy required to remove an electron from its orbital; it is the energy required to overcome the Coulombic attractive force. The binding energy for the single electron in the lowest shell of a hydrogen atom is 13.6 eV. When electron cascades occur as a result of an electron from a higher orbital falling to a vacated lower orbital, discrete sets of emitted characteristic radiation are produced and named according to the orbital into which the electrons fall upon releasing energy. The series falling into the innermost orbital, the K-shell, is called the Lyman series. The series for electrons falling into the second shell, the L shell, is the Balmer series, and so on for the Paschen series and the Brackett series, and so forth. For hydrogen, the released radiation is in the visible to ultraviolet range. As the number of protons (Z) in the nucleus and the number of orbital electrons increase, the atom becomes more compact. The electrostatic attraction of the greater number of protons in the nucleus becomes stronger and the orbitals reside closer to the nucleus. Because the cumulative charge of the nucleus increases and the distance, r, decreases, the binding energy, the amount of energy required to remove the electron, increases and the emitted characteristic radiation passes into the x-ray range. The released x-rays become an important factor for biological damage (Figure 2-2).

2.3. Interactions Involving Neutral Particles

2.3.1 *Massless Neutral Particles*

When an energetic photon — a quantum particle of electromagnetic radiation — enters an absorbing material, it transfers its energy to the absorber atoms through one of three mechanisms. Low-energy photons (order of magnitude: 10 eV–100 keV) transfer energy via the photoelectric effect described by

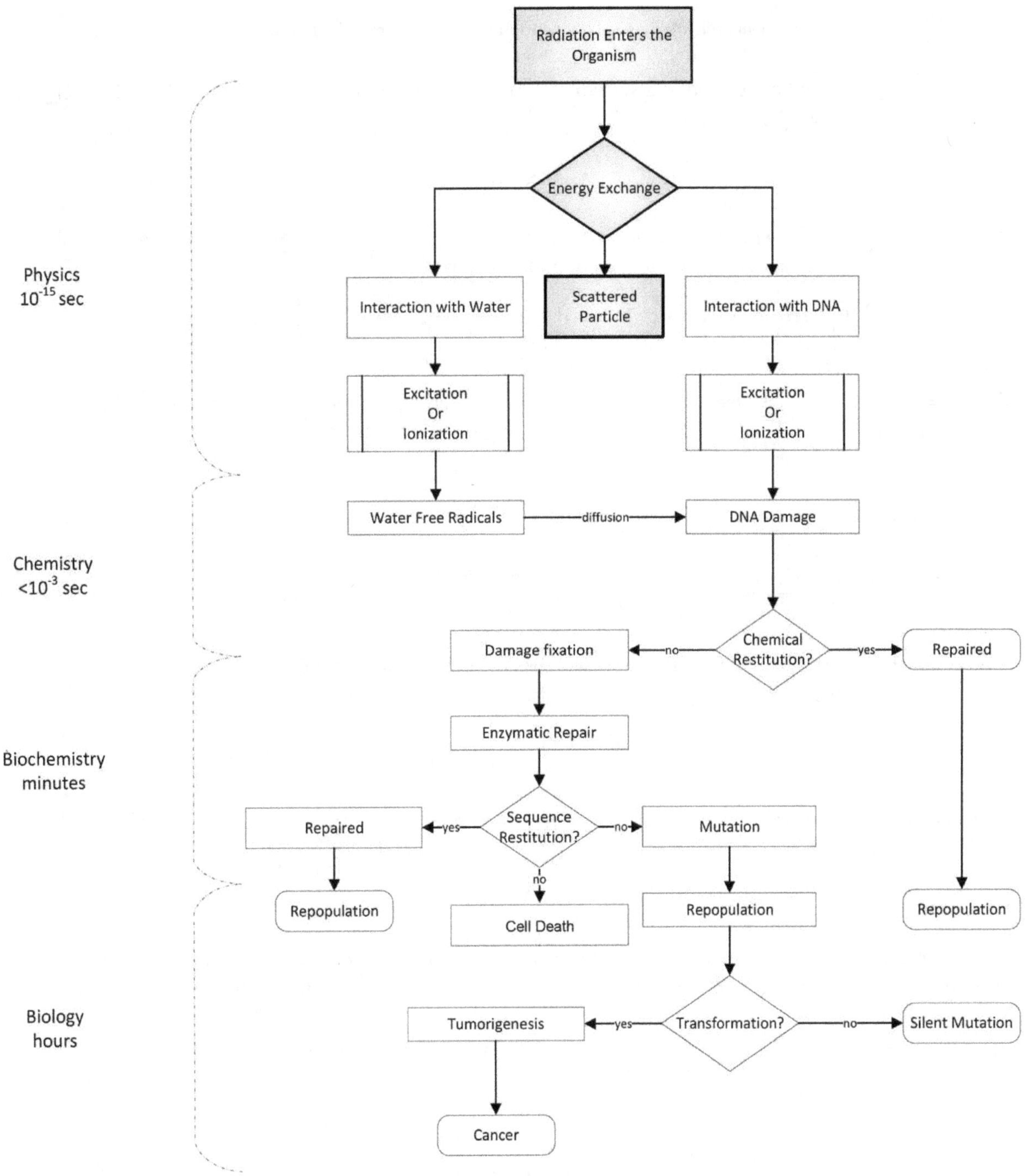

Figure 2-2 Radiation biology in a nutshell. This flowchart will reappear periodically throughout this book to indicate the relationships relevant to the topic of discussion. A gray box indicates "you are here." The disciplines noted on the left provide the relative kinetics of each domain. The final domain — biology — can occur over any duration from approximately one hour to several decades. This chapter describes the exchange of energy from the incoming radiation to the biological absorber.

Albert Einstein (for which he won the Nobel Prize in Physics in 1922). In the range of 10 keV–100 MeV (primarily 100 keV–10 MeV), most photons transfer energy to atoms via Compton scattering described by Arthur Holly Compton (for which he won the Nobel Prize in Physics in 1927). Above 1 MeV, photons may interact with the nucleus and eject a positron/electron pair from the impacted atom by a mechanism called pair production (for which Patrick Maynard Stuart Blackett won the Nobel Prize in Physics in 1948).

Quantum Mechanical Description of the Structure of Hydrogen

The hydrogen atom can be modeled as a system of two interacting point charges with the potential energy of the bound electron given by

$$V(r)=\frac{e^2}{r}$$

where r is the distance between the proton and the electron (the radius). The three-dimensional Schrödinger wave equation in spherical coordinates for the bound electron is

$$\frac{1}{r^2}\frac{\partial}{\partial r}\left(r^2\frac{\partial\psi}{\partial r}\right)+\frac{1}{r^2\sin\theta}\frac{\partial}{\partial\theta}\left(\sin\theta\frac{\partial\psi}{\partial\theta}\right)+\frac{1}{r^2\sin^2\theta}\frac{\partial^2\psi}{\partial\phi^2}+\frac{8\pi^2\mu}{h^2}[E-V(r)]\psi=0$$

where $\psi=\psi(r,\theta,\phi)$ and μ is the electron mass. The partial differential equation for the wave function is replaced by three ordinary differential equations (ODE) with a single variable:

$$\psi(r,\theta,\phi)=R(r)\,\Theta(\theta)\,\Phi(\phi)$$

and multiplied through by $r^2\sin\theta/(R\Theta\Psi)$ to obtain

$$\frac{\sin^2\theta}{R(r)}\frac{d}{dr}\left(r^2\frac{dR(r)}{dr}\right)+\frac{1}{\Phi(\phi)}\frac{d^2\Phi(\phi)}{d\phi^2}+\frac{\sin\theta}{\Theta(\theta)}\frac{d^2\Theta(\theta)}{d\theta^2}+\frac{8\pi^2\mu}{h^2}r^2\sin^2\theta\left[E-V(r)\right]=0.$$

Because the second term is a function of ϕ, whereas the other terms are independent of ϕ, the second term must equal a constant. By setting the constant equal to $-m^2$ we get the ODE:

$$\frac{d^2\Phi(\phi)}{d\phi^2}=-m^2\Phi(\phi).$$

Upon substitution, rearrangement, and by multiplying through by $\sin^2\theta$, the wave equation becomes

$$\frac{1}{R(r)}\frac{d}{dr}\left(r^2\frac{dR(r)}{dr}\right)+\frac{8\pi^2\mu r^2}{h^2}\left[E-V(r)\right]=-\frac{1}{\sin\theta\,\Theta(\theta)}\frac{d^2\Theta(\theta)}{d\theta^2}+\frac{m^2}{\sin^2\theta}=0.$$

The terms on the left are functions of r. The terms on the right are functions of θ. Both sides of the equation must be equal to the same constant. By setting the constant equal to β, we can obtain two more ODEs:

$$\frac{1}{\sin\theta}\frac{d^2\Theta(\theta)}{d\theta^2}-\frac{m^2}{\sin^2\theta}\Theta(\theta)+\beta\,\Theta(\theta)=0$$

and

$$\frac{1}{r^2}\frac{d}{dr}\left[r^2\frac{dR(r)}{dr}\right]+\frac{\beta}{r^2}R(r)+\frac{8\pi^2\mu}{h^2}\left[E-V(r)\right]R(r)=0.$$

The eigenvalue solutions to the three ODEs provide three quantum numbers used to describe the possible electron density configurations of a hydrogen atom.

A bit of mathematical magic happens here and for the details the student is referred to any good text on quantum mechanics, but the essence is that logic requires m to be an integer value and the second ODE has normalized solutions only if $\beta=\ell(\ell+1)$, where ℓ (the angular momentum) is a positive integer. Furthermore, m must be bounded by ℓ such that

$$m=0,\pm1,\pm2,\ldots,\pm\ell.$$

The third ODE has normalized solutions only if

$$E_n = \frac{2\pi^2 \mu e^2}{h^2 n^2} \text{ where } n = 1, 2, 3, \ldots$$

The integer n is the principle quantum number.

For more detail, see Shultus and Faw (2007).

Let's examine these interactions in two steps. First, a photon "collides with" an atom. Then, some portion of the photon's energy is transferred to the atom. The two events are not equivalent; it is the transfer of energy that is critical to biological effects. Here is an allegory: when John calls to say he had a car accident, you ask, "Are you okay?" The event and the consequences of the event are connected, but dissimilar. They are two different properties and the probability of each is different, but interdependent. So, first, we examine the *collision* of photon and atom. Photoelectric interactions involve relatively low-energy photons and bound electrons. In this case, a photon intercepts an orbital electron; the photon vanishes and an energetic electron escapes (Figure 2-3). If the incident photon possesses more energy than the binding energy of the innermost orbital electron, it preferentially passes all its energy to that innermost electron (K-shell). At the subatomic scale, physical interactions favor matching energies — the more closely the energy of the bound electron matches the incoming photon, the more likely the interaction. Therefore, most photoelectric events involve the K-shell electron. Because the atom has lost an electron, it carries a positive charge and thus is "ionized," and an ion pair has been created: a negatively charged electron and a positively charged atom. It was precisely this phenomenon that caused Einstein to postulate the particle-like behavior of electromagnetic radiation, the "photon."

Unlike the photoelectric phenomenon that involves only bound electrons (generally K-shell), unassociated (free) or loosely bound outer shell electrons interact with incoming photons through a different mechanism. Compton scattering is an inelastic collision between a photon and an electron resulting in a scattered photon and a recoil electron (Figure 2-4). Because the binding energy of the electron (less than approximately 1.0 keV) and the kinetic energy of the incident photon (greater than approximately 10 keV) are mismatched, only a portion of the photon energy is passed to the electron. The scattered photon has less energy (envisioned as a wave, it has lower frequency, ν, and longer wavelength, λ) than the incoming photon; less by the amount that the electron carries off. An ion pair is created.

The final mechanism of interest through which photons interact with matter becomes accessible at photon energies above 1.022 MeV. Pair production occurs when a photon enters the nuclear field of an atom and the energy ($E = mc^2$) is converted into two particles of equal mass and equal but opposite

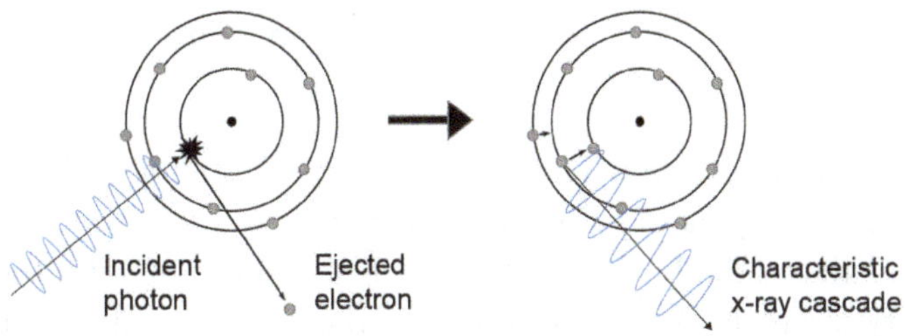

Figure 2-3 Schematic representation of the photoelectric effect. An energetic photon collides with an orbital electron. The photon disappears. The electron escapes its orbital. Ejection of an inner shell electron will provide an opportunity for higher shell electrons to fall into a lower shell releasing characteristic x-rays.

Photoelectric and Compton Interactions Explained by Electron Wave Functions

The likelihood (cross section) of photoelectric vs. Compton interaction correlates with the energies of the participant particles, and this can be understood using perturbation theory. We are accustomed to thinking of electromagnetic (EM) radiation presenting as a wave; it is less typical to think of EM in terms of particles — that is, photons. The reverse is true of electrons: we are accustomed of thinking of electrons as particles, but they can be equally well described as waves. For example, as a particle, the K-shell electron of potassium has a binding energy of 3.6 keV. Represented as an energy wave, the innermost orbital electron has a wavelength of 0.34 nm. X-ray wavelengths range from less than 0.001 nm (12 MeV x-rays produced by a linear accelerator) to 10 nm (defined on the EM scale). So, incident x-rays with wavelengths of 0.34 nm to about 0.034 nm have a high probability of interacting with the potassium K-shell electron wave.

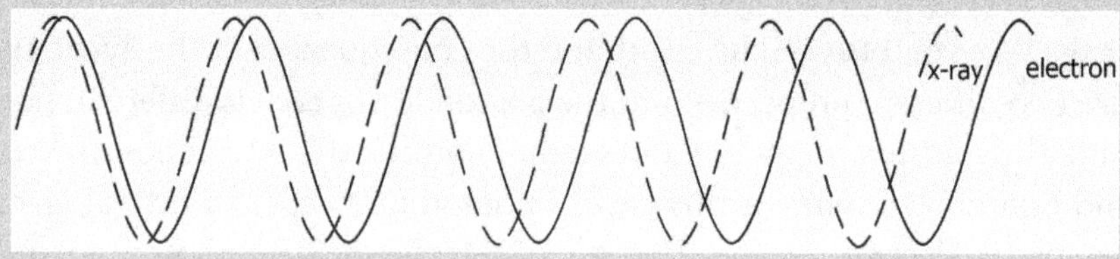

X-rays of lower energies represented by wavelengths longer than 0.34 nm do not possess and therefore cannot transfer sufficient energy to the innermost potassium electron to allow it to escape the positive nuclear influence. Incoming x-rays of higher energy fail to interact as efficiently with tightly bound electrons, such as our potassium K-shell example. Thus, the probability for photoelectric interaction decreases with increasing x-ray energy.

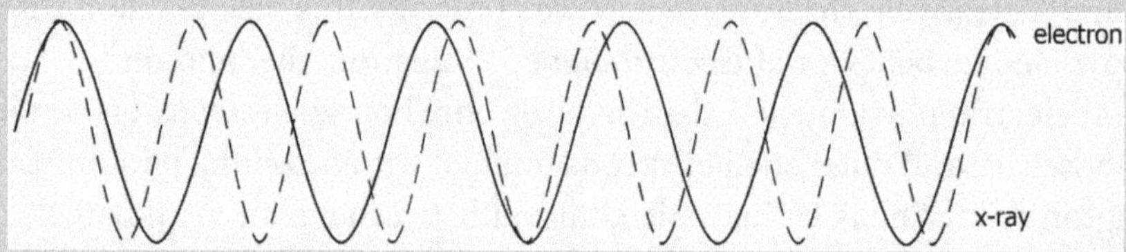

On the other hand, high-energy x-rays are better described as quantized. From about 0.1 nm up to about 2.5×10^{-4} nm, the high frequency, short wavelength x-ray is so compact compared with the electron wavelength, it inelastically scatters. For longer electron wavelengths, the mismatch becomes greater. Therefore, for very low kinetic energy unbound electrons, or loosely bound outer shell electrons, Compton scattering becomes more probable, and the interaction is better described by particle behavior (photons and electrons).

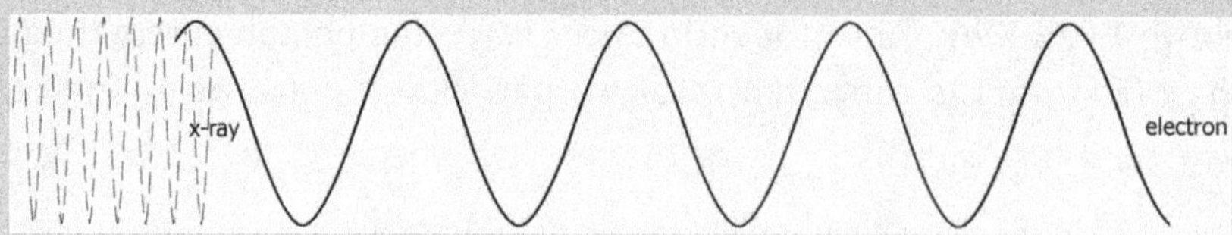

charge: a positron and an electron. The nucleus performs a function not unlike a catalyst, but it absorbs an undetermined amount of energy and gains an undetermined amount of momentum. The escaping particles lose energy through Coulombic interactions (see Section 1.5.3) with the atoms of the absorber until they reach a velocity of nearly zero, at which point the positron will pair up with an electron, its identical but oppositely charged entity, and annihilate. Two 0.511-MeV photons escape the annihilation departing in opposite directions. An ion pair and two photons are created. Because the mass energy

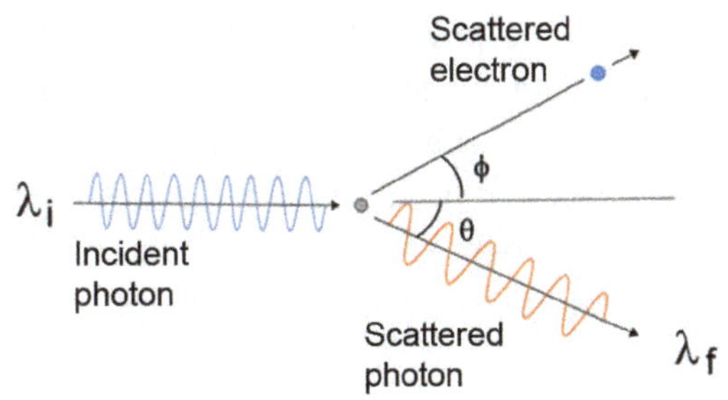

Figure 2-4 Illustration of Compton scattering. An energetic photon (described by wavelength λ_i) enters from the left and collides with a free electron. Part of the photon's energy is imparted to the electron that scatters with a velocity according to $E = ½mv^2$. The photon departing the collision has reduced energy (increased wavelength) $\lambda_f = \lambda_i - ½\, mv^2$.

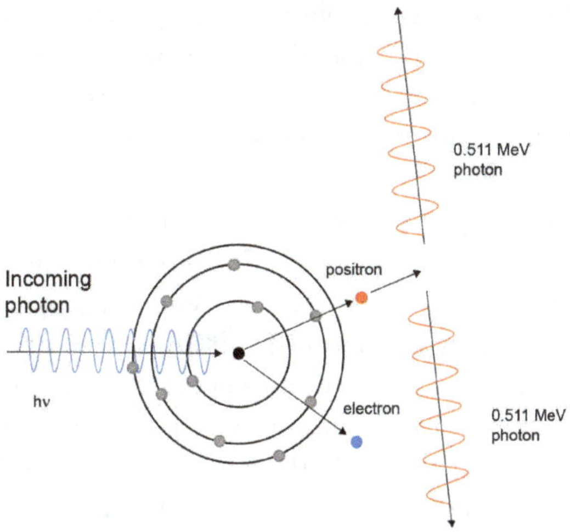

Figure 2-5 Representation of pair production. A photon with a threshold energy of 1.022 MeV enters the nucleus and vanishes. The nucleus then expels a positron and an electron. When the ejected antimatter positron stops, it interacts with an electron and annihilates emitting two 0.511-MeV photons, equivalent to the rest mass of the two particles.

equivalent ($m = E/c^2$) of an electron is 0.511 MeV, the incoming photon must have at least 2×0.511 MeV available to create the electron/positron pair (Figure 2-5).

The next logical question might be, "What is the likelihood that an incoming photon will interact through each of the three mechanisms?" Clearly, a photon of less than 1.022 MeV will have no chance (a probability of 0) of inducing pair production because it has insufficient energy to be converted into the mass of an electron/positron pair, but what about those with energies above 1.022 MeV? How often will they transfer energy through pair production and how often will they transfer energy through Compton scattering or photoelectron creation? These probabilities are called "cross sections" and have the units of area, that is, length squared — the larger the cross section, the higher the probability. For example, imagine a cross section is equal to a variable area (πr^2) of an infinitely inflatable particle that is in the path of an approaching photon. The larger the area (inflation), the greater the chance is that a photon will collide with the particle. *The cross section is not an actual physical area, it is a means of visualizing the probability of an interaction.* Each of the three interactions for photons in matter can be expressed in terms of a cross section: τ for photoelectric interactions, σ for Compton scattering, and κ for pair production. These are generally referred to as "partial attenuation coefficients" because the total of all three cross sections ($\tau + \sigma + \kappa$) defines the probability that a given photon will collide while passing through an absorber and thereby attenuate. Dividing the partial attenuation coefficients by the density of the absorbing material yields the "mass partial attenuation coefficients": τ/ρ, σ/ρ, and κ/ρ. These mass partial attenuation coefficients are convenient because the cross sections will then depend on the atomic and electronic characteristics of the absorber. This is an important point: the transfer of energy from photons (x-rays and γ-rays) is dependent upon the energy of the photons and the characteristics of the material through which they pass.

The partial mass attenuation coefficient for the photoelectric process, τ/ρ, varies with the energy of the incident photon as approximately $(h\nu)^{-3}$, where h is Plank's constant and ν is the frequency of the waveform of the photon. According to this relationship, photoelectric interactions rapidly become less probable as the incoming photon energy increases (see Figure 2-7). The electronic cross section per gram varies with the atomic number of the absorber as approximately Z^3 for high Z materials and $Z^{3.8}$ for low Z materials (not shown). In other words, the probability of a photoelectric interaction increases rapidly with decreasing photon energy and increases rapidly with increasing atomic number. You already intuitively understand this. In a radiographic film (Figure 2-6), bones attenuate x-rays more than soft tissue because diagnostic x-ray machines emit relatively low-energy x-rays (100 keV–300 keV) and bones have higher Z (Ca) than tissue (H, O, and C).

With respect to photoelectric energy attenuation, the partial mass attenuation coefficient (τ/ρ) is relatively easy to determine. There is a finite probability that a photon will interact with a bound electron (first event), and if it does, all the photon energy is transferred to the electron (second event). The attenuation coefficients for Compton scattering are more difficult to determine because although the probability of interaction may be known (first event), some of the energy of the incoming photon is attenuated but not all. The deflected photon retains energy that must be considered in order to evaluate the loss of energy to the electron (Figure 2-4). In other words, the *interaction cross section* does not equal the *attenuation coefficient.* Fortunately, Klein and Nishina calculated the electronic differential cross section for Compton scattering — the probability of photons scattered into a unit solid angle per electron:

$$\frac{d_e\sigma_t}{d\Omega} = \frac{e^4}{2m_0^2c^4}\left[\frac{1}{1+\alpha\left(1-\cos\theta\right)}\right]^2 \times \left[1+\cos^2\theta + \frac{\alpha^2\left(1-\cos\theta\right)^2}{1+\alpha\left(1-\cos\theta\right)}\right], \tag{2.3}$$

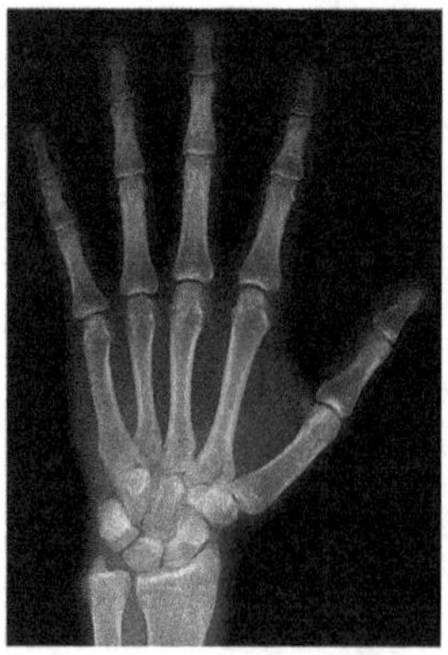

Figure 2-6 Radiographic film of a human hand. Relatively low-energy x-rays are attenuated through photoelectric interactions that have a greater mass partial attenuation coefficient in bone than in tissue. Unattenuated x-rays pass through and darken the film. Here, the initial energy is constant and the Z of the absorber changes. Radiograph by Mikael Haggstrom, MD, reproduced with permission through CC0.

where θ is the scattering angle and α is the ratio of energy of the incoming photon (hν) to the rest energy of the outgoing electron (m_0c^2). The Klein–Nishina formula can be rewritten in terms of the classical Thompson scattering formula to yield

$$\frac{d\sigma}{d\Omega}=\frac{r_0^2}{2}\left(1+\cos^2\theta\right)\left[\frac{1}{1+\alpha\left(1-\cos\theta\right)}\right]^2\times\left[1+\frac{\alpha^2\left(1-\cos^2\theta\right)^2}{\left\{1+\alpha\left(1-\cos\theta\right)\right\}\left(1+\cos^2\theta\right)}\right], \tag{2.4}$$

where r_0 is the classical electron radius. The details of this derivation are not necessary for our purpose, which is to find the electronic cross section (first event, with the free electron) for Compton scattering. Integrating Equation 2.4 over all scattering angles produces the expression for the *total Compton scattering coefficient*:

$$\sigma_e=2\pi r_0^2\left[\frac{1+\alpha}{\alpha^2}\left(\frac{2(1+2\alpha)}{1+2\alpha}-\frac{\ln(1+2\alpha)}{\alpha}\right)+\frac{\ln(1+2\alpha)}{2\alpha}-\frac{1+3\alpha}{(1+2\alpha)^2}\right]. \tag{2.5}$$

Now, we have an expression that defines the probability of photon *scatter* (collision). Following collision, all the incident energy is sometimes transferred to the electron (θ = 180°) and sometimes it is all retained in the scattered photon (θ = 0), but most of the time it is partitioned between the recoil electron and the scattered photon (Figure 2-4). Before we can resolve the partial mass attenuation coefficient (energy transfer to the electron), we need to determine the percentage of energy transferred in each case.

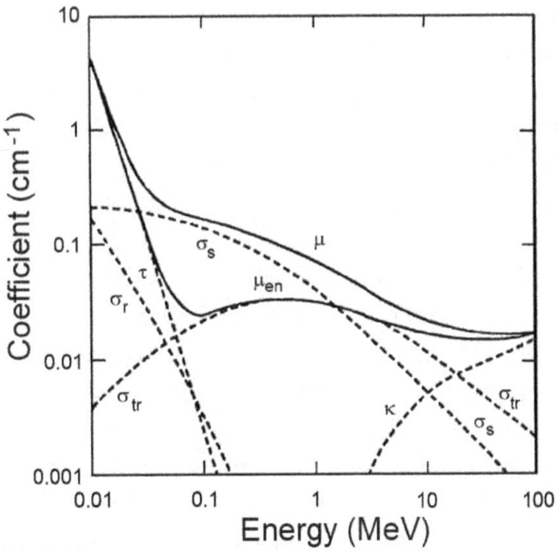

Figure 2-7 Photon interactions in water. The partial attenuation coefficients are shown for each of the photon interactions with water at increasing photon energy. The probability of a photoelectric event (τ) dominates at energies less than 50 keV and varies according to approximately (hν)$^{-3}$. Compton photon scattering probability (σ) dominates from about 100 keV to about 25 MeV. The probability of pair production (κ) initiates at 1.022 MeV and increases rapidly with incoming photon energy; it dominates above 50 MeV. The total attenuation coefficient (μ) indicates the sum of the individual probabilities. Reproduced with permission of Wiley, from *Atoms, Radiation, and Radiation Protection*, Turner, J. E., 3rd edition, copyright 2007, permission conveyed through Copyright Clearance Center.

Multiplying the expression for the electronic differential cross section (Equation 2.4) by an expression for fractional energy transferred to the electron provides the differential energy transfer coefficient

$$\frac{d_e\sigma_{tr}}{d\Omega}=\frac{r_0^2}{2}(1+\cos^2\theta)\left[\frac{1}{1+\alpha(1-\cos\theta)}\right]^2\times\left[1+\frac{\alpha^2(1-\cos^2\theta)^2}{\{1+\alpha(1-\cos\theta)\}(1+\cos^2\theta)}\right]\left(\frac{\alpha(1-\cos\theta)}{1+\alpha(1-\cos\theta)}\right). \quad (2.6)$$

Integrating Equation 2.6 over all angles yields the energy *transfer* coefficient.

$$\sigma_{tr}=\frac{3}{4}\sigma_0\left[\frac{2(1+\alpha)^2}{\alpha^2(1+2\alpha)}-\frac{1+3\alpha}{(1+2\alpha)^2}+\frac{(1+\alpha)(1+2\alpha-2\alpha^2)}{\alpha^2(1+2\alpha)^2}-\frac{4\alpha^2}{3(1+2\alpha)^3}-\left(\frac{1+\alpha}{\alpha^3}-\frac{1}{2\alpha}+\frac{1}{2\alpha^3}\right)\ln(1+2\alpha)\right] \quad (2.7)$$

The expression is ponderous, but nonetheless we now have a means of calculating the *cross section for energy transfer to the recoil electron* (or attenuation), given a Compton scattering event. Figure 2-7 shows the relationship between the probability that a photon will scatter (σ_s) and the probability that energy will be transferred to the electron (σ_{tr}). The probability that energy will be retained by the scattered photon (σ_s) is simply the difference between the total Compton scattering coefficient (σ) and the energy transfer coefficient (σ_{tr}). The probability of interaction and the probability of photon scatter fall off with increasing energy and are virtually independent of the Z of the absorber (not shown). In water, the probability of transferring energy to the free electron increases with energy to about 200 keV and then declines. As we shall see later, it is the transfer of energy to the electron (σ_{tr}) that determines local energy absorption (μ_{en}), and so the cross section for absorption (Figure 2-7) looks very much like the cross section for energy transfer (σ_{tr}) where Compton scattering dominates.

Clearly the mass attenuation coefficient for pair production is equal to zero for all photon energies less than 1.022 MeV because there is insufficient photon energy to convert into the masses of the positron and the electron. For incoming photons with energies above 1.022 MeV, the process is much like photoelectric attenuation because the photon is converted into a positron/electron pair in the vicinity of the nucleus; all the energy is attenuated. The photon vanishes and a positron/electron pair materializes, and so, the attenuation coefficient is determined by the cross section for the pair production event. This cross section increases rapidly above the threshold energy (Figure 2-7) and approaches a limit around 100 MeV incoming photon energy in water. Pair production, like photoelectron production, depends on the photon energy and the absorbing material; the cross section for pair production varies as approximately Z^2. The process is more likely to occur in bone than in soft tissue.

An illuminating graphical representation of the coefficients for each of the three interactions described is presented in Figure 2-7. Water is an acceptable surrogate for biological tissue because adult humans are approximately 60% water: blood and kidneys consist of 83% water, muscles are 76% water, although adipose tissue contains only 10% water. Figure 2-7 presents, on a log-log scale, the partial attenuation coefficients for photoelectric (τ) and pair production (κ) events. The probability of a photoelectric event dominates at energies less than 50 keV and varies as approximately $(h\nu)^{-3}$. Compton scattering dominates from about 100 keV to about 25 MeV. The graph breaks down Compton scattering into the energy transfer coefficient ($\sigma_{tr} = \sigma\, ^{T}/h\upsilon$) — the total probability of scattering multiplied by the average fraction of the incident photon energy (expressed as the initial kinetic energy of the scattered electron) — and the energy scattering coefficient ($\sigma_s = \sigma\, ^{h\upsilon'}/h\upsilon$) — the total probability of scattering multiplied by the average fraction of the incident photon energy (expressed as the kinetic energy of the scattered photon). The probability of pair production initiates at 1.022 MeV and increases rapidly with incoming photon energy. The total attenuation coefficient (μ) indicates the sum of the partial attenuation coefficients, both energy transfer to electrons and scattered photons; the probability that a photon will collide. The utility of μ is discussed in Section 2.3.1.1. Also indicated in the graph is the total energy absorption coefficient (μ_{en}): the total probability that energy will be deposited locally.

A few important concluding remarks: we have not yet discussed quantitative energy absorption in such a way that we can begin to "follow the energy" into and through a biological system. Thus far, we have discussed the mechanisms through which photon energy is attenuated, and the probability that each of these mechanisms might occur. We also did not mention a process called triplet production wherein photon energy is deposited into an orbital electron and a positron/electron pair also is manifested. Thus, two electrons and a positron carry off the photon energy. This interaction requires a minimum incoming photon energy of 2.04 MeV, and it is rare in low Z absorbers such as tissue. Another photon scattering mechanism is "coherent scattering," or Rayleigh scattering. During this interaction, no energy is converted into kinetic energy and so it is of no interest to the radiation biologist. The incident photon causes the orbital electrons of the absorber to oscillate and thereby radiate electromagnetic radiation. The result is a change in the direction of the incoming photon without loss of energy. The partial mass attenuation coefficient for coherent scattering (σ_{coh}/ρ) is included in Figure 2-7.

2.3.1.1. *Macroscopic implications for massless particles*

Let's step back and examine the process of photon attenuation on a macroscopic scale (Figure 2-8). Say you have a slab of homogeneous absorbing material. A beam of x-rays impinging normally on an absorbing slab with intensity I will be attenuated, will scatter, or will pass through the slab without interacting at an intensity I′. A detector placed in the optimal position will detect only x-rays that have passed through the absorber without interacting; the uncollided radiation. The relationship between the incident radiation and the uncollided radiation is

$$I' = Ie^{-\mu x} \tag{2.8}$$

where x is the thickness of the slab and μ is the linear attenuation coefficient: the probability per unit distance traveled before a particle interaction (units of cm^{-1}). The term $e^{-\mu x}$ represents the portion of photons that do *not* collide before leaving the slab, as I′ represents the exiting uncollided photons.

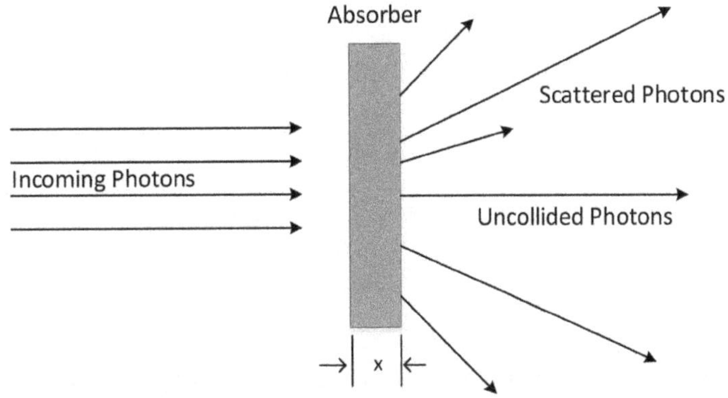

Figure 2-8 Transmission of photons through a homogeneous absorbing material. Photons enter from the left, attenuate or scatter within the attenuating material. Those that are not attenuated — the photons from characteristic x-rays, Compton scattering, annihilation photons, and the uncollided photons — exit the far side of the absorber. In the case of an appropriately sized detector placed far beyond the attenuator, only uncollided photons are collected. This is termed "good geometry."

Because the detection system is arranged under conditions of "good geometry," the scattered photons are not collected by the detector and so the linear attenuation coefficient accounts for the loss of intensity by all mechanisms, including scatter, excitation, and ionization resulting in characteristic x-rays, and annihilation photons. In Figure 2-7, the attenuation coefficient is designated μ to indicate that it includes attenuation by both energy transfer to an electron and photon scatter (photoelectric, Compton, and pair production). Our discussion of mass attenuation coefficients considers single photons, each having a discrete energy. When we consider the fluence of many photons, the beam rapidly diverges into a spectrum of energies, each photon having an energy resulting from that photon's collisional history. We know from our examination of individual microscopic interactions that the cross section for photoelectric events with low-energy photons is much greater than the cross section for Compton scattering or pair production at higher energies (Figure 2-7). In other words, low-energy x-rays are preferentially attenuated. The consequence is that as x-rays penetrate, the energy distribution is shifted to higher energies. The beam becomes less intense but more penetrating because the linear attenuation coefficient for low-energy photons is greater than the linear attenuation coefficient for higher energy photons. In Figure 2-9, the photon energies are represented in the visible spectrum with blue indicating the higher energies and red indicating the lower energies. The *number* of penetrating photons (arrows) is given by Equation 2.8; the *energy spectrum* is represented by the colored arrows.

An important and useful concept for radiation protection is the half value layer (HVL). This is the thickness of a given material that will reduce the intensity of an x-ray beam by half. The health physicist can speak in terms of HVLs of shielding material, for example. Notice that this value is readily derived from Equation 2.8

$$\frac{I'}{I} = \frac{1}{2} = e^{-\mu HVL} \rightarrow \ln 2 = \mu HVL \rightarrow HVL = \frac{0.693}{\mu} \tag{2.9}$$

where μ is the effective linear attenuation coefficient for the spatially relevant energy spectrum. However, it is important to remember that because low-energy x-rays are preferentially attenuated, the second HVL will be thicker than the first, and the third thicker than the second and so on. This phenomenon of "beam hardening" affects dose deposition. An intuitive grasp of beam hardening will serve the reader well.

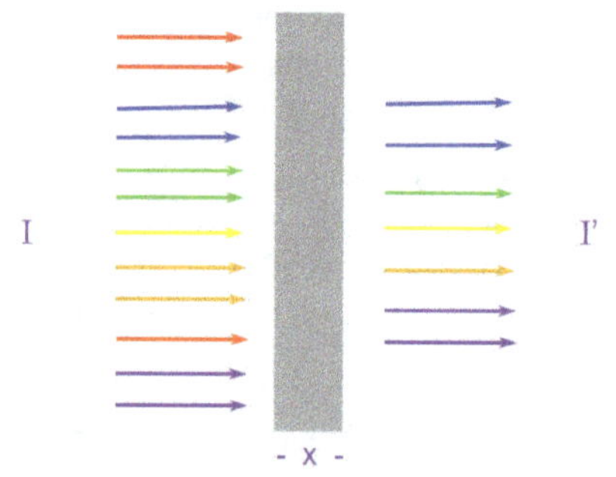

Figure 2-9 Representation of an x-ray spectrum impinging on a slab of absorbing material. Photon energy is represented in the visible spectrum with red photons (arrows) having lower energy and blue photons (arrows) having more energy. The relationship between the incident beam and the uncollided exiting beam is $I' = I\,e^{-\mu x}$. The number of photons passing through the slab uncollided is reduced from the number of photons impinging on the slab and furthermore, the number of lower energy photons is reduced to a greater extent than the higher energy photons.

Because the radiation biologist is more concerned with what happens inside the absorbing material (person) than the characteristics of the escaping radiation, notice that the relationship for the collided radiation, the radiation that does *not* pass through without interacting, can be determined by

$$I_x = I_0 - I' = (I_0 - I_0 e^{-\mu x}) = I_0(1 - e^{-\mu x}). \tag{2.10}$$

2.3.2. *Massive Neutral Particles — Neutrons*

Thus far, we have considered the interactions between massless, neutral particles (photons), and the absorbing material through which they pass. There is also a massive neutral particle of interest — the neutron. Neutrons exhibit similar attenuation characteristics as they pass through an absorber with intensity falling off exponentially with depth (thickness, x). However, neutrons do not interact with orbital electrons, they instead interact with the *atomic nuclei* of the attenuator. This difference in energy transfer has profound biological effects that are manifested in a "quality factor" or "radiation weighting factor" that we will examine later.

Neutrons interact with atomic nuclei through five mechanisms including three types of scatter: elastic scatter, inelastic scatter, and nonelastic scatter. The remaining two disparate interactions are neutron capture and spallation. The cross sections for each of these interactions depend on the energy of the neutron and the nuclear composition of the target atom. Furthermore, because energetic neutrons interact with the nucleus of the atoms composing the absorber, molecular chemistry is not relevant; the individual cross sections for each elemental species simply can be added to obtain the total cross section for a given neutron energy. The compositional percentage is a multiplicative factor. For example, the neutron cross section for water (H_2O) would be two times the cross section for hydrogen plus the cross section for oxygen. Neutron cross sections are not intuitive and must be looked up on a chart of the nuclides. Figure 2-10 provides a depiction of the relative thermal neutron cross sections for three prominent biological nuclides: hydrogen, carbon, and oxygen. Figure 2-11 shows the variation in total scattering and total absorption cross sections with increasing neutron energy for the same three elements. To understand the contribution of each of the five mechanisms, we will examine each of the mechanisms in the order of their prominence with energy.

Let's examine scattering first. Neutrons with kinetic energies of up to about 14 MeV experience scattering. Hydrogen (specifically, the single proton that comprises the nucleus of hydrogen) presents the largest cross section for elastic scattering in biological absorbers, with significant cross sections also for oxygen (O), carbon (C), and nitrogen (N) in descending order (Figure 2-10, N not shown). Between 250 keV and 14 MeV, about 85% of the neutron energy transferred is due to scatter interactions with hydrogen. Elastic scatter retains all the incoming neutron's energy as the neutron collides with and departs the nucleus.

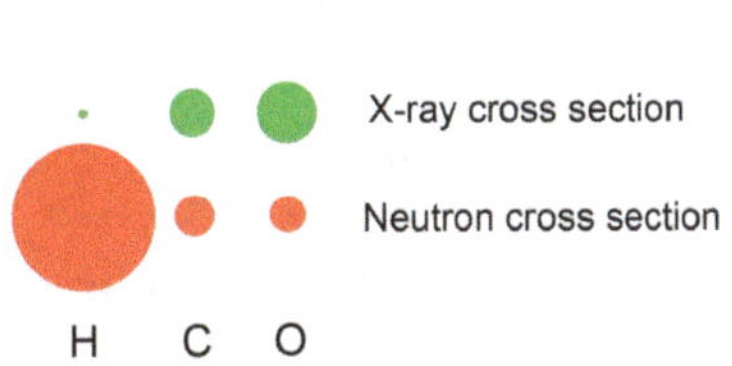

Figure 2-10 Relative attenuation cross sections of hydrogen (H), carbon (C), and oxygen (O) for x-rays (green) and thermal neutrons (red). While the probability of x-ray interactions increases with increasing Z, the nuclear cross section for neutron interactions is not intuitive and must be obtained from a chart of the nuclides.

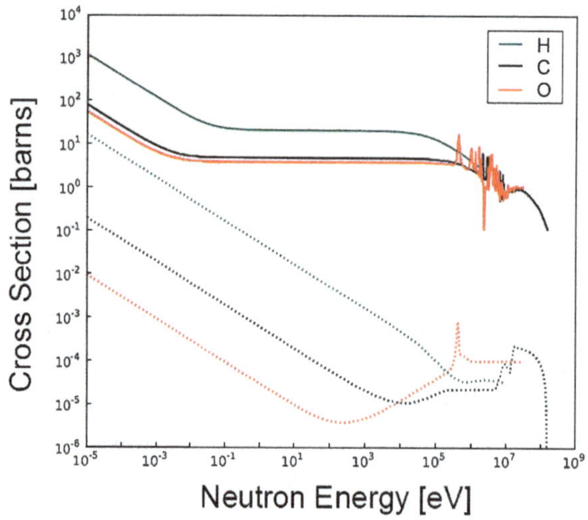

Figure 2-11 Neutron scattering (solid line) and absorption (dotted line) cross sections of common biological elements. Resonance structures can be seen in the MeV energy range. The data was obtained from the database NEA N ENDF/B-VII.1 using JANIS software and plotted using mathplotlib. Reproduced and modified with permission from ProkopHapala through CC BY-SA.

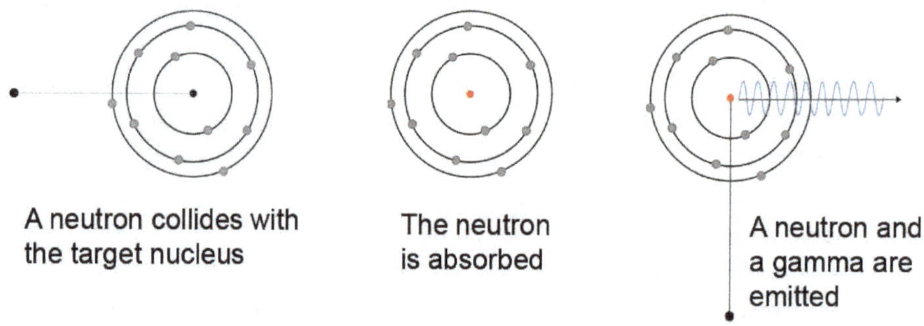

Figure 2-12 Inelastic scattering loses energy to the excited nucleus and the scattered gamma. The massless, uncharged gamma and the massive, uncharged neutron will not transfer energy locally.

The larger nuclei — carbon, oxygen, and nitrogen — are more likely to participate in inelastic (Figure 2-12) or nonelastic (Figure 2-13) scattering. When a neutron is inelastically scattered, the neutron is transiently taken up by the target nucleus and is released coincident with a gamma photon; the initial neutron energy is partitioned between the two emission products and any retained excitation of the target nucleus. The reaction for nitrogen looks like this:

$$ {}^{14}_{7}\text{N}(n,\, n')\,{}^{14}_{7}\text{N}. \tag{2.11}$$

The incoming neutron energy range for inelastic scattering requires that the threshold energy be adequate for conservation of momentum and energy.

Nonelastic scattering results in the emission of a novel charged particle, often a proton or alpha particle. Because the binding energy for the emitted particle determines the threshold interaction energy, larger target nuclei favor nonelastic scattering over inelastic scattering. An example for scattering from carbon would be:

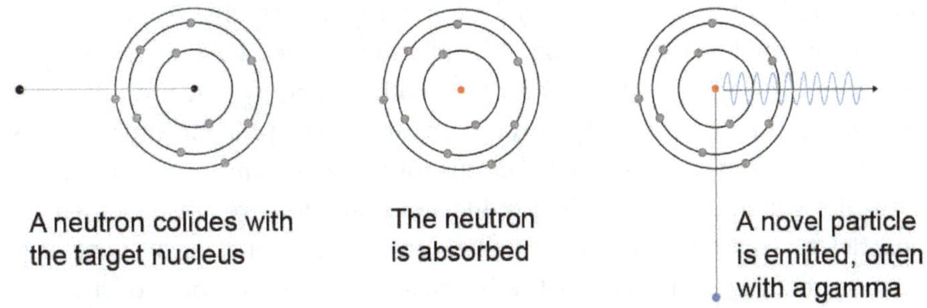

Figure 2-13 Nonelastic scattering is similar to inelastic scattering except that a novel particle such as a proton or alpha particle is emitted following the absorption of the neutron.

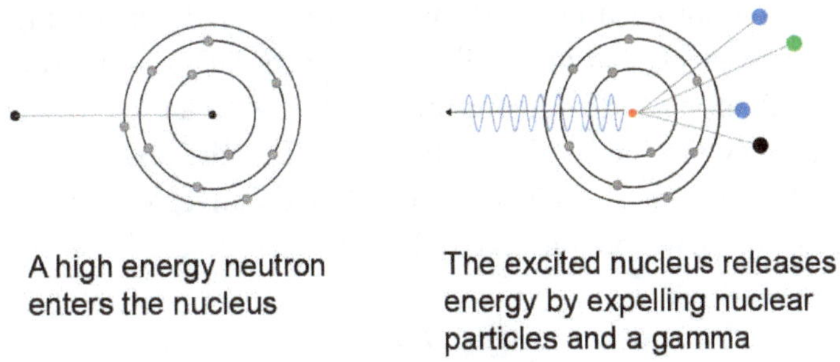

Figure 2-14 Spallation results from the capture of an energetic neutron. A high energy neutron enters the nucleus and transfers its excess energy to several particles, expelling them along with gamma radiation.

$$^{12}_{6}\mathrm{C}(n,\ \alpha)^{9}_{4}\mathrm{Be}. \tag{2.12}$$

Because one nonelastically scattered product possesses charge, that part of the energy is absorbed locally, whereas energy allotted to escaping neutrons and gammas leave the local environment. So, neutron scattering results in charged and uncharged energetic products depending on the neutron energy and the target nucleus binding energy. The types of discharged particles and their energies will determine the biological effects, so we will return to these mechanisms several times.

The cross section for neutron capture is significant for near thermal neutrons and again overwhelmingly favors interactions with hydrogen nuclei (Figure 2-10). Thermal energy neutrons possess energies well below an electron volt (~0.025 eV), so one can assume that neutron interactions with hydrogen consist of neutron capture at very low energies and elastic scattering at low energies (<14 MeV). Thermal neutrons drift into the hydrogen nucleus and bind to the proton:

$$^{1}_{1}\mathrm{H}(n,\gamma)^{2}_{1}\mathrm{H}. \tag{2.13}$$

There are also noteworthy cross sections for carbon, oxygen, and nitrogen, for example, the reaction for nitrogen:

$$^{14}_{7}\mathrm{N}(n,p)^{14}_{6}\mathrm{C}. \tag{2.14}$$

Notice that radioactive isotopes can be generated through neutron capture, resulting in the release of charged particles.

If the neutron captured by the nucleus enters the nuclear field with kinetic energy above 100 MeV, the nucleus is likely to fragment. The cross section for spallation (FIgure 2-14), as this fragmentation is designated, increases from 100 MeV to 400–500 MeV. Such high-energy neutrons do exist in outer space but are not often encountered by earthlings except under catastrophic circumstances such as nuclear reactor meltdown. The neutron capture cross sections for hydrogen and carbon are on the order of 10^{-5} barns, even less for nitrogen and oxygen. The products of spallation are novel elements, ions, and penetrating neutral particles — gammas and neutrons — and so the penetrating neutral particles can be considered "incoming radiation" for each of the interactions examined previously, with suitable cross sections for the interactions discussed.

2.4. Interactions Involving Charged Particles

Whereas neutral particles — photons and neutrons — interact with absorbing materials through stochastic processes, that is, cross-sectional probabilities, charged particles interact with the environment through Coulombic forces. Photons and neutrons must collide with atomic components of the absorber. Charged particles continuously interact with all charges within the absorber — electrons, nuclei, and ions — and the strength of the interactions is determined by the inverse square of the distance between the particles. This is Coulomb's Law for stationary charged particles,

$$F = k_e \frac{q_1 q_2}{r^2} \tag{2.15}$$

where k_e is approximately 9×10^9 N m^2 C^{-2}, q_1 and q_2 are the charges of two interacting particles, and r is the distance between the charged particles. The sum of the forces between a particle of interest and all other charged particles in the environment defines the cumulative electrostatic force on a particle. An electron will interact with electrons in the absorber according to

$$F_{\text{coul}} = \frac{2Ze^2}{4\pi\varepsilon_0 r^2}. \tag{2.16}$$

where e is the electron charge and ε_0 is the permittivity of free space. The forces on an energetic charged particle are a bit more complicated, but the situation can be imagined to be like a ball in space under the influence of surrounding gravitational objects. The ball is pulled toward other objects depending on their distances and their masses. The charges within the absorber tug and push the energetic particle depending on the distance, the strength of the charge and the sign of the charge. Recall that all three x-ray interactions (photoelectric, Compton, and pair production) produce energetic electrons (Figures 2-3, 2-4, and 2-5). We are primarily concerned with these electrons, as they are responsible for the deposition of dose resulting from x-rays. Common charged particle therapies include proton and carbon ion beams, and alpha particles are emitted by radioisotopes. Nuclear events emit both x-rays (resulting in energetic electrons) and charged particles. Comprehending the deposition of energy within a biological absorber requires internalizing the fundamental physical difference between the way neutral particles and charged particles transfer energy. Photons transfer energy *stochastically* and charged particles transfer energy *continuously.*

Most often, electrons lose kinetic energy to the environment through interactions with orbital electrons resulting in ionization or excitation. However, electrons also can interact with the positively charged nuclei. If an electron manages to penetrate the repulsive electron shells, the attractive force of

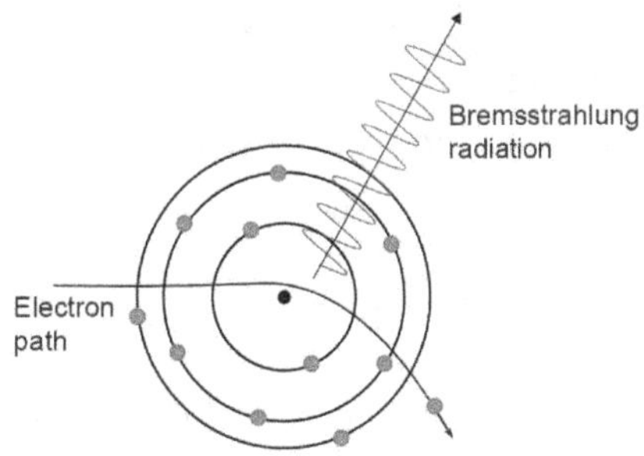

Figure 2-15 Bremsstrahlung radiation. An electron penetrates the negatively charged orbital shells and accelerates under the influence of the positively charged nucleus. The change in the velocity vector causes a release of energy in the form of a photon. Depending on the degree of deflection, a photon of any energy from 0 MeV to the initial kinetic energy of the electron may be created.

the nuclear protons will cause the electron path to be diverted resulting in a change in the velocity vector and therefore a negative acceleration, or deceleration (Figure 2-15). This deceleration results in reduced velocity and thus a loss of kinetic energy ($\frac{1}{2}mv^2$). The lost energy manifests as a photon proportional to the square of the velocity vector change. Photons released in this manner are referred to as "radiative energy loss" or "Bremsstrahlung" (strictly translated from German: braking radiation). The greater the electron's deviation from the linear path, the greater the radiative loss, and therefore Bremsstrahlung results in a continuous spectrum of photon energies. Radiative loss results from Coulombic forces between an energetic particle of charge ze, and a nucleus of charge Ze, creating acceleration proportional to the square of the product of the charges divided by the energetic electron mass, m_e, $\left(Zze^2/m_e\right)^2$. From this relationship, it becomes clear that the greater the Z of the material and the smaller the mass of the energetic particle, the more probable Bremsstrahlung becomes. The useful x-ray radiation produced in diagnostic x-ray tubes and medical accelerator thin targets is Bremsstrahlung. The phenomenon is less important in soft biological tissues (low Z) but remains greater for electrons than protons or other heavy particles.

2.5. Energy Transfer

2.5.1. *Massless Neutral Particles*

Now, we are ready to begin investigating the transfer of energy from radiation to biological material. Photoelectric energy transfer deposits all the energy of the incoming photon into an inner shell orbital electron. The photon vanishes and an energetic electron escapes (Figure 2-3) carrying away the energy of the incoming photon minus the binding energy expended in freeing the electron. The expression for the energy transferred from a photon to an electron during a photoelectric interaction is thus:

$$E_e = h\nu - \mathrm{BE} \tag{2.17}$$

where $h\nu$ is the quantum energy of the incoming photon and BE is the binding energy of the escaping electron. All the energy of the incoming photon is transferred to the electron during this event. Ejection of an inner shell electron provides an opportunity for electrons in higher energy shells to occupy lower energy shells, releasing characteristic x-rays. The radiation biologist must account for the characteristic radiation as an additional x-ray source.

Compton scattering results in a scattered photon and a recoil electron produced by an inelastic collision between a high-energy photon and an outer shell or "free" electron (Figure 2-4). The scattered photon has less energy than the incoming photon, less by the amount that the electron carries off. The energy of the recoil electron is therefore

$$E_e = h\nu - h\nu' \tag{2.18}$$

where $h\nu$ is the energy of the incoming photon and $h\nu'$ is the energy of the scattered photon. Expressed in relativistic terms, the kinetic energy of any electron is

$$E_e = total\,energy - potential\,energy = m_0c^2\left[\frac{1}{\sqrt{1-{v^2}/{c^2}}}-1\right] \tag{2.19}$$

where the potential energy is the rest mass of the electron. Because momentum must be conserved, the recoil angle of the electron (ϕ) is determined by the scatter angle of the photon. Referring to Figure 2-4, the momentum of the incoming photon must equal the sum of the horizontal vectors of the scattered photon and the recoil electron:

$$\frac{h\nu}{c} = \frac{h\nu'}{c}\cos\theta + \frac{m_0 v}{\sqrt{1-{v^2}/{c^2}}}\cos\phi. \tag{2.20}$$

Similarly, the vectors in the vertical direction for the recoil electron and scattered photon also must be equal:

$$\frac{h\nu'}{c}\sin\theta = \frac{m_0 v}{\sqrt{1-{v^2}/{c^2}}}\sin\phi. \tag{2.21}$$

The final expressions for the energies of the recoil electron and the scattered photon can be derived by substitution into Equation 2.18, elimination of the velocity (v), and rearrangement (solving in terms of the angle of deflection of the scattered photon, θ),

$$E_e = h\nu\,\frac{\alpha(1-\cos\theta)}{1+\alpha(1-\cos\theta)} \tag{2.22}$$

and

$$h\nu' = h\nu\,\frac{1}{1+\alpha(1-\cos\theta)} \tag{2.23}$$

where α is the ratio of the incoming photon energy to the rest mass equivalent energy of the electron:

$$\alpha = \frac{h\nu}{m_0c^2}. \tag{2.24}$$

The expression for α reveals an important insight. Intuitively, the result of the interaction depends on the energy of the incoming photon relative to the particle with which it collides. Imagine a very energetic photon corresponding to a large α. What happens to the term $\alpha(1 - \cos\theta)$? How does that affect the scattered photon energy following the collision? The electron energy? Now, imagine a relatively low-energy incoming photon. How is this situation different? Because photons travel relatively large distances before interacting — on the scale of centimeters in water — most of the scattered photons will leave the volume of interest, whereas the electrons will deposit energy locally. Again, we concern ourselves with the energetic electron, but the scattered photon constitutes an additional x-ray source.

When the incoming photon of Figure 2-4 collides with an outer shell electron and is backscattered in the direction of the incoming photon, $\theta = 180°$ and $\phi = 0°$ and the maximum energy is imparted to the electron that travels straight forward. This, then, is the maximum amount of energy that is transferred to an electron:

$$E_{e(\max)} = h\nu \frac{2\alpha}{(1+2\alpha)}. \tag{2.25}$$

The maximum energy transferred is *not* equal to the incoming photon energy ($h\nu$) as it is during photoelectron creation.

Pair production is the third photon interactive process that can occur when a photon of sufficient energy enters the nuclear field of an atom and is converted into two particles of equal mass and equal but opposite charge: a positron and an electron (Figure 2-5). The nucleus absorbs an undetermined amount of energy, and so the energies of the escaping particles must be represented as spectra as they were in the case of isotopic emission of beta particles during decay (Figure 1-9). The escaping charged particle's energy, the kinetic energy of the newly created particle pair (i.e., their velocity), dissipates through Coulombic interactions with the atoms of the absorber until they reach nearly zero kinetic energy. At this point, the positron aligns with an electron — its identical but oppositely charged entity— and annihilates. Because each of these particles claims a rest mass equivalent of 0.511 MeV, two photons, each having an energy of 0.511 MeV, escape the annihilation, departing 180° in opposite directions to preserve angular momentum. In truth, the center of mass retains an average momentum of 0.009 m_0c^2 causing a distribution of vectors of about 0.5° around the mean of 180°. Furthermore, about 2% of the time, a positron with persistent kinetic energy annihilates resulting in annihilation photons exceeding 0.511 MeV and deviating from the 180°separation. These last two minor points are of interest to imaging physicists but can be neglected by most radiation biologists. Therefore, for our purposes, the energy transferred from the incoming photon to the positron/electron pair, plus that retained by the nucleus, is equal to the incoming photon energy less the energy required to create the masses,

$$E_{tr} = h\nu - 1.022 \text{ MeV}. \tag{2.26}$$

The energy passed to the electron and the positron is absorbed locally through Coulombic interactions. The annihilation photons have long pathlengths and escape the local volume (see Figure 2-18). The photons resulting from characteristic x-ray emission, Compton scattering, and positron annihilation continue penetrating the absorbing material, becoming the incoming photons engaging in photoelectric events, Compton scattering and pair production. Lower energy photons preferentially interact through photoelectric mechanisms and the higher energy photons interact through Compton scattering or pair production (if they possess at least 1.022 MeV energy). In turn, the released and scattered photons engage in photoelectron creation, Compton scattering, and perhaps pair production, and so on. Because the scattered photon energy is less than the incident photon energy, the scattered photons are eventually reduced to sufficiently low energy to be converted into photoelectrons. And so, pair production produces energetic electrons and 0.511-MeV photons; Compton scattering produces energetic electrons and scattered photons; and photoelectric interactions produce energetic electrons with coincident characteristic x-ray emission. Excitation of orbital electrons — where sufficient energy is transferred to raise an electron to a higher energy orbital, but not enough to allow it to escape — results in characteristic x-rays. The released 0.511-MeV photons and the scattered photons continue to interact until the remaining energy is sufficiently low to engage in a photoelectric event. The energy *transferred* is the maximum energy available for absorption, but it is greater than the energy *absorbed* because of the escaping photons (characteristic x-rays, scattered photons, annihilation photons, and Bremsstrahlung radiation).

2.5.2. *Massive Neutral Particles*

Elastic collisions between two massive bodies retain the total kinetic energy following the event; there is no net conversion of kinetic energy into other forms such as heat or noise. In the case of neutron scatter, energy is transferred from the collisional neutron to the recoil nucleus. In the case of hydrogen, the recoil proton — a heavier charged particle — will interact with the absorber to deposit the collisional energy locally. The scattering is nearly isotropic, and the energy transferred can be derived by

$$E_{\mathrm{tr}} = E_{\mathrm{n}} \frac{4M_{\mathrm{a}}M_{\mathrm{n}}}{(M_{\mathrm{a}}+M_{\mathrm{n}})^2}\cos^2\theta . \tag{2.27}$$

Here, the initial energy of the collisional neutron is E_{n}, the angle of deflection of the nucleus is θ, the mass of the nucleus is M_{a} and the mass of the neutron is M_{n}. The average energy transferred to a target nucleus is obviously,

$$\overline{E}_{\mathrm{tr}} = E_{\mathrm{n}} \frac{2M_{\mathrm{a}}M_{\mathrm{n}}}{(M_{\mathrm{a}}+M_{\mathrm{n}})^2} . \tag{2.28}$$

When the collision between a neutron and a nucleus is inelastic, neither escaping particle carries a charge and so neither deposits energy locally. Recall that the energetic neutron is absorbed by the nucleus and a lower energy neutron escapes. The excited nucleus emits a gamma with an energy equal to the remainder of the energy. As in the case of coherent (Raleigh) scattering, the interaction is of little interest to the radiation biologist except that the released gamma will interact later through photoelectric, Compton scattering, or pair production events.

If the neutron collision is nonelastic, protons or alpha particles may be released from the impacted nucleus (Equation 2.14). These heavier charged particles travel short distances under the influence of Coulombic forces and deposit their energy locally. Because of their large mass, the alpha particles lose energy extremely quickly over very short distances. This large energy deposition over a short range results in severe biological damage and therefore, nonelastic neutron collisions are of paramount concern for understanding neutron radiation biology. Nonelastic neutron collisions release a spectrum of particle energies that depends on the type of particle emitted, the recoil of the nucleus, and sometimes energy given off from the excited nucleus as gamma emission.

For thermal and near thermal neutrons ($E_{\mathrm{n}} \approx 0.025$ eV), neutron capture accounts for nearly all the energy transferred to tissue. The escaping gamma resulting from capture by the hydrogen nucleus (Equation 2.13) will have an energy of 2.2 MeV provided by the reduction in mass achieved during capture. This gamma will not deposit energy locally however, the proton ejected during capture by nitrogen (Equation 2.14) does deposit 0.58 MeV energy locally. So, although the latter is not predominant, this reaction provides greater local energy transfer.

2.5.3. *Massive Charged Particles*

All energetic charged particle lose energy to their environment through Coulombic interactions. Because atoms contain many electrons per nucleus, the majority of Coulombic interactions occur between energetic charged particles and bound electrons. Consider the following situation: a particle with charge equal to ze and mass M is traveling at a velocity v (possessing kinetic energy ½ mv^2), approaches an electron of mass m_0 and charge e. Figure 2-16 illustrates this situation. The force exerted on the electron depends on the distance between the two charged particles, r. When $r = \text{b}$, the distance between the two particles (the "impact parameter") is at a minimum. The force between the particles will be:

$$F = \frac{kze^2}{r^2} \tag{2.29}$$

where k is the constant of proportionality, 8.9875×10^9 N m^2 C^{-2}.

Protons are several orders of magnitude more massive than electrons but carry the same charge. Because they are significantly more massive, their track structure — the path they follow and the pattern of energy absorption along that path — is more linear and simpler to describe than that of electrons. So, let us begin with a description of the energy absorption physics of protons and later describe the energy absorption from electrons. The proton energy (velocity) loss under the influence of the interparticle force (Equation 2.29) can be derived thusly. The momentum imparted to the electron can be found by

$$\Delta p = \int_{-\infty}^{+\infty} F_y dt = \int_{-\infty}^{+\infty} \frac{kze^2}{r^2} \cos\theta \; dt. \tag{2.30}$$

An expression for the classical electron radius has been derived from classical coherent scattering theory,

$$r_0 = \frac{ke^2}{m_0 c^2} = 2.81794 \times 10^{-15} \, \text{m}. \tag{2.31}$$

We can use this expression and rewrite Equation 2.30 in terms of the angle θ:

$$\Delta p = \frac{kze^2}{vb} \int_{-\pi/2}^{+\pi/2} \cos\theta \; d\theta = \frac{2zke^2}{vb} = \frac{2zr_0 m_0 c^2}{vb}. \tag{2.32}$$

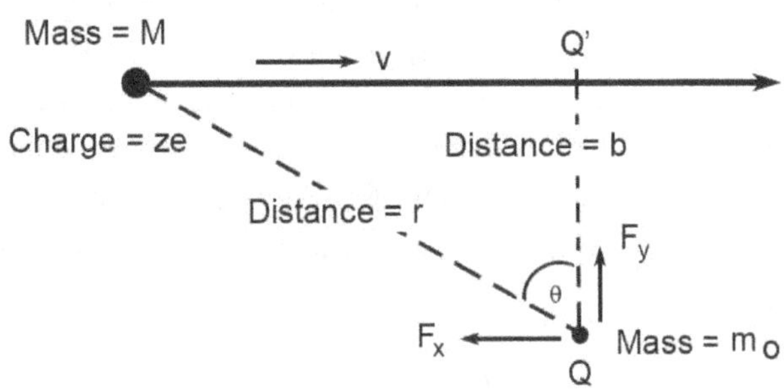

Figure 2-16 Formulation schematic for the Coulombic interaction between a massive, charged particle passing an electron. *F* is the force acting on the charged particle (electron) of mass m_0.

The energy lost from the heavy particle will thus be:

$$\Delta E(b) = \frac{(\Delta p^2)}{2m_0} = \frac{2z^2 r_0^2 m_0 c^4}{v^2 b^2} = \frac{z^2 r_0^2 m_0 c^4 M}{b^2 E} \tag{2.33}$$

because $v^2 = 2E/M$. The variables in this formulation are z, M, b, and the kinetic energy of the moving particle, E. For protons and electrons, $z = 1$. The energy loss depends strongly on the inverse of the impact parameter b, and the inverse of the velocity of the incoming particle, as both are squared. The source of the radiation determines the initial velocity of the particle and the difference between that and the residual particle energy determines the energy loss (ΔE). Because of the inverse relationship, the *greater* the velocity of the particle, the *less* energy it loses due to interactions with electrons. Notice that the momentum transfer to the bound electron depends on time (Equation 2.30). A fast particle has insufficient time to interact with the bound electron. As the particle loses velocity, it exchanges increasing energy quanta with each interaction until it nearly comes to rest, at which time it loses maximal energy (Equation 2.33). By our gravity analogy, the attraction increases as the energetic particle slows.

So, how are electron and proton projectiles different? Equation 2.33 indicates that energy loss depends directly upon the mass of the particle and the square of the charge. The difference between protons ($z = 1$) and electrons ($z = 1$) is due therefore only to mass. A collision between particles of like mass will result in the deflection of both particles. It is not possible to determine the degree of deflection; this can only be approximated mathematically for large numbers of collisions. Nonetheless, the energy loss resulting from each collision remains dependent upon the impact parameter, the velocity of the incoming particle, and the mass. A single 200 keV energetic electron arising from an ionization event — a photoelectron, a scattered electron, the electron or positron of an ejected pair, an Auger electron, or a beta particle — can give rise to approximately 15,000 subsequent ionizations or excitations. Clearly, electron energy loss to biological tissue constitutes the primary mode of radiation injury. In the context of accelerated heavier and multiply charged particles such as alpha particles, spallation products, or accelerated carbon ions, each energy loss event is significantly greater (proportional to M and Z^2) than for electrons or protons, and the heavy particle is not deflected. Imagine these collisions as classical interactions between balls of different masses. A two-electron collision would look like a golf ball rolled at another golf ball. A collision between a carbon ion and an electron would look like a bowling ball rolled at a golf ball. In the latter case, the golf ball is deflected, but the bowling ball continues along its path without deviating.

2.6. Kerma, Cema, and Dose

2.6.1. *Kerma*

Initially, for gamma and x-ray radiation, energy is transferred from an incoming photon to an electron in the biological absorber. The energy lost from these neutral particles results in the excitation of orbital electrons to higher energy orbitals, or ionization of an atom resulting in an energetic electron and a positively charged ion. This process is *energy transfer* and it is measured in kerma (**k**inetic **e**nergy **r**eleased per unit **ma**ss). Kerma (K) is defined as the transfer of energy per unit mass; it is also the photon fluence (Φ) multiplied by the probability of interaction (mass attenuation coefficient, μ/ρ) times the average energy transferred per event,

$$K = \frac{dE_{tr}}{dm} = \Phi \frac{\mu}{\rho}\left(\overline{E}_{tr}\right). \tag{2.34}$$

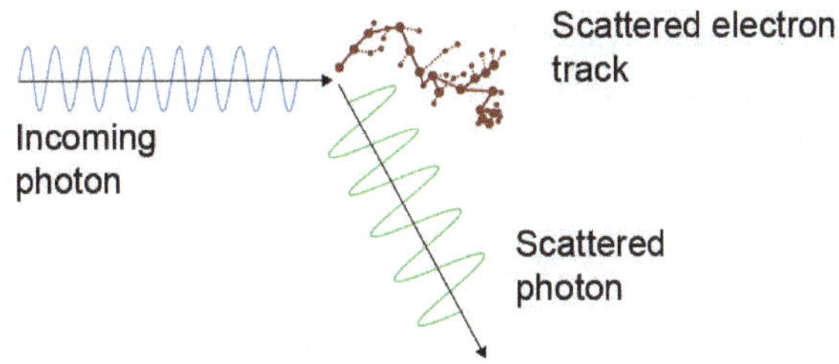

Figure 2-17 Detail of a Compton scattering event. An incoming photon ($E = h\nu$) scatters off a free electron. The energized electron continues along a tortuous path as it interacts with local bound electrons through Coulombic forces, ejecting secondary electrons with which it interacts. The loss of energy from the primary electron to the secondary electrons increases rapidly near the end of the path as the velocity slows to zero. Some of the secondary electrons have sufficient energy to eject additional electrons; these secondary electrons are called "delta electrons." The charged particle interactions were modeled by Monte Carlo calculation. The scattered photon ($E = h\nu'$) leaves the volume of interest.

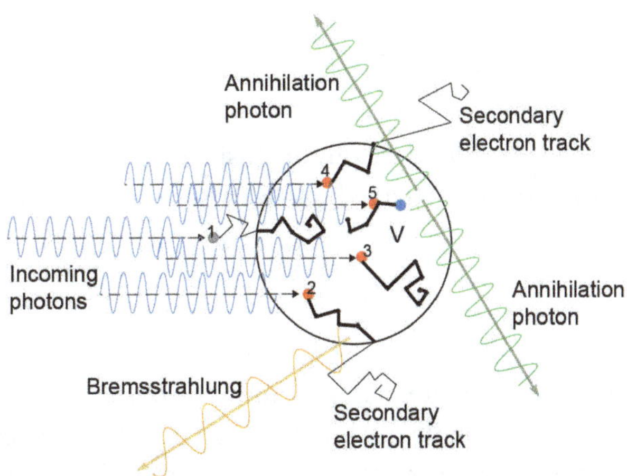

Figure 2-18 Illustration of the mechanisms of energy loss, transfer, and absorption in volume, V. Kerma (K) results from the kinetic energy transfers (red dots) to electrons 2, 3, and 4, and the nucleus (positron/electron pair) 5, but not electron 1 because it is not within the volume of interest (V). $K = \Sigma(E_i) = E_2 + E_3 + E_4 + E_5$. The energy lost from K includes the annihilation photons and Bremsstrahlung that escape the volume of interest. Dose is deposited (energy is absorbed) along the heavy black sections of the secondary electron tracks (black lines). The primary electron 1 enters the volume with a residual energy (cema), inducing subsequent dose deposition.

The units of kerma are energy per mass: joules per kilogram (J/kg). Kerma provided a special radiation biology term: the gray (Gy); 1 Gy = 1 J/kg. This is important: kerma is *not* the energy lost from an incoming photon, which may include excitation, a scattered photon, or nuclear absorption. The *energy transferred*, E_{tr}, includes only the *kinetic energy gained by electrons* encountered by photons (the energy transferred from the incoming photon in Figure 2-17 to the scattered electron). In Figure 2-18, the energy transferred at "1" does not contribute to the kerma because the transfer of energy from the photon to the electron does not occur within the volume of interest. The electrons, to which the energy is transferred, may lose some energy if they interact with a nucleus, losing some of their energy as Bremsstrahlung. In fact, if the primary electron suffers radiative energy loss at any point along its path, this energy is no longer available for subsequent ionizations. Imagine your uncle gave you $10 in quarters, but before you had the opportunity to spend it, you put a quarter into the parking meter, dropped a quarter into a Red

Kettle, and lost a quarter down the sewer grill. Now you have only $9.25 to spend. Thus, kerma is usually divided into two components: collisional kerma (K_{col}) and radiation kerma (K_{rad}). If the fraction of energy lost is designated "*g*," then the collisional kerma is

$$K_{col} = K(1 - \bar{g}). \tag{2.35}$$

2.6.2. *Cema*

Kerma pertains only to neutral particle energy transfers — *indirectly ionizing radiation.* Radiation sources also produce charged particles. All charged particles expend energy through Coulombic interactions. Charged particle radiation is therefore *directly ionizing radiation.* Cema (converted energy per unit mass) defines the average amount of energy converted in a small volume from charged particle collisions with atomic electrons. The unit for cema is J/kg, as before.

$$C = \frac{\overline{E}_c}{dm} \tag{2.36}$$

Cema involves the energy transferred in electronic collisions by *incoming* charged particles. Kerma involves the energy imparted to *outgoing* charged particles such as the ionized electrons after photon collisions.

2.6.3. *Dose*

Radiation dose is the energy absorbed in a volume of interest and is measured in joules per kilogram (J/kg), as it is for kerma and cema. One joule per kilogram is a gray (Gy) as before. The medium absorbs energy through excitation and ionization events (ΔE; Equation 2.33), as well as pair production and thermal energy deposition that result from collisions along the secondary electron track. The secondary electron tracks are represented as black lines in Figure 2-18. Because dose is defined as the change in energy due to absorption per unit mass, it is functionally the photon fluence multiplied by the probability of interaction (mass attenuation coefficient) times the average energy per absorption event (ionization, excitation, thermal transfer). Dose is relevant for both indirectly (photon) ionizing events and directly (charged particle) ionizing events.

$$D = \frac{dE_{ab}}{dm} = \Phi \frac{\mu}{\rho}\left(\overline{E}_{ab}\right) \tag{2.37}$$

2.7. Stopping Power

The loss of kinetic energy per unit distance along the charged particle track (dE/dx) is called "stopping power." The energy lost (ΔE) is calculated by Equation 2.33. More frequently the "mass stopping power" ($1/\rho\ dE/dx$) is employed, where ρ is the density of the absorbing material. The pattern of energy deposition is stochastic and cannot be predicted for an individual particle. Nonetheless, stopping power allows us to describe the characteristics of the track structure at various energies, in different absorbers. Remember, it is the deposition of energy in biological tissue that causes damage. The track structure predicts the precise character of the damage and that in turn predicts the biological response. A good analogy for stopping power might be friction. A hockey puck propelled on ice will travel a long way before it comes to rest. Ice does not have much friction — or in our analogy, stopping power. However, if that same hockey puck is taken out to the street and struck with the same impact (i.e., the same energy imparted), it would travel a much shorter distance. The blacktop has more friction — or stopping power. The resistive force for charged particles is the electrostatic interaction with the orbital electrons of the absorber (previously the gravity of astronomical bodies). The hockey puck loses energy through heat; the secondary electron loses energy through heat, ionization, and excitation within the atomic milieu. We examine the stopping power along the black tracks (both thick and thin) of Figure 2-18. It is not a dosimetric quantity and so it is not necessarily confined to the "volume of interest." Tracks 2 and 4 have residual energy at the point where they leave the volume of interest, and that residual energy can be calculated using stopping power. But notice in Figure 2-18 that energy is lost from the electron traversing track 2 to Bremsstrahlung and track 5 to annihilation photons. When calculating dE/dx, we must consider these losses just as we did when we calculated collisional and radiative kerma. The Bethe-Bloch equation for electron collisional mass stopping power looks like this:

$$\frac{S_{\text{col}}}{\rho} = \frac{1}{\rho}\frac{dE}{dx} = \frac{\mathrm{N_A} Z}{A}\frac{\pi r_e^2 2m_0c^2}{\beta^2}\left[\ln\left(\frac{\tau^2(\tau+2)}{2\left(I/m_0c^2\right)^2}\right) + \mathrm{F}(\tau) - \delta - \frac{2C}{Z}\right] \tag{2.38}$$

where $\mathrm{N_A}$ is Avogadro's number, r_e is the classical electron radius, I is the mean excitation energy, τ is the ratio of the electron kinetic energy to the rest energy equivalent of the electron mass (m_0c^2), δ is a density correction, and C/Z is a shell correction factor. The remainder of the terms are as usual: Z is the atomic number of the absorber, A is the atomic mass number, β is the ratio of the velocity of the particle to the speed of light (v/c). The mean excitation energy cannot be calculated from atomic physics; the values have been empirically determined and can be looked up on tables (e.g., ICRU Report 49). This term represents the portion of transferred energy that is expended in exciting electrons to higher orbitals — this transfer does not result in ionization but does result in the collapse of the electrons back into their original orbitals emitting characteristic x-ray radiation. The term τ is similar to α; both indicate the energy of the incoming particle (charged in the first case and neutral in the latter) normalized to the rest energy of the electron (m_0c^2). The function of τ, ($\mathrm{F}(\tau)$), depends on the energetic charged particle species. For an electron, $\mathrm{F}(\tau)$ looks like this:

$$\mathrm{F}(\tau) = 1 - \beta^2 + \frac{\tau^2/8 - (2\tau+1)\ln 2}{(\tau+1)^2}. \tag{2.39}$$

Notice that the numerators of the first two terms in Equation 2.38 are composed of constants with fixed values, except for Z. If we substitute "k" for the value of those constants, and we suppress the second order terms, we find that

$$\frac{S_{\text{col}}}{\rho} \propto \mathrm{k}\frac{Z}{A}\frac{1}{\beta^2}. \tag{2.40}$$

In other words, to a first approximation, the mass stopping power depends on the absorber: the ratio of the number of electrons to the nuclear mass (Z/A). For low Z elements, the number of protons equals the number of neutrons. So, in general, Z/A equals ½ for biological materials. The denominator, β^2 (v^2/c^2), reflects the particle's kinetic energy (½mv^2). A high energy particle, traveling at relatively high velocity, will not lose much energy because β is large (and squared). As the particle loses energy (velocity), the rate

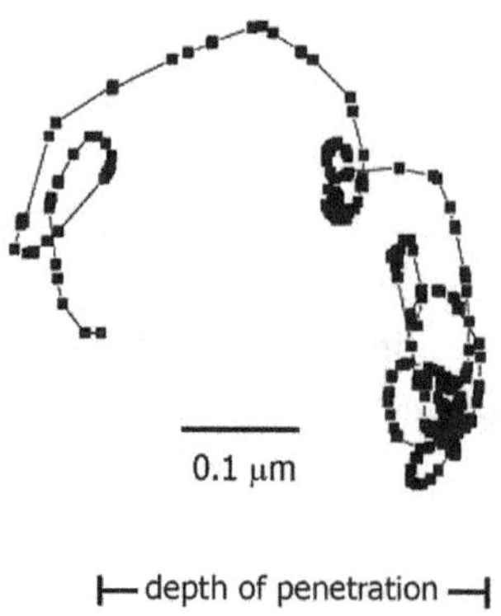

Figure 2-19 Monte Carlo simulation of a 10-keV electron track in water. Black lines represent the primary electron track initiating at the left of the frame. The pathlength (*l*) is measured along the electron track, whereas the range and *dx* are measured linearly to the furthest depth of penetration. Reproduced with permission of the Oxford University Press, from Electron track simulation using ETMICRO, Kim, E. H., *Radiation Protection Dosimetry*, 2006; 122: 53–55.

of energy loss (dE/dx), becomes rapidly greater. This phenomenon can be seen in Figure 2-19 where the end of the electron path exhibits a cluster of energy loss events. The bracketed terms of Equation 2.38 influence the stopping power at energies above three times the mass of the energetic charged particle: 3×0.511 MeV = 1.5 MeV for electrons (the influence of τ). The delta term is a density correction factor. As an energetic electron passes atomic electrons, the repulsive force polarizes the atoms slightly. This shields atoms further away such that the energy exchanges with those more distant atoms decrease. The δ-effect is greater for solids and liquids than for gases because in the latter the atoms are further apart, and it is therefore called a density correction factor. The δ-effect suppresses the slowly increasing dE/dx at high energies. The fraction C/Z affects the low energy extreme of the stopping power; it is the shell effect correction. According to the Bohr model, the electrons orbit around the nucleus at velocities sufficient to prevent collapse. When the projectile electron slows to velocities less than the orbital velocity of the K-shell electron, the cross section for interaction vanishes. The K-shell electron is no longer available for excitation (the I-value is overestimated). The factor C/Z corrects for this overestimation as evidenced at low energies by a fold-over on a graph of Stopping Power versus Energy that prevents the stopping power from shooting off to infinity at velocity = 0. The fold-over is not apparent in Figure 2-20 as it occurs below 0.01 MeV.

At this time, it is important to note the dependence on the absorbing medium — both Z/A and I increase as Z increases. This increase is relevant for metals, and also for hard biological materials such as bone. Increased dE/dx is important for calculating dose deposited in soft tissue immediately in front of and just distal to bone in patients being treated with radiation. This phenomenon increases dose to bone marrow and affects tissue near metal implants. For a detailed derivation of the Bethe-Block equation for electrons, the reader is referred to Attix (2004) or any equivalent radiological physics text.

From the expression for stopping power (Equation 2.38), it should be possible to predict the distance that a particle will travel before stopping, provided the particle type, energy, and absorbing material are known. By our analogy, if you knew how hard the puck was struck, you could predict if it would reach the goal or stop short. This distance is called the particle range ($\mathfrak{R}$) and it is simply the inverse of the stopping power. The distance traveled is calculated as the linear distance, not the distance along the electron path, which can be tortuous. Depending on the degree to which the energetic electrons are deflected at each collision, the depth of penetration may vary from something extremely short to something equal to the total pathlength. Figure 2-19 illustrates a Monte Carlo simulation for a 10-keV electron traversing water. For electrons, the pathlength is almost always greater than the range, and in some cases, the furthest depth of penetration may be beyond the terminal stopping point (Figure 2-19). In addition, energy is deposited in discrete events (for an electron, the average energy deposition event is 20 eV), making residual energy calculations difficult. The mathematical solution to these dilemmas is the continuous slowing down approximation (CSDA),

$$\mathfrak{R}_{\mathrm{CSDA}} = \int_0^{E_{\max}} dE \frac{1}{S(E)} \text{ or } \frac{1}{\rho}\mathfrak{R}_{\mathrm{CSDA}} = \int_0^{E_{\max}} dE \frac{1}{S(E)/\rho}. \tag{2.41}$$

Equation 2.41 assumes that energy is being lost from the energetic primary electron continuously (rather than stochastically) as it traverses the absorbing medium. The result is that $\mathfrak{R}_{\mathrm{CSDA}}$ is slightly less than true $\mathfrak{R}$ (by ≥0.2%). The values of $\mathfrak{R}_{\mathrm{CSDA}}$ have been calculated for various energies and absorbers and are tabulated in many references. They can be found on the National Institute of Standards and Technology (NIST) website for protons (PSTAR) and electrons (ESTAR), and for an exhaustive number of absorbers.

For electrons, Bremsstrahlung (radiative loss) is significant even in biological tissues. The radiative stopping power for electrons looks like this:

$$\frac{S_{\mathrm{rad}}}{\rho} = \left(\frac{1}{\rho}\frac{dE}{dx}\right)_r = \frac{\mathrm{N_A}}{137}\left(\frac{e^2}{m_0c^2}\right)^2 \frac{Z^2}{A}\left[E + m_0c^2\right]\bar{B}_r. \tag{2.42}$$

where $\frac{1}{137}\,(e^2/m_0c^2)^2 = 5.80 \times 10^{-28}$ cm²/atom, Z/A is approximately ½ for biological tissues, $\mathrm{N_A}$ is Avogadro's number, and m_0c^2 is 0.511 MeV. $\bar{B}_r$ is a slowly varying function of Z and electron energy equal to roughly 6 at 1 MeV, increasing to 15 at 100 MeV. This formulation clearly demonstrates a dependence on the medium Z as indicated previously, and a dependence on the energy of the electron. Figure 2-20 illustrates the radiative, collisional, and total stopping powers with respect to the initial energy of the incident electron in aluminum

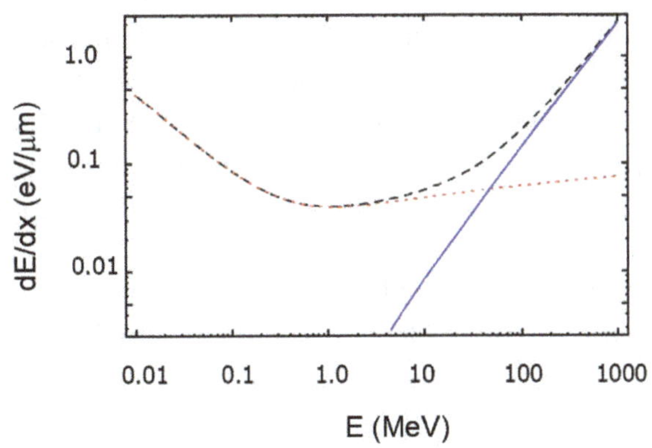

Figure 2-20 Relative contributions of collisional (red dashed) and radiative (solid blue) energy loss in aluminum. The contribution from Bremsstrahlung is not impactful below and the collisional loss is virtually constant above three times the mass of the particle (1.5 MeV for an electron). The total stopping power is the sum of the collisional and radiative components (black dashed line). The fold-over at very low energies is not apparent in the energy range depicted. Reproduced with permission of the American Institute of Physics, from Collisional and collective effects in two-dimensional model for fast-electron transport in refluxing regime. Volpe *et al.*, *Physics of Plasmas*, 2013; 20. https://doi.org/10.1063/1.4771586.

($Z_{Al} = 13$; $Z_{water} = 10$). Notice that below 1.5 MeV, radiative stopping power is negligible. Recall that the energetic electron must have sufficient energy to penetrate the electron shells before interacting with the nucleus.

Because ionization and subsequent energy absorption resulting from x-rays and γ-rays is the principle precursor to radiation damage, we began our discussion of stopping power by examining electron interactions. Proton radiation is more straightforward because of the increased mass; the charge magnitude is precisely the same. If a proton and an electron are initially at the same energy and the range of either is known, the unknown range can be approximated by:

$$\mathfrak{R}_{CSDA} = \mathfrak{R}^{P}_{CSDA}\frac{m_0}{M_0} \tag{2.43}$$

where $\mathfrak{R}^{P}$ indicates "proton range" and M_0 or m_0 indicate the rest mass of the proton and the rest mass of the electron, respectively. The ratio of masses, 0.511 MeV/931.5 MeV = 5.486×10^{-4}, reduces the range of the electron compared with the range of the proton an impressive distance, largely a reflection of the electron/electron scattering that is evident in the tortuous electron path. Compare Figure 2-19 that shows a Monte Carlo simulation of a single electron track in water with Figure 2-21 showing a Monte Carlo simulation of a single proton track in water. The residual energy in Figure 2-21 is initially approximately 9 keV. Protons do not exhibit significant Bremsstrahlung in low Z materials, so S_{rad}/ρ is negligible. While the energetic electrons are repelled by the orbital electrons, the protons are attracted, so the electron interactions "push" electrons out of their orbits and protons "pull" electrons out of their orbits. Heavier ions behave in analogous fashion and we will examine some of these in greater detail when we discuss radiation quality, but an understanding of the differing physics explained in this chapter exhibited by photons, electrons, and protons provides a solid foundation for the implications of dose and biological response.

Now, let's step back again and look at a macroscopic expression of dose. When a linear accelerator produces a radiation beam and that beam is intercepted by a water tank — a slab of water — the deposition of dose in the water can be measured by a translational dosimetry device (Figure 2-22). By this mechanism, we can examine the inside of the slab in Figure 2-8. A comparison of the depth–dose curves — the deposition of dose in water at increasing depth — for x-rays, electrons, and protons is shown in Figure 2-23.

Because photons (x-rays) transfer energy stochastically –that is, the photon creates ion pairs through photoelectric interactions, Compton scattering, or pair production, with a *probability of interaction* equal to the sum of the individual cross sections, μ (Figure 2-7)–the *dose* (a reflection of μ_{en}, Figure 2-7) is greatest near the surface of the water where the x-rays first enter the absorbing medium. As photons transfer their energy, fewer photons remain (Figure 2-8) and the number of collisions decreases with depth, less energy is transferred, and the dose decreases. Thus, the ^{60}Co depth–dose curve of Figure 2-23 (the broken dotted line) falls off with depth.

The proton curve of Figure 2-23 (solid line) looks quite different. The entrance dose is about 30% of the maximum deposited dose; the maximum dose presents deep into the water, with the depth depending

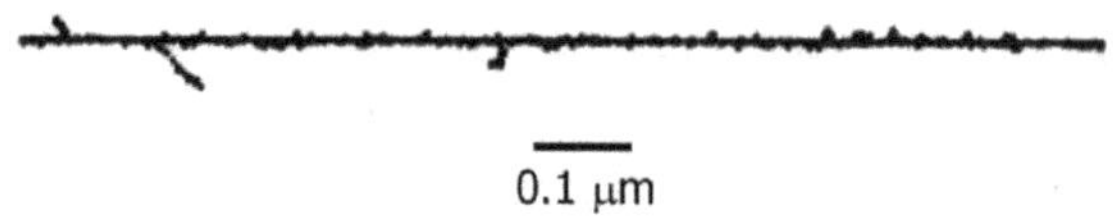

Figure 2-21 Monte Carlo of a simulated 9-keV proton track in water. This calculated representation of a proton path (*dE/dx* = −40 keV/mm) demonstrates the linearity of the proton track structure. Branches indicate delta electron tracks. Heavy ion range is approximately equal to the greatest depth of penetration. Reproduced with permission from "Development of Monte Carlo track structure simulations for protons and carbon ions in water." Thiansin Liamsuwan, Doctoral Thesis, Stockholm University, 2012.

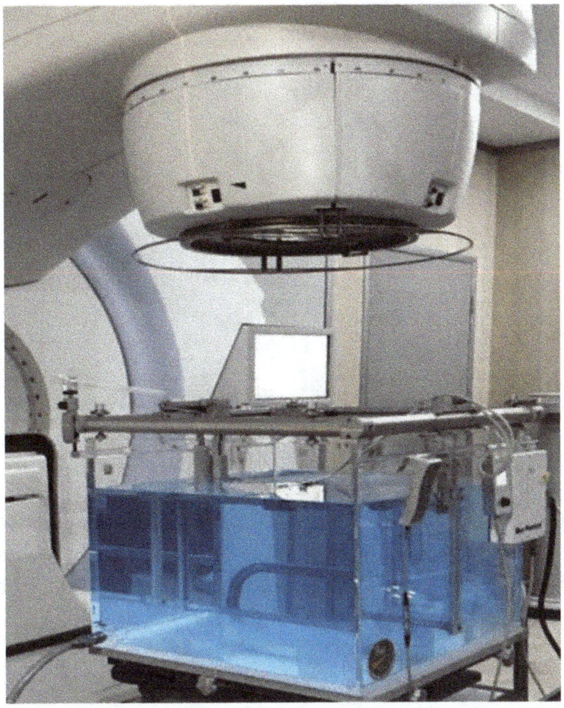

Figure 2-22 Setup for measurement of dose in water using a clinical linac. The dosimeter travels downward in the water as dose is recorded at increasing depth. Reproduced with permission from Measurement of Percentage Depth Dose (PDD) for 6 MeV in water phantom and homogeneous actual planning. Samar Essa, *Iraqi Journal of Physics,* 2018; 16:1–6.

on the initial energy of the proton. Also, the deposition of energy falls off rapidly between 23 and 24 cm of water (for a 200-MeV initial energy proton beam). These characteristics reflect the loss of energy (dE/dx) from the primary proton to secondary electrons (plus excitation events) due to Coulombic interactions. The sharp rise in dose reflects the rapid increase in stopping power at low residual energy; as the proton slows, the energy loss events become more frequent and larger (Equation 2.40). This phenomenon is referred to as the Bragg peak. At depths beyond the Bragg peak, the protons retain no residual energy; the velocity drops to zero, and the protons have reached maximum range. All protons with the same initial energy stop at the same depth as predicted by $1/S_{coll}$. The slight slope of the Bragg peak's distal edge ($\Delta x \approx 1$ cm) reflects energy straggling — the statistical variation in proton energies caused by small variations in track structure among the protons. If the distal slope is continued linearly (a line is superimposed on the slope) to the x-axis, a very small tail becomes apparent at the end. This tail reflects a very small amount of radiative loss experienced by protons. Robert Wilson first noted in 1946 that proton dose deposition might be beneficial if applied to radiation therapy.

One might expect the electron depth–dose curve (dashed line) to be similar to the proton depth–dose curve because both electrons and protons are singly charged particles responding to electrostatic forces (Equation 2.38). In fact, there are similarities. The electron curve falls to zero dose at a predictable depth, as does the proton curve ($\mathfrak{R}$). The dose also does not decrease with depth, as does the ^{60}Co curve, but nonetheless there is no obvious Bragg peak at the end of range. The electron track structure differs from the proton track structure because of the difference in the masses of the charged particles. The electron path is torturous (Figure 2-19 compared with Figure 2-21). Although $\mathfrak{R}_{CSDA}$ predicts that the ranges are equal for electrons having the same initial energy, the depths of furthest penetration are in fact variable due to lateral electron scattering. If we calculate the depth of dose curve using a CSDA and neglect primary electron scattering, the Bragg peak persists. Figure 2-24 compares CSDA depth–dose

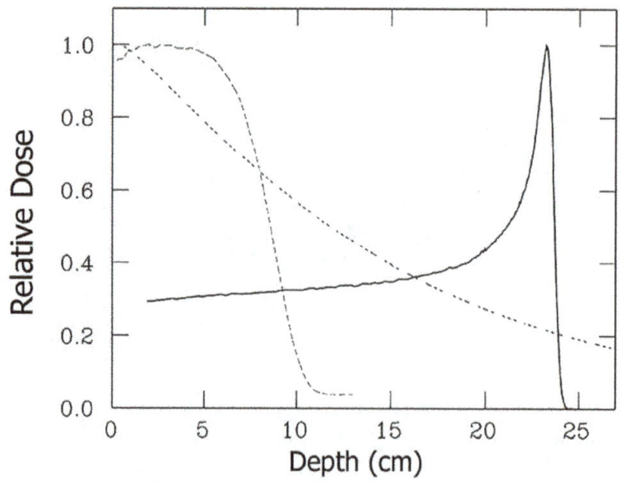

Figure 2-23 Comparison of photon, electron, and proton depth doses. Dose deposition was measured in water using broad beam geometry and an ion chamber dosimeter. Shown are ^{60}Co (<1.25 MeV>) x-ray (... ...), 20-MeV electron (---) and 200-MeV proton (–) energy absorption with distance in water. The charged particles exhibit a clearly defined range resulting from their mass stopping powers (*dE/dx*) while the x-ray dose reflects the stochastic nature of $e^{-(\mu_{en}x)}$ absorption. The proton Bragg peak is apparent at the terminal range ($\mathfrak{R}$) because the track structure is relatively linear.

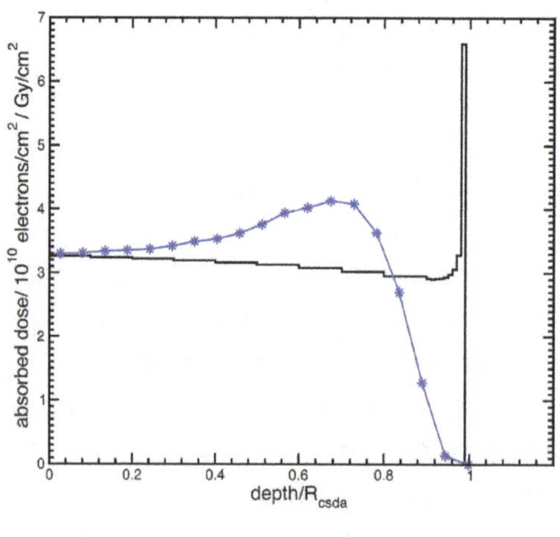

Figure 2-24 Comparison of 20 MeV electron plots of dose per fluence vs. depth normalized to range, using the CSDA. The curve results from calculations including electron scatter (* indicates points with straggling included). The histogram portrays the relationship without scattering in the calculation. The shape of the Bragg peak is an artifact of energy binning. Reproduced with permission of the National Research Council of Canada, from Lecture notes: Electron Monte Carlo Simulation. Alex F. Bielajew and DWO Rogers, Institute for National Measurement Standards.

calculations: the curve is calculated for a beam of 20-MeV electrons and the histogram is calculated for the same energy electrons with no multiple scattering. For details, the reader is referred to Kase *et al.* (Rogers & Bielajew, 1990). The distal falloff in Figure 2-23 is less steep than for protons because of increased electron straggling, and there is a significant Bremsstrahlung tail at the furthest depths (this is diminished in Figure 2-24 because radiative loss was not included in the calculation). If the distal slope is extended linearly to the axis, the point of intercept is the projected range of the electrons, discounting radiative losses. These depth–dose curves exemplify cases particular to radiation therapy and although the universe of radiation biology is more expansive, a grasp of the translation of microdosimetry to a practical case such as therapy will prove useful for extrapolating basic principles.

2.8. Summary

- For the purposes of basic radiation biology, the Bohr model of atomic structure is sufficient.
  - An electron moves in a circular orbit about the nucleus obeying the laws of classical mechanics.
  - Instead of an infinity of orbits, only those orbits whose angular momentum (L) is an integral multiple of $h/2\pi$ are allowed.
  - Electrons radiate energy only when moving from one allowed orbit to another.
- Photons transfer energy to an absorber in one of the three ways:
  - Photoelectric interactions.
  - Compton scattering.
  - Pair production.
- There exists a probability of interaction (a cross section) for each energy transfer mechanism for any specific energy in any given absorber.
- Water is often used to represent a biological absorbing material.
- The relationship between the incident radiation and the uncollided radiation escaping an absorber is $I' = Ie^{-\mu x}$.
- The HVL is the thickness of a given absorber that reduces the incident photon intensity by half.
  - The subsequent HVLs are affected by beam hardening — the preferential absorption of low-energy photons.
- The neutron cross section depends not on the electron structure, but rather on the atomic nuclear structure and it is not intuitive.
  - Relevant to biology: hydrogen, carbon, oxygen, and nitrogen have relatively large cross sections for neutron interaction.
- Neutrons scatter off nuclei through three mechanisms:
  - Elastic scattering.
  - Inelastic scattering.
  - Nonelastic scattering.
- Spallation occurs at high energies.
- Energy is transferred to the absorbing material through the creation of energetic secondary electrons.
- Charged particles (electrons or heavy ions) interact with absorber electrons through Coulombic interactions that depend on the charge and also the mass and velocity of the energetic particle.
  - This interaction is described by the Bethe-Bloch equation for Stopping Power:

$$\frac{S_{\text{col}}}{\rho} = \frac{1}{\rho}\frac{dE}{dx} = \frac{\mathrm{N_A} Z}{A}\frac{\pi r_e^2 2m_0c^2}{\beta^2}\left[\ln\left(\frac{\tau^2(\tau+2)}{2\left(\mathrm{I}/m_0c^2\right)^2}\right) + \mathrm{F}(\tau) - \delta - \frac{2C}{Z}\right]$$

- Electrons radiate energy in the form of photons when they interact through Bremsstrahlung with the positive charge of the nucleus.
  - The proportion of energy converted to Bremsstrahlung depends heavily on the Z of the absorber (Z^2) and the energy of the electron.

- Kerma is the energy *transferred* to a unit mass; it represents the initial kinetic energy of photon scattered electrons.
  — The collisional kerma (K_{col}) does not include the energy lost to Bremsstrahlung (~*g*).
- Cema is the energy *converted* in a small volume; it represents charged particle collisions with atomic electrons.
- Dose is the energy *absorbed* in the volume of interest.
- Electron tracks are tortuous because the masses of the colliding particles are similar.
  — The range is calculated using the CSDA.
- Heavy ion tracks are approximately linear because the mass of the energetic particle is much larger than the electron.
- The majority of dose deposited by photons and electrons is near the surface (e.g., patient's skin), whereas the majority of dose deposited by heavy ions is at the end of their range.

2.9. Problems

1. Assume that a hydrogen atom undergoes transition from $n = 3$ to $n = 1$. What is the energy, momentum, and frequency of the emitted photon?
2. What are the possible electron and x-ray emission energies from a potassium (K) atom if a 3.4 keV photon transfers energy via photoelectric interaction?
3. Using the graph of Figure 2-7, what is the probability (partial attenuation coefficient) for each of the three types of photon/electron interactions at 50 keV (τ, σ, and κ)? At 2.0 MeV? At 40 MeV?
4. Develop a formula for the maximum electron kinetic energy of a Compton scattered electron. For each of the following initial photon energies, compute the maximum kinetic energy that the Compton scattered electron may have.
   Photon energies keV: 5, 50, 500
   MeV: 1, 10, 100
   What can you conclude about the relationship of the incoming photon energy to the maximum scattered electron energy?
5. Calculate $E_{e(max)}$ and $h\nu_{min}$ for incident photon of 5.11 keV and 5.11 MeV. What is the percent energy transferred to the electron?
6. A photon of wavelength 0.20 nm is incident on a slab of carbon, and its wavelength is shifted by 0.0100%. What is the photon scattering angle and what is the maximum energy that the scattered electron could gain?
7. In the accompanying figure, (a) if "A" represents photons, how would you calculate (provide formula) the *number of* collided photons (scattered or absorbed) in "B"? (b) If "A" represents electrons, how would you calculate the energy deposited along "C"? (c) If "A" represents neutrons, how would you calculate the neutron intensity that should be measured in a detector setup according to good geometry, exiting "B"? (d) If "A" represents photons, how would you find the HVL of "B"?

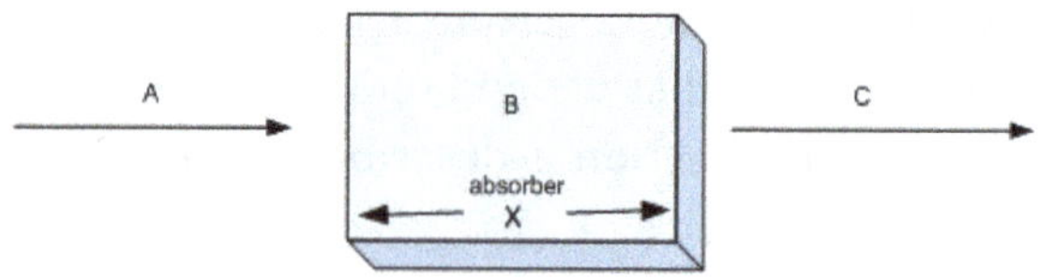

8. The intensity of a beam of neutral particles decreases by a factor of 4 while passing through a slab of material 3.96 cm thick. What is the linear attenuation coefficient for this material?
9. The radiation emitted by a ^{60}Co source has an HVL of 1.2 cm in lead. If a 5-cm thick cast block is fashioned from lead, what percent of ^{60}Co radiation will be transmitted through this block?

10. A parallel beam of neutrons is incident on a 5-cm thick slab of absorber at a 60° angle (see figure). The intensity (fluence, ϕ) of the incident beam is 10^{11} particles cm^{-2} s^{-1}. The intensity of the uncollided radiation exiting the slab is 10^8 particles cm^{-2} s^{-1}. What is the linear attenuation coefficient of the material for neutrons?
11. Sources 1, 2, and 3 are composed of the same radioisotope and have the same activity (see figure). The count rate at X from source 1, with the shield in place, is 14 cpm (counts per minute). The shielding has an optical thickness (μx) of 4.0. Source 1 is 2 m from the point of interest (X) and sources 2 and 3 are 0.8 m from the point of interest (each). Determine the total count rate at the point of interest (in counts per minute).

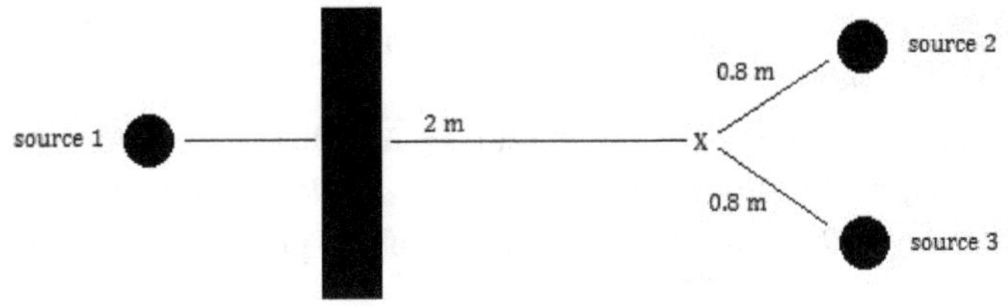

12. For 10-MeV photons, the mass attenuation coefficient for air is 0.204 m^2 kg^{-1}. The average energy transferred (E_{tr}) is 7.37 MeV and the average energy absorbed (E_{ab}) is 7.10 MeV. Assuming a fluence of 10^4 photons traverses a volume of air thickness 1.5 g/cm^2, compute (a) the number of scattering events, (b) the energy appearing as kinetic energy of electrons in the medium, (c) the energy removed from the incident beam, and (d) the amount of the original incident energy that reappears as Bremsstrahlung.
13. A 7-MeV neutron undergoes elastic scattering off a nuclear proton. The proton is scattered at an angle of 60° from the incoming neutron trajectory. What is the energy transferred to the nucleus?
14. If a 200-MeV proton ejected an orbital electron carrying 1.2 MeV kinetic energy through a Coulombic interaction, what must the impact parameter (b) have been? For the same impact parameter, what initial kinetic energy would the ejected electron have if the incident particle had been a 200-MeV electron? Compare the energies of the ejected electrons in each case.
15. Estimate the range of 12-MeV electrons in soft tissue and in air if the density of air is approximately 0.0013 g/cm^3. Assume that the Stopping Power (2.0 MeV cm^2/g) is the same for soft tissue and air.
16. Use the NIST "STAR" database [https://www.nist.gov/pml/stopping-power-range-tables-electrons-protons-and-helium-ions] to find the density, Stopping Power (in MeV/cm) and CSDA range (in cm) for (a) a 10-Mev proton and for (b) a 10-MeV electron in (1) tissue (skeletal muscle) and (2) cortical bone.

2.10. Bibliography

Attix, F. H., 2004. *Introduction to Radiological Physics and Radiation Dosimetry.* Weinheim: Wiley-VCH Verlag GmbH & Co.

Berger, M. J., Coursey, J. S., Zucker, M. A. & Chang, J., 2017. *Description of the ESTAR Database.* [Online] Available at: https://physics.nist.gov/PhysRefData/Star/Text/method.html [Accessed 26 February 2019].

Einstein, A., 1905. Concerning an heuristic point of view toward the emission and transformation of light. *Annalen der Physik,* pp. 132–148.

Geiger, H. & Marsden, E., 1909. On a diffuse reflection of the α-particles. *Proceedings of the Royal Society of London,* pp. 495–500.

Griffiths, D. J., 1995. *Introduction to Quantum Mechanics.* New Jersey: Prentice-Hall.

Hubbell, J. H., 1977. Photon mass attenuation and mass energy-absorption coefficients for H, C, N, O, Ar, and Seven Mixtures from 0.1 keV to 20 MeV. *Radiation Research,* 70(1), pp. 58–81.

ICRU, 1984. *Stopping Power and Ranges for Protons and Alpha Particles.* [Online] Available at: https://icru.org/home/reports/stopping-power-and-ranges-for-protons-and-alpha-particles-report-49 [Accessed 19 February 2019].

Klein, O. & Nishina, Y., 1929. Über die streuung von strahlung durch freie elektronen nach der neuen relativistischen quantendynamik von Dirac. *Zeitschrift für Physikalische Chemie,* p. 52. 853–869.

Rogers, D. W. O. & Bielajew, A. F., 1990. Monte carlo techniques of electron and photon transport for radiation dosimetry. In: *The Dosimetry of Ionizing Radiation, Vol. III.* San Diego: Academic Press, pp. 427–533.

Shultus, J. K. & Faw, R. E., 2007. *Fundamentals of Nuclear Science and Engineering.* Boca Raton: Taylor and Francis Group.

Staub, S. E., Bethe, H. & Ashkin, J. H. A., 1953. *Experimental Nuclear Physics: Volume I.* New York: Wiley.

Turner, A. E., 2007. *Atoms, Radiation, and Radiation Protection.* 3rd ed. Weinheim: Wiley-VCH GmbH & Co.

Wilson, R. R., 1946. Radiological use of fast protons. *Radiology,* 47(5), pp. 487–491.

Chapter 3

Radiation Chemistry

3.1. Introduction

Having assimilated a thorough understanding of the nature of radiation and the structure of matter, as well as the mechanisms of energy transfer from energetic particles (neutral and charged) to bound electrons, we are ready to examine the impact of energy transfer on biological chemistry. The most abundant biological chemical is water. It is reasonable, and correct, to assume that within a biological system, the probability of energy transfer to water is therefore highest (see Figure 3-1). Although there are many other potential biochemical targets, evidence suggests that the only other truly important target is deoxyribonucleic acid (DNA) (see Figure 3-1). Why? DNA is the largest macromolecule (target) in the cell. Furthermore, chemical alteration of DNA is likely to cause disruption of somatic and genetic functionality. For simpler organisms, such as bacteriophages and viruses, a quantitative correlation has been shown between DNA damage and biological disfunction. For higher organisms, various types of DNA damages reflect loss of function. Finally, successful DNA repair relates closely to cell division capacity; cells that lack repair competency (generally through synthetic genetic manipulation) are extremely susceptible to radiation-induced cell death. Energy transferred to water becomes important only in the vicinity of DNA; short-lived products of radiolysis have the potential to damage DNA. If water comprises 70% of the volume of the cell nucleus and DNA comprises 5% — fair approximations — an electron is 14 times more likely to transfer energy (assuming an average energy transfer of 50 eV) to water in the vicinity of DNA than to the DNA. We will therefore examine the transfer of energy to water first.

3.2. Radiolysis of Water

The lysis or splitting of water by radiation is called radiolysis. Recall that the energy transferred from incoming radiation (kerma or cema), or energy transferred from a secondary electron through Coulombic interaction, may cause ionization of a bound electron or it may excite the bound electron to a higher energy state. Imagine that the atom is not isolated but is covalently bonded to other atoms in the form of a molecule, water, for example. Water is an amazing molecule with several unique physical characteristics that support life in fundamental ways. Of particular interest here is the bipolar nature of a water molecule. Composed of one oxygen atom and two hydrogen atoms, the two hydrogen electrons complete the outer shell of the oxygen by orbiting the oxygen nucleus and their respective hydrogen nucleus (a single proton) in a figure eight pattern. The orbital path causes the electrons to spend more time circling the oxygen, leaving the protons "hanging out" somewhat isolated much of the time. This electronic structure also pushes the protons off the equatorial plane to form a stable configuration resembling a pudgy arrowhead. Thus, the molecule is "bipolar," with a relatively negative head and relatively positive base. This polarity enables all kinds of interesting biology. Regarding the current discussion, water molecules can cage a freed electron by orienting with their relatively positive heads toward the negatively charged electron. This configuration, referred to as an aqueous electron (e^-_{aq}), extends the lifetime of the electron's dissociation to approximately 2×10^{-4} seconds. The extended lifetime enables a greater drift radius and thereby increases the potential for electrons to reach and interact with DNA.

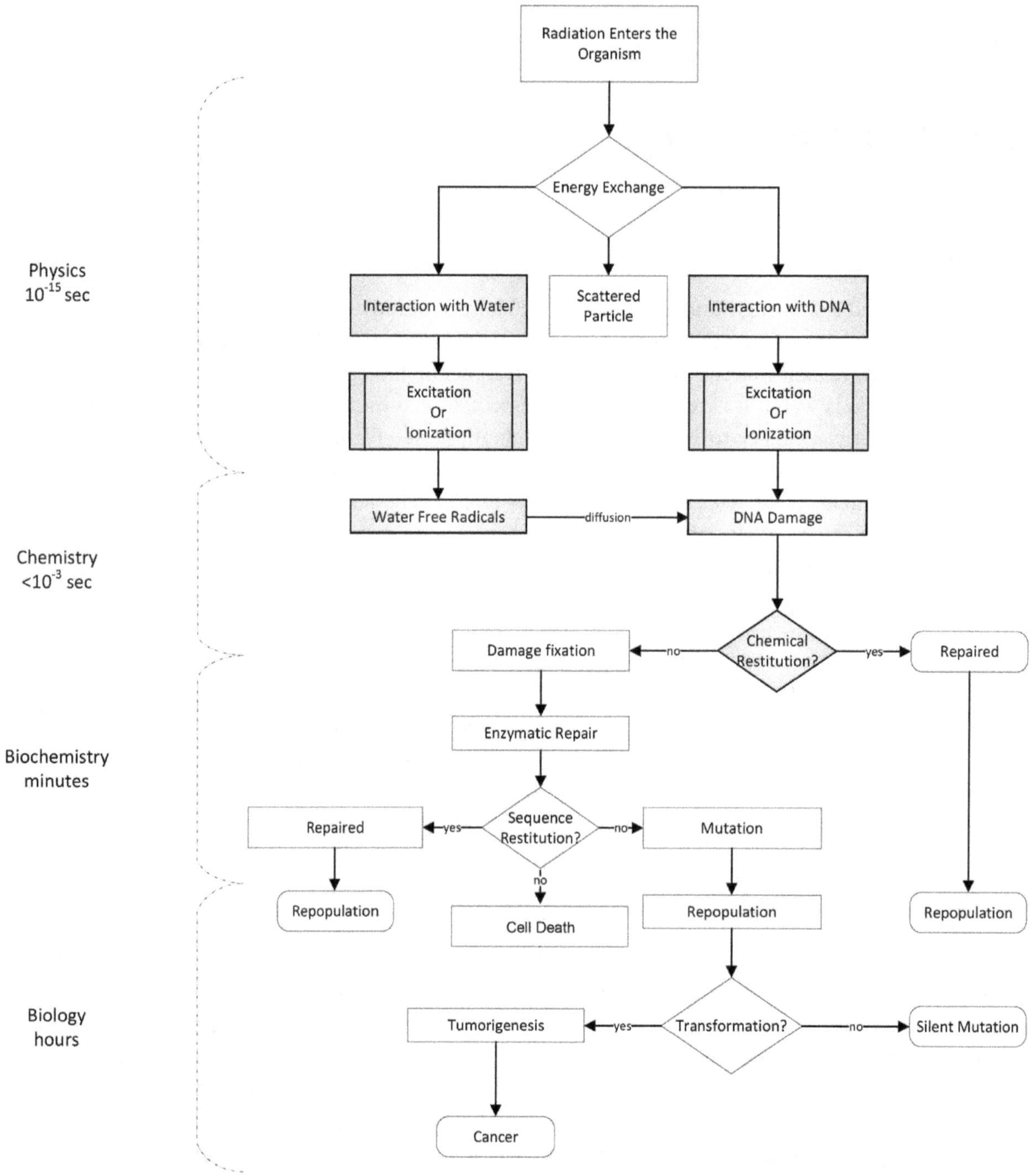

Figure 3-1 This chapter describes the final physical interactions between energetic charged particles and biological absorbers. The two absorbers of interest to radiobiologists are water and DNA. The discussion then continues to examine the subsequent chemistry resulting from the energy transfer. The pertinent processes discussed in this chapter are boxed in gray.

If a passing energetic electron transfers sufficient energy to a water molecule's L-shell electron to ionize the molecule — in other words, if the energy transfer ejects the electron — the resulting ion pair will consist of an electron and a positively charged water molecule (Figure 3-2). On the other hand, if insufficient energy is transferred to eject the electron, but sufficient energy is transferred to raise the electron to a higher energy orbital, the covalent bond is weakened and the excited hydrogen can separate

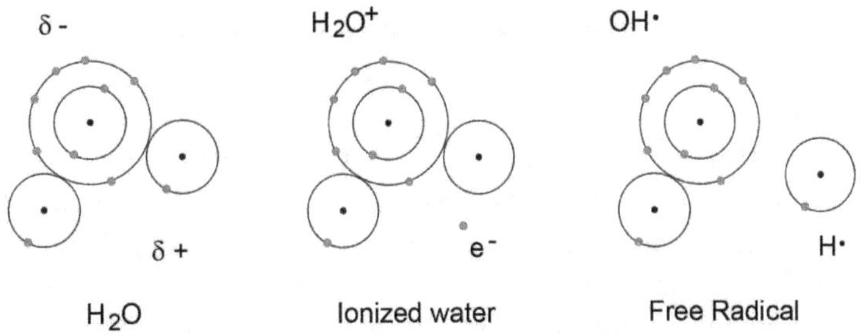

Figure 3-2 Bohr characterization of a water molecule in the relaxed state (left), ionized by the transfer of energy through Coulombic interaction (center), and fragmented into two radical species by the excitation of an electron following energy transfer.

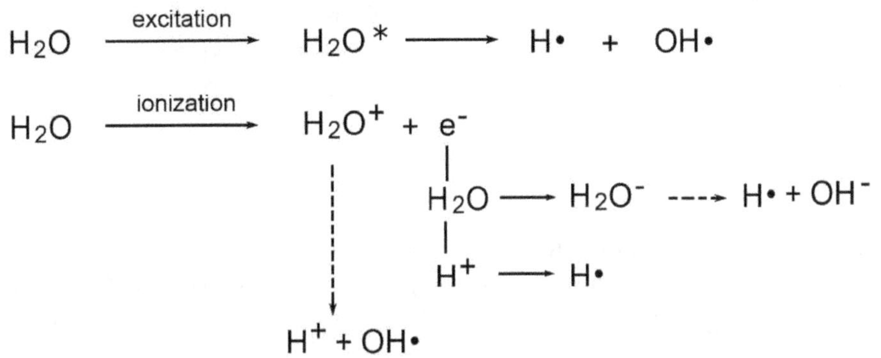

Figure 3-3 Radiolysis of water. Energy transferred to a bound electron in water may excite that electron to a higher energy level. This allows a hydrogen atom to dissociate from the water (H•), leaving an uncharged, highly reactive hydroxyl radical (OH•, upper reaction). Energy greater than the binding energy, transferred to a bound electron, may allow that electron to escape (i.e., it becomes a secondary electron), ionizing the water molecule ($H_2O^+ + e^-$, lower reaction). The proton (H^+) dissociates from the water ion leaving a hydroxyl radical (OH•). The secondary electron may bind to a water molecule, which then dissociates into a hydrogen atom (H•) and a negatively charged hydroxyl (OH^-). Or the electron may associate with a proton to form a hydrogen atom (H•).

from the water molecule, leaving a hydroxyl molecule (OH), and a hydrogen atom (H). This point is important: neither of these products carries a charge — they are neutral. The magnitude of the positive charge of each nucleus is precisely balanced by the negative charges of the associated electrons. These "radicals" are extremely reactive, not because of charge, but because each has an unfilled outer shell. A full hydrogen K-shell requires two electrons, and the filled oxygen L-shell requires eight electrons. This explains the molecular forms of hydrogen (H_2) and oxygen (O_2). Radicals will seek out other molecules with which to bond and share electrons.

What happens to the ion pair, H_2O^+ and e^-? If it retains sufficient kinetic energy, the electron may become a delta electron and carry on ionizing more water molecules. Alternatively, it may become an aqueous electron, or it may diffuse until it attaches to a water molecule to form H_2O^-. The now unnecessary hydrogen will depart, leaving a hydrogen radical (H•) and a charged hydroxyl molecule (OH^-). Or the electron may find a dissociated proton (H^+) and form a hydrogen radical (H•). From where did the proton come? The water ion (H_2O^+) releases it. But notice in Figure 3-3 that release of the proton results in a hydroxyl radical (OH•). For emphasis, we will use the "dot notation" to highlight the uncharged reactive radicals: H• and OH•.

Additional Free Radical Biochemistry

Radiation biochemistry is not limited to water radiolysis products. Although water radiolysis generates the most critically examined, published, and understood set of DNA-damaging reactants, several other radical sources and sinks can be traced to radiation damage, some targeting DNA and others affecting cellular functionality. The following reactions, principally involving superoxide radical and nitric oxide radical, are representatives of additional important radiation biochemistry. The backward Fenton reaction provides a sink for OH• radical. A similar reaction employing Cu ions also reduces hydroxyl radical burden. Superoxide dismutase famously increases radiation resistance by reducing the concentration of superoxide radical and preventing the generation of both OH• and HO_2•. When reacted, nitric oxide plus superoxide neutralize both DNA-damaging radicals.

Superoxide Reactions

$O_2 + e^- \longrightarrow O_2^{\cdot -}$

$O_2^{\cdot -} + H^+ \longrightarrow HO_2^{\cdot}$

$O_2^{\cdot -} + H_2 \xrightarrow{\text{superoxide dismutase}} H_2O_2$

$O_2^{\cdot -} + H_2O_2 \longrightarrow O_2 + OH^{\cdot} + HO^-$

(Haber-Weiss Reaction)

Fenton Reaction

$H_2O_2 + Fe^{2+} \longrightarrow Fe^{3+} + OH^{\cdot} + HO^-$

Nitric Oxide Reactions

$NO^{\cdot} + O_2^{\cdot -} \longrightarrow ONOO^-$

$NO^{\cdot} + O_2 \longrightarrow NO_2^{\cdot}$

To affect biology, the products of radiolysis must interact with DNA. Reactants must be created close enough to the DNA to have the potential to reach it before interacting with a different target. Given that the radicals are relatively similar in size, they diffuse at similar rates. For example, the hydrogen diffusion rate constant is 8×10^{-5} cm^2 s^{-1} and the hydroxyl diffusion rate constant is 2.3×10^{-5} cm^2 s^{-1}. Thus, the drift radius will primarily depend on the stability of the reactant. The less reactive radicals persist longer. Reaction kinetics depend on the relative yields (i.e., concentrations) of each of the radiolysis products and their potential reactants. Nearly 75% of the radiolytic products are either electrons or hydroxyl radicals (37% each). Hydrogen radicals account for another 9%. The remaining products are chiefly H_2O_2 and H_2 with insignificant minor reactants. The hydrogen radical has an average lifetime of 1×10^{-6} seconds and the hydroxyl radical has an average lifetime of 2.7×10^{-6} s (Attri *et al.*, 2015); an aqueous electron has an average lifetime of 2×10^{-4} seconds. The hydrogen and hydroxyl radicals thus have a drift radius of about 200 nm. In addition to drift radius, the pattern of radical creation is important as it relates to DNA damage. Imagine the difference between closely clustered energy transfer events within the hydration (water) layer surrounding and directly in contact with the DNA molecule versus sparsely transferred energy events 100–200 nm distant from the DNA. Each event designated by a block in Figure 2-19 represents about six, 20 eV energy transfers. Where that track occurs in relation to a DNA strand and whether the initial events or final events happen close to the DNA has a major impact on the potential damage. We will examine track structure, or microdosimetry, later. Two other reactants are noted in Figure 3-3: H^+ and OH^-. These ions are ubiquitous in aqueous environments, determine pH, and when formed in equal amounts reassociate to form water molecules without impacting biochemistry in any way. They are an innocuous energy sink for the initial deposition of radiation.

3.3. The Structure of DNA

For completeness, let's review the structure of DNA. The secondary structure of DNA is composed of four nucleosides. Each of these nucleosides is composed of a sugar ring — a deoxyribose — and one of the

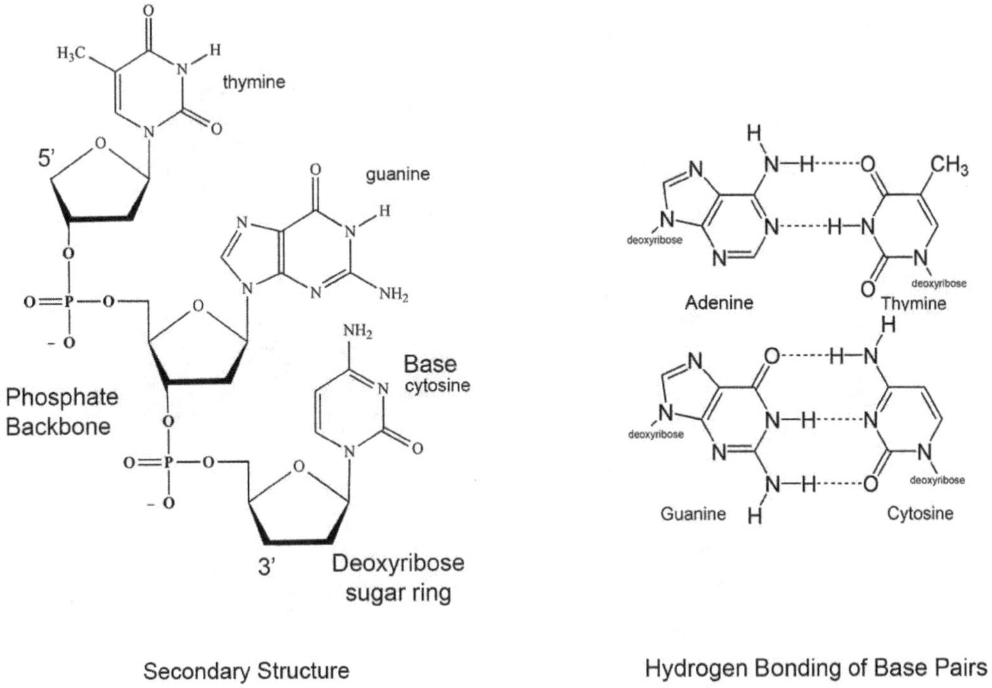

Figure 3-4 The secondary structure of DNA is a repeated pattern of deoxyribose sugars linked covalently by phosphate groups to form a chain. Each sugar moiety is linked to one of four nucleotide bases, A, T, G, or C. When two of these chains are aligned antiparallel, a hydrogen in the base can hydrogen bond to the oxygen or nitrogen of the opposing base, forming a double strand composed of the two chains.

four nitrogen-containing bases: adenine (A), cytosine (C), guanine (G), or thymine (T). The three prime (3′) hydroxyl of one deoxyribose ring is linked to the five prime (5′) hydroxyl of the next deoxyribose ring by a covalently bound (~400 kJ/mol) phosphate molecule (PO_4); the nucleoside plus phosphate structure is referred to as a nucleotide. Through this mechanism, an infinite chain of nucleotides can be constructed (Figure 3-4). A unique characteristic of the base structures is that when correctly oriented, pairs can hydrogen bond. The purines (A and G) hydrogen bond to the pyrimidines (C and T). Hydrogen bonds are relatively weak (~40 kJ/mol), but like Velcro, many individual bonds add up to a stable configuration. The chemistry of this DNA helical structure is significant, as the weakness of hydrogen bonding between the two strands allows the two strands of DNA to "unzip," that is, unwind, without requiring much energy. This elegant secondary structure of antiparallel, hydrogen-bonded chains of nucleotides minimizes potential energy in an aqueous environment by twisting gracefully into the familiar double helix with a pitch of about 11 nucleotides (3.4 nm) and an axial diameter of about 2 nm (Figure 3-5). To compact the nearly 2 m of DNA composed of 3 billion base pairs that is present in each human cell, 147 base pairs of the helix are twice wrapped around a nucleosome: a protein core composed of dimers of histones H2A, H2B, H3, and H4. A fifth histone (H1) acts as a clasp to fasten the DNA to the nucleosome. A spacer, or "linker" DNA of 38 base pairs extends between nucleosomes. When nuclear material is mildly denatured in situ, this structure looks like "beads on a string" under high magnification electron microscope examination. The axial diameter is about 10 nm. Physical characteristics, primarily electrostatic interactions in the physiological medium, wind the beads on a string into a compact 30-nm diameter fiber that packages the nucleosomes into protuberances normal to the axis (z direction) in the x direction and the y direction. The 30-nm fiber loops into loose, irregular coiling hung on a protein scaffold. It is thought that both

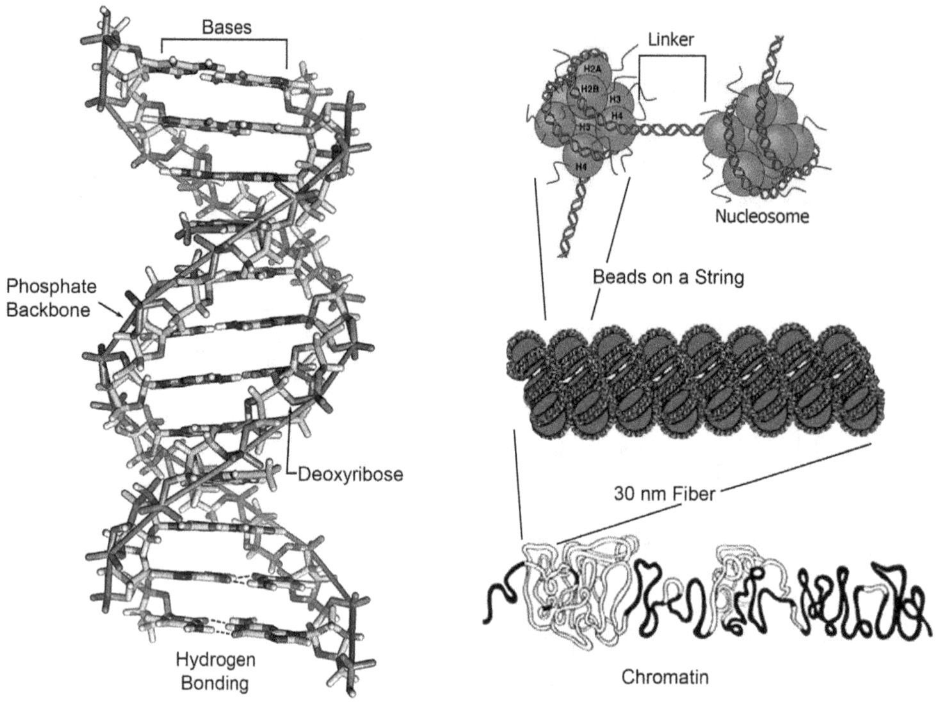

Figure 3-5 Higher order structures of DNA. The tertiary structure of DNA is the familiar double helix or ribbon model. The phosphate backbone spirals around the hydrogen bonded base pairs with a pitch of approximately 3 nm or about 11 base pairs per turn. In situ, the double helix wraps around a nucleosome composed of dimers of H2A, H2B, H3 and H4. The linker DNA of about 80 base pairs allows the structure to twist into a 30 nm diameter cord with nucleosomes positioned normal to the axis. Tension from coiling and a protein scaffold further compact the structure into chromatin that takes two forms: loosely compacted euchromatin (thin open lines) and tightly compacted heterochromatin (thick black lines). Individual panels reproduced with permission through CC BY-SA.

the 30-nm fiber and the protein scaffold control represented in Figure 3-5 the level of compaction of the condensed DNA, or "chromatin." Chromatin can take two forms: active euchromatin open to transcription and very compact heterochromatin that is transcriptionally inactive. Although the discussion of DNA damage habitually refers to interactions between DNA and neutral radiation, charged particles, or free radicals, it is essential to keep in mind that the actual structure being affected is a very compact form of DNA: chromatin that can be further wound and compacted into individual chromosomes.

3.4. Indirect Damage of DNA

We have established that aqueous electrons, hydroxyl radicals, and hydrogen radicals can react chemically with DNA provided that the radicals reach the DNA before interacting with another reactant. The lifetime of these radicals is on the order of 10^{-6} seconds. The reaction rates of these radicals with DNA are OH: 3×10^8 mol/s, e_{aq}: 1.4×10^8 mol/s, and H: 8×10^7 mol/s, making the hydroxyl radical more reactive than the aqueous electron or hydrogen. However, the hydroxyl radical diffuses an average of only 3.5 nm during its lifetime, so it must be created very close to the DNA molecule. Three types of reactions with radicals are possible: (1) extraction of a hydrogen, (2) dissociative reactions, and (3) addition reactions.

A hydrogen extraction reaction (also called oxidation) with DNA may remove a hydrogen atom to produce molecular hydrogen (H_2) by bonding with a hydrogen radical

$$[\text{DNA} - \text{H}] + \text{H}\cdot \rightarrow \text{DNA}\cdot + \text{H}_2 \tag{3.1}$$

or form water by bonding with a hydroxyl radical

$$[\text{DNA} - \text{H}] + \text{OH}\cdot \rightarrow \text{DNA}\cdot + \text{H}_2\text{O}. \tag{3.2}$$

For example, the four bases shown in Figure 3-4 each contain hydrogens covalently bonded to a nitrogen. These hydrogens participate in the hydrogen bonding that holds the two DNA strands together. A free radical may associate with the nitrogen-bound hydrogen, providing an alternative electron to fill the outer shell. This reaction removes the bound hydrogen leaving a reactive nitrogen that is attached to a base, a base that is part of the DNA strand. Thus, the entire DNA molecule becomes a radical (DNA•), just as the excited water molecule in Figure 3-2 becomes a hydroxyl radical (OH•) when the hydrogen departs. In fact, any hydrogen within the DNA macromolecule is susceptible to hydrogen extraction; the reactivity of a given site is strongly dependent on location within the DNA molecule. The type of damage resulting from hydrogen extraction determines the subsequent repair mechanisms and the potential for physiological toxicity. As in our example, the bases may become reactive. Almost equally susceptible are the deoxyribose sugar moieties. Hydrogen extraction from either of these entities results in a reactive nucleoside. Hydrogen extraction may also occur at the 3′ or 5′ carbon sites, affecting the deoxyribose–phosphate backbone integrity. Interestingly, a radical formed in a base also can be transferred to the deoxyribose–phosphate backbone. Either of these reactions results in a reactive DNA nucleotide.

The hydroxyl radical may also remove an azanide (NH_2) from a base by a dissociative reaction,

$$[\text{DNA} - \text{NH}_2] + \text{H}\cdot \rightarrow \text{DNA}\cdot + \text{NH}_3. \tag{3.3}$$

Similar results are obtained through ammonia cation (NH_3+) interaction with an aqueous electron,

$$[\text{DNA} - \text{NH}_3^+] + e^-_{aq} \rightarrow \text{DNA}\cdot + \text{NH}_3. \tag{3.4}$$

These dissociative reactions again affect the bases of the nucleotides but as before, the radical can be displaced from the base to the deoxyribose–phosphate backbone. The chemical results are as before with hydrogen extraction.

Finally, the carbon double bonds within the ubiquitous base ring structure are vulnerable to attack by the hydroxyl free radical. The hydroxyl radical reaction rate implies that this addition reaction (also called reduction) may be favored above the previous H• and e^-_{aq} reactions. The hydroxyl radical is added to the ring at one of the carbon sites,

$$[\text{R} - \text{CH}] = [\text{CH} - \text{R}] + \text{OH}\cdot \rightarrow [\text{RCHOH} - \text{CH}\cdot - \text{R}]. \tag{3.5}$$

where R represents a "reactive species." The secondary structure for this reaction with thymine would look like Figure 3-6.

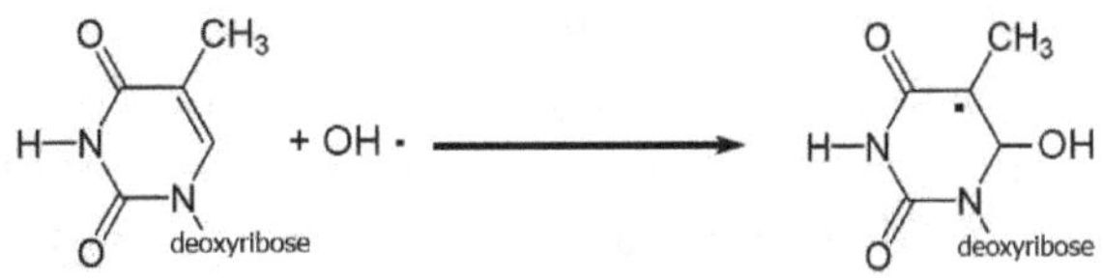

Figure 3-6 Formation of a thymine radical through an addition reaction. The hydroxyl radical binds at the double bonded carbon, adding an OH group and resulting in an unpaired electron at the adjacent carbon.

What happens to the radical DNA sites that result from these three reactions? There are two possible outcomes. Restitution returns the DNA to its native form at the expense of another reactive molecule that becomes a radical. The transient damage becomes irrelevant:

$$DNA\cdot + [\ R - H] \rightarrow [DNA - H] + R\cdot\ . \tag{3.6}$$

Alternatively, the chemical disruption becomes "fixed," or stabilized through oxidation,

$$DNA\cdot + O_2 \rightarrow [DNA\cdot\text{–peroxide}], \tag{3.7}$$

where a peroxide is any compound with the structure R-O-O-R (as in hydrogen peroxide, H_2O_2). Fixation is the initiation of the process that results in biological radiation damage. The activation energy, or the reverse reaction rate constant, becomes sufficiently large following fixation that the damage can be repaired only through biological enzymatic intervention. Obviously therefore, oxygen plays an important role in radiation DNA damage.

3.5. Direct Damage of DNA

The intuitive reader has probably wondered, "if incoming radiation can excite water molecules to form OH• and H•, and ionize water molecules to form H_2O^+ and e^-, could similar stochastic and Coulombic events directly excite and ionize DNA?" Yes, radiation can directly affect the DNA chemistry. However, direct damage is more difficult to observe experimentally because DNA is hydrated (surrounded by solvent shell water molecules), as it must be to maintain the familiar helical structure. Thus, hydrated DNA inevitably suffers from radiolysis-induced indirect damage. These experiments are generally conducted using low doses of photon irradiation and short polymers of nucleotides, either crystallized or suspended in an anhydrous film, often at extremely low temperatures (~4 K). For this reason, conclusions drawn from these experiments must be considered carefully and extrapolation to physiological conditions found in cells should be made cautiously. Nonetheless, empirical evidence suggests that 1/3 of the damage inflicted upon DNA by photon radiation (x-rays and γ-rays) results from direct damage, whereas 2/3 (70%) of the damage results from indirect damage.

Recall that the most reactive radiolysis product, HO•, results from electronic excitation, and the second most reactive product, e^-_{aq}, results from ionization (Figure 3-3). In the case of direct damage to DNA, it appears that excitation is of minor importance in the generation of persistent damage. This may reflect the stability of chemical ring structures that readily distribute absorbed energy. Therefore, ionization events dominate direct damage (Figure 3-7). In addition to ionization of DNA moieties (bases, deoxyribose rings, and phosphate backbone links), ionization of water molecules in the tightly bound solvent shell (approximately two water molecules per nucleotide) rapidly transfers freed electrons and holes to the DNA. The resulting DNA lesions are indistinguishable from those formed de novo in the DNA and therefore are included in our definition of direct damage. These transfers do not involve water radicals or aqueous electrons and are distinct from the pathway that we have designated as "indirect." The free electrons are relatively localized (short lived), and the rate of intra-strand electron transfer along the DNA strand is greater than inter-strand exchange between two DNA strands. These exchanges are akin to solid-state electronic configurations where an electron hole is created by an electron transitioning from a tightly bound state to a weakly bound state. Proton transfers, on the other hand, readily occur across the hydrogen bonds between the two DNA strands. The rates of electron hole transfer and proton transfer depend on the adjacent bases. Cytosine–Guanine pairs are more susceptible to electron hole and

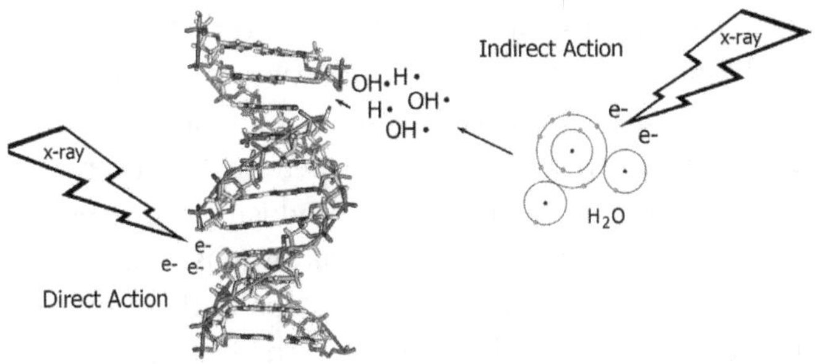

Figure 3-7 Indirect and direct DNA damage by photon radiation. Most often (70%), photon radiation deposits energy in water, ionizing or exciting the molecules, releasing H• and OH• free radicals, and aqueous electrons (see Figure 3-2). These products react with DNA, disrupting the structure. This radiation action is indirect (right). Less often (30%), photon radiation directly ionizes the DNA molecule, disrupting the covalent bonds of the structure. The local release of energetic electrons that subsequently ionizes DNA is also a direct effect (left).

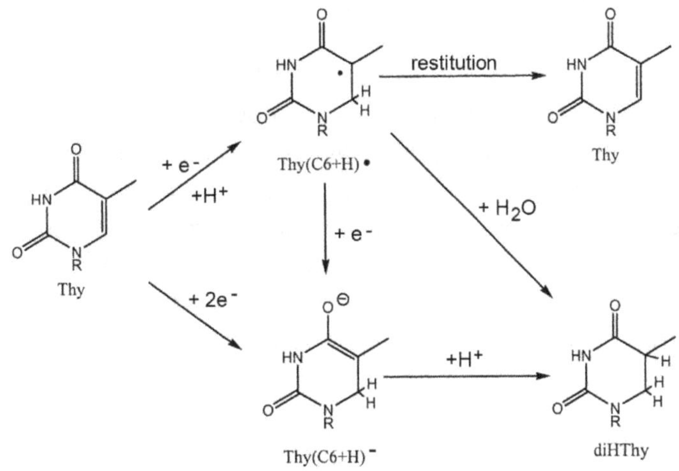

Figure 3-8 Direct damage of thymine. Thymine (far left) is reduced and then protonated in the upper reaction to yield a radical derivative of thymine, or doubly reduced in the lower reaction to form an ion. Alternatively, an electron is captured by the Thy(C6+H)• radical, (center reaction). If a water molecule is added to the Thy(C6+H)• radical, or if the ion is protonated, a stable aberrant base is created. This scheme illustrates three possible direct damage pathways. Reproduced with permission of the Radiation Research Society from Swarts *et al.* (2007).

proton transfer than Adenine–Thymine pairs, and base damage is three times more probable than backbone damage. So, the details are complicated. One of the more interesting aspects of radiation DNA chemistry suggests that unlike radiolysis, creation of the ultimate fixed (persistent) DNA damage may require more than one radical reaction. A mechanistic model offered by Swarts proposes three possible reactions: (1) a radical driven single-track pathway, (2) a radical driven multiple-track pathway, and (3) a pathway for which there is no radical precursor, at least no radical that can be scavenged by a radical trap under experimental conditions. To provide a concrete example of direct base damage, let's use thymine again. Figure 3-8 presents all three possible pathways and is supported by experimental evidence (Swarts *et al.*, 2007). Suppose a thymine base captures an electron released from an ionized hydration water molecule or an ionized neighboring base. The addition of the electron "reduces" thymine, producing an ionized radical: Thy•$^-$. This very reactive species will quickly capture a proton (H^+) to produce the neutral radical: Thy(C6+H)•. Three reactions are now possible. The radical base can release the newly acquired hydrogen radical ($H^+ + e^-$) and return to its native thymine structure, that is, restitution. A water molecule may be added to the radical, fixing the damage through addition of an oxygen (diHThy). Or, as in the first reaction, an electron ion may be added to convert the radical into an ion (Thy(C6+H) $^-$) followed by the addition of a proton (H^+) with the same result (diHThy) as before. It is also possible for thymine to absorb two free electrons generated in a densely ionizing event, producing the same intermediate ion created in the previous reaction pathway (Thy(C6+H)$^-$). This last pathway is interesting because it leads to base damage not resulting from radical formation. The initial steps of the proposed pathways in Figure 3-8 require the addition of one or two electrons. In chemistry, the base is said to be reduced; the base has become reduced through the addition of a hydrogen atom ($H^+ + e^-$). Following photon energy transfer, the pyrimidines (C and T) are preferably reduced leading to persistent base damage. On the other hand, following energy transfer, purines (A and G) release a hydrogen atom; they form radicals through "oxidation." The deoxyribose sugar DNA backbone rings also can be preferentially oxidized. Therefore, it is quite possible for a radical to be transferred from a DNA nucleotide base to the deoxyribose–phosphate backbone of the DNA molecule. This chemistry implies that the terms "restitution" and "fixation" are perhaps a bit naive. Rather, oxygen facilitates "damage transfer," a sequence of radical relocations within the DNA component structures.

Figure 3-9 illustrates the types of damages that may result from direct energy deposition in the deoxyribose–phosphate backbone of DNA and Figure 3-10 provides the reaction details. In each of the illustrated cases, a hydrogen is removed (the deoxyribose is oxidized) leaving a reactive carbon component of a deoxyribose radical. Removal of hydrogen from carbons 1 or 2 results in the loss of the associated base. Removal of hydrogen from carbons 3 or 4 disrupts the phosphate bond and fragments the deoxyribose ring. In the first case (carbon 1), the base is replaced by a double bond to oxygen. In the second case (carbon 2), the deoxyribose ring is broken open. Oxidation of carbons 3, 4, and 5 (not shown) produce disruption of the deoxyribose–phosphate backbone and fragmentation of deoxyribose. The structural

Nitric Oxide DNA Damage

Free radicals resulting from radiolysis of water transiently disrupt DNA structure; the chemical alteration can become stable following the addition of oxygen to produce DNA••-peroxide. Another radical, nitric oxide (NO•), effects similar levels of biological detriment following radiolysis-induced DNA scission, although the chemistry differs from O_2 reactivity. NO• is produced biochemically from arginine by NO synthases (NOS) and nominally functions as a neurotransmitter. Nitric oxide reacts with guanine radicals to produce 8-xanthine or azaguanine, and with adenine radicals to produce 8-aza-adenine or hypoxanthine.

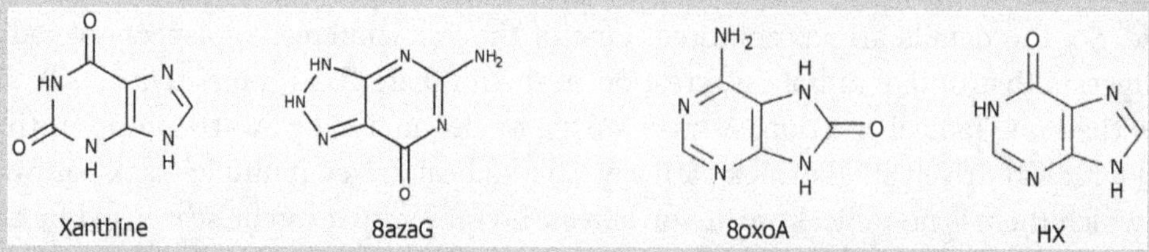

Both O_2 and NO• react with DNA radicals (DNA•) to form products that require biological enzymatic repair, but while reactions with oxygen form products that are radicals (Equation 3.7), the nucleic acid NO• products are not radicals. It is possible that NO• may be more effective at diminishing OER than O_2, by restoring stable DNA lesions under hypoxic conditions. Interestingly, the processing of nonreactive, NO• stabilized base appears to rupture both DNA strands resulting in a double strand break, whereas O_2 reaction products tend to produce simple base damage resulting in single strand breaks (Figure 3-9).

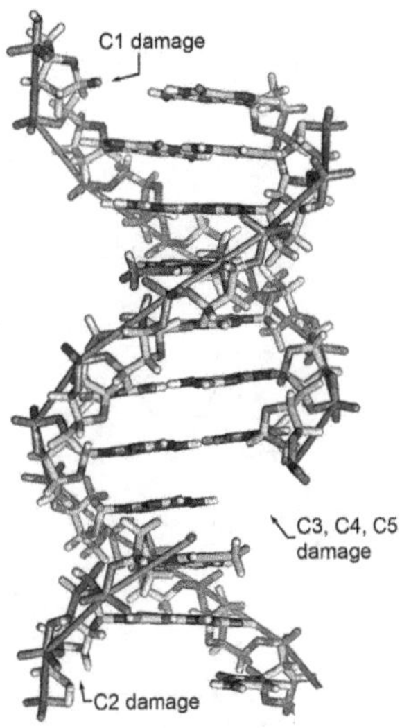

Figure 3-9 A representation of structural damage resulting from the oxidation of A or T, and the reduction of G, C, or deoxyribose. Extraction of the carbon 1 (C1) hydrogen removes the associated base. Extraction of the carbon 2 (C2) hydrogen opens the deoxyribose ring and dissociates the bound base. Although the removal of a hydrogen bound to carbons 3, 4, or 5 (C3, C4, or C5) yields different products, all these reaction result in the loss of the deoxyribose ring, one or two of the PO_4 moieties and the associated base.

differences among these products are critical. Bases that exhibit a radical structure — that is, bases that are oxidized or reduced through indirect interaction with products of water radiolysis, or bases that are oxidized or reduced through direct deposition of energy — become chemically altered (as in Figure 3-8) or are removed (as in Figure 3-10 at carbon 1 or carbon 2). These altered or missing bases must be recognized and replaced to retain the functional integrity of the DNA. We shall examine these mechanisms in Chapter 4. Oxidation of the deoxyribose–phosphate backbone through indirect or direct action results in a strand break (Figure 3-10 at carbon 3 or carbon 4). Strand breakage is fundamental to the cellular response to radiation. Much of the remainder of this book will concern strand breakage.

The chemistry presented in this chapter is far from an exhaustive representation of all possible reactions. The interested student is encouraged to explore the references at the end of the chapter as well as many other excellent resources representative of this fascinating topic. The information presented herein is intended to provide sufficient illustrative examples and depth that the reader can conceptualize the physical and chemical processes leading to DNA damage. Figure 3-9 illustrates the structural damage to DNA that is detailed in Figure 3-10. Future incorporation of simplified DNA graphic representations will be helpful as our perspective expands. This chapter serves as a useful resource to assist in the interpretation of those graphics.

3.6. Summary

- DNA is the germane biological target of radiation energy transfer.
- The abundance of water in biological systems makes water a significant radiation energy sink/absorber.
- Radiolysis of water into free radicals (H•, OH•) and aqueous electrons (e^-_{aq}) potentiates DNA damage.
  - To induce damage, products of radiolysis must be released within 200 nm of DNA.
  - Free radicals are produced through excitation and chemical reaction.

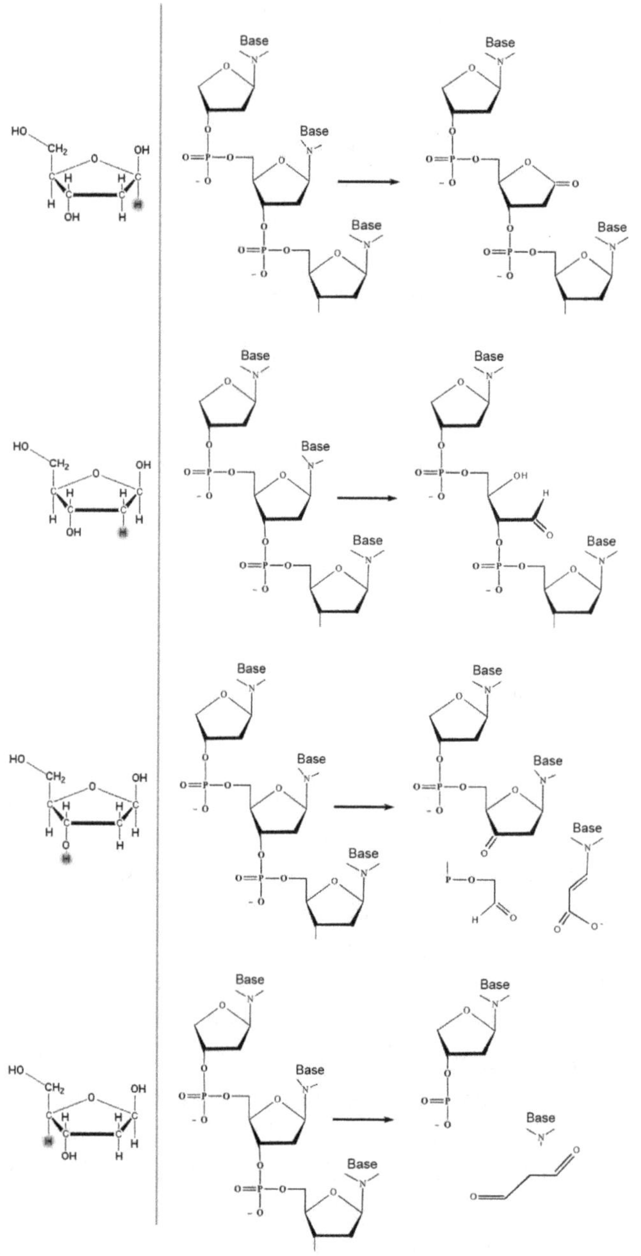

Figure 3-10 Oxidation of deoxyribose. The left panel indicates which hydrogen is removed to create a deoxyribose radical. The right-hand panel compares intact DNA with the final product of oxidation. From top to bottom. Removal of hydrogen from carbon 1 results in the loss of the associated base. Removal of hydrogen from carbon 2 breaks open the deoxyribose ring and releases the associated base. Removal of hydrogen from carbon 3 disrupts the phosphate bond and fragments the deoxyribose ring. Removal of hydrogen from carbon 4 disintegrates and releases both the sugar and base structures. The adjacent 5′ phosphate-ribose structure is also chemically disrupted.

- Excitation of water leads to H• and OH• radicals.
- Ionization of water leads to secondary electrons, aqueous electrons, and positive water ions. Further reactions can lead to H• and OH• radicals.
- Radiation damage of DNA is the result of indirect damage by radiolysis products or direct damage through the deposition of energy into DNA chemistry.
- The structure of DNA has three components: four bases, a deoxyribose sugar ring, and phosphate backbone.
- The alpha helix form of DNA is condensed into a 10-nm diameter "beads on a string" structure through winding around histone cores and further compacted by coiling into a 30-nm fiber.
- The cellular form of DNA is chromatin, a looping 30-nm fiber supported by a protein scaffold that exists in two forms: euchromatin that is transcriptionally active and heterochromatin that is hyper condensed and inactive.
- Chemical disruption of the DNA structure leads to functional loss and biological damage.
- Pyrimidines are preferably reduced, and purines are preferentially oxidized.
- Deoxyribose rings in the DNA are preferentially oxidized.
- A radical can be transferred from a DNA nucleotide base to the deoxyribose–phosphate backbone.
- Base chemical alteration, base removal, and strand breakage define the three types of DNA damage that must be recognized and repaired by cellular enzymic repair mechanisms so that functionality can be maintained.

3.7. Problems

1. A Fricke dosimeter uses the ferrous ion, Fe^{2+}, to scavenge free radicals. Oxidation of Fe^{2+} to Fe^{3+} through the addition of a proton is commonly known as "rust." The relevant Fricke oxidation reactions in water are:
   a. $H\cdot + O_2 \rightarrow HO_2$
   b. $HO_2\cdot + Fe^{+2} \rightarrow HO_2^- + Fe^{3+}$
   c. $HO_2^- + H^+ \rightarrow H_2O_2$
   d. $OH\cdot + Fe^{2+} \rightarrow OH^- + Fe^{3+}$
   e. $H_2O_2 + Fe^{2+} \rightarrow OH^- + Fe^{3+} + OH\cdot$
   How many molecules of ferric iron (Fe^{3+}) are produced per hydrogen radical (H•) released by radiolysis? Show the sequence of reactions.
2. Ionizing radiation deposits energy in a molecule and releases an electron through photoelectric interaction, Compton scattering or pair production, or through Coulombic interactions with an energetic charged particle. Using the oxidation reaction at C3 of the deoxyribose ring (the removal of a hydrogen), show how ionization results in oxidation. Use the Bohr model and show only the necessary atoms.
3. Using a structural diagram, show how a radical can be transferred from a base to the deoxyribose–phosphate backbone.
4. The composition of one DNA strand is 0.30 A and 0.24 G. What can be said about the composition of T and C on the same strand? What can be said about the composition of bases of the complementary strand?
5. A mutant bacteriophage is missing several base pairs; it is a "deletion mutant." If the length is 15 μm rather than 17 μm, how many base pairs are missing from the mutant?

3.8. Bibliography

Annunziato, A. T., 2008. DNA packaging: nucleosomes and chromatin. *Nature Education*, 1(1), p. 26.

Attri, P. *et al.*, 2015. Generation mechanism of hydroxyl radical species and its lifetime prediction during the plasma-initiated ultraviolet (UV) photolysis. *Science Report*, 5, p. 9332.

Chatterjee, A. & Holley, R. W., 1993. Computer simulation of initial events in the biochemical mechanisms of DNA damage. *Advances in Radiation Biology*, 17, pp. 181–226.

Chiu, S.-M. *et al.*, 1986. DNA-protein cross links in nuclear matrix. *Radiation Research*, 107, pp. 24–38.

Dorfman, L. M. & Adams, G. E., 1972. *Reactivity of the Hydroxyl Radical in Aqueous Solution*. Gaithersburg: National Bureau of Standards.

Kapiszewska, H. H., Wright, W. D., Konings, A. W. T. & Roti-Roti, J. L., 1989. DNA supercoiling changes in nucleoids from irradiated L5178Y-S and -R. *Radiation Research*, 119, pp. 569–575.

Kornberg, R. D., 1977. Structure of chromatin. *Annual Review of Biochemistry*, 46, pp. 931–954.

Lea, D. E., 1962. *Actions of Radiation on Living Cells*. 2nd ed. London: Cambridge University Press.

Roti-Roti, J. L., Wright, W. D. & Taylor, Y. C., 1993. DNA loop structure and radiation response. *Advances in Radiation Biology*, 17, pp. 227–259.

Samuni, A. & Czapski, G., 1981. Radiation induced damage in Escherichia coli B: the effect of superoxide radicals and molecular oxygen. *Radiation Research*, 76, pp. 624–632.

Schrodinger, E., 1944. *What is Life? The Physicist's Approach to the Subject - With an Epilogue on Determinism and Free Will*. Cambridge: Cambridge University Press.

Schwarz, H. A., 1974. Recent research on the radiation chemistry of aqueous solutions. *Advances in Radiation Biology*, 1, pp. 1–30.

Swarts, S. G. *et al.*, 2007. Mechanisms of direct radiation damage in DNA, based on a study of the yields of base damage, deoxyribose damage and trapped radicals in d(GCACGCGTGC)2. *Radiation Research*, 168, pp. 367–381.

Ward, J. F., 1988. DNA damage produced by ionizing radiation in mammalian cells: identities, mechanisms of formation and repairability. *Progress in Nucleic Acid Research and Molecular Biology*, 35, pp. 95–125.

Watson, J. & Crick, F., 1953. A structure for deoxyribose nucleic acid. *Nature*, 171, pp. 737–738.

Chapter 4

The Biochemistry of DNA Repair

4.1. Introduction

Chapter 3 presented several justifications to support the hypothesis that DNA represents the single most biologically important recipient of radiation damage. This assertion can be validated on two purely physical bases: DNA presents the largest cellular biomolecule (target), and the free radical products of water radiolysis — water being the substance of greatest cellular volume — react with the macromolecule to damage DNA structural chemistry. Even more persuasive justifications are biological: chemical damage produced by radiation leads to changes in DNA sequences that manifest as altered cellular and physiological impairment. In this chapter, we will show that the four DNA bases, A, T, G, and C, comprise a triplet code that is transcribed to ribonucleic acid (RNA, a single-stranded derivative of DNA possessing an oxygen at C2 of the ribose ring) which is commonly translated into protein. Proteins maintain cell and tissue structure and function, and thus, rigorous fidelity of DNA is necessary for cell survival and organism health as well as genetic perpetuation. We will discuss several types of DNA damage: single-strand breaks (ssb), double-strand breaks (dsb), base and nucleotide damage, and we will show how these breaks contribute to chromosomal aberrations leading to organism disfunction that results from cell death or cell transformation. To prevent these untoward outcomes, evolution has provided mechanisms that prevent and repair these various classes of DNA damage. This chapter will begin a discussion of these mechanisms by examining the processes available at the cellular scale.

4.2. A Brief Review of Molecular Biology

The historical central dogma of molecular biology goes something like this: DNA is *transcribed* to RNA, which in turn is *translated* into protein (Figure 4-1). In addition, DNA is *replicated* during cell division allowing for an exact copy of the genome to be distributed to each of the daughter cells. More current dogma acknowledges that the human genome contains about 20,000 encoded genes comprising only 1%–2% of the DNA sequence. This fraction of the DNA embodies a triplicate-based code that dictates protein products supporting cell growth, metabolism, and physiology. Because RNA sequence analysis (RNAseq) has shown that over 80% of the human genome is transcribed into RNA, approximately 20% of the genome must be "nontranscribed" DNA that may have regulatory or other unidentified functions. Furthermore, some of the transcribed RNA is never translated into protein (e.g., tRNA and rRNA which will be described later, and other regulatory RNA such as ribozymes, microRNA [miRNA], and long noncoding RNA [lncRNA]). Compared to DNA, RNA is quite fragile. It is single-stranded (though it folds into double-stranded structures), susceptible to chemical degradation and enzymatic destruction, and short-lived with a median mRNA half-life of 10 hours (fast decaying mRNA transcripts exhibit half-lives of < 2 hours). The transient nature of RNA provides a mechanism for rapid initiation of mRNA transcription and subsequent protein translation, and then equally rapid termination of mRNA transcription and

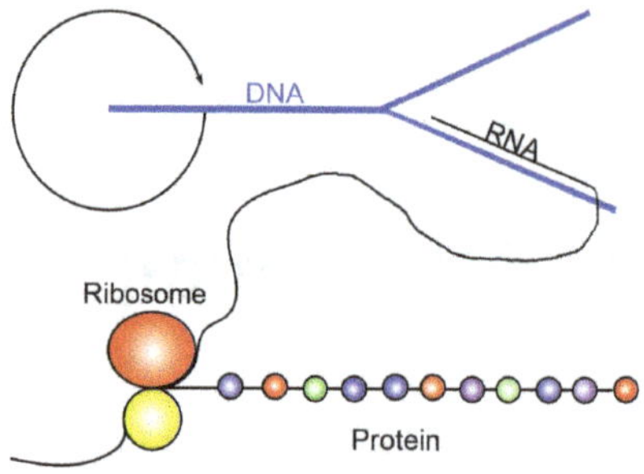

Figure 4-1 The central dogma of molecular biology. During the cell cycle in S phase, the DNA is replicated to create two identical copies (left). DNA is continuously being transcribed into RNA (right) that is subsequently translated within ribosomes to amino acid sequences that comprise proteins (bottom).

protein translation. RNA is as susceptible to radiation damage as DNA (Figure 4-2), but it is a small target and the loss of a small fraction of RNA is of virtually no consequence as it can be rapidly replaced by the upregulation of mRNA transcription.

DNA replicates during DNA synthesis or S phase of the cell cycle. With notable rare exceptions, our tissues need to continually repair and replenish their cells throughout our lifetime. As we develop from embryo to adult, tissues need to differentiate into various types of cells required for organ and tissue function, and increase in number so tissues/organs can increase in volume. Normal friction removes cells from the surfaces of skin, blood vessel walls, and the lining of our digestive system. Cells in most tissues must divide to increase in number and to replace this loss. This cell division occurs in tissue stem cells that differentiate into various functional cell types required for tissue maintenance. The process by which cells split into two daughter cells is regulated by the *cell cycle* (Figure 4-3). Importantly, because each daughter cell requires a complete library of genetic instructions, DNA must be precisely replicated during DNA synthesis or S phase and equally distributed to each of the two daughter cells. Under *in vitro* conditions (in cell culture), cell growth and division may be continuous. *In vivo* (in tissues), however, the duration between cell cycles is tissue-type dependent with most cells not actively dividing. Most cells rest in "Gap 0" (G_0), content to carry out various housekeeping activities without cell division, but remain metabolically quite active. When induced through stimulus by various growth factors and nutrients, a cell enters the cycle at "Gap 1" (G_1). During this phase, the cell increases in size and prepares for division and the DNA presents largely as euchromatin. Enzymes and other proteins are actively transcribed, translated, and produced. Cell building blocks such as amino acids for proteins, fatty acids for cell membranes, and nucleotides and ribonucleotides for DNA and RNA are chemically synthesized. In culture, the G_1 phase persists for approximately 10–40 hours depending on the cell type. Once the structures are organized and the component concentrations are sufficient, the cell enters the "Synthesis" (S) phase. During S phase, the DNA replicates and the various DNA segments on each chromosome must be relaxed, uncoiled, and unzipped as they are duplicated. In culture, S phase lasts about 8–10 hours. After S phase, the cycle enters another shorter growth phase, "Gap 2" (G_2), during which the cell prepares for mitosis. DNA-specific mitotic cofactors such as centrioles and kinetochores assemble, and scaffolding components accumulate over a period of 2–3 hours. Finally, "Mitosis" (M) occurs. By far, the most complicated and studied phase of the cell cycle, mitosis takes only about 1 hour (Figure 4-4). Mitosis is described in subphases: prophase, metaphase, anaphase, and telophase. During prophase, the DNA condenses into chromosomes and sister chromatids (DNA segments, one maternal and one paternal, containing the same genes) pair. Enzymes called endonucleases nick the DNA severing the strands, and sections of DNA exchange between chromatids. The motivation for this is unknown, but if the

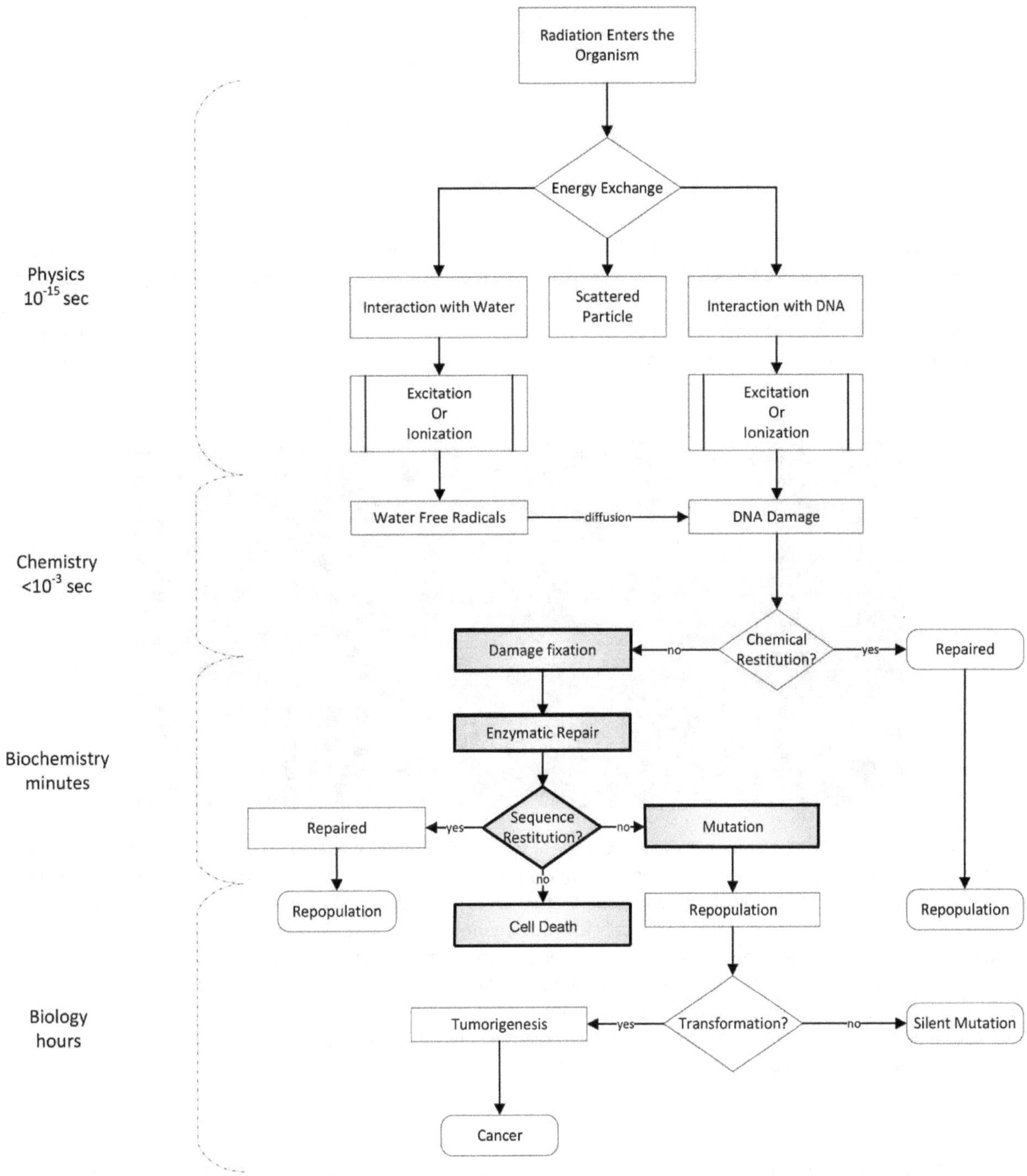

Figure 4-2 This chapter is concerned with the consequences of radiolytic DNA damage. If the radiation-induced lesion is stabilized through oxidation, the cell must recognize the damage and repair it or suffer the biological consequences. If the damage is recognized and successfully repaired, the cell continues to function normally, including metabolic functions and cell division. If the damage is not successfully repaired, the cell may suffer from genetic mutation or experience cell death.

endonucleases are blocked, mitosis will not proceed. Studying this process provides valuable insights into DNA repair. During metaphase, spindle fibers emanating from the centrosomes at the cellular poles attach to the kinetochores of each chromosome. The nuclear membrane dissolves. The chromosomes align along an equatorial plate with one of the replicated chromosomes on the north side of the plane

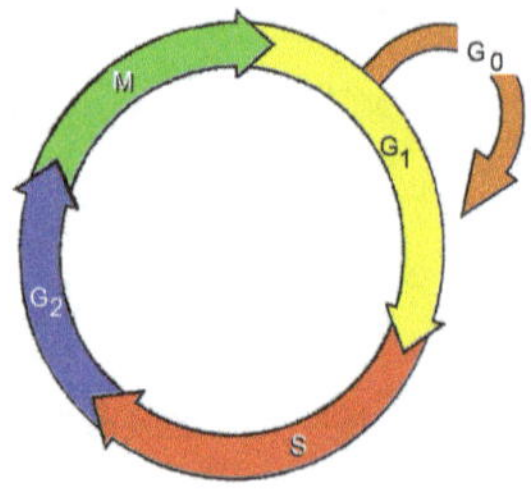

Figure 4-3 The somatic cell cycle. In general, cells of an organism spend most of their time in G_0. They are not mitotically active. When induced, they enter the cell cycle at G_1, a period of growth and protein production comprising about half of the cell cycle. DNA replication (synthesis [S]) follows, a nearly equivalently lengthy process. During G_2, structures and proteins required for cell division are produced. The phase is short, lasting only couple of hours. Mitosis (M) is the rapid but complicated process of cell division that takes only about an hour.

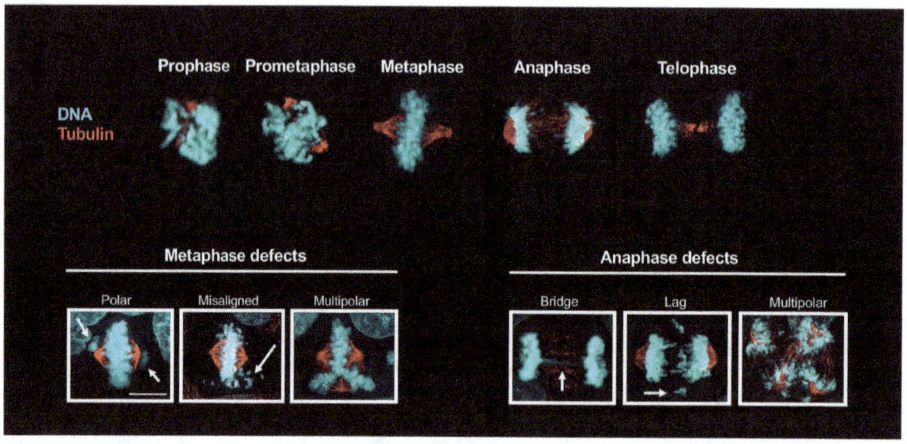

Figure 4-4 Mitosis, the M phase of the cell cycle. Sequence at top illustrates the five phases of mitosis. Chromosomes condense during prophase (blue) and spindle poles composed of microtubules begin to form (red). The sister chromatids then align on the metaphase plate as spindle microtubules are properly attached to kinetochores. During anaphase and telophase, the sister chromatids are separated resulting in two genetically identical daughter cells. Bottom left: mitotic defects during metaphase include chromosomes stuck at the spindle pole (polar chromosomes), misaligned chromosomes, and multipolar spindles. Bottom right: mitotic defects during anaphase include bridge and lagging chromosomes, which are often induced by ionizing radiation, and multipolar anaphases in which the cell attempts to divide in three or more directions. These mitotic errors lead to aneuploid progeny that may affect cell survival. Arrows indicate abnormally located chromosomes. Scale bar is 10 μm. This figure was graciously provided by Pippa F. Cosper and Beth A. Weaver, University of Wisconsin, Madison.

and the other replicated chromosome directly opposite on the south side of the plane. During anaphase, the chromosomes migrate along the spindle fibers to opposite poles, assuring that identical DNA is distributed to each half of the cell. The cell constricts along the equatorial plane and pinches off forming two independent cells during telophase. Each of the daughter cells reorganizes, forms a new nuclear membrane and the DNA relaxes to form the typical configurations of euchromatin and heterochromatin. Changes in chromatin structure during the cell cycle affect both the vulnerability of DNA to specific types of damage and the potential for repair. In addition, biochemistry has provided "checkpoints" where the cycle can be paused or disrupted should the precious genome present signs of damage or disorder.

DNA replication during S phase is a critical process that warrants a more detailed investigation (Figure 4-5). Because DNA is compacted, topoisomerases are required to relax the structure. These

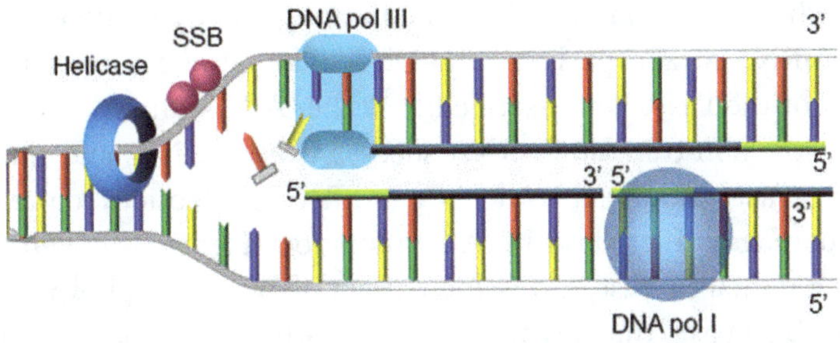

Figure 4-5 DNA replication. During the S phase of the cell cycle, the DNA is replicated in a semiconservative manner. A helicase unwinds the helix and breaks the hydrogen bonds; strands are held apart by single strand binding (SSB) proteins. DNA pol III binds to the template strand and follows the helicase, continuously adding nucleotides as dictated by the template strand and linking the phosphates to the deoxyribose C3. The antiparallel strand is completed by DNA pol I, also adding complementary nucleotides 3′ to 5′ in short Okazaki fragments that are annealed by ligases. Reprinted with permission of the copyright holder, Buzzle.com; from A primer on DNA and DNA replication by Larry H. Bernstein, MD, FCAP. https://pharmaceuticalintelligence.com/ 2014/07/29/a_primer_on_dna_and_dna_replication/.

enzymes nick the phosphate backbone to release tension and allow the DNA strands to relax, untwist, and unwind. Helicases bind to the open end and slide along the strand breaking the hydrogen bonds to create a translocating open wedge of two single-stranded DNA templates. Unwinding is essential for DNA accessibility and synthesis. An indication of the importance of this function is that there are 31 helicases specific to DNA replication and 64 helicases devoted to RNA transcription. Single-strand binding proteins stabilize the open structure preventing regeneration of the hydrogen bonds and enabling DNA polymerases to read the template strand and add nucleotides complimentary to the template bases. DNA is always replicated from the 3′ to 5′ direction (C3 of deoxyribose to C5 of deoxyribose). The 3′ strand can be replicated continuously by DNA pol III; the antiparallel 5′ strand is problematic. DNA pol I binds to the antiparallel strand at the replication fork. The polymerase walks toward the 5′ end, adding as many nucleotides as possible until it reaches duplex DNA and releases. As the helicase opens a new segment of template, pol I binds at the opened replication fork and again adds nucleotides until it reaches duplexed DNA and falls off. The short segments of double-stranded DNA are called "Okazaki fragments," named in honor of their discoverers Professors Tsuneko and Reiji Okazaki at Nagoya University. Lastly, a ligase anneals the phosphate backbone between fragments. The process is semiconservative; each of the two daughter DNA molecules contains one original strand and one newly synthesized strand. Because the fidelity of the newly synthesized strand is critical for the maintenance of the genetic code, DNA polymerases have an exonuclease subunit responsible for recognizing and removing aberrant nucleotide pairs as the polymerase walks along the template strand.

Sections of DNA that code for proteins are called "genes." They are transcribed into mRNA. Each gene has an initiation codon ("start transcribing here") and a stop codon. When a cell requires a protein, for example, when an erythrocyte needs hemoglobin, RNA polymerase along with several cofactors bind to the promoter site for the hemoglobin gene. Although the signaling for specific gene transcription is complicated biochemically and beyond the scope of this text, the interested reader is encouraged to examine the references noted in the Bibliography. The RNA polymerase breaks the DNA hydrogen bonds, creating a transcription bubble in the DNA that travels along 3′–5′ until the stop codon is reached. In a process analogous to DNA polymerase, RNA polymerase adds RNA nucleotides (in RNA, C2 of the ribose sugar is oxygenated and the base uracil replaces thymine) complementary to the template DNA strand. This polymerase also anneals the phosphate backbone at C3 to form single strands of messenger RNA (mRNA).

Because DNA prefers to hydrogen bond to DNA (the binding energy is slightly greater for DNA–DNA than DNA–RNA), the mRNA strand peels off the DNA maintaining a constant transcription bubble length. Only one strand of the DNA is transcribed; it is the 3′–5′ strand that can be continuously processed. Genes also code for "non-translated" RNA such as tRNA, rRNA, and ribozymes. Just like DNA polymerases, RNA polymerases proofread both DNA and RNA nucleotides to prevent coding errors.

Recall the types of damage inflicted on DNA by radiation (Figure 3-9). During replication, when a polymerase encounters a damaged base, a missing base, a deoxyribose phosphate strand break, a mismatched base or a dimerized base, the polymerase proofreading function prevents the continuation of synthesis. An exonuclease cannot remove an integrated deoxyribose phosphate molecule, but the interruption of synthesis recruits an endonuclease that nicks the phosphate backbone and removes several nucleotides from the template strand. The strand gap triggers renewed polymerase activity. Simple phosphate backbone breaks resulting from either radiation chemistry or endonuclease/polymerase activity are mended by ligases. This point is important: cells have inherent mechanisms evolved to create and resolve DNA damage as part of normal biological functionality. These processes are leveraged to resolve damage caused by radiation.

The transcribed mRNA is translated, and protein is assembled by ribosomes through an intriguing mechanical process. The genes that code for tRNA prescribe sequences that induce stabilizing intrastrand RNA base pairing, folding the structure into a roughly three-dimensional "T" shape. Here, we concern ourselves with two critical tRNA regions, although other important structural details will be omitted. The most distal loop from the terminus contains a triplet "anticodon" sequence. A tRNA-ligase (aminoacyl-tRNA synthetase) "reads" the anticodon and attaches the designated amino acid (aa) onto the exposed 3' CCA tail of tRNA. Thus, tRNA deciphers each set of three anticodon bases into only one specific amino acid. The codes for translation of DNA into amino acids is provided in Table 4-1. For

Table 4-1 The genetic code. DNA triplet sequences in the leftmost columns are transcribed into mRNA that is translated by tRNA to the amino acid listed to the right of each triplet codon. The code exhibits considerable redundancy.

DNA Code	Amino Acid	DNA Code	Amino Acid	DNA Code	Amino Acid	DNA Code	Amino Acid
TTT	Phe	TCT	Ser	TAT	Tyr	TGT	Cys
TTC	Phe	TCC	Ser	TAC	Tyr	TGC	Cys
TTA	Leu	TCA	Ser	TAA	STOP	TGA	STOP
TTG	Leu	TCG	Ser	TAG	STOP	TGG	Trp
CTT	Leu	CCT	Pro	CAT	His	CGT	Arg
CTC	Leu	CCC	Pro	CAC	His	CGC	Arg
CTA	Leu	CCA	Pro	CAA	Gln	CGA	Arg
CTG	Leu	CCG	Pro	CAG	Gln	CGG	Arg
ATT	Ile	ACT	Thr	AAT	Asn	AGT	Ser
ATC	Ile	ACC	Thr	AAC	Asn	AGC	Ser
ATA	Ile	ACA	Thr	AAA	Lys	AGA	Arg
ATG	Met	ACG	Thr	AAG	Lys	AGG	Arg
GTT	Val	GCT	Ala	GAT	Asp	GGT	Gly
GTC	Val	GCC	Ala	GAC	Asp	GGC	Gly
GTA	Val	GCA	Ala	GAA	Glu	GGA	Gly
GTG	Val	GCG	Ala	GAG	Glu	GGG	Gly

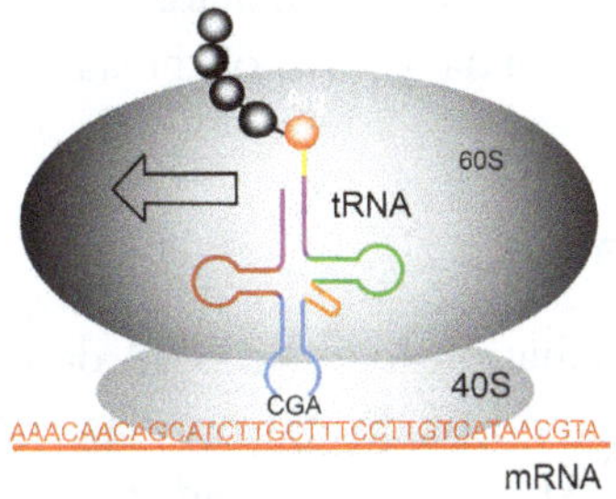

Figure 4-6 RNA translation mechanism. The ribozyme binds tRNA when the anticodon compliments the mRNA triplet code. The 60S subunit covalently bonds the code-specified amino acid to the growing peptide. The 40S subunit slides the mRNA through inserting charged tRNA into the larger subunit and releasing discharged tRNA in an orderly manner.

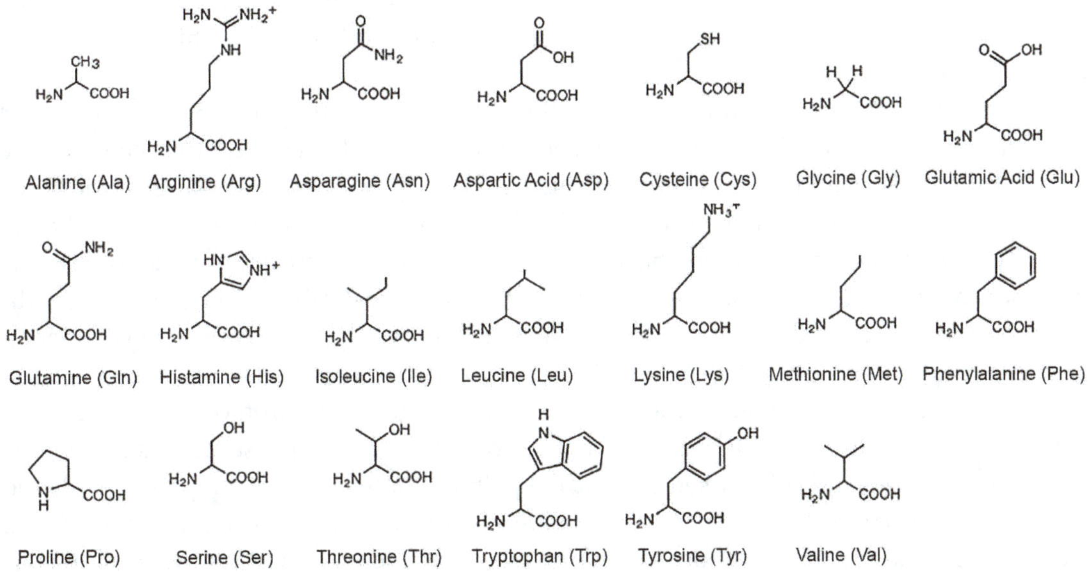

Figure 4-7 Amino acid secondary structures. Amino acid chains determine the protein tertiary structure and folding critical to protein function. Protein shape derives from the minimization of potential energy and takes into consideration amino acid charge, bulkiness, acidity, and hydrophobicity.

example, the triplet DNA codon TTT (phenylalanine) transcribes into AAA mRNA. The tRNA, with an anticodon UUU and terminal-bound phenylalanine, hydrogen bonds to the AAA mRNA sequence. The loaded, or charged, tRNA with the appropriate amino acid attached binds to the larger subunit of a ribosome. Ribosomes are cytosolic ribozymes composed of rRNA and protein. The larger subunit (60 S, where S is a Svedberg unit of density) contains three RNA strands — 120 nucleotides, 160 nucleotides, and 4700 nucleotides; these are bound to approximately 49 proteins. Three sequential binding sites receive tRNA (site 1), link the amino acid to the terminal end of the immature protein through action of a peptidyl transferase (site 2), and release the uncharged tRNA from the ribosome (site 3). The smaller subunit (40 S) contains a 1900 nucleotide strand of rRNA and approximately 33 proteins. The mRNA strand attaches to the smaller subunit of the ribosome. Binding of mRNA facilitates assembly of the two ribosomal subunits. The 40 S subunit ratchets the mRNA, from right to left in Figure 4-6, carrying the tRNA from site 1 to site 2 to site 3. Humans have 20 common amino acids, the three letter symbols of which are presented in Figures 4-7. Several of these amino acids can be modified to affect protein functionality including DNA repair. Individual amino acids have physical characteristics such as charge or aromatic structure that influence the tertiary form of the assembled protein.

4.3. DNA Damage and Repair

For the purposes of radiation biology, radiation-induced DNA damages that sever the DNA backbone are categorized as ssbs or dsbs. The oxidation of deoxyribose as the result of direct or indirect damage leading to a break in the phosphate backbone (ssb) was addressed in Chapter 3. We have also seen that polymerases and enzymes can create ssb when compromised base pairing is encountered during replication and transcription. How do dsb occur? Figure 4-8 shows how two independent ssbs in complementary strands of DNA within 6–10 base pairs of separation can occur when a single energetic electron track transfers energy as it passes, or when two independent electron tracks each break one strand. Each of the independent ssb may be the result of direct damage, or indirect damage. Alternatively, a single event depositing a large amount of energy within the helix can break both strands. This last mechanism is generally a result of direct damage.

When ssb and dsb are successfully repaired, the cell continues to grow, metabolize, and divide without impact to the organism. If the DNA damage is not successfully repaired, the cell may die or the cell will survive but the genome may be mutated (Figure 4-2). Therefore, DNA repair is critical to radiobiological outcome. Because of the importance of DNA damage repair to the cell, tissue, and organism, many alternative pathways exist to accomplish the DNA repair task. Some are unique; some are redundant, and some overlap. Nonetheless, predominant generalities persist. DNA ssbs repair quickly; they are chemically uncomplicated and a template for repair is readily available. DNA dsbs repair more slowly because of the break complexity or loss of DNA basepair template instructions (compared in Figure 4-9). "Seek and hybridize" activities to an unspoiled DNA copy of the sister chromosome may be required, and therefore DNA repair may be delayed until an appropriate cell cycle phase such as S or G_2 phase is reached. Repair enzymes and cofactors in DNA dsb repair are more complex compared to ssb and base damage repair. To visualize the relative repair activities, let's provide some ballpark figures. Irradiation of a cell with 1 Gy of x-rays induces about 40 dsb, 10^3 ssb, and 10^4 damaged bases. Figure 4-10 compares a generalized graphical representation of the repair kinetics for ssb and dsb. Actual kinetics vary according to the repair mechanisms and cell types (as well as species) under consideration. Nonetheless, ssb repair is considerably faster than dsb repair in all cases.

Let's begin by examining ssb repair. A chemically altered or missing base impacts the DNA double helix hydrogen bonding. Base excision repair (BER) recognizes and corrects this simple base damage (Figure 4-11). BER initiates when triggered by one of many proofreading mechanisms, an example of which is 8-oxoguanine-DNA glycosylase (OGG1) that recognizes the 8-oxoguanine form of damaged

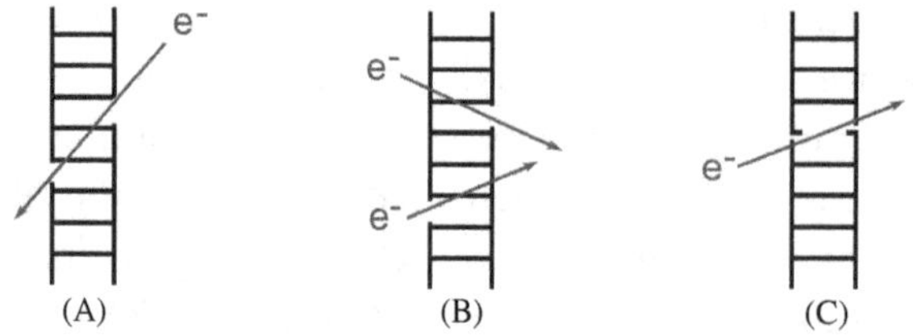

Figure 4-8 DNA scission. **(A)** Double-strand breaks (dsb) occur when a single electron track deposits energy directly or indirectly into both strands, within 6–10 base pairs, or **(B)** two independent electron tracks each deposit energy into a single complementary strand. **(C)** A single large energy transfer within the DNA helix can also cause a dsb.

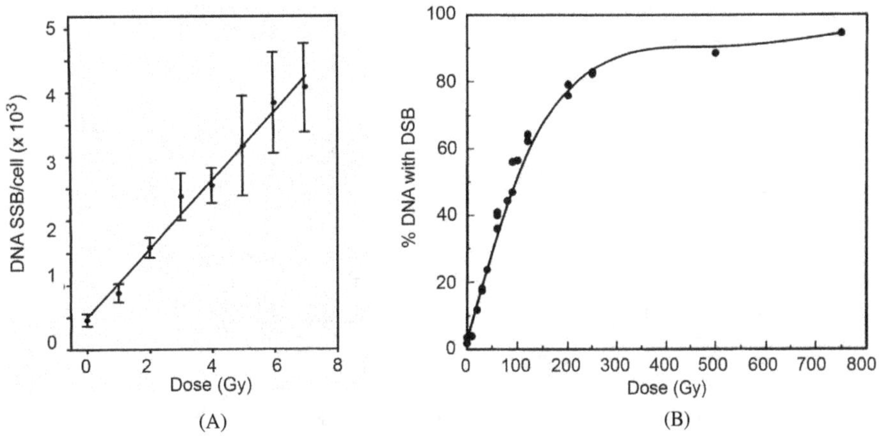

Figure 4-9 Comparison of the induction of ssb and dsb. Panel (**A**) examined the induction of ssb in mononuclear blood cells exposed to a ^{137}Cs source. Panel (**B**) examined the induction of dsb in yeast DNA (doses required to create strand breaks in the smaller yeast nucleus are much greater than doses required to create strand breaks in mammalian nuclei). Both experiments were performed at low temperature to prevent repair. The induction of ssb is linear with dose because ssbs are created by a single event. The induction of dsb has a linear component at lower doses and a curved component at higher doses because there are two mechanisms for creating dsb: a single hit (linear as in ssb; Figure 4-8A and C) or two separate hits (Figure 4-8B). The likelihood of two hits within 10 bp increases with dose. Reproduced with permission of the Radiation Research Society from (A) J. G. Hengstler *et al.*, (2000); (B) D. D. Ager *et al.*, (1990).

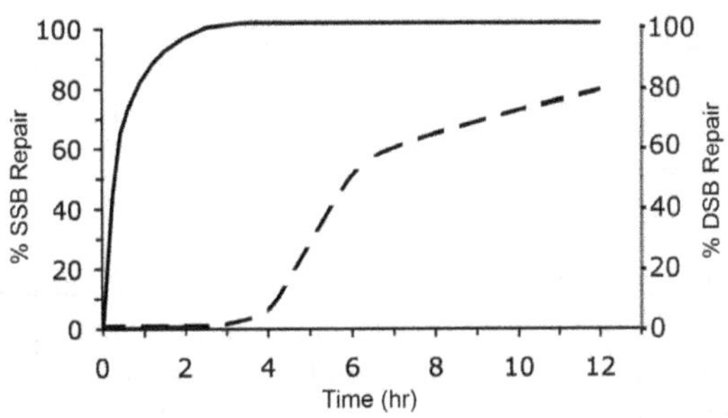

Figure 4-10 Comparison of ssb and dsb repair kinetics. ssb repair (solid line) proceeds rapidly, completing within approximately 2 hours, whereas dsb repair (dashed line) tends to be biphasic and requires approximately 24 hours to complete. Experimental results representing actual processes deviate somewhat from this representation according to cell line, species, and experimental protocols.

guanine (Kovtun *et al.*, 2007). OGG1 tracks along the DNA strand inserting an amino acid probe that twists each G base out of the helix core as OGG1 slides by. When OGG1 encounters an 8-oxoguanine, it hangs up and cannot proceed. Once a damaged base is recognized by OGG1 or any of the other proofreading proteins, a glycosylase (either a subunit of the proofreading enzyme or one that is recruited to the site by a conformational change of the proofreading protein) removes the damaged base. An AP (apurinic or apyrimidinic) endonuclease nicks the phosphate backbone (adds an OH to the diphosphate) at the C3 of the affected nucleotide deoxyribose, and a displacing polymerase (e.g., pol β) uses the opposing template strand to insert several (2–10) paired nucleotides. A "flap endonuclease" nicks the phosphate backbone to release the displaced nucleotides, and a ligase anneals the nicks (catalyzes the phosphodiester bonds between nucleotides). More serious types of base damage distort the orderly base stacking of the double helix and create bulges. These physical distortions are recognized by a different

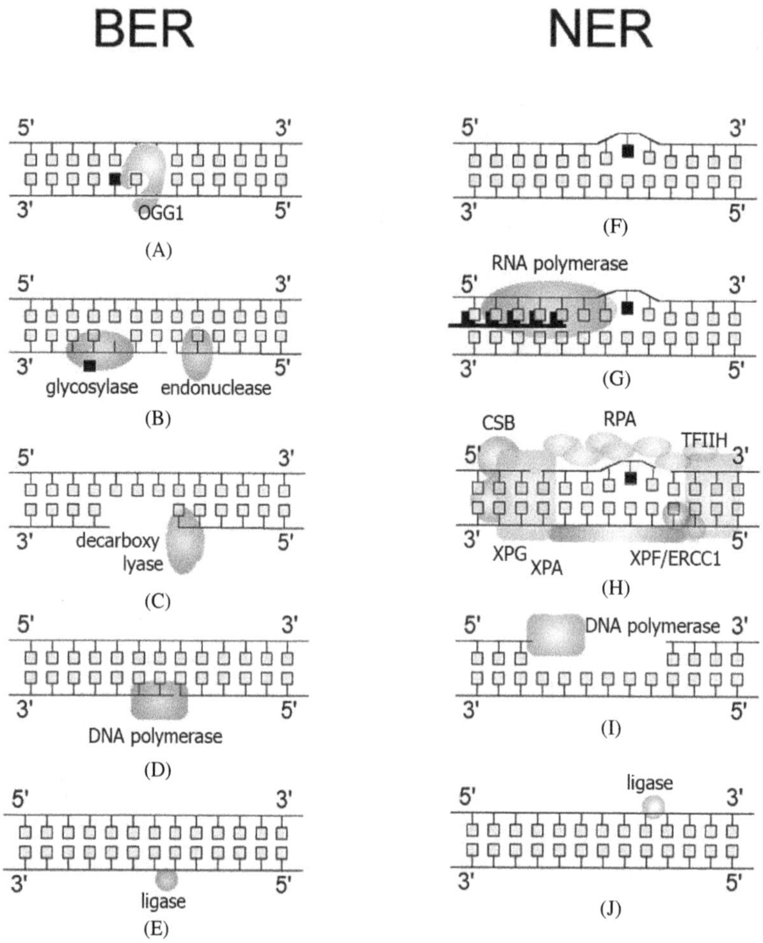

Figure 4-11 The ssb repair. BER (A–E) acts on damaged and missing bases. **(A)** A proofreading protein scans the DNA and hangs up at a damaged base. **(B)** A glycosylase removes the base and an endonuclease nicks the backbone. **(C)** A lyase removes several nucleotides. **(D)** A DNA polymerase uses the template strand to replace the missing nucleotides. **(E)** A ligase anneals the newly synthesized sequence. Nucleotide excision repair (F–J) acts on **(F)** bulky base damage, such as base dimers. **(G)** RNA polymerase encounters bulky damage. **(H)** Specialized proteins release the polymerase and recruit a stabilizing complex of cofactors including RPA and endonucleases that nick the strand and remove several nucleotides 5′ and just 3′ of the damaged base. **(I)** The DNA strand and cofactors release and DNA polymerase replaces the missing nucleotides. **(J)** Ligase seal the nick in the phosphate backbone.

set of editor proteins triggering nucleotide excision repair (NER). An example of NER occurs during transcription when the RNA polymerase hangs up on bulky damage (Figure 4-11). The polymerase recruits two proteins (CSA and CSB) that in turn bind transcription factor 11 H (TF11H). TF11H unwinds the DNA and enables binding of a 10 subunit complex that opens 17–32 nucleotides. The polymerase, CSA, CSB, and TF11H are released. Another protein, replication protein A (RPA) wedges the structure open. Endonucleases at either end of the structure (XPB and XPD) nick the damaged strand and release it. A DNA polymerase fills the gap and a ligase anneals the nicks at either end. BER and NER are "homology dependent" — these processes use the opposing DNA strand as a template to dictate the replacement of the missing and damaged bases. "Error prone repair" replaces standard polymerases with polymerases that have larger DNA binding sites enabling the enzymes to slip past damage, inserting random nucleotides into the de novo strand. Error prone synthesis exhibits low fidelity when replicating

DNA; it is roughly homology dependent. Mismatch repair (MMR) targets inappropriate pairs (purine–purine or pyrimidine–pyrimidine) created by error prone repair. Mismatch damage is not caused by radiation but rather inaccurate replication during the S phase or repair of radiation damage. These examples are representative of many variations and this discussion is merely illustrative and presented without a great deal of detail. Students are encouraged to explore the amazing multitude of DNA repair mechanisms that is beyond the scope of this text.

The dsbs are more difficult to repair faithfully. In this case, the complementary strand may be damaged or missing. Nonetheless, dsb can be repaired through either a homology-independent or a homology-dependent manner (Figure 4-12). Fortunately, the severed strands of DNA do not float off and dissociate; DNA is bound to nucleosomes and to a fibrous scaffold. Nonhomologous end-joining (NHEJ) is the simpler, faster mechanism of dsb repair but as it is nonhomologous, it can lead to severe mutation. Before video, movie film was composed of a series of still frames that advanced past a projection lamp at a speed high enough for the human brain to create intermediates and see smooth movement. When film severed, it was spliced by trimming the ends square, scuffing the two end frames and gluing them

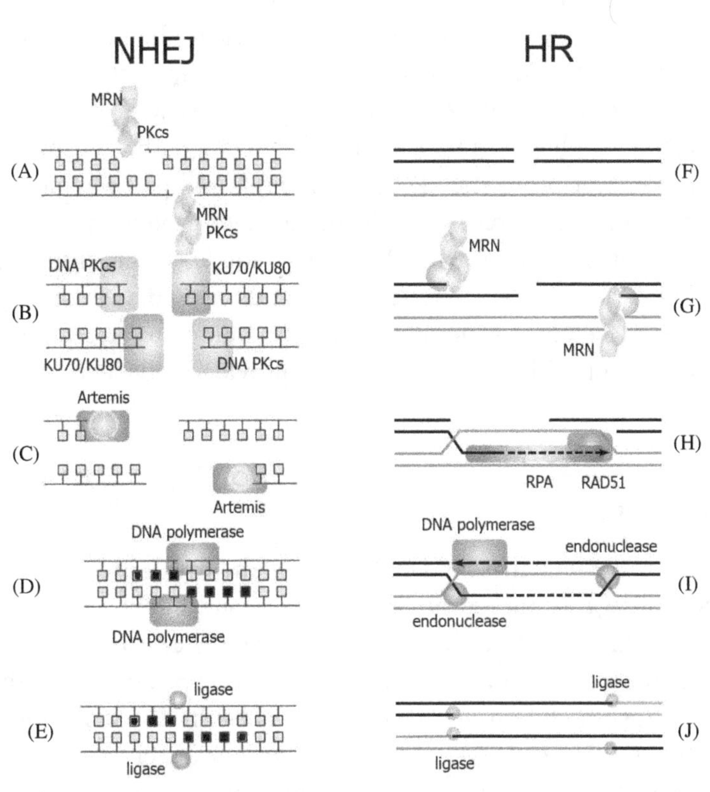

Figure 4-12 The dsb repair. Examples of nonhomologous end joining (NHEJ, A–E) and homologous recombination (HR, F–J) are represented schematically. **(A)** A dsb where both strands suffer backbone damage within 10 base pairs is recognized by the MRN complex. **(B)** Ku70–Ku80 dimers replace MRN and prepare the ends for resection. **(C)** Artemis replaces Ku70–Ku80 and removes several bases, 5′–3′, leaving overhanging "sticky ends." **(D)** A polymerase uses the template strands to replace missing nucleotides and the shortened strands are pulled together. **(E)** A ligase anneals the residual nicks. **(F)** Postsynthesis, a redundant copy of the complete genome exists. **(G)** MRN recognizes the broken ends and recruits Sae2, an exonuclease that removes several nucleotides, 5′–3′. **(H)** A protein/DNA filament is formed with RPA and a cofactor, and this structure searches chromosomes for matching sequences. A polymerase then replaces the missing nucleotides using the invaded DNA as a template. **(I)** The DSBR pathway copies the complementary strand of the invaded DNA. Endonucleases nick the contiguous strands to release fragments of DNA. **(J)** Ligases anneal the strands. Here, crossover has occurred.

together one on top of the other. NHEJ is achieved in a similar way (Figure 4-12). Exonucleases resect the two strands from the 5′ ends leaving opposing "sticky ends" — two single strands of DNA. If the ends are homologous, the strands can be overlapped and annealed as in the case of microhomologies. More often, a DNA polymerase (pol β) replaces the resected nucleotides and the nicks are annealed by a ligase. The resulting missing nucleotides can cause a frameshift mutation. NHEJ primarily repairs dsb presynthesis during G_1, before the cell has replicated its DNA and therefore a secondary template is unavailable. If dsb damage occurs during or postsynthesis (late S, G_2, M), there exists a replicated copy of the predamaged site on the sister chromosome. In that case, homologous recombination (HR) becomes the preferred option (Figure 4-12). DNA is again resected from the 5' ends by a complex of cofactors, the MRN complex (Mre11, Rad50, and Nbs1), plus Sae2. But in this case, the sticky end "invades" matching, replicated DNA. Briefly described, a protein (RPA) coats the sticky end and Rad51 (in mitosis) or Dmc1 (in meiosis) form a filament of nucleic acid and protein that searches for DNA sequences similar to that of the sticky end. The helicase Sgs1 separates the replicant helical strands. HR proceeds by one of two pathways: double-strand break repair (DSBR) or synthesis-dependent strand annealing (SDSA). During DSBR, the second sticky end also duplexes with the unwound homologous DNA. A polymerase fills in the missing nucleotides on each strand. Endonucleases (Exo1 and Dna2) nick one of the strands to release the section of resolved duplex DNA. Depending on which strands are nicked, the product may be recombinant or not. During SDSA, the invading strand is extended using the complementary DNA and is released as the polymerase advances. The newly synthesized strand recombines with the damaged complementary strand, provides a template for the polymerase, and the SDSA pathway finishes with ligation. This pathway does not yield crossover products. SDSA is the preferred pathway during mitosis and is also active during meiosis, though DSBR dominates during gamete production. As in NHEJ, if the break occurs between repeated sequences, another HR pathway, single-strand annealing (SSA), becomes an option. Here, following resection of the 5' strand, the strands realign, overlapping repeated sequences. The resulting deletion of nucleotides generally results in mutation. Again, these examples are not exhaustive or discussed in detail. They are intended to provide a conceptual understanding of these repair processes.

4.4. Genetic Mutations

Figure 4-2 indicates that the consequences of DNA damage and repair are threefold: the damage is successfully repaired; the damage is inaccurately repaired resulting in mutation; or the damage is unsuccessfully repaired leading to cell death. In the first case, the affected cell returns to the cell cycle and continues to function normally. Relevant to the second case, exposure to ionizing radiation increases the natural mutation rate (one mutation per gene per 10^6 cells) by about 10^3-fold. The mutations may occur in a nontranscribed section of DNA such that alteration of the sequence may be irrelevant. A change in DNA coding sequence may not result in a change in protein primary structure because the code is redundant (Table 4-1). A change in DNA sequence may result in the silent replacement of an amino acid (the substitution does not result in a protein structural or functional change). Furthermore, most mutations are recessive; they are not expressed unless both parental copies of the gene are affected. Some mutations are beneficial, for example, if they provide a growth advantage or improve the biological activity of the protein encoded by the mutated gene. The introduction of genetic variability may be the evolutionary pressure that enforces crossover during meiosis. Most serious, expressed mutations are deleterious to the cell and result in cell death, removing the mutation from the gene pool. However, in rare cases, multiple mutations in several different regulatory networks within a cell causes cell transformation that can lead to carcinogenesis. In the third case, the damage to the site may be sufficiently

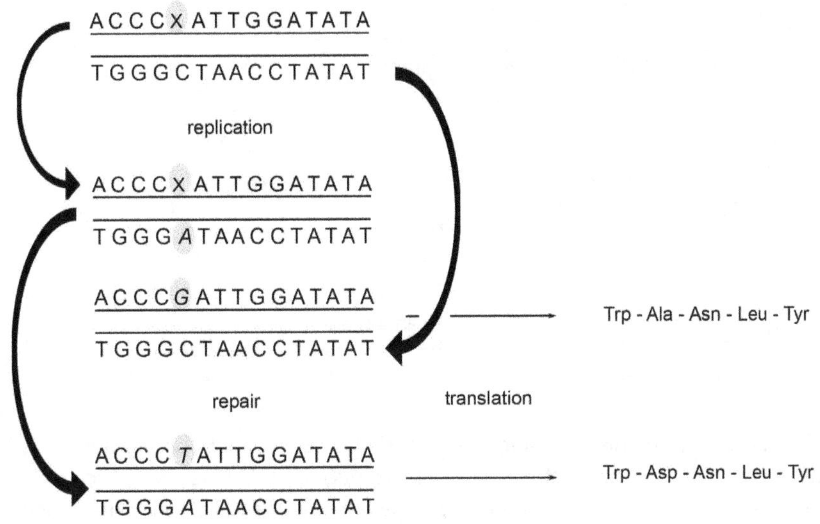

Figure 4-13 Point mutation. DNA is damaged causing a base to be altered, lost, or replaced with a mismatched base (X). When the 5′–3′ (upper) strand is replicated, a random base or base complementing the erroneous base, is inserted into the newly synthesized strand. The lower strand (3′–5′) is faithfully replicated. Eventually, the damaged strand is repaired and the base complementary to the incorrect nucleotide is inserted. In this example, the undamaged template leads to a translation with alanine (Ala) at the damaged codon. The replicated damaged template leads to asparagine (Asp) at the damaged codon.

severe that repair is not possible, or the damage may be compound (either by the number of chromosomes damaged or the by number of damaged sites per chromosome) and complete resolution is improbable. Under these circumstances, the cell can undergo programmed cell death (PCD) or cell lysis, or senescence. We'll come back to types of cell death a little later.

What is mutation and what types of mutation can occur? The simplest form of mutation is a point mutation (Figure 4-13). Suppose a damaged or missing base persists or an incorrect base is inserted during repair, and the mistake is not recognized by a proofreading mechanism. During replication, the unaffected strand is faithfully patterned pairwise in a novel strand. Each A designates a T to be placed in the new strand, each C designates a G, and so on. However, the damaged strand induces an erroneous base insertion into the replication strand at the point of the mistake. A damaged or missing base results in a random insertion; an incorrect base is faithfully read and the appropriate (but incorrect) nucleotide is inserted into the replicated strand. If a random base is inserted into the new strand, customary proofreading mechanisms later replace the missing or damaged base with the appropriate (but incorrect) base pair. This gene now transcribes into a different RNA sequence than prior to the damage and the genome has been mutated. Point mutations tend to occur during replication of unrepaired damage or following failed homology dependent repair. At first glance, point mutations seem rather innocuous leading to null or minor alterations in protein function. However, what would happen if the point mutation occurs within a stop codon? An elongated mRNA transcript would produce a severely elongated protein product. Point mutations in initiation sequences (DNA sequences that bind transcription polymerase complexes) can be equally troublesome if they impede transcription of the gene.

Frameshift mutations (Figure 4-13), wherein one or more base pairs (not including multiples of three) are removed from the DNA, can result from dsb, particularly in the case of NHEJ. These omissions create serious disruptions effectively removing genes from the genomic library and producing novel products that may well be harmful. Figure 4-14 illustrates the impact of a frameshift mutation. Deletions of multiples of three base pairs result in the loss of one or more amino acids from the translated protein.

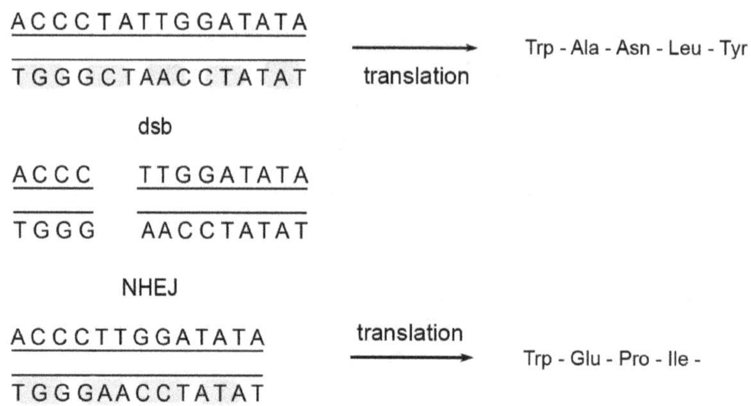

Figure 4-14 Frameshift mutation. A pathway leading to a frameshift mutation might be a double strand break (dsb) repaired by NHEJ. The resulting deletion in the case illustrated here of two nucleotides alters the triplicate code for translation. The polypeptide created from the undamaged DNA is completely different from the polypeptide manufactured following the lost bases.

4.5. Chromosomal Damage

We again remind ourselves that the DNA we have been discussing in our examination of radiation-induced DNA damage and repair exists not as a simple alpha helical structure. The double helix is wound around histones (beads on a string) and twisted into a 30-nm fiber and attached to a fibrous scaffold to form chromatin. This packaging forms a structural mechanism governing enzyme recruitment and function. For example, the genes and sequences contributing to the densely compacted heterochromatin where the DNA is not transcribed, differ depending upon cell type differentiation; the heterochromatin sequences in a muscle cell are different from the heterochromatin sequences in a nerve cell. Whether transcription is controlled by the compaction of the chromatin or the compaction of the chromatin is a result of transcriptional inactivity is not clear. Nonetheless, the packaging of DNA into euchromatin and heterochromatin presents a barrier to enzyme-related activities such as transcription, replication, and repair. The situation can be appreciated when relative size is considered. The DNA double helix axial diameter is approximately 2 nm; a "bead" (the histone core around which the DNA is wound twice) diameter is approximately 6.3 nm. The longer axis of the Ku 70–Ku 80 dimer is approximately 12 nm. The Ku dimer physically cannot access the ends of a dsb unless the chromatin is unpackaged, and the histone core is released. Chromatin remodeling into a relaxed, accessible structure takes about 13 seconds (see Figure 3-5). Phosphorylated SIRT6 (Sirtuin 6, a stress-activated protein) migrates to the dsb and recruits poly-ADP-ribose polymerase 1 (PARP1) to the site. PARP1 synthesizes poly-ADP-ribose chains that bind ALC1 (a helicase). The helicase accomplishes about half of the total structural relaxation allowing the attachment of MRE11 (part of the MRN complex) at the dsb. Next, H2AX histones, a variant of H2A that accounts for about 10% of the H2A dimers in the cores, are phosphorylated along an expanse of about two million base pairs surrounding the dsb. A complex composed of RNF8 (ubiquitin-protein ligase), CHD4 (chromodomain-helicase-DNA-binding protein), and NuRD (nucleosome remodeling deacetylase) binds to phosphorylated H2AX (γ-H2AX) and releases the DNA from the nucleosomes, completing the structural relaxation. So, some repair enzymes attach, and repair processes begin before the release of DNA from the nucleosomes, while the ends are held in proximity. It is easy to see how, following release and relaxation, HR crossover and recombination may occur between opened sections of chromatin. Following completion of repair, chromatin is remodeled and eventually returned to its pre-repair state of condensation.

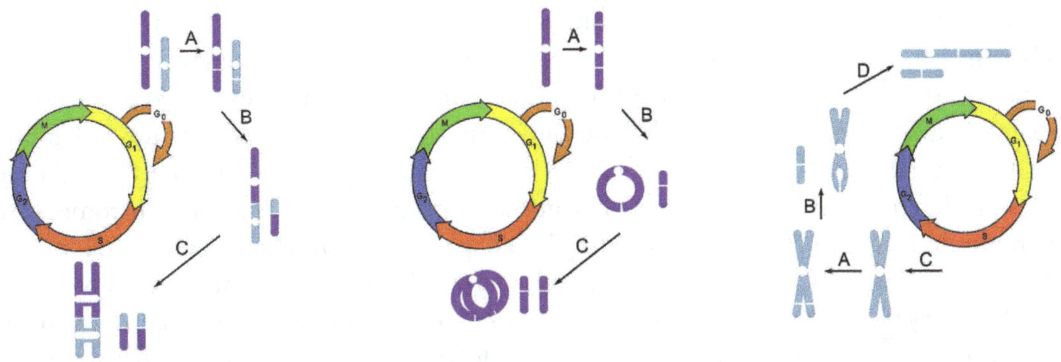

Figure 4-15 Formation of chromosomal abnormalities. At (**A**), the cell is exposed to radiation inducing two dsb. At (**B**), the breaks are repaired but the ends of one dsb anneal to the ends of an alternate dsb. At (**C**), the genetic material replicates producing homologous chromosomes. At the end of mitosis (**D**), the kinetochores release the duplicate chromatid pairs, now presenting two centromeres within one chromosome. In the left panel, the dsb arises presynthesis in G_1. Replication of the dicentric chromosome during S results in a dicentric crosslinked pair. The center panel shows the fate of a single chromosome that received a dsb at either end. The exposed ends join with one another to form a circle. When replicated, two circles are joined together at the duplicated centromeres. The panel at the right follows a postsynthesis chromosome pair each of which experiences a dsb. The ends join to form a loop that opens during anaphase when the centromeres separate. In each scenario, one or two acentric pieces of chromosome are also formed.

Chromosomes form during cell division (meiosis or mitosis). This supercompacted structure facilitates accurate sorting of genetic material into identical daughter cells. Damage to the chromatin can be readily visualized in these structures and so we speak of "chromosomal damage." Don't be fooled; the genetic alterations persist regardless of chromatin packaging status. Resolution of dsb produces three possible outcomes: the break is repaired; the break fails to rejoin; or broken ends resort and rejoin to ends of a different segment of DNA (crossover, reassortment, or recombination). Manifestations of the latter two situations become apparent following chromosome formation. Suppose a cell is exposed to ionizing radiation while in G_0 or G_1. As the cell progresses through the cell cycle (G_1), dsb repair is initiated. If the ends fail to rejoin, the strands separate into two pieces of chromatin of unequal size. When the cell enters S phase, both pieces of chromatin replicate. At mitosis, the chromatin condenses into chromosomes. Two shorter-than-normal chromosomes form, as do two un-rejoined pieces (which will condense into chunks of chromosome nonetheless). If the dsb resolves by annealing to a different strand of broken chromatin, several odd structures can result depending on the relationship of the recombined DNA to the centromere, a DNA sequence that links sister chromatids. Figure 4-15 illustrates the formation of two presynthesis chromosomal abnormalities (i.e., the dsb occurs in G_0 or G_1) and one postsynthesis abnormality (i.e., the dsb occurs late S or G_2).

Because mitosis is elaborate, we describe it in four stages (Figure 4-4). During prophase, the chromatin condenses into chromosomes, the nuclear membrane breaks down, kinetochores bind to the centromeres, and microtubules connect the kinetochores to the cell poles. The kinetochores are duplexed protein structures that attach to the centromeres of each sister chromatid and to the microtubules. During metaphase, microtubule protein motors pull one duplexed kinetochore (and centromere) of the replicated pair toward the "north pole" and one duplexed kinetochore toward the "south pole." The tension forces the chromosome pairs to the metaphase plate, a plane defined by the cell "equator." It is during anaphase that the kinetochore pairs dissociate, and microtubule motors separate the identical chromosomes into different hemispheres. In the case of dicentric chromosomes (possessing two pairs of centromeres, or four kinetochores), a single chromosome may be pulled in two directions — one kinetochore to the "north" and one to the "south." The next phase, telophase, wherein the cell elongates and pinches in at the equator, is compromised by the jammed microtubules associated with the dicentric chromosomes. Cytokinesis cannot occur. If two ends of a single chromosome join following a dsb, as in the center panel of Figure 4-15, the chromosome forms a circle. Replication creates two interlocked circular DNA strands called dicentric rings. The circumstances portrayed in the far right panel of Figure 4-15 yield a loop and then a dicentric chromosome in anaphase, when the centromeres release. The illustration in Figure 4-15 may be misleading because chromosome structure does not exist except during M phase, following condensation of the chromatin. The drawing is intended to be illustrative, though a laboratory technique, premature chromosome condensation (PCC), does allow us to visualize these structures during G_1, S, and G_2. The structures following "C" shown in the first two panels of Figure 4-15 can be seen in chromosomal spreads, as well as the structures following "B" in the third panel (rightmost). If the kinetochores separate successfully, dicentric chromosomes and rings also can be seen. In each case, one or more acentric chromosome segments are formed when the severed ends of chromosomes anneal. Because these segments lack centromeres and kinetochores, they are not redeployed by the microtubules and diffuse freely about the two cell hemispheres. At cytokinesis they segregate randomly, resulting in some daughter cells with duplicate genetic material and some daughter cells lacking these segments. Cells that lack efficient HR DNA repair exhibit increased chromosomal rearrangement and aneuploidy. A chromosomal spread of cells in mitosis allows us to visualize the condensed chromosomes (see Question 6). A useful display of chromosomes is an organization called a karyotype (Figure 4-16). In this format, the chromosomes are artificially ordered from largest

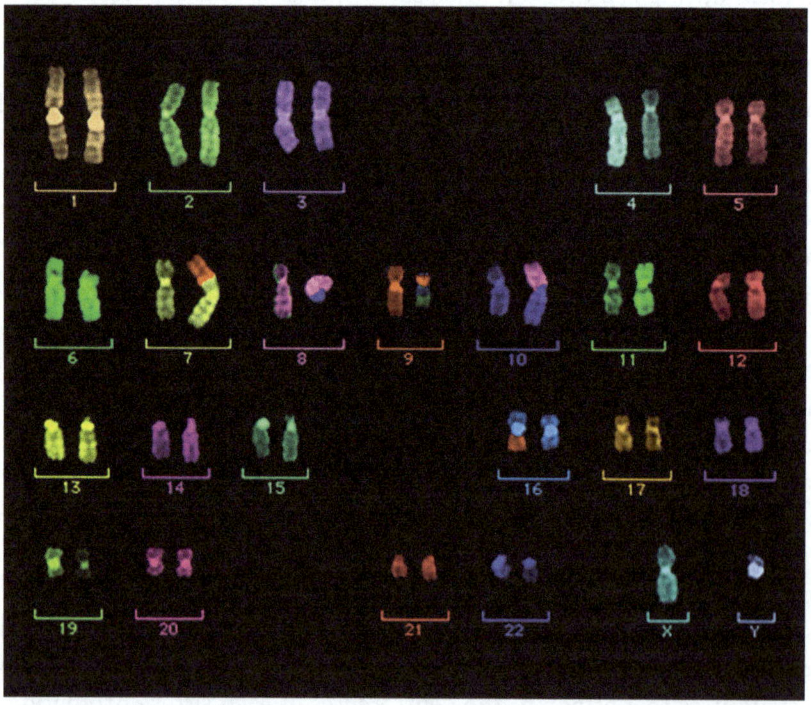

Figure 4-16 Karyotype of human lymphocyte chromosomes postirradiation with 0.3 Gy of γ-rays. The chromosomes are hybridized with 6 different fluorophores and viewed through combinations of filters to achieve 23 different colors. Translocation of the long arm of chromosome 9 to chromosomes 7 and 16, reciprocal translocations of chromosome 8 and chromosome 10, and reciprocal translocation of chromosome 16 to chromosome 9 can be seen. Reprinted with permission of the Radiation Research Society from Karyotypes of human lymphocytes exposed to high-energy iron ions. M. Durante *et al*, *Radiation Research*, 2002; 158: 581–590.

(chromosome 1) to smallest (chromosome 22). In Figure 4-16, the chromosomes have been "painted" with a fluorescent label that uniquely identifies each chromosome with a different color. Chromosomal rearrangements can then be readily identified.

4.6. Cell Death

Radiation biologists have a unique perspective on cell death because of the technique used to determine percent cell survival (see Chapter 6). An *in vitro* cell survival assay depends on each postirradiation surviving cell to divide enough times to create a visible colony of progeny. A cell that cannot divide does not form a visible colony and thus is not counted. It therefore did not "survive" by this functional definition. Clinically, suppose a physician wants to kill all the cells in a tumor with radiation; any cells that persist but cannot divide do not contribute to tumor regrowth. Mission accomplished.

Cells die, that is, completely cease to function biologically, and break down into component parts, by several mechanisms. Mitotic catastrophe is the primary cell death mechanism following ionizing radiation. Because unrepaired dsb can result in mutation, the cell cycle has checkpoints that suspend the cycle until dsb are fully resolved. If the cell enters mitosis despite chromosomal abnormalities (Figure 4-15) the microtubule motor assembly hangs up in anaphase and the microtubules disassemble. These structural malfunctions are likely caused by excessive accumulation of cyclin B1/cdc2 complex and result in abnormal chromosomal distributions, aneuploidy and polyploidy. The end result of these processes is that chromosomes do not segregate properly, which culminates in cells with multiple nuclei following cell division with membranes fully formed. The chromosomes can also be degraded, possibly by cytoplasmic nucleases prior to sequestration within the nuclear envelopes. Actual cell decomposition is often delayed, sometimes allowing multiple additional rounds of the cell division. Three manifestations of mitotic catastrophe have been identified: cell death resulting in a unique, chaotic degradation of cellular components characterized by pulverized DNA, vacuoles and blebbing; caspase-dependent PCD, or caspase-negative necrotic death.

Apoptosis is the most orderly form of PCD. It is best identified by the regular fragmentation of the genome into increasingly smaller segments with time (Wyllie, 1980), although morphological changes include blebbing and cell shrinkage. An endonuclease, Caspase-Activated DNase (CAD), cleaves DNA in the linker regions of the beads on a string structure, producing fragments in multiples of linker-nucleosome-linker (Figure 4-17). Eventually, every linker is cleaved, and every fragment is the length of one linker-nucleosome-linker segment (~180 base pairs), resulting in a single band at the bottom of the gel. At the same time, executioner caspases degrade cellular proteins involved in cell energy production, DNA repair, RNA synthesis, protein synthesis, and maintenance of the DNA nuclear and cytoskeletal scaffolding that keeps cell structures organized. This produces the characteristic morphological indicator of apoptosis: cellular blebbing that resembles a bunch of grapes. These cellular blebs, or "apoptotic bodies" can be phagocytosed by macrophage scavenger cells, preventing local toxicity or immunological response that might damage neighboring cells.

Caspase-independent PCD pathways include anoikis, cornification (eyes only), excitotoxicity or Wallerian degeneration (nerves only), and ferroptosis. During anoikis, cells detach from the extracellular matrix thereby losing attachment-dependent functionality and signaling. As the name implies, ferroptosis is dependent upon iron and is exemplified by the presence of lipid peroxides. Furthermore, there is an immunogenic form of apoptosis that is of special interest because irradiation triggers this pathway. The production of reactive oxygen species (oxygen radicals) stimulates release of heat-shock proteins that are presented on the cell's plasma membrane. There they stimulate immune cells leading to a CD8+ T-cell response and dendritic cell attraction. The cells following this pathway also excrete adenosine

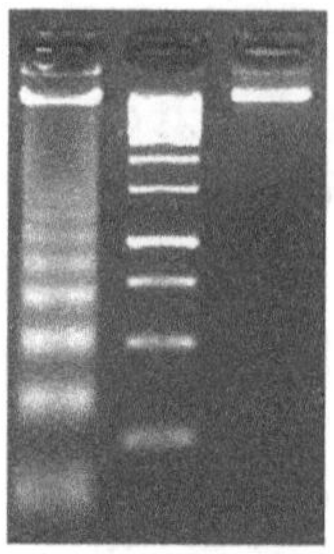

A B C

Figure 4-17 Agarose gel electrophoresis with ethidium bromide staining. (**A**) The systematic degradation of DNA from cells undergoing Apoptosis shows regular bands of DNA fragments. (**B**) A marker ladder with known DNA fragment sizes is used to assess the lengths of the fragments. (**C**) Intact DNA from normal, healthy cells is too large to migrate through the gel.

triphosphate (ATP) that attracts monocytes. Pertinent to oncology, immunogenic apoptosis can induce an antitumor response through the local activation of T-cells and attraction of dendritic cells. Because the immune system naturally polices organisms and destroys precancerous and metastatic cells before they can attach and form tumors, by recruiting T-cells and dendritic cells to irradiated sites, immunogenic apoptosis may be a useful secondary effector for tumor resolution.

Cells also die via chaotic mechanisms devoid of discernable regulated patterns. If deprived of oxygen, cells experience ischemic death characterized by a reduction in the intracellular pool of ATP. ATP drives ion pumps in the membrane (by donating a phosphate) and thus depletion results in an ionic imbalance causing cell swelling. The endoplasmic reticulum (ER) and Golgi dilate, the mitochondria condense, the chromatin clumps, and the cytoplasm forms blebs that pinch off. Subsequent phagocytosis is accompanied by an inflammatory response likely because of ATP leakage. As tumors become too large for oxygen to diffuse from superficial vessels to the core, the innermost cells become ischemic. This is therefore a common form of cell death deep within tumor masses.

The most catastrophic form of chaotic cell death, which leads to the demise of most cell types exposed to high doses of radiation (with the exception of lymphocytes that die after low doses of radiation), is necrosis. During necrosis, the cell swells until the membrane ruptures and the cellular contents spill into the intercellular space. The subsequent severe inflammation adversely impacts the tissue. There appears to be a slightly more programmed form of necrosis called necroptosis that may be a detour mechanism when apoptosis signaling becomes blocked.

A particularly intriguing mechanism embracing cell component salvaging may be hijacked under certain circumstances to result in cell decomposition and eventual death. Manifested primarily during embryonic development and starvation stress, autophagy ("self-eating") represents an ancient cell survival mechanism. Autophagy recycles aged or damaged cell organelles, proteins, and other cellular components, restoring them to basic cellular building blocks such as amino acids, fatty acids, sugars, and DNA or RNA precursors. In general, degradation is exclusive to the cytoplasm and is not expressed in the nucleus unless cell death is imminent. Autophagy is characterized by the formation of double-membrane cellular vacuoles (bubbles) that engulf damaged organelles in a predictable sequence, terminating in binding and fusion with a lysosome followed by degradation of cytoplasmic structures. During embryonic development, autophagy is responsible for the resorption of transient structures, such as finger webbing. In many cancers, autophagy is thought to support cancer cell growth under suboptimal conditions. Nonetheless, biochemistry suggesting autophagic activity following radiation insult has been noted by several researchers. It appears that hyperactive autophagic reaction or prolonged autophagic reaction may be responsible for radiologic cell death in some instances, although the triggering mechanism for this pathway has not been fully described.

Finally, because of our functional definition of cell survival, we include cell senescence in this discussion of cell death. Senescence is a semi-irreversible process leading to secession of cell division. It differs from G_0 quiescence in that the cell very rarely reenters the cell cycle. The state is characterized by severely shortened telomeres (noncoding, repetitive DNA caps). Senescence is a protective mechanism that prohibits replication of cells exhibiting aneuploidy, acentric chromosomes, and unrepaired dsb; conditions that lead to transformation and carcinogenesis. Senescence provides for continuation of structural integrity, specialized functionality such as secretion, and networking, while preventing the replication of damaged DNA. It is neither a form of cell death, nor a form of cell survival according to the functional definition.

The biochemical signaling pathway post-irradiation leading to cell cycle arrest, DNA repair, or cell death initiates with recognition of DNA damage (Figure 4-18). Activation of the primary signal takes place at the site of a dsb; it initiates when the MRN complex binds to a dsb. In addition to the MRN repair activities that we have already discussed, MRN binding to DNA recruits ataxia telangiectasia-mutated (ATM) protein kinase to the dsb site. Tip60 acetyltransferase is recruited to the ATM and is activated by trimethylated H3 in the nucleosome core (methylation of histones is an early response to radiation insult). Then, Tip60 acetylates ATM, activating the ATM kinase activity.

Table 4-2 Modes of cell death.

Motif	Description	Morphology
Mitotic catastrophe	Cell death resulting from premature entrance into mitosis, bypassing the G_2 cell cycle checkpoint; often exhibits aneuploidy or polyploidy and chromosomal aberrations	Giant cells, spontaneous premature chromosome condensation, and chromosome fragmentation
Apoptosis	Programmed cell death proceeding through systematic cleaving of DNA into progressively smaller fragments with containment of cellular components within micellular bodies eliminating inflammatory response	Cell shrinkage, reduced nuclear cross section, membrane blebbing
Immunogenic apoptosis	Apoptosis response plus release of heat-shock proteins that stimulate immune cells; also release ATP that attracts monocytes	As in apoptosis but distinguished by the presence of immune cells
Anoikis	Cellular detachment from the extracellular matrix resulting in apoptotic behavior	Detached, rounded cells with blebbing and small nucleus
Ischemic death	Depletion of ATP results in an ionic imbalance causing cell swelling; phagocytosis is accompanied by an inflammatory response	Cell swelling, ER and Golgi dilation, mitochondrial condensation, chromatin clumping, and blebbing
Necroptosis	The cell swells until the membrane ruptures, the cellular contents spill into the intercellular space resulting in severe inflammation	Cell lysis with the presence of immune cells
Senescence	Irreversible nonmitotic state leading to eventual cell death	Normal morphology
autophagy[a]	Aberrant cellular recycling through lysosomal degradation of cytoplasmic structures, resulting in abnormal protein aggregation, and damaged organelles leading to cell death.	Vacuoles, blebbing, and disproportionally large nucleus

[a]Autophagy appears to play a role in the tabulated forms of cell death, however, the extent of its primary involvement in radiation-induced cell death remains controversial and is under active investigation.

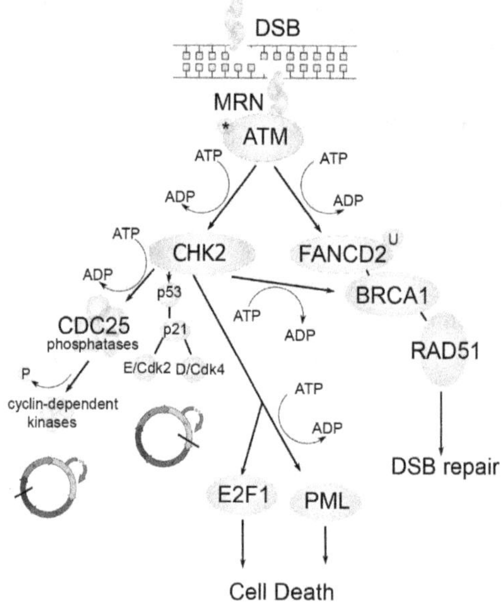

Figure 4-18 Biochemical signaling pathway for resolution of dsb. MRN binding to DNA recruits ATM. ATM phosphorylates FANCD2 and CHEK2. FANCD2 complexes with BRCA1. CHEK2 phosphorylates BRCA1. FANCD2 and BRCA1 complex with RAD51 leading to HR. CHEK2 phosphorylates CDC25 phosphatases that dephosphorylate cyclin-dependent kinases blocking entry into mitosis at G_2. CHEK2 stabilizes p53; p53 binds to p21 and p21 binds to cyclin E/Cdk2 and cyclin D/Cdk4 causing G_1 arrest. CHEK2 phosphorylates E2F1 and PML, both of which are implicated in apoptosis.

Kinases transfer phosphorous from ATP to a target molecule. ATM phosphorylates FANCD2 and CHEK2. Ubiquinated FANCD2 complexes with BRCA1, and CHEK2 phosphorylates BRCA1. FANCD2 and BRCA1 complex with RAD51 to promote HR dsb repair. Meanwhile, CHEK2 also phosphorylates CDC25 phosphatases responsible for dephosphorylating (inactivating) cyclin-dependent kinases (CdK), preventing them from initiating mitosis (creating a checkpoint at G_2). CHEK2 stabilizes p53 that in turn binds to p21. Then, p21 binds to cyclin E/Cdk2 and cyclin D/Cdk4 (inactivating them) to cause G_1 arrest. Finally, CHEK2 phosphorylates E2F1 and PML, both of which are implicated in apoptosis. In short, ATM through CHEK2 and FANCD2 is the primary signaling pathway controlling dsb repair, cell cycle regulation and cell death. The entire biochemistry is more complex and nuanced than represented here. The positive and negative controls and overlapping pathways reveal a biochemical elegance that is difficult to imagine. The reader is encouraged to explore this topic in greater depth.

4.7. Summary

- The central dogma of molecular biology states that DNA is replicated during cell division, translated to RNA, and the RNA is transcribed to protein.
- The human genome contains about 20,000 genes that are coded by only 1%–2% of the DNA.
- Over 80% of the human genome is transcribed into RNA of some description.
- DNA caries the genetic code in triplets of purines and pyrimidines. The code is redundant.
- Not all RNA is translated. Examples of functional RNA are tRNA, rRNA, regulatory RNAs, and ribozymes.

- The transient nature of RNA makes it an excellent control point for regulating the quantity of specific proteins and thereby cellular metabolism.
- There are 20 amino acids that comprise all proteins.
- Most cells in an organism are quiescent, residing in G_0. They enter the cell cycle at G_1 when induced by an external growth signal.
- The cell cycle begins at G_1, a growth phase followed by S when DNA replication (or synthesis) takes place. This is followed by another growth and organizational phase, G_2. Cells then enter mitosis (M), or cell division, culminating in the creation of two duplicate cells.
- Enzymatic cleavage of DNA causing dsb is required for meiosis to proceed. Most breaks are rejoined, but some chromosomal exchange always takes place during cell division.
- The cell cycle has "checkpoints" in G_1 and G_2. The cycle can be paused for DNA repair at these checkpoints by inactivating cyclin proteins.
- The level of DNA compaction affects the accessibility to repair enzymes.
- DNA polymerases proofread the DNA and have exonuclease activity capable of removing damaged nucleotides.
- A gene is a segment of DNA that gets transcribed into a protein or functional RNA. Each gene has an initiation sequence to which polymerases bind and a stop codon.
- RNA polymerases proofread the DNA sequence and recruit repair enzymes when bulky damage is encountered.
- Ribosomes match the tRNA anticodon to the mRNA code for a specific amino acid, translating the mRNA into a sequence of amino acids that comprise a protein.
- The ssbs occur when a charged particle directly deposits energy in the deoxyribose phosphate backbone either ionizing or exciting electrons and causing chemical damage. Strand breaks also occur indirectly, when free radicals oxidize or reduce the deoxyribose phosphate backbone.
- The dsbs occur when a single energetic charged particle deposits energy (either directly or indirectly) in two events, damaging the deoxyribose phosphate backbone of both helical strands of DNA within 10 base pairs. Or, dsb occurs when two independent energetic charged particles each deposit energy damaging the deoxyribose phosphate backbone of paired strands of DNA, within 10 base pairs. Or, a single energetic charged particle may deposit sufficient energy in a single event to damage both strands.
- The three possible results of dsb are successful repair, mutation, or cell death.
- The ssbs are easily repaired by BER or NER. Because these are homology-dependent mechanisms, errors are seldom created during ssb repair. Most errors are subsequently repaired via proofreading mechanisms.
- The dsbs are repaired by non-homologous (NHEJ) or homologous (HR) methods. HR is dominant postsynthesis, when two copies of the genome are present.
- Mutation is an alteration in the gene sequence. Translesion synthesis, NHEJ, and misrepaired base damage leads to mutations. Most mutations are inconsequential (result in a redundant code for the same amino acid), silent, recessive, or result in cell death.
- Point mutations are less severe than frameshift mutations.
- Ionizing radiation increases the natural rate of mutation by a factor of 10^3.
- Chromosomal damage is an alteration of the chromatin sequence that can be seen when the genetic material is condensed into chromosomes during cell replication (meiosis or mitosis). It arises because of improper rejoining of chromatin fragments produced through dsb.
- Chromosomes and chromatin must be unpackaged to expose the alpha helical structure of DNA before repair and remodeling can take place.

- Aberrant chromosome structures reveal whether the dsb occurred pre- or postsythesis.
- Dicentric chromosomes lead to failed cell division at telophase.
- Acentric fragments lead to random sorting of genetic material into the two daughter cells.
- Because cell survival is determined by the ability of cells to divide, viable cells are assumed dead if they cannot divide.
- Several processes lead to cell death including mitotic catastrophe, apoptosis, autophagy, anoikis, immunogenic apoptosis, ischemic death, and necrosis.
- Senescence occurs when a cell is capable of functioning (metabolizing) but can never enter mitosis. It is different from quiescence (G_0) and is recognizable by extremely shortened telomeres.
- ATM is the primary inducer of DNA damage response. ATM is recruited to the site of a dsb by the MRN complex. ATM phosphorylates CHEK2 and FANCD2 producing two cascades leading to DNA repair, cell cycle arrest, and/or cell death.

4.8. Problems

1. What is the central dogma of molecular biology? Name at least three genetic processes that violate the dogma.
2. The following RNA sequence is transcribed from DNA: 5′-GAAGGGCGCUUCCAUUCCCGCCGGAA-3′, where uracil (U) in RNA replaces thymine (T) in DNA. Predict at least three configurations that this single strand might form through hydrogen bonding (with at least 4 bp per bound site).
3. The image on the right represents a sequencing gel. The 5′ end of each fragment is labeled with ^{32}P. The radiation from the label reveals the molecular mass as a band on a film exposed to an acrylamide gel. What is the sequence of this fragment?

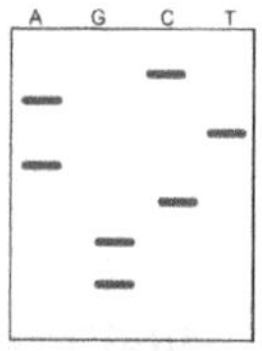

4. Most DNA damage results from (direct or indirect) action and represents (ssb or dsb)?
5. The image on the right shows a "comet assay." Cells are irradiated on ice to prevent repair. Ethidium bromide, that binds to DNA and fluoresces under ultraviolet light, is added. The cells are embedded in agarose gel and subjected to an electrical field.
   (A) What is the disk labeled "control" in the image?
   (B) Explain why the irradiated cells are different.

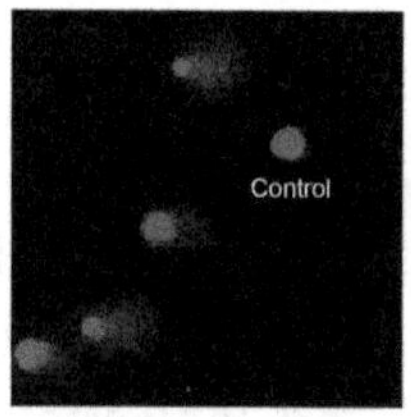

6. Identify the two aberrant chromosomes in the chromosomal spread to the right. Was the damage incurred presynthesis or postsynthesis?

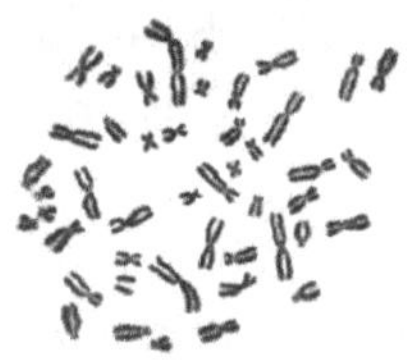

7. For the schematic shown here, what type of damage has occurred? What type of repair will likely be required to rectify the lesion? How will this repair proceed, chemically?

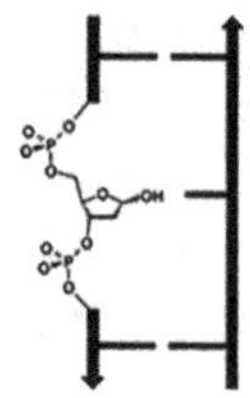

8. Cells ("stem cells") are recruited from quiescence (G_0) into the cell cycle to replace senescent cells that have expired through PCD. Thus, the average lifetime, or turnover rate, of a cell type will determine the number of those cells cycling at any given time. What is the average lifetime of each of the following cell types (look them up)? Given that radiation damage induces mitotic catastrophe, PCD and senescence during the

cell cycle, arrange the cell types in order of radiation sensitivity (most sensitive first). What do you suppose might happen to irradiated, quiescent and senescent cells?
Red blood cells
White blood cells
Skeletal (bone)
Small intestines
Fat cells
Skin cells
Colon crypt cells
Cardiomyocytes
Lung alveoli
Taste buds
CNS neurons

9. Suppose a point mutation occurred at a specific DNA locus coding for proline. Two mutant variants were isolated, one with serine at the locus and one with leucine. Further mutation produced phenylalanine at the locus. What are the possible codons at this locus; wild-type, ser, leu, and phe?
10. The following polypeptide sequence was isolated from a hypothetical cell line:
Cys-Val-Arg-Trp-Ala-Gln-Asp-Ser-Arg-Val-Ala-Gly-Pro-Gln-Ala-Pro.
The culture was subjected to UV radiation and a mutant was isolated that now presented the following polypeptide:
Cys-Val-Arg-Trp-Leu-Arg-Ile-Pro-Gly-Trp-Leu-Asp-Pro-Arg-Pro-Gln
What type of mutation occurred (point or frameshift)? Provide the DNA sequence for the wild-type gene and the mutated gene.
11. From the caption of Figure 4-18, what can you surmise is required for the cell cycle to progress from G_2 to M?
12. The following figure appeared in a publication: Predicting Radiosensitivity with Gamma-H2AX Foci Assay after Single High-Dose-Rate and Pulsed Dose-Rate Ionizing Irradiation, authored by van Oorschot *et al.*, appearing in the journal *Radiation Research*, volume 185, pages 190–198, in 2016.

    The authors of this paper were interested in finding a differential response when cells were irradiated with a single (high dose rate) burst versus when cells were irradiated with a (slower dose rate) pulsed delivery. For the purposes of this question, do not be concerned with that result, that is, panel C is not important. At what timepoint do each of the mutants express maximum phosphorylation of H2AX? With respect to dsb repair, what is the implication for the wild-type (L^+R^+) reduction in H2AX phosphorylation after 24 hours (panel B)? Comparing the mutants L^+R^- and L^-R^+ with the wild-type at the 24-hour timepoint (B), what can you conclude about the NHEJ and HR pathways? Why does the L^-R^- mutant phosphorylation state at 24 hours resemble the L^-R^+ mutant? What necessary information is missing from both figure and legend (hint: what is different about panel C)? If this were your experiment, you might say, "I wonder what happens at 72 hours." What is your hypothesis for a follow-up experiment examining the same cells under the same conditions at 72 hours? Sketch the hypothetical bar graph.

VAN OORSCHOT *ET AL.*

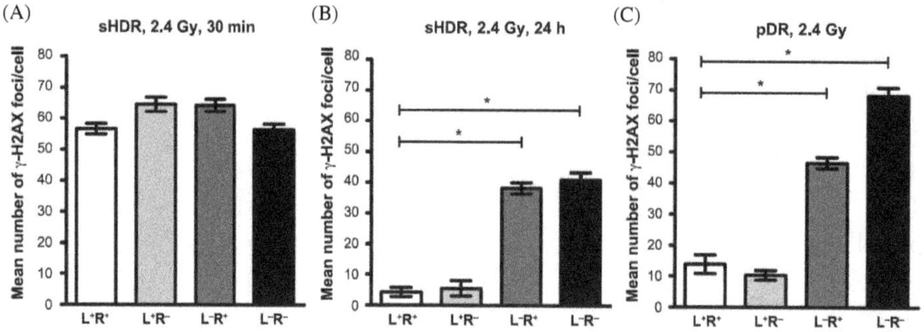

Gamma-H2AX foci in mouse embryonic fibroblasts (MEFs) after 2.4 Gy irradiation. L^+R^+: repair-proficient cells (wild-type), L^+R^-: MEFS with mutation in Rad54 (HR-deficient), L^-R^+: MEFS with mutation in LigIV (NHEJ deficient), and L^-R^-: MEFS with mutation in Rad54 and LigIV (both HR and NHEJ deficient). Panel A: Mean foci numbers 30 min after sHDR irradiation. Panel B: Mean foci numbers 24 h after sHDR irradiation. Panel C: Mean foci numbers after pDR irradiation. A minimal of 100 cells per condition are counted per experiment (n = 3); error bars are ±SEM.

13. The following figure appeared in a publication: Induction and Repair of DNA DSB as Revealed by H2AX Phosphorylation Foci in Human Fibroblasts Exposed to Low- and High-LET Radiation: Relationship with Early and Delayed Reproductive Cell Death, authored by F. Antonelli *et al.*, appearing in the journal *Radiation Research*, volume 183, pages 417–431, in 2015.

    What does the number of phosphorylated foci per cell indicate? What happens at T = 0? What happens at approximately 30 minutes? Why do the curves decrease with time? Suggest a hypothesis that explains why the curves do not return to 0 foci/cell.

INDUCTION AND REPAIR OF DNA DSB

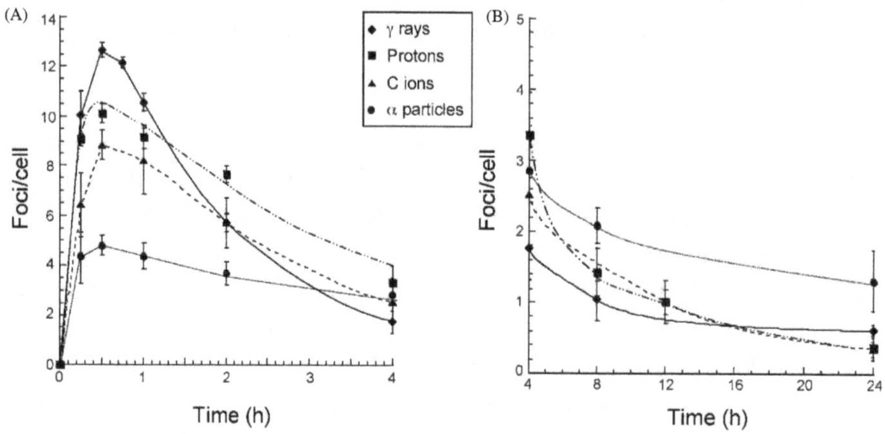

Phosphorylation-dephosphorylation kinetics of H2AX foci induced in AG01522 cells after irradiation with 0.5 Gy of γ rays (solid diamond), protons (solid square), carbon ions (solid triangle) and alpha particles (solid circle). H2AX kinetics are reported up to 4 h (panel A) and from 4–24 h (panel B) after irradiation. At least eight independent experiments were performed for γ rays, five for protons, four for carbon ions and seven for α particles. Net data are reported, after subtraction of the control values. The error bars represent the standard error of the mean.

4.9. Bibliography

Ager, D. D. *et al.*, 1990. Measurement of radiation-induced DNA double-strand breaks by pulsed-field gel electrophoresis. *Radiation Research,* 122(2), pp. 181–187.

Alberts, B. *et al.*, 2002. *Molecular Biology of the Cell.* 6th ed. New York: Garland Publishing.

Gewirtz, D. A., Holt, S. E. & Grant, S., 2007. *Apoptosis, Scenescence and Cancer.* Totowa: Humana Press.

Grand, R. J. A. & Reynolds, J. J., 2019. *DNA Repair and Replication: Mechanisms and Clinical Significance.* Boca Raton: CRC Press, Taylor & Francis Group.

Griffiths, A. J. *et al.*, 2015. *Introduction to Genetic Analysis.* 11th ed. New York: W.H. Freeman and Company.

Haber, J., 2014. *Genome Stability: DNA Repair and Recombination.* Milton Park: Garland Science, Taylor & Francis Group.

Hall, E. J. & Giaccia, A. J., 2019. *Radiobiology for the radiologist.* 8th ed. Philadelphia: Wolters Kluwer.

Hengstler, J. G. *et al.*, 2000. Induction of DNA single-strand breaks by ^{131}I and ^{99m}Tc in human mononuclear blood cells in vitro and extrapolation to the *in vivo* situation. *Radiation Research,* 153, pp. 512–520.

Kerr, J., Wyllie, A. & Currie, A., 1972. Apoptosis: a basic biological phenomenon with wide-ranging implications in tissue kinetics. *British Journal of Cancer,* 26(4), p. 239–257.

Kovtun, I. *et al.*, 2007. OGG1 initiates age-dependent CAG trinucleotide expansion in somatic cells. *Nature,* 447(7143), pp. 447–452.

Pack, L., Daigh, L. & Meyer, T., 2019. Putting the brakes on the cell cycle: mechanisms of cellular growth arrest. *Current Opinion in Cell Biology,* 60, pp. 106–113.

Wyllie, A., 1980. Glucocorticoid-induced thymocyte apoptosis is associated with endogenous endonuclease activation. *Nature,* 284(5756), pp. 555–556.

Chapter 5

Higher Order Biology and Radiation Damage

5.1. Introduction

Chapter 4 interrogated intracellular biochemistry with a focus on the impact of radiation damage to DNA fidelity, chromosomal structure, and resultant cell survival. In this chapter, we will examine the radiogenic phenomena beyond simple molecular biology, consequences of radiation damage that network with more complex levels of biological organization. Beyond prompt DNA damage and repair, irradiated genomes can become unstable, inducing cells to express mutations or mortality only after several divisions. In addition, we will also explore DNA modifications that can influence chromosomal structure and how radiochemistry can impact the DNA modification state. Furthermore, cells are not independent; they are interconnected to form tissues so that the cellular response to radiation damage reacts to biological mechanisms that promote tissue preservation and organ functionality. When radiation-induced damage repair succeeds, both cells and tissue triumph. However, when cell or tissue damage repair falters, organism well-being motivates some type of response. If the genetically damaged cell is required for tissue integrity, cells may enter senescence; if the genomic threat to the organism is unacceptable, cells preferentially die. Also, cellular radiogenic injury generates tissue-level and organism-level consequences. For example, organ dysfunction resulting from cell death may reflect the kinetics of a subset of differentiating populations that express at times dependent on the number of differentiation steps and the duration of each step (e.g., villi development from crypt cells). This has additional relevance in germ tissues that develop into functional units or even the whole organism after being exposed to radiation. In the latter case, surviving fertilized but genetically damaged germ cells will replicate mutated sequences into every cell of the new organism with highly complex consequences. It is also important to realize that radiation-induced changes in cells and tissues can result from not just DNA damage or gene rearrangement, but also from DNA gene expression regulatory mechanisms such as DNA methylation and chromatin packing. Examination of these "extra-genomic" or "epigenetic" regulatory mechanisms provides context for further analysis of radiation-induced complex multiscale phenomena.

5.2. Genomic Instability

The most simplistic level of radiation biology considers the molecular construct of DNA and the chemical/structural alterations introduced by ionizing radiation. Changes in molecular integrity may impact chromosomal structure and thereby disrupt gene function. The genome therefore represents a higher functional unit composed of DNA organized into chromosomes. Genomic instability arises from radiation or chemical insult and occurs nearly without exception in cancer,[1] so it behooves us to understand the phenomenon. In the lab, genomic instability emerges thusly. A gently irradiated plate of nonconfluent cells may double as usual over 24 hours. However, the next 24 hours may produce not twice as many

[1] Genomic instability may include mutations in genes involved in DNA repair or manifest as defects such as broken, missing, rearranged, or extra chromosomes.

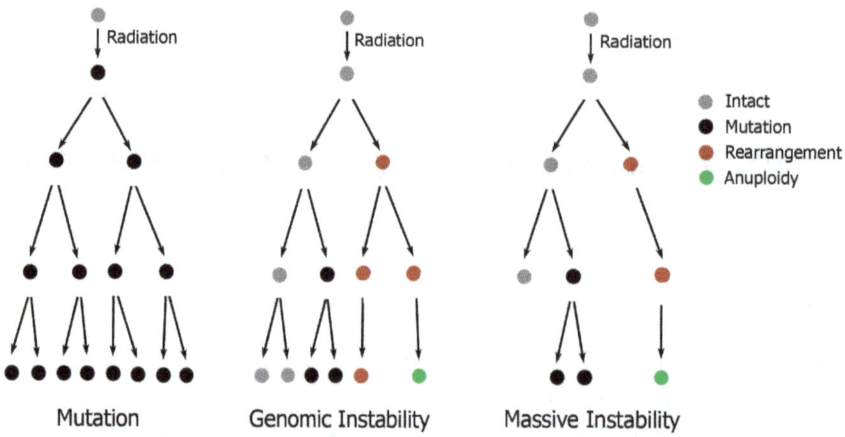

Figure 5-1 A comparison of genetic mutation with genomic instability. In this example, irradiation of a healthy cell introduces a heritable mutation in DNA sequence, epigenetic modification, or chromosome structure (left). Alternatively, irradiation of a healthy cell may produce no apparent damage in this cell (first generation), however, genetic alterations may appear later — in this case, a rearrangement appears in the second generation and a mutation appears in the third generation. Accumulating damage may lead to cell death (center). If the induced genomic instability is severe (right), increased, delayed cell death is anticipated.

cells, but perhaps only 50% more and following the next 24 hours, the cell number may barely increase or even decrease. Chromosomal spreads (see Chapter 4, problem #6) from these cell cultures would reveal increasing frequency of lethal chromosomal aberrations over time, in the absence of additional radiation exposure. In general, genomic instability can be attributed to a loss of cell cycle control. This loss of control and the associated DNA damage tolerance commonly correlates with persistent damage to any of the genes encoding any of the proteins involved in the complex biochemical processes responsible for cell cycle checkpoint activation (Figure 4-18), permitting cells to progress through division despite existing chromosomal damage. The loss of growth control (continuous cell cycling) combined with the loss of cell cycle control (disabled cell cycle check points that allow for repair) generates cells with limitless, increasing chromosomal abnormalities. From the simplistic perspective of molecular biology, an increase in DNA damage tolerance leads to decreased cell death, allowing for a greater and greater accumulation of mutations. Figure 5-1 illustrates the conceptual differences between the processes of genomic instability and genetic mutation as described in Chapters 3 and 4. In this figure, each dot represents a cell; the top row of gray dots indicates cells exhibiting nascent, nonmutated genes. If these normal cells are exposed to a mutagen — radiation in this case — the progeny, represented in the second generation present genetic or epigenetic alterations in the case of mutation (first pedigree), or may appear genetically unharmed as in the cases of provoked genomic instability (the second and third pedigrees). In these latter two cases, genetic changes may present in the progeny, or may not appear until the third generation or later (unrepresented in the figure). In the figure, gene rearrangement similar to that illustrated in Figure 4-16 (chromosome 7) emerges in one of the two second generation progeny cells. This rearrangement carries through to the third generation, which produces one daughter cell retaining the rearrangement, one daughter giant cell that fails to undergo cytokinesis (aneuploidy), and one cell that does not survive. The apparently unharmed cell of the second generation of this pedigree produces an apparently unharmed daughter and a daughter with an unprovoked genetic mutation. If the genomic instability is severe, as in the third pedigree, the high frequency of lethal configurations induces increased apoptosis or mitotic death. We will encounter the ramifications of genomic instability several times as we continue our exploration of radiation biology.

5.3. Epigenetic Mutation

In Chapter 4, we mentioned that ATM, the progenitor cell life cycle determination (Figure 4-18), is activated by trimethylated histone H3, and that methylation of H3 is an early response to radiation insult. Methylation of H3 is one example of epigenetic modification: structural or chemical modification of chromosomes that does not affect the DNA sequence (Figure 5-2). Nonetheless, epigenetic modification affects gene expression in complex and relatively poorly understood ways. Furthermore, epigenetic modification can become "mutated" and inherited — sometimes for a few generations and sometimes permanently. Because epimutations influence the expression of genes (including DNA repair genes), and because inappropriate methylation leads to the loss of cell life cycle control (via ATM signaling), it is not surprising that epigenetic mutations cause increased genomic instability. Gene silencing (inactivation) through mutation of a gene promotor region may result from DNA misrepair, as we have seen, but inaccurate remodeling of the original epigenetic status (aka "clearance" of repair sites) may also cause gene silencing. In fact, more gene silencing epigenetic mutations are identified in the DNA repair genes of biopsied cancer cells than DNA sequence mutations. For this reason, functional assays such as RNA transcription may be better indicators of altered genetic expression than simpler assays that detect DNA sequence mutations.

Methylation is a unique epigenetic marker in that both cytosine and nucleosomal histones represent the targets of methylation (Figure 5-3). In the case of DNA base methylation, a hydrogen atom of cytosine is replaced with a methyl molecule (CH_3). Gene promotor regions are typically hypomethylated with characteristic CpG islands (prevalent in promotor regions) consistently unmethylated. In other words, when the DNA sequence immediately upstream of a gene lacks methyl adducts, the gene is actively

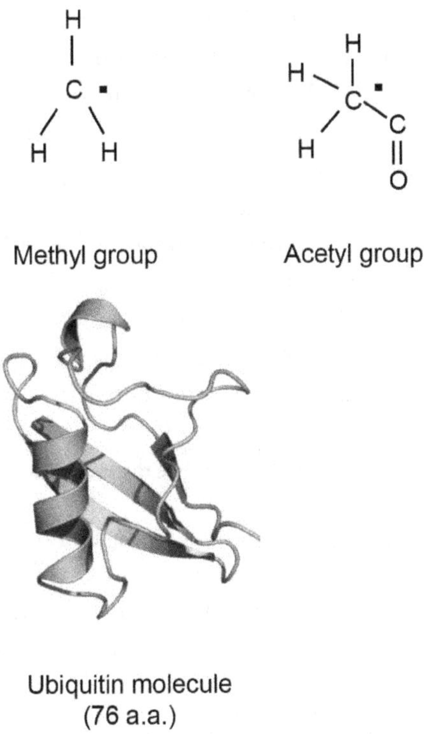

Figure 5-2 Common epigenetic adducts. The methyl group can be added to DNA at cytosine residues or to histones H3 or H4 at lysine residues. Acetyl is slightly larger, adding a carbon and oxygen to the methyl group. Ubiquitin is enormous by comparison, composed of 76 amino acids, most of which are considerably larger than the other two adducts.

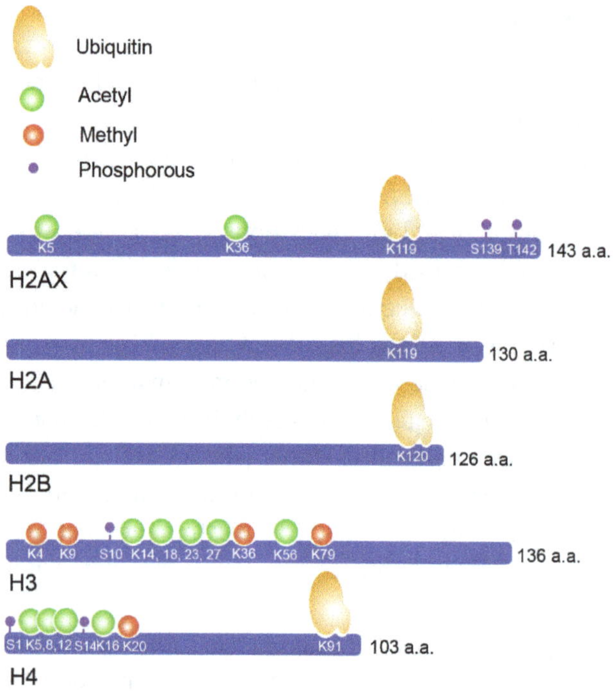

Figure 5-3 Representation of histone epigenetic modification. Bars represent histones and are identified below each bar. Amino acid modification sites are labeled within the bar, below the epigenetic marker; K = Lys, S = Ser, and T = Thr. The epigenetic adducts are identified in the key provided at the top of the figure. Not drawn to scale.

transcribed. Hypermethylation of promotor DNA sequences represses transcription. An example of the repercussions of radiation-induced epigenetic remodeling reflects DNA hypermethylation of CpG islands in tumor suppressor gene promoters. This hypermethylation correlates with inactivation of tumor suppressor genes and it is frequently observed in chronic inflammation and precancerous lesions. Furthermore, abnormal cytosine methylation may be passed to progeny cells. During DNA replication, a methyltransferase, DNMT1, copies the methylation pattern from the template strand to the *de novo* strand. This process has been dubbed "epigenetic templating" to differentiate it from complimentary assembly of the genetic code during replication. How important is epigenetic patterning? Maintenance of the epigenetic pattern may be responsible for cell differentiation (the determination of which genes are transcribed, and which genes are silenced) or as in the case of DNMT1 is essential for appropriate X-chromosome condensation during embryonic development.

How could DNA methylation influence the promotion of cancer?

An example is provided by the microhomology-mediated end joining (MMEJ) pathway. This minor error-prone repair pathway for DNA dsb repair utilizes 5–25 complimentary base pairs to realign broken strands. The noncomplimentary flaps on each strand are removed by an endonuclease and the strands are ligated, resulting in the loss of several bases from the sequence. Resulting frameshift mutations increase the likelihood of cell transformation. Hypomethylation of the gene that transcribes the flap endonuclease (FEN1) induces overexpression of the enzyme, thereby shifting repair kinetics in favor of the MMEJ repair pathway. The shift in kinetics results in a larger proportion of the repair activities occurring via the MMEJ pathway. Both hypomethylation of the FEN1 gene and overexpression of the FEN1 enzyme have been identified in breast, prostate, stomach, pancreas, and lung cancers, and neuroblastoma.

Methylated histones (specifically methyl bound to H3 at Lys 9 and at Lys 27, and H4 at Lys 20) are associated with condensed chromatin, although in this case correlation does not indicate causation — that is, the relationship between H3/H4 methylation and the formation of heterochromatin does not indicate necessarily that histone methylation promotes condensation. Although the global phenomenon of silenced gene promotor hypercondensation in all differentiated cells strongly suggests a trend that cannot be ignored, HP1 (a protein that blocks transcription) specifically binds to methylated H3 at Lys 9, indicating that binding of inhibitory proteins may be an additional mechanism of epigenetic control. In an apparent contradiction, trimethylation of H3 at Lys 4 is strongly associated with actively transcribed genes. This modification, however, binds a nucleosome remodeling factor (NURF) that facilitates nucleosome sliding along the DNA. Similarly, trimethylation of H3 at Lys 36 recruits mismatch repair proteins. These examples of biochemical docking indicate that the purpose of methylation may be the creation of binding sites rather than direct chromosome condensation. The evidence suggests that two mechanisms are possible: either the methylation-induced condensation state of chromatin determines gene accessibility, or the methylation of histones provides efficient binding of promotor cofactors . . . or both.

In addition to methylation, several other epigenetic modifications of the histone cores are possible, the most common being phosphorylation, acetylation, and ubiquitinoylation. The phosphorylation of H2AX at Ser 139 and at Thr 141 enables DNA double-strand break repair by recruiting an enzyme complex that releases the DNA from the nucleosomes (Chapter 4). H2AX is a variant of H2A that accounts for about 10% of the H2A dimers in nucleosome cores. The phosphorylation of H4 at Ser 1 may play a role in DNA dsb rejoining by binding casein kinase 2 (CK2) during nonhomologous end-joining (NHEJ). H4 also may be phosphorylated at Ser 14. Phosphorylation of histone H3 at Ser 10 represents a paradox; it is associated with relaxed chromatin during transcription and replication but highly condensed chromatin during other stages of mitosis. The difference most likely corresponds to the status of additional epigenetic markers, specifically methylation and acetylation (at Lys 14, Lys 18, Lys 23, Lys 27, and Lys 56). So, the response may depend on some pattern of modifications still not well understood and under investigation.

Acetylation generally indicates open, active chromatin. Nucleosomal lysine has a positively charged tail (NH_3^+; Figure 4-7) that electrostatically attracts the negatively charged DNA phosphate backbone. Acetylation neutralizes the tail, reducing the attraction between the nucleosome and the entwined DNA helix. Alternatively, a biochemical mechanism of action for acetylation proposes that acetylation may create a binding site for enzymes involved in transcription including structural remodeling proteins. The existence of acetylated lysine binding sites in several transcription complex enzymes supports this scheme. Of course, as before, both mechanisms, biophysical and biochemical, may be in play.

Ubiquitin is a relatively common biochemical marker that designates proteins as targets for a variety of processes including degradation, cellular sequestration, modification of activity, and protein interaction. Ubiquinated FANCD2 complexes with BRCA1 in a pathway leading to DNA dsb repair (Figure 4-15) and H2AX is polyubiquinated at Lys 63 in response to a DNA dsb, helping to recruit BRCA1 to the damage site. Again, evidence exists for both biophysical and biochemical chromatin remodeling schemes. Ubiquitination alters the chromatin structure; ubiquitin is nearly as large as the histone to which it attaches. Nonetheless, it provides a binding site for transcription modifying enzymes, either inhibitory or activating.

One final comment that further obfuscates but demonstrates the richness of the roles and devices of epigenetic regulation: different combinations of markers are likely to signal different responses. Acetylation at one site may induce a different effect from acetylation at another site, and acetylation next to methylation likely elicits a different outcome from either modification alone. Although much remains to be discovered about epigenetics, it is essential to keep in mind that epigenetics is at least as powerful in determining cellular transformation and the heritability of altered gene expression as is DNA mutation. For this reason, the impact of radiation chemistry within the realm of epigenetics deserves serious consideration.

5.4. Cell Division

Cell division eclipses radiation biology at several crucial intersections: replication of DNA promotes repair fidelity through proofreading, cell cycle stage influences radiosensitivity, and persistent DNA dsb damage during cell division gives rise to chromosomal aberrations. Therefore, a thorough appreciation of cell division processes informs several radiation biology fundamentals.

5.4.1. *Meiosis*

Most cells replicate through mitosis (Figure 5-4). These are somatic cells; they possess 46 chromosomes (23 pairs) composed of perhaps 50,000 genes,[2] a prescribed set of which are expressed to define a specific tissue phenotype. Nonterminally differentiated somatic cells have the potential to divide many times, and so we say that they *cycle through mitosis.* On the other hand, a small number of germline cells, gametes (eggs or sperm), are haploid and contain 23 chromosomes composed of only one copy of each gene enabling the zygote, following fertilization, to carry one paternal and one maternal copy of each gene. Gametes are pluripotent; germ cells can be induced to develop into any type of tissue determined by the eventual pattern of gene expression. To reduce the number of chromosomes from 46 to 23,

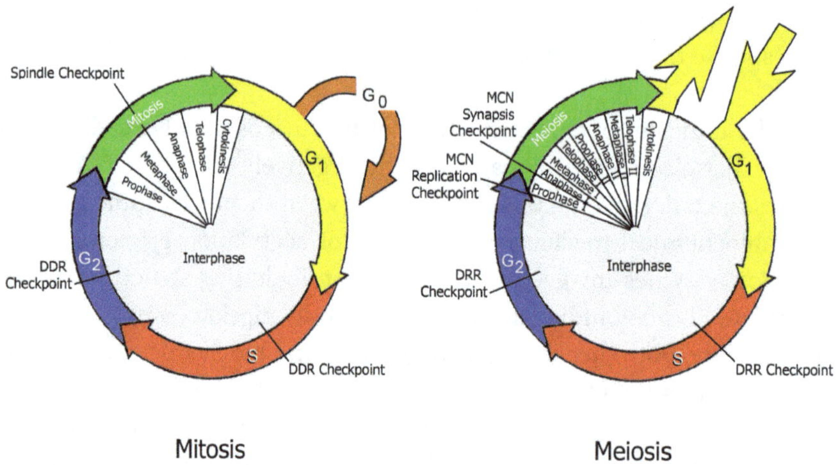

Figure 5-4 A comparison of mitosis and meiosis. Somatic cells undergo cell division according to the cycle on the left to produce two genetically identical daughter cells. While some cells cycle continuously, most cells *in vivo* reside in a quiescent phase, Gap 0 (G_0) until they are recruited into the cycle by growth factors. DNA damage response (DDR) checkpoints sense DNA dsb and stall the cycle until the damage can be cleared. The spindle checkpoint responds not to dsb but rather incomplete spindle attachment. Germ cells undergo dual divisions according to the cycle on the right to produce four haploid progeny. The cycle is completed only once per cell and it retains DDR checkpoints during synthesis (S phase) and the largely vestigal Gap 2 (G_2) phase. During prophase I of meiosis, the MCN surveys several genomic constructs including dsb and regulates the temporal flow through prophase I, stalling when necessary, blocking when necessary, and inducing cell death when necessary.

[2] There are approximately 20,000 protein encoding genes, plus regulatory mRNA and micro-RNA encoding genes.

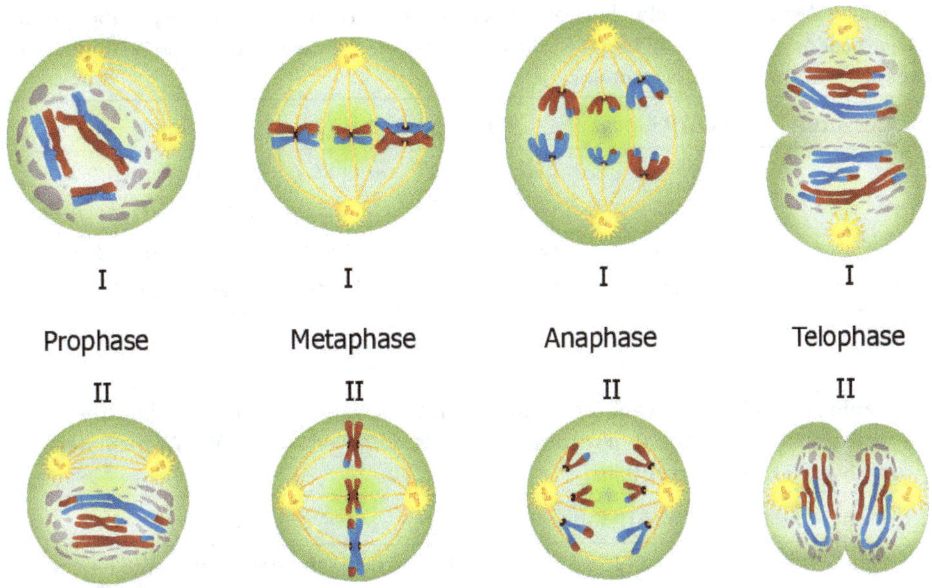

Figure 5-5 Schematic representation of meiosis. The first cell division (top) is characterized by interlocking the replicated chromosome pairs through crossover chiasmata during prophase I, alignment of the pairs at the equatorial plate during metaphase I, and anaphase I dissolution of chiasmata and separation of chromosome pairs to form a haploid genome. The second cell division is characterized by separation of the sister chromatids at anaphase II. Reproduced and modified with permission through CC BY-SA 4.0, https://commons.wikimedia.org/w/index.php?curid=49630204. Figure by Ali Zifan based on *Campbell Biology* (10th Edition). Jane B. Reece and Steven A. Wasserman.

germline cells undergo a second cell division that distributes only one copy of the genome to each daughter cell. The "two-cycle" cell division is called *meiosis.* A diploid primary oocyte produces three haploid polar bodies and one haploid ovum; a primary spermatocyte produces four haploid sperm. Germline cells complete this cycle only once.

Germ cells enter meiosis at prophase I. Following meiotic S phase, the replicated identical chromatids are conjoined through sister chromatid adhesion just as in mitosis; however, during prophase I, the homologous chromosomes pair up and undergo "recombination" where strands of the paired chromosomes interlace between pairs. One or more of these exchanges per chromosome pair form chiasmata that link the homologous pairs and allow them to associate across the equatorial metaphase plate (Figure 5-5). At anaphase I, the chiasmata dissolve, homologous pairs separate, and the pairs segregate to opposite hemispheres while the sister chromatids remain conjoined. At metaphase II of the second cell division, the sister chromatid cohesion dissolves and the chromatids separate as in mitosis. The obligatory homologous recombination requires the activity of enzymes that produce DNA dsb, enzymes that resolve DNA structural mismatch, and enzymes that ligate the recombined chromosomes. The mechanistic similarities between crossover and DNA dsb repair are obvious. From the perspective of radiation biology, a DNA dsb indicates damage that requires repair and thus checkpoints represent a damage control mechanism — the DNA damage response (DDR) pathways in mitotic S phase and G_2 (Table 5-1) are retained in meiotic cells. As in mitotic DDR, during meiotic cell division ATM is activated (phosphorylated) in response to blunt-ended DNA dsb (Figure 4-18) and a similar serine/threonine kinase, ATR (ATM- and Rad3-Related; not shown in Figure 4-18) is activated in response to replication protein A (RPA)-coated single-stranded DNA (Figure 4-12). Both kinases activate subsequent cyclin-dependent kinases that promote cell cycle blocks or cell death. On the other hand, the

Table 5-1 Mitotic cell cycle checkpoint activation by DDR. Activation by trimethylated H3 proceeds through ATM/ATR phosphorylation followed by ATM/ATR-mediated phosphorylation of target proteins.

Point of Arrest	ATM/ATR Kinase	Target	Action
Late G_1	p53	Cyclins D and E	Inhibits cyclin-kinase binding
Early S	Chk2	Cyclins E and A	Mitochondrial release of Cytochrome C
Late S	MRN	(unknown)	Replication fork metabolism?
Late G_2	Chk1	Cyclin B	Inhibits CDC25 activation

Table 5-2 Meiotic prophase I checkpoint activation by MCN. Checkpoint activation by ATM/ATR phosphorylation is followed by ATM/ATR-mediated activation of target proteins.

Point of Arrest	ATM/ATR Kinase	Target	Action
Late S	Chk2	Spo11	dsb delay pending recombination
Early Prophase I	Chk2	Cdc25	Blocks dsb processing via Ntd80
Late Prophase I	Chk2	HORMAD	Synaptic stabilization

meiotic checkpoint network (MCN) choreographs normal, requisite DNA strand scission during crossover, chromosomal reassortment, and DNA rejoining (Table 5-2). Radiation-induced DNA dsb repair during meiosis hijacks the MCN pathway. With respect to both pathways, MCN and DDR, kinases are activators and cyclins are regulators. Kinases flip on and off through phosphorylation and dephosphorylation like lightning bugs flickering at dusk. Individual cyclin concentrations rise and fall, ebb and flow, like the notes of a symphony.

During meiotic cell division, the first MCN checkpoint ensures that replication has been completed before the cell cycle enters prophase I. MCN stalls progression from S phase to prophase I by coopting replication fork cofactors to inhibit enzymatic formation of dsb; strand exchange required for chiasmata formation and appropriate chromosome pair segregation in anaphase I is blocked. The inhibition of DNA dsb formation ensures that only replicated chromosomes may crossover. Following S phase replication completion, the release of kinases from replication fork cofactor sequestration initiates chromosomal morphogenesis, enzymatic dsb formation, and DNA strand invasion. These activities mark progression into prophase I. Then, during prophase, MCN regulates DNA repair by controlling the generation of 30 base single strand overhangs at dsb resection sites. By inhibiting inter-strand repair completion, MCN promotes the perpetuation of chiasmata. Finally, MCN checks for completed synapsis of DNA to spindle fibers. The meiosis-specific unsynapsed chromosome structure that triggers MCN-induced delay has been difficult to identify because the loss of MCN factors causes unresolved dsbs that trigger cell death (Figure 4-18), but investigations into mouse meiosis indicate that under conditions of incomplete synapsis, ATR facilitates phosphorylation of H2AX, condensing chromatin and recruiting silencing factors. Finally, MCN regulates exit from prophase I through Cdc5 kinase activation. To prevent exit from prophase I, MCN inhibits Ntd80 transcription factor. When Cdc5 kinase is activated, inhibition of Ntd80 is reversed and cells rapidly progress to metaphase I. So, ATR regulates DNA dsb-response differently during S phase and during prophase. The decision for which response should be activated results from ATR interactions with replication fork cofactors (Figure 4-5), dsb mediators (Figures 4-11 and 4.12), or unsynapsed structures.

We have seen that mitosis represents a highly radiation sensitive phase of the cell cycle. Does MCN activity improve germ cell radiation resistance during meiosis? The presence of radiation-induced DNA dsb interferes with normal MCN checkpoint meiotic interruption. At the replication checkpoint, radiation-induced dsb circumvent the checkpoint mechanism that prevents enzymatic strand breakage by SPO11. The coincident replication stress induces ATR phosphorylation of Chk1 leading to cell death. This MCN pathway is independent of and suppresses the ATM/Chk2 DDR pathway leading to cell death (Figure 4-18). Furthermore, MCN promotes the stability of chiasmata by inhibiting dsb repair completion, thereby amplifying radiation-induced chromosomal aberrations arising during anaphase I, many of which prove lethal. Here, ATM phosphorylates CHK2 activating the CHK2-dependent p53 family of proteins leading to cell death (Figure 4-18). Thus, despite MCN dsb formation and repair activities insinuating that meiosis might be less radiation sensitive than mitosis, the reverse is true. From the perspective of species genetic integrity, this strategy makes sense.

Because the propagation of genetic damage from the fertilized zygote to every cell of the embryo potentiates such extreme prospective harm, germ cells exhibit hypersensitivity to radiation resulting in mortality. Lethal chromosomal aberrations prevent the second division of meiosis, and so we find that following reproductive organ radiation exposure, fertility diminishes (germ cell number is reduced). Persistent genomic defects in gametes that evade cell death express more frequently in oocytes than in sperm. And so, MCN provides additional safeguards against embryonic defect. MCN instigates signaling cascades beyond prophase I that obstruct formation of robust oocytes, sperm, and zygotes. Because it would be inappropriate to create and track irradiated gametes in humans, we must rely on animal studies and verify pertinence to humans. In *Drosophila,* persistent DNA dsb damage leads to ATR/Chk2-dependent modification of an mRNA helicase required for translation of a developmental gene responsible for dorsoventral patterning of the fly's eggshell. In several fungi, either persistent DNA dsb or defective synapsis result in MCN-dependent blocks to spore formation. In humans, aneuploidy — an inappropriate number of chromosomes within a cell — is the primary cause for early human fetal mortality. A grasp of radiation protection is greatly enhanced by appreciating the differences between mitotic cell division and meiotic cell division, and by keeping the destinies of these various mutated progeny in mind.

5.4.2. *Cell Cycle Synchronization*

As we saw in Chapter 4, when cycling cells are irradiated, checkpoints halt division until DNA damage repair completes (clears). In tissue culture, where all cells are continuously dividing (unless the culture has reached confluence), cycle synchronization is easily and rapidly achieved collecting the cells at checkpoints prior to releasing them to cycle in synchrony. *In vitro* cycle synchronization can be achieved by chemical DNA synthesis inhibition (treatment with hydroxyurea, aphidicolin, or thymidine) that blocks cells at the G_1/S boundary or cells can be blocked at mitosis (G_2/M) by treatment with mitotic spindle inhibitors such as nocodazole, colchicine, and vinca alkaloids. Synchronization may also be achieved by physical means (irradiation or mitotic shake off[3]). When irradiated, cultured human cells bunch in M, G_1, or preferentially, G_2. A subsequent additional exposure to radiation will block virtually all cells in G_1, and they will cycle in synchrony for several divisions following that exposure. *In vivo,* the capacity for synchronization is more complicated. Most cells rest in G_0 and won't "arrest" unless they reenter the cycle and any persistent DNA dsb are detected. Depending on the tissue, cells can remain quiescent (in G_0) for a long time, often years. Cell cycle synchronization is not a major concern in these tissues. However, some

[3]Cells growing in culture round up in mitosis and can be collected by shaking the flask and collecting the mitotic cells on ice, a method developed by Toyozo Terashima and Leonard J. Tolmach.

tissues — for example, epithelial cells of the small intestine — cycle almost as continuously as cells in culture. Human cancer cells also cycle continuously. In these cancer cells, the G_1/S phase cell cycle DNA damage checkpoint is often inactive because of mutations in a tumor suppressor gene (p53) and mutation or over-expression of G_1/S cell cycle regulatory proteins (e.g., cyclin D and regulatory binding partners — the cyclin-dependent kinases cdk4 and cdk6). Therefore, when cancer cells are irradiated, they preferentially block at the G_2/M checkpoint. Cell cycle synchronization in rapidly cycling tissues/tumors becomes a radiobiological concern because some phases of the cell cycle are more sensitive to radiation, and some are more resistant. Cells are more sensitive to radiation-induced cell death when exposed during M or G_2. Why might that be? Because DNA is not actively transcribed or replicated during M, the proofreading functions of the DNA and RNA polymerases do not initiate repair. Damage persists; telekinesis fails. Likewise, G_2 is a postreplication gap phase during which DNA-specific cofactors, such as centrioles and kinetochores assemble, and scaffolding components accumulate. Thus, during G_2, proofreading activity is diminished. The accumulated, unrepaired DNA damage perceived at the G_2 checkpoint preferentially leads to programmed cell death, or cells that pass through the G_2 checkpoint experience mitotic catastrophe during M. On the other hand, cells irradiated in late S phase are radiation resistant. Recall that synthesis denotes DNA replication. If DNA dsb damage occurs precisely when DNA polymerases are actively replicating the genome, the damage is promptly detected by the proofreading subunits of DNA polymerase and repaired using the highly accurate homologous recombination repair (HRR) pathway. During early S phase, replication activity is ramping up, so resistance is not as notable. During G_1, transcription is aggressively producing proteins necessary for growth that will roughly double the cell volume. Thus, some portions of the genome are being actively proofread. Therefore, the radiation sensitivity of the G_1 and early S phases lies between the two extremes. When small doses of radiation are delivered at 24-hour intervals, as happens clinically during a process called fractionation, rapidly cycling cell populations may become synchronized and subsequent doses may be delivered regularly during sensitive or resistant phases. This can have a profound effect on tissue response. Tissues that are actively replenishing may experience synchronization and become more radiation responsive or less radiation responsive if the dose occurs when the majority of cells are in early M-phase or late M-phase. Radiation oncologists, medical physicists, and radiation biologists must be aware of whether the normal tissue they are exposing is susceptible to radiation-induced cell synchronization.

5.5. Tissue Kinetics

With the rare exception of damage so severe that cell maintenance/metabolic activity is disrupted resulting in cell lysis, only dividing cells are susceptible to perpetuated radiation damage — that is, death or transformation. As a rule of thumb, the impact of radiation correlates with the percentage of dividing cells in a population. This principle has historically guided the philosophy of radiation therapy known as the "Law of Bergone and Tribondeau" (e.g., rapidly dividing cancers are more sensitive than surrounding slower cycling normal tissue) and radiation protection (e.g., the most radiation sensitive organs are those that replenish regularly). Although this principle represents a good rule of thumb, a more insightful understanding requires interrogation of higher levels of organization, as we shall see.

Cell cycle kinetics are determined by the rates of passage through G_1, S, G_2, and M, with M being reasonably consistent throughout various cell types. The average rate of cell division in an asynchronous population can be determined by clocking mitosis (M), a stage that is unique due to chromosome condensation and cell fission. An easy way to determine cell cycle intervals is to add tritiated thymidine (^{3}H-TdR) to the cell medium or experimental system (tissue or organism). Cells take up the ^{3}H-TdR and incorporate it into the nascent DNA strand during replication (S phase), radioactively labeling one strand of the DNA helix of every chromosome. The ^{3}H-TdR is then removed (the medium is exchanged with fresh medium without ^{3}H-TdR, or the ^{3}H-TdR is flushed with unlabeled thymidine). When the cells cycle to M, the number of cells with radioactive chromosomes can be compared with the number of unlabeled cells. The proportion of labeled cells is the labeling index (LI),

$$\mathrm{LI} = \lambda \frac{T_{\mathrm{S}}}{T_{\mathrm{C}}}, \tag{5.1}$$

where T_{S} is the duration of synthesis and T_{C} is the duration of the cell cycle. The term λ accounts for unequal cell numbers throughout the cycle. The number of cells in the population following mitosis is twice the number of cells just prior to mitosis, and because mitosis is entered at different time points for various members of the population, this term is difficult to determine. However, in most cases, λ can be assumed to be ln2, or 0.693, because of the exponential nature of doubling and distribution. T_{S} can be determined either by examining the length of time it takes for the activity of cells to continuously increase from zero to maximum (in a synchronized population), or by determining the mitotic index (MI),

$$\mathrm{MI} = \lambda \frac{T_{\mathrm{M}}}{T_{\mathrm{C}}}, \tag{5.2}$$

where T_{M} is the duration of mitosis. Combining the two equations, 5.1 and 5.2, allows for the solution to any of the duration variables.

However, the LI and MI assume that the entire cell population is actively dividing — that none of the cells are quiescent in G_0. A more generalized solution considers the rate of change in the mass of the cell population,

$$\frac{d(\mathrm{mass})}{dt} = \frac{\ln 2}{\mathrm{MDT}}[\mathrm{mass}], \tag{5.3}$$

where MDT is the mass doubling time. Here, the mass can be a cell population, a tissue, an organ, or an organism. Provided that the defined system is closed, that cells cannot be added to the system nonmitotically or removed by surgery or cell death, the MDT represents the cell cycle time.

Open systems exhibit type-specific kinetics partially determined by the cell cycle but dependent upon other parameters as well. For example, imagine a petri dish of cells. The dish fills at a rate reflecting how many cells are dividing (some may be contact inhibited) and how long it takes to complete a mitotic cycle. This system is defined only by the initial number of cells and the rate at which the number increases. But suppose that some of the cells die because of nutrient deficiency, old age (yes cells do age due to telomere shortening and damage to mitochondria), infection, or some other adversity. Now the rate at which the dish fills is reduced by the rate at which cells are removed (die). Cell cycle kinetics determine radiation sensitivity only under ideal *in vitro* growth conditions where the system is closed. In all other circumstances, system kinetics dominate.

Let's model the systems under consideration as a bucket (Figure 5-6). If the bucket holds all the water that ever will be, it is a static, closed system (Figure 5-6A). In tissue culture, this is equivalent to a petri dish of confluent cells where all cells are contact inhibited (in G_0). The heart (myocardium) is a tissue of this type, or the neurons of the central nervous system (CNS), if we exclude the hypothalamus that

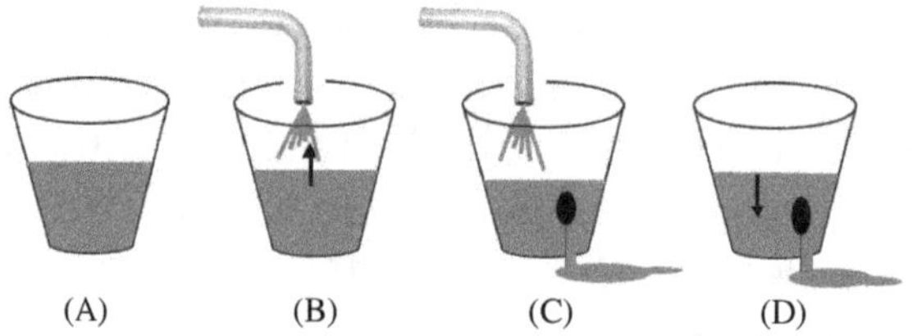

Figure 5-6 Tissue kinetics can be modeled as buckets of water. (**A**) A closed system is represented by a bucket with neither input nor output. Heart tissue exemplifies this model; cells neither divide nor die during the organism's postembryonic lifetime. (**B**) An increasing system adds to the volume in the bucket. Tumors represent this kinetics model; they grow uncontrollably, continuously increasing the cell number. (**C**) In a steady state system, water flows into the bucket at the same rate as it flows out. Regenerating tissues such as skin, blood, bone, and vasculature reflect this kinetics model. (**D**) A decreasing system is one where water flows from the bucket until it is empty. Oocyte maturation, life cycle, and death follow this kinetic model.

appears to slowly regenerate. If the petri dish is not confluent and the cells are dividing, there is a flow of cells into the system. The number is increasing. Now in our model, the bucket has a hose with a variable flow rate (Figure 5-6B). This system remains closed, the growth can be analyzed using cell cycle kinetics or MDT, but growth may not be constant; the change in number of cells (N) has a variable rate: dN/dt. An idealized tumor reflects this model. The majority case, tissues in steady state, can be modeled by a bucket with a hose and a hole (Figure 5-6C); water flows in at the same rate as it flows out ($dN_{in}/dt = -dN_{out}/dt$). Most healthy tissues reflect this kinetics, for example skin cells, bone marrow cells, or red blood cells. Here, stem cells divide adding to the population and aging cells die leaving the population. Differentiation schemes can also be described by this model, for example an erythroblast differentiates into a proerythrocyte, which differentiates into an erythrocyte (red blood cell). Looking at the population of proerythrocytes, erythroblasts flow in as they become proerythrocytes and flow out as the proerythrocytes become erythrocytes. Furthermore, some systems, like proerythrocytes, have two sources of influx; in addition to the differentiation of erythroblasts into proerythrocytes, proerythrocytes divide. In this case, the model must consider two variable flow rates: dN_{in}/dt and $dN_{mitosis}/dt$. The final kinetics model (Figure 5-6D) resembles a bucket with a hole and no hose. In this situation, the bucket empties. Oocytes in female ovaries follow this kinetics model. In general, females are born with a maximum number (~2 million) of oocytes. Once puberty is achieved, these oocytes mature serially and leave the ovaries through the fallopian tubes until no more oocytes remain. The eggs are either fertilized and become embryos (leaving the population of oocytes) or they die. Each of these examples are idealized. In truth, systems B and D in Figure 5-1 are extremes of system C, where the flow in or the flow out is minimized ($dN_X/dt = 0$). In A, both the flow in and the flow out equal 0. Tissue kinetics are important when interpreting the effects of radiation on organs. Radiation affects both cell cycle kinetics (the rate at which cells add to the tissue) and cell death (the rate at which cells are removed from the system). However, the volume representing noncycling cells remains unchanged — in this scenario, $dN_{in}/dt \neq dN_{out}/dt$, so the system does not reach a steady state. There is still one more consideration. Cells in G_0 are not senescent; they are quiescent and capable of entering, and usually do enter the cell cycle eventually. Thus, there is another rate of increase due to cells entering the cycle from G_0: dN_{G0}/dt. One might expect that the same percentages of the induced G_0 population, compared with the population of dividing cells, will repair successfully, repair unsuccessfully or die through the usual mechanisms, once they enter mitosis. That intuition is incorrect. Cells in G_0 are more resistant to radiation than anticipated. Why? Likely because quiescent cells while not growing and dividing are metabolically active. They need to replace expended proteins, transcribe nontranslated RNA, carry out mitochondrial functions, and so on. To do this, cells in G_0 must transcribe DNA, and this process involves RNA polymerases that manifest proofreading activities. As we and others have shown,[4] cells in G_0 have adequate time to repair much of their DNA damage prior to entering the cell cycle. In our model, the variable flow of water into the bucket is a complicated, time-dependent sum of all the rates of all the amplifying components against which we must estimate the rate of flow out of the bucket due to only radiogenic cell death.

5.6. Matters of Scale

Biology is more than complicated; it is highly complex. Complicated systems can be examined, understood, and reproduced. They can be broken down into component parts to be studied and then reassembled. A transistor's function can be analyzed in isolation and then placed into a circuit with totally predictable results. Complex systems have some mysterious set of organizational rules that sporadically produce unpredicted results. This is sometimes described as, "the whole being greater than the sum of the individual components." Nonetheless, we often study these biological systems by investigating the smaller scale components and then examining the way these components behave in progressively more complex systems, continuously evolving our understanding of the hidden rules. In other words, we investigate using various scales of biology, beginning with the least complex and advancing cautiously to the most complex. This method is referred to as a "bottom up" technique. It is justified by the belief that local rules determine global behavior. For example, bird flock flight patterns can be modeled by the nearest neighbor principle: each bird responds to the movement of the bird closest to it (much less so to the next one out). This is complex behavior modeled on local principles. Alternatively, some behaviors are modeled best from the top down. Here, the guiding principle professes that global behavior influences local behavior. For example, in ecology, landscape features determine animal dispersal.

We began our exploration of radiation biology in Chapter 3 by discussing the interactions of radiation with DNA and the associated hydration layer. This system included only the macromolecule (DNA) and the solvent (water). In solution (in a test tube; on the benchtop), DNA can be influenced structurally by adding nucleosomes and manipulating the ionic concentration of the solvent. In its simplest forms, the solvent can be distilled water or isotonic saline (0.9% sodium chloride dissolved in sterile distilled water). DNA in solution can be damaged by radiation either directly or indirectly, and it can be cut with nucleases and annealed with ligases — if the experiments are performed in buffered solutions that have the correct salt concentrations, pH, and cofactor ions such as Ca^{2+} and Mg^{2+} that optimize the enzyme three-dimensional (3D) structure and activity. Using this system, DNA ssb and DNA dsb can be interrogated, genetic damage can be investigated, and fixation or restitution of damage (chemical processes) can be appreciated. This system represents the molecular scale. At the molecular scale, radiation sensitivity is determined by chemical kinetics, macromolecular size, and DNA structure. The number of components is minimal, and the system is limited, but this does not exclude the extrapolation of experimental results to more inclusive systems.

Isotonic salt concentrations are not only impactful at the molecular scale; the same solvent chemistry is required for repair of radiation-induced DNA damage in cultured cells, as well. As we and others have shown,[5] if you irradiate radiation-resistant normal mammalian or cancer cells with a range of x-ray doses, cell survival rates indicate that repair of the DNA strand breaks is robust. However, if you irradiate the same resistant cells with the same radiation doses and then immediately treat the cells with a

[4] See publications by Mendonca, Dethlefsen, Little, Alpen and others.

[5] See results and conclusions by Elkind, Dewey, Wheeler, Ed Alpen, Mendonca, and others.

hypertonic (0.5 M NaCl) saline solution, the cells shrink, cell killing at each radiation dose increases, and DNA dsb repair is nearly completely inhibited or misrepaired. The cell, then, represents the next more inclusive experimental system. Within an appropriately accommodating environment (nutrient medium, temperature, gas pressure, and humidity), cultured mammalian cells provide supporting biochemistry enabling examination of cell cycle checkpoints, progression to replication, apoptosis, and mitotic catastrophe. Enzymatic repair, mutation, transformation, epigenetics, replication, and transcription take place only when components and mechanisms provided by the cell are available. An example of the interrogation of radiation biology on the cellular scale can be exemplified by the DNA repair research performed by Löbrich (Löbrich *et al.*, 1995). He and his coworkers measured the rate of the unidentified (at the time) dsb "misrepair" pathway(s). He removed DNA from irradiated, cultured fibroblasts and cut it with a restriction endonuclease, NotI. This nuclease produces a 3.2 megabase pair (Mbp) fragment, a 2 Mbp fragment, and a 1.2 Mbp fragment. The three NotI scission site sequences reside at three specific locations. A radiation-induced dsb within any of these DNA sequences prior to extraction produce two fragments because the binding site for the endonuclease has been eliminated. If the two fragments rejoin correctly prior to DNA extraction, the original fragment length is restored (Figure 5-7). But if the two fragments anneal to other broken ends of other dsb fragments, novel lengths of DNA appear. The hybridized fragments were separated using pulsed field gel electrophoresis (PFGE; isolating incorrect rejoining) and conventional electrophoresis (displaying all rejoining) and the percentage of incorrect rejoining was calculated at about 25%. Furthermore, the experiment determined that faithful rejoining completed in a few hours (with a halftime of 2 hours), but misrepair transpired over several hours. Although the experimental analysis was carried out on the lab bench, the phenomena under investigation, DNA repair/misrepair, occur internal to the cell *in vitro*. This is the cellular scale. At this scale, radiation sensitivity is determined by cell cycle kinetics and repair enzyme competence, that is, cellular components discussed in Chapter 4.

In Chapter 8, we will begin to investigate the continuity of cellular radiation biology in tissues and organs. In general, tissues are composed of a single cell type and supporting stromal components. Organs are composed of several tissues — for example, in addition to epithelial tissue, skin includes fibroblasts, sebaceous glands, hair follicles, and sweat glands (Figure 5-8). Organs are a more inclusive system.

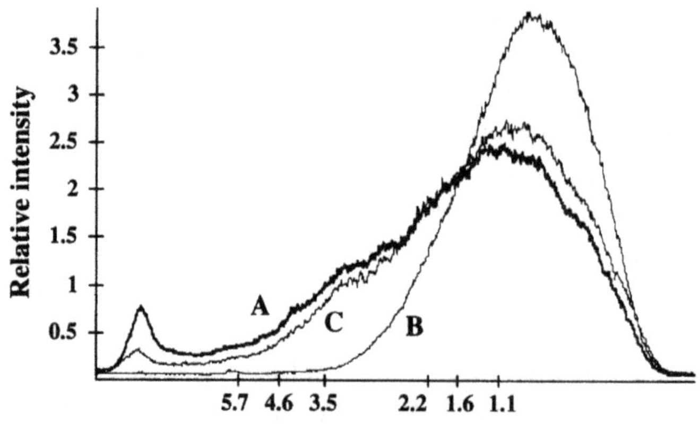

Figure 5-7 Normalized intensity distributions of NotI restriction DNA fragments. Distributions were obtained by measuring the ^{14}C signal from labeled DNA. (**A**) NotI fragment lengths from unirradiated cells. (**B**) Fragments from x-ray irradiated cells. (**C**) Fragments from x-ray irradiated cells following 24 hours of repair. The fragment size distribution is shifted following irradiation to less than the native 3.2 Mbp–1.2 Mbp. Following repair, the distribution closely resembles the original distribution. Reproduced with permission of National Academy of Sciences USA, Copyright, Lobrich *et al.* (1995).

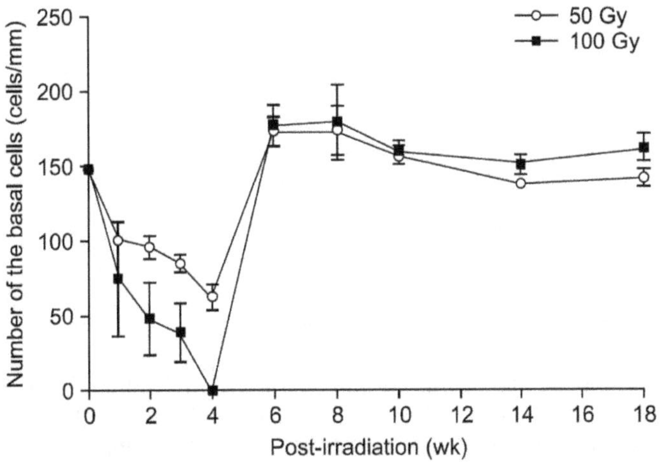

Figure 5-8 Loss of skin basal cells following β-radiation exposure. Pig skin biopsies were stained and the number of stem cells counted. The higher dose resulted in increased stem cell death. However, the kinetics — the rate at which stem cells were removed from the organ and the rate at which the pool was replenished — was consistent at both doses. Radiation wounding results from diminished new cell production while the loss of cells due to aging remains constant. Reproduced with permission of The Korean Society of Veterinary Science from Kim J-S *et al.* (2015).

Tissue-level controls dictate passage into the cell cycle, from G_0 to G_1. Mitotic initiation is influenced by extracellular matrix composition and secretion of growth factors (e.g., epidermal growth factor [EGF], vascular endothelial growth factor [VEGF], and transforming growth factor alpha [TGFα]). This higher level of communication controls cellular metabolism resulting in organ function. An example of higher scale communication might be found in the lungs. Alveoli exchange CO_2 and O_2, defining the function of the lungs. Weakened alveolar epithelium must be replaced to ensure lung function, and so damaged epithelial cells release growth factors, causing surrounding cells to enter the cell cycle. Similarly, kidney function is assessed by examining the competence of the glomeruli — the blood filtering structures of the kidney — through blood creatinine levels. We call these organ structures, like alveoli and glomeruli, "functional subunits" (FSU). Organ failure arises from FSU damage which reflects cell death; thus, FSU failure can be predicted by examining *in vitro* dose escalation studies. However, the radiation sensitivity of the organ depends on the number and organization of FSU, and therefore is more complicated than a simple extrapolation from *in vitro* cell survival studies. Often, organ failure can be predicted only based on the probability of sufficient FSU damage. Similarly, because carcinogenesis and cataract formation arise (Figure 5-9) from cell transformation, these organ scale outcomes must be based on the probability of mutation progressing to transformation (and thereafter to carcinogenesis in the case of cancer). We will delve into these topics in depth a little later.

At the most complex level, tissues and organs comprise an organism — for example, a human being. On this scale, molecules, cells, tissues, and organs all interact. Genetics and genomics express systemwide, individualizing every organism and creating unique organism-wide responses. Hormone secretion and inflammatory cells including monocytes and macrophages respond to any number of internal and external environmental influences. Age influences cells, tissues, and organisms in both dependent and independent ways. Conscious decisions influence the environment — what we eat, where we live, our level of activity, and the medications we take. Every human being has a unique environmental history. At this level, it is apparent that the global human network is complex. Because each person is unique, statistics require enormous numbers of subjects to achieve significance. Complexity is one reason research often prefers to

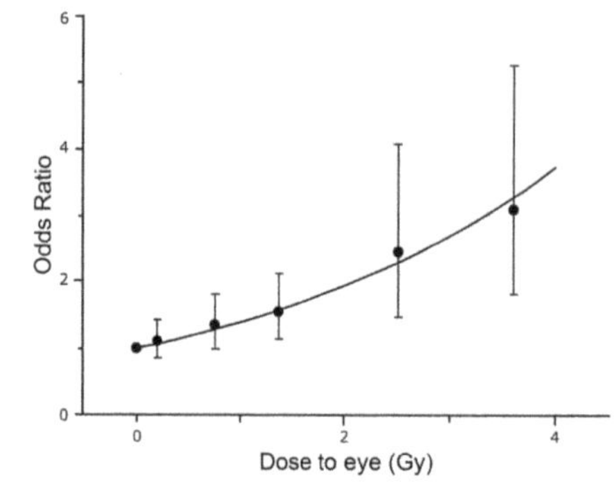

Figure 5-9 Probability of atomic bomb survivors requiring cataract surgery. The odds ratio (a comparison of survivors who required surgery to the collective population) indicates that greater dose correlates with greater probability of an individual developing one or two cataracts. Cataract occurrence cannot be predicted — only the *probability* of occurrence can be estimated. Individual variations in tolerance increase uncertainty with dose. Reproduced with permission of the Radiation Research Society from Neriishi *et al.* (2007).

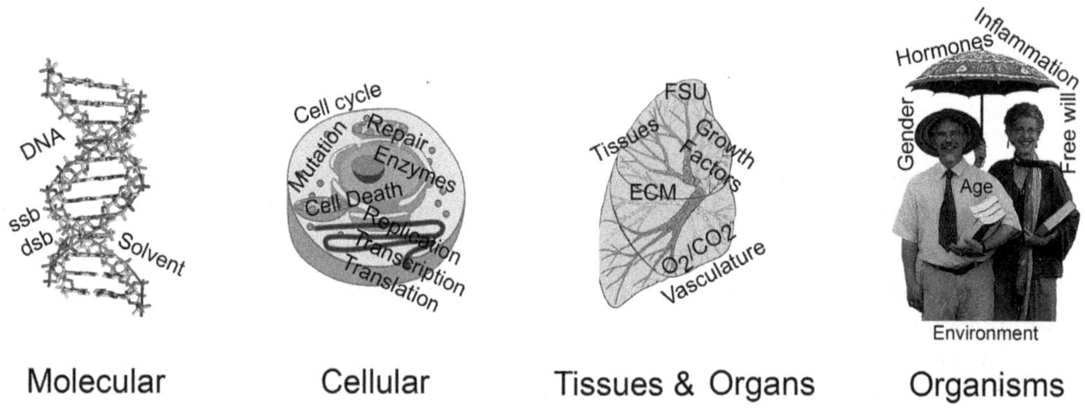

Figure 5-10 Matters of scale. When performing experiments and discussing results, it is important to keep matters of scale in mind. The larger the system, the more complex the biology.

use animal models, or computational models. Animals are selected for carefully controlled genetic identity, environmental history, age, sex, and so on, to manage as many variables as possible in the test population. This is the organism scale. Multiple factors determine radiation response including the volume of damage (number of organs involved), tissue kinetics, clearance kinetics of internalized radioactivity, tissue connectivity, and physiological status. Caution must be taken when extrapolating lessons learned at smaller scales (Figure 5-10). Each scale should be considered a unique system to be characterized according to the behavior at that scale. The correct perspective for inspection depends on the problem at hand, and so the research hypothesis must be carefully worded with the results cautiously interpreted.

This chapter is intended to introduce higher order factors that impact radiation biology. Although we will predominantly progress unidirectionally from smallest scale to largest, going forward the remainder of this text will often pass fluidly among scales to embrace the big picture. Every effort will be made to remind the reader that underlying complexities are in play and to review the essential details of the specific biological circumstances. In every case, references have been provided to guide interested students to more in-depth analyses of complex biological behavior.

5.7. Summary

- Most cells reproduce through mitosis; gametes reproduce through meiosis.
- Meiosis consists of two divisional cycles and produces haploid cells.
- During meiosis, sister chromatids (replicated DNA) are held together by adhesion molecules and are attached at their centromeres.
- The chromosome pairs are connected through chiasmata composed of invading complimentary DNA strands of the paired chromosomes.
- Chromosome pairs separate during anaphase I. Dissolution of chiasmata requires dsb formation.
- Mitosis employs two DNA damage repair (DDR) checkpoints at G_1 and G_2, a spindle integrity checkpoint at the metaphase/anaphase interface and DNA integrity checkpoints during S phase.
- Meiosis retains the DDR checkpoint mechanisms and adds a meiosis-specific, MCN that is active primarily during prophase I. Both DRR and MCN rely on ATM and ATR activation through phosphorylation.
- MCN also influences replication during late S phase and post meiotic-gamete/zygote development.
- MCN delays chiasma formation until replication has completed by sequestering dsb signaling proteins to replication cofactors.
- Once replication protein factor concentration diminishes, MCN stabilizes the chiasmata and stalls prophase I progression by inhibiting DNA dsb resolution.
- MCN stalls transition from prophase I to anaphase I until all synapses between kinetochores and spindle fibers are stable. The trigger component(s) and pathway remain under investigation.
- The results of somatic cell mutation are different from the results of germ cell mutation. The former results in transformation leading to tumorigenesis and the latter results in fetal abnormalities, teratogenesis or latent cancer.
- Cells can be synchronized such that they pass through the cycle in unison.
- Radiation can synchronize cells by blocking progression at cell cycle checkpoints.
- Cells are more sensitive to radiation in G_2 and M.
- Cells are most resistant to radiation in late S.
- Cell cycle rates can be measured by the LI and the MI.
- Radiation affects cycling cells and quiescent cells differently as reflected in mutation rates and cell death.
- Quiescent cells repair through transcription proofreading and mostly die only after reentering the cycle.
- The mass doubling time can be applied to open or closed systems but will reveal cell cycling time only if the system is closed.
- Tissue kinetics can be modeled as the flow into and the flow out of a reservoir. There can be several different contributing rates of input and several different rates of output.
- Genomic instability results from an increase in DNA damage tolerance, allowing for a greater and greater accumulation of mutations.
- Chromatin is modified epigenetically through methylation, phosphorylation, acetylation, and ubiquitination.
- Both cytosine (DNA) and lysine (histone protein) bind methyl.
- Trimethylation of H3, which happens in response to radiation, induces ATM activation and thereby cell cycle regulation, cell death, and dsb repair.
- Ubiquitin is a protein composed of 76 amino acids that binds to lysine residues on all four histones.

- Epigenetic modification of chromatin alters the condensation state which either inhibits or provides access for transcription, or epigenetic modification may enhance transcription cofactor binding.
- Epimutations can be passed from one generation to the next, just as DNA mutations.
- With respect to cell transformation, epimutations are at least as important as DNA mutations.
- Because biology is complex, different system scales must be examined and comprehended through specific perspectives.
- Each system scale has special rules guiding the kinetics of that system: chemical, cell, tissue, and organism.

5.8. Problems

1. Two experiments were conducted to determine the cell cycle parameters of a given synchronized cell line. The results of those experiments determined that MI = 0.94 and LI = 7.7. If $T_S = 9.8$ hours, how long does it take this cell type to complete mitosis (M phase)? What is the relationship between λ and the total length of the cell cycle?
2. The accompanying figure represents experimental results acquired by Sinclair and Morton (1966) They synchronized V9 hamster cells at M (t_0) and either incubated them with ^{3}H-TdR (solid circles) or also exposed them to x-ray radiation (open circles) at various times postsynchronization.

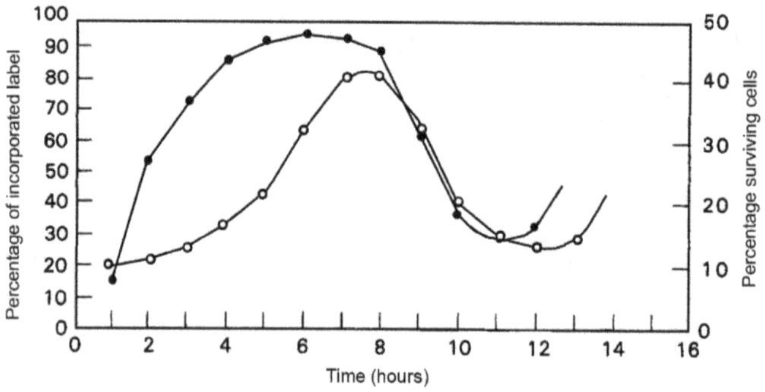

   Why does the percentage of incorporated label increase from 1 hour to 5 hours following synchronization? Why does the incorporated label level off at 5–8 hours and then rapidly decrease between 8 and 10 hours? The greatest percentage of cells survived following irradiation at what time postsynchronization? Does this make them relatively radiation sensitive or radiation resistant? At what time points of x-ray radiation exposure do the fewest cells survive? By considering the data from both experiments, during which phases of the cell cycle are cells radiation sensitive and during which phases are cells radiation resistant?
3. What are the differences between the centromere and the kinetochore?
4. A typical karyotype displays the chromosomes from a mitotic spread in order of decreasing size and numbers them 1–23. This organization is useful for visualizing chromosomal translocations. A Robertsonian translocation fuses entire chromosomes as portrayed below.

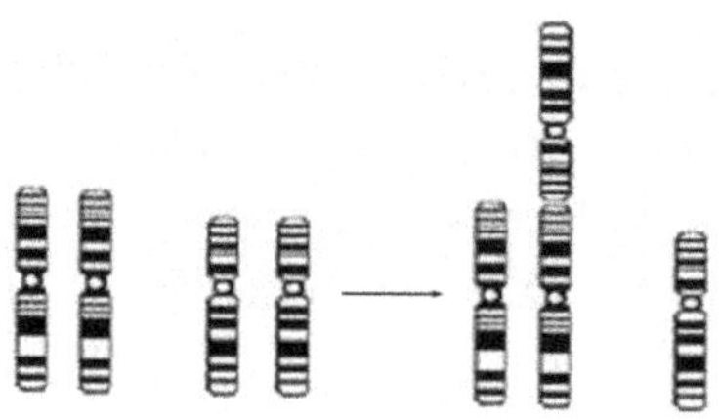

Show the haploid result of meiosis — chromosomal content of the gametes — with respect to these chromosomes. What would you expect the effect to be on a fetus resulting from the fertilization of these gametes: fetal abnormalities or cancer or both? Explain your answer.

5. If Michael inherited a novel mutation from his mother (one that his mother did not have), at what stage did this mutation likely occur (type of division and phase name)?
6. An analogy for the role of cyclin in cell cycle regulation would be more like which of these alternatives? Why?
   a. A signal emitted periodically
   b. A row of dominoes that fall sequentially
   c. An on/off switch
7. Suppose five cells, each at a different stage of the cell cycle, are exposed to radiation. Where will each cell stall in the cycle? What happens if they are exposed again at a time $\frac{1}{2}T_C < t < \frac{2}{3}T_C$?
8. This figure represents the number of platelets per milliliter drawn from a subject exposed to approximately 3 Gy of x-rays. Samples were drawn at the times indicated following exposure.

   Explain the shape of the curve in terms of our kinetic model using buckets, hoses, and holes. Explain the shape of the curve in terms of biology (platelets).

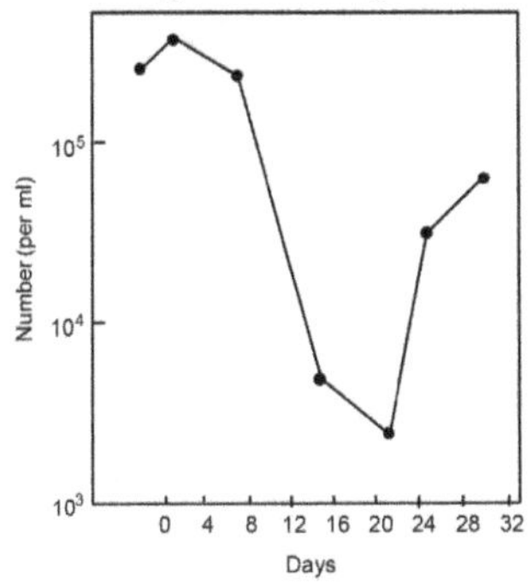

9. Explain why a Northern blot (that tracks RNA transcription) would be a better indicator of genomic mutation than a Southern blot (that tracks changes in DNA sequence).

5.9. Bibliography

Alberts, B. *et al.*, 2015. *Molecular Biology of the Cell.* 6th ed. New York: Garland Science.

Csikasz-Nagy, A. *et al.*, 2006. Analysis of a generic model of eukaryotic cell-cycle regulation. *Biophysical Journal,* 90, pp. 4361–4379.

Harvey, Z. H., Chen, Y. & Jarosz, D. F., 2018. Protein-based inheritance: epigenetics beyond the chromosome. *Molecular Cell,* 69(2), pp. 195–202.

Herodin, F. *et al.*, 2012. Assessment of total — and partial — body irradiation in a baboon model: preliminary results of a kinetic study including clinical, physical, and biological parameters. *Health Physics,* 103(2), pp. 143–149.

Jeggo, P. A., 1998. DNA breakage and repair. *Advanced Genetics,* 38, pp. 185–218.

Kim, J.-S. *et al.*, 2015. β-irradiation (^{166}Ho patch)-induced skin injury in mini-pigs: effects on NF-κB and COX-2 expression in the skin. *Journal of Veterinary Science,* 16(1), pp. 1–916.

Löbrich, M., Rydberg, B. & Cooper, P. K., 1995. Repair of x-ray-induced DNA double-strand breaks in specific Not I restriction fragments in human fibroblasts: joining of correct and incorrect ends. *Proceedings National Academy of Science,* 92, pp. 2050–12054.

Neriishi, K. *et al.*, 2007. Postoperative cataract cases among atomic bomb survivors: radiation dose response and threshold. *Radiation Research,* 168, pp. 404–408.

Russo, A. *et al.*, 2015. Genomic instability: crossing pathways at the origin of structural and numerical chromosome changes. *Environmental and Molecular Mutagenesis,* 56, pp. 563–580.

Schmidt-Ullrich, R. K. *et al.*, 2000. Signal transduction and cellular radiation responses. *Radiation Research,* Volume 153, pp. 245–257.

Sinclair, W. K. & Morton, R. A., 1966. X-ray sensitivity during the cell generation cycle of cultured chinese hamster cells. *Radiation Research,* 29(3), pp. 450–474.

Subramanian, V. V. & Hochwagen, A., 2014. The meiotic checkpoint network: step-by-step through meiotic prophase. *Cold Spring Harbor Perspectives in Biology,* October, 6(10), pp. 1–26.

Thompson, L. H. & Schild, D., 2001. Homologous recombinatorial repair of DNA ensures mammalian chromosome stability. *Mutation Research,* 477, pp. 131–153.

Withers, R. H., Taylor, J. M. G. & Maciejewski, B., 1988. Treatment volume and tissue tolerance. *International Journal of Radiation Oncology, Biology and Physics,* 14(4), p. 751–759.

Wood, R. D., Mitchell, M., Sgouros, J. & Lindahl, T., 2001. Human DNA repair genes. *Science,* 291, pp. 1284–1289.

Zhang, Z., 2007. Concepts, measurements and scientific problems of biocomplexity. *Integrative Zoology,* 2, pp. 100–110.

Chapter 6

Modeling Radiation Biology

6.1. Introduction

Radiation biology seeks to understand the fundamental impacts of radiation on biological systems pursuant of predicting the outcomes of radiation exposure. Although Roentgen identified machine produced radiation (from a Crooke's tube) more than a century ago in 1895, serious investigations into radiation biology were somewhat lacking. Roentgen's contemporary, Pierre Curie, noted that radiation induced developmental abnormalities in tadpoles, and it was often reported in the literature that radiologists developed skin lesions, cancer, and sometimes lost fingers. Nonetheless, systematic study of radiation biology didn't emerge until around 1905, and radiation biophysics only appeared during the second decade of the twentieth century. Quantitative radiation biophysics requires a method to measure dose and an understanding of the physics of radiation as well as biological chemistry and cellular physiology. Advancements in radiation biophysics had to wait for several fields of inquiry to provide the necessary foundation. Following the atomic bombing of Hiroshima and Nagasaki, several resourceful physicists turned their attention to the biological effects of radiation, and through their collaboration with outstanding biologists and biochemists, insights into the fundamentals of radiation biology escalated dramatically. The most seminal piece of work was published by a former experimental nuclear physicist, Douglas E. Lea, in 1947 (Lea, 1947). Try to imagine this moment in time, when dose could not be directly measured, and DNA was unknown. How could the effects of radiation be predicted?

Prediction is the realm of computational modeling. The simplest form of computational modeling is a straight line. If one wishes to purchase a car, and one puts aside $100 a week, determining when the car can be purchased is an easy matter of graphing dollars saved against time. When the line reaches the price of the car on the ordinate, the number of weeks on the abscissa indicate the purchase date. If the savings are collecting interest, the interest can be calculated and added to the principle. The straight line becomes curvilinear, but again, when it reaches the price of the car, the ordinate discloses the purchase date. Simple. Good models have five main attributes:

- A model should be clear in its design and help us better understand something.
- A model should identify important relationships.
- A model should predict accurately and be consistent with evidence.
- A model should improve communication and help explain complex relationships.
- A model should be useful and help guide experimental design or thinking.

Models can become extremely complicated, for example, models of the weather. Hurricane paths are predicted by several computational models including the European Center for Medium-Range Weather Forecasting (ECMWF), the Global Forecast System (GFS), the National Weather Service's Geophysical Fluid Dynamics Laboratory model (GFDL), and the new National Weather Service's Hurricane Weather Research Model (HWRF). It can be interesting to compare how well each of these models performs as a newly formed hurricane advances on the continent and to consider the differences in computational philosophy or mathematical formulation. We will examine some models of radiation induced biology (e.g., cancer) in Chapter 7. Here, we will introduce the development of predictive models for cell survival following irradiation. To predict radiobiological outcome, all that is needed is a set of equations that describe the biological response to known quantities of radiation in a controlled environment.

6.2. Modeling Cell Survival

Models are generated for observable systems, allowing model testing against experimental or fortuitous (naturally occurring) results. To investigate cell survival, the least complex scale is that of the cell; not multicellular organisms or tissues. Remember, the correct perspective for inspection depends on the problem at hand. Isolating and sustaining cells required the development of sterile cell culture techniques, a process that is ongoing today for many types of cells. The first momentous step toward that capability was the development of the "petri dish" (1887). This apparatus is simply two sterile cylindrical glass dishes with the top plate being slightly larger than the bottom so it can be placed on top and form a closed system with reduced air exchange. The reduced air exchange discourages external spores or bacteria in the air from invading into the inner chamber and keeps the cells sterile and the growth media from evaporating. For many early experiments, growth medium was placed in the upright, smaller bottom dish, and the cells to be tested such as yeast, plant root cells, and bacteria would be transferred into the dish where it would divide and grow. The plate could be irradiated and examined to determine the reduction in population growth caused by cell death. It was not possible until the 1920s to induce mammalian cells to thrive under culture conditions (Figure 6-1). Temperature and atmospheric restrictions (e.g., O_2 and CO_2

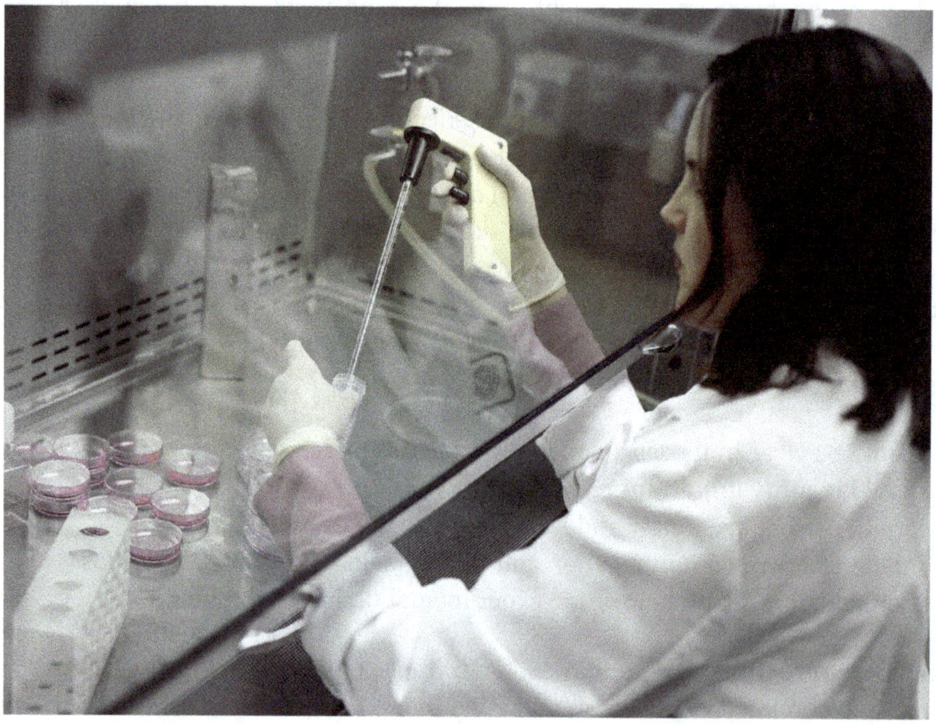

Figure 6-1 Culturing cells for *in vitro* experiments. The laminar flow hood is a sterile environment enabling the opening of petri dishes without fear of contamination. Cells are "fed" liquid medium containing proteins and chemicals that support cell division and metabolism. Old medium is removed and replaced with fresh medium several times a week. The medium contains a color indicator to monitor pH. Reproduced with permission from Kevin McCormack, California Institute for Regenerative Medicine.

percentages) had to be established, as well as adherent surfaces and nutrient requirements. Finally, mouse fibroblasts were successfully cultured for several passages (the process of removing the cells from the petri dish, diluting them in growth medium, and replating them at a lower density). In 1943, the first fibroblast cell line was established. A permanent cell line generates cells that have become immortal — they can be passaged an infinite number of times. Most modern cell lines, especially human cell lines, are derived from cancer cells dissociated from patient's tumors such as carcinomas and sarcomas. Human and mouse fibroblasts remain the most easily manipulated and maintained cells. Again, plates of eukaryotic normal or cancer cells can be irradiated and examined to determine cell survival.

6.2.1. *Lea's Target Theory*

Cell survival modeling uses the biophysical processes set in motion by the exposure of cells to ionizing radiation (Figure 6-2). For experiments conducted *in vitro*, two major assumptions are made: the average damage induced by radiation exposure is not based on the number of cells damaged but rather the damage induced in each cell, and repopulation is an indicator of cell survival. The first assumption works well except at extremely low doses of charged particles such as protons, alpha particles, and carbon ions where only some cells may be hit. The second assumption remains acceptable because, if a cell loses the ability to divide but the cell survives, it becomes senescent. The progeny of this senescent cell is not biologically impactful in terms of cell growth, because a mitotically incompetent cell will not divide uncontrollably, mutations will not be perpetuated, and genomic instability will not occur. Therefore, the inability of cells to divide is sometimes referred to as "reproductive death," a more accurate description than cell death. Recall that it is during mitosis (or meiosis) that cell cycle controls lead to cell death, and not the other way around (see Chapter 4, Section 4.6). Furthermore, because a single surviving, mitotically competent cell has the capacity to give rise to a colony of cells, all cells in the colony are clones of the original cell. Thus, cell survival is often referred to as "clonogenic survival"; again, a more accurate description, as it is insignificant whether the original cell survives, so long as it gives rise to a colony of at least 50 cells. Nonetheless, "cell survival" and "cell death" are common usage in the literature, and so we will continue to use these terms in this text. We begin our discussion of cell survival modeling by stating some assumptions that summarize chapters 2 through 5. Reproductive death is assumed to be the result of a multistep process:

- Energy is deposited (in the form of ionization or excitation) in a critical volume of the cell.
- The deposition of energy in a critical volume will lead to molecular lesions in the cell.
- The expression of these molecular lesions causes the cell to lose the ability to divide.

The first important model predicting cell survival following exposure to ionizing radiation appeared during the zenith of radiation biophysics around the mid-twentieth century — following World War II. In 1947, Douglas Lea published his groundbreaking description of the target theory of radiation (Lea, 1947), tragically the same year he died in a car crash at the age of 37. At the time, it was not possible to culture mammalian eukaryotic cells; prokaryotic cells (such as bacteria) were grown on nutrient gels in petri dishes on the benchtop. Recall from Chapter 2 that the quantity of radiation delivered to a target is a rather poor predictor of the amount of energy that the material will absorb. Only some of the initial energy will be transferred, and of that, only some of the energy will be absorbed. Dose resembles temperature — a measure of the energy deposited in a mass. Nonetheless, the mechanisms for the deposition of energy, and the probability of each interaction were well known (Figure 2-7). Fortunately, Lea worked with Louis Gray, another physicist, for whom the SI unit of dose (Gy) was named. So, using the

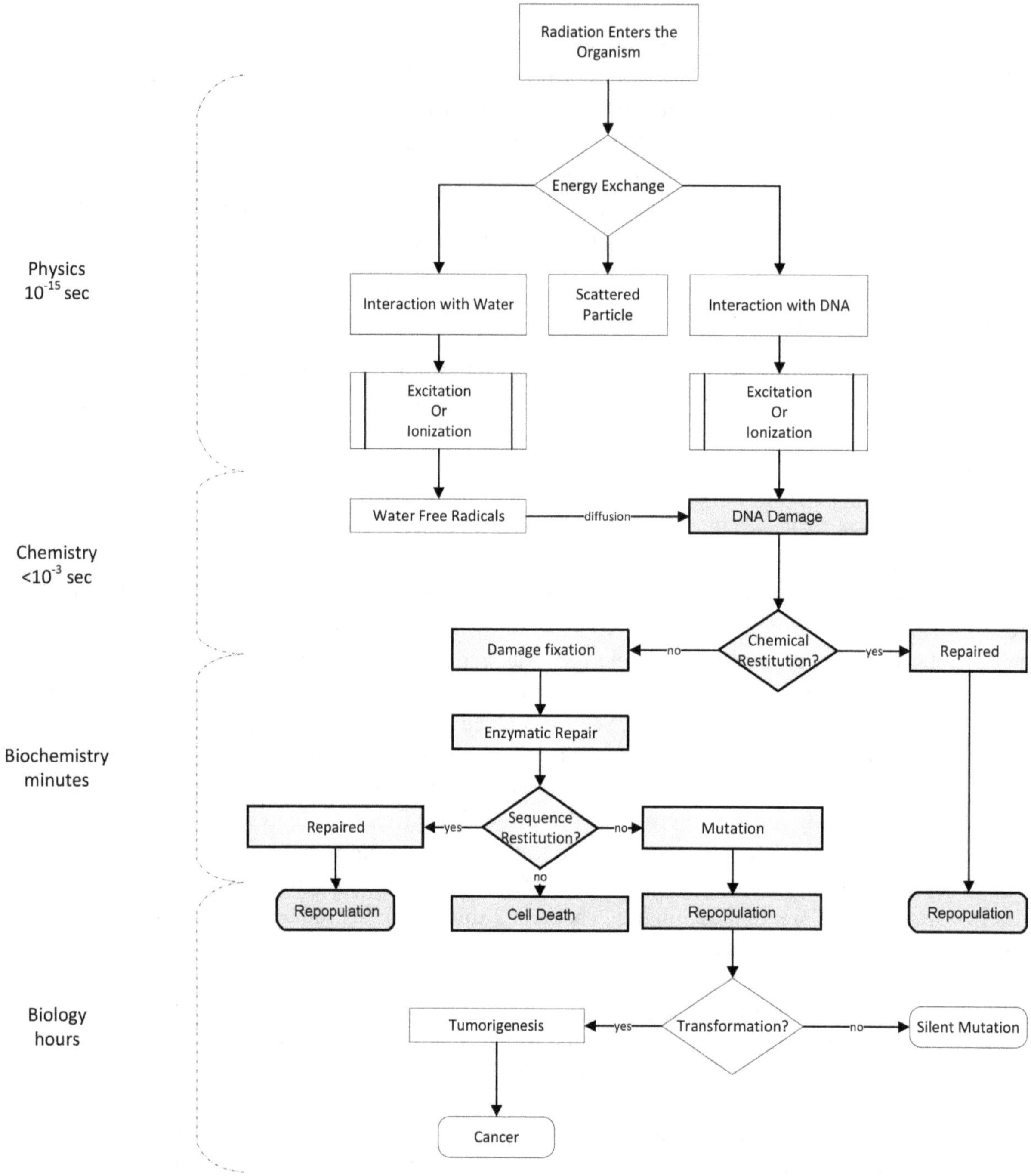

Figure 6-2 Cell survival following DNA damage is modeled by observing two possible endpoints: cell death or cell division (repopulation). Cell division is assumed to indicate cell survival. The pathways leading to the cell death or repopulation endpoints are employed to guide agreement between the model prediction and the experimental outcome.

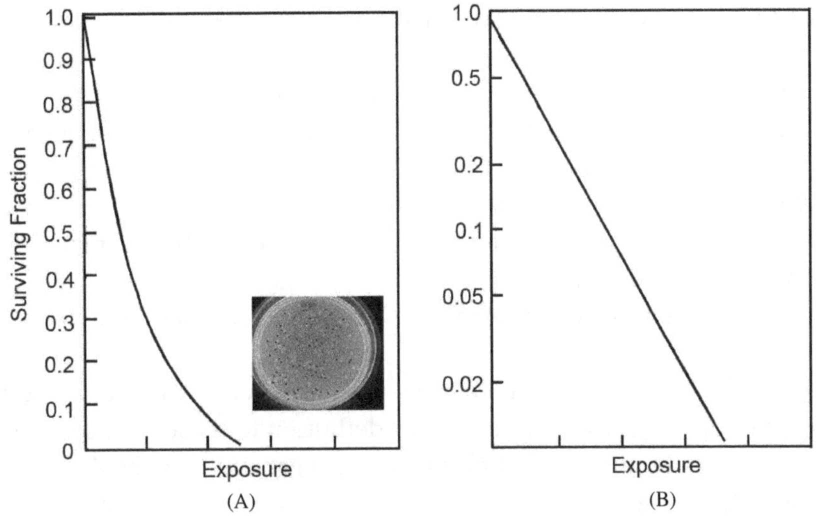

Figure 6-3 Cell survival curves representing (some) bacteria, haploid yeast, or human spermatogonia. (**A**) The number of counted colonies per number of cells plated (survival fraction) is plotted for each radiation exposure delivered. Inset: Clonal colonies on agar. The black dots are marker on the dish cover denoting a counted colony. (**B**) When plotted on a semi-logarithmic graph, the relationship between surviving fraction and exposure is linear. This suggests an exponential relationship. The fraction of surviving cells at any exposure can be predicted from the graph.

information and techniques available at the time, Lea developed his theory based on probability theory and the experimental evidence provided by the irradiation of bacteria and haploid yeast.

Exposure of selected types of bacteria to x-ray radiation produced survival curves similar to the ones presented in Figure 6-3. These experiments are performed something like this. Cells (bacteria) are suspended in isotonic solution; a sample is counted under a microscope to determine the concentration and diluted such that 1 ml spread onto the growth medium in a petri dish will produce a sparse distribution of cells (perhaps 100 cells/ml). The cells are permitted to stabilize for a short time, but not divide. Groups of triplicate samples (three dishes) are subjected to either no radiation or progressively higher totals of radiation exposure. This process is referred to as a "dose escalation." The cells are incubated (permitted to rest under normal growth conditions) for as long as it takes the bacterium to divide and form visible colonies which are then counted. Three dishes at each exposure are averaged and the average number of colonies per plate are plotted. To normalize the data, the number of colonies is divided by the number of cells originally added to the plate and presented as the survival fraction. Here, again as a reminder, each colony is assumed to have arisen from a single surviving (reproductively competent) cell.

Here is what Lea understood to be true when he developed his "target theory." Deposition of energy is a discrete and random process (stochastic) both in time and space. In each cell, there exists a physically describable (but not yet identified) *target* for radiation action. There may be multiple targets in a single cell and the inactivation of *n* targets leads to the loss of reproductive capacity. The exact number of targets is not known. There are no conditional probabilities for interaction of radiation with the individual targets; they are independent — the order of inactivation is not important. After the first target is inactivated, the inactivation of the second target is neither more nor less likely. Thus, Lea made the following assumptions.

1. The population of cells is exposed to x-ray radiation in a manner unlikely to produce multiple ionizing events in a single target, if the target volume is small.
2. The cell contains one or more sensitive volumes or targets of size ν.
3. *Active events* can produce the necessary biological damage (excludes ion recombination, radical recombination, and chemical restitution).
4. The *active events* occurring within volume, ν, are called *hits*.
5. The total cell volume (of all exposed cells) is V (the product of the average cell volume and the number of cells in the radiation field).
6. D is the density of *active* events; the number of events that occur per unit volume.

The nature of an *active* event can be understood from our discussion in Chapter 2, Section 2.3. An *active* event is clearly the ionization or excitation (limited by definition to those events resulting in fixed DNA damage). Let's do a "back-of-the-envelope" approximation to get a feel for the density of these events. If the average secondary electron energy transfer is estimated at 60 eV, a relatively small delivered energy of 100 ergs/g (1 cGy) would result in about 4×10^3 energy transfer events per cell in Equation 6.1. D is proportional to dose and for convenience can be thought of as a stand-in for dose. Now, notice that the probability of an *active* event is simply the ratio of the target volume to the total exposed volume (ν/V), because the total number of *active* events is the number of *active* events per volume multiplied by the volume (DV), and the number of *active* events occurring in a target is the number of *active* events per target volume multiplied by the target volume ($D\nu$). This provides a nice visual: the probability of an *active* event is just a comparison of the target volume size to the irradiated volume size. This probability is designated ρ (i.e., $\rho = \nu/V$) and is called the *hit probability*. For convenience and simplification, DV can be designated $\mathcal{D}$, the total number of events in the irradiated volume. We can express the probability that a cell will be *hit* h times (p) as:

$$p(\rho,h,\mathcal{D}) = \rho^h (1-\rho)^{(\mathcal{D}-h)} \binom{\mathcal{D}}{h} \tag{6.1}$$

where ρ^h is the probability that a cell will be hit h times, and $(1 - \rho)^{(\mathcal{D}-h)}$ is the probability that the remaining events will *not* be *hits*. Why? The probability of one *hit* ($h = 1$) per cell is ρ. The probability of two *hits* ($h = 2$) per cell is $\rho \times \rho$ (or ρ^2). The probability of three *hits* ($h = 3$) per cell is $\rho \times \rho \times \rho$ (or ρ^3); and so on. Because *hit probabilities* are less than one, the exponent causes the term to become increasingly smaller for greater numbers of *hits*. Now, since the total probability of all outcomes is unity (i.e., the probability of heads *or* tails occurring is 1, or 100%), then the probability of an event *not* being a *hit* is $1 - \rho$. It is the probability of all outcomes (1) excluding the probability of a *hit* (ρ). So, this probability is also less than one. The total number of events, $\mathcal{D}$ (Figure 6-4) is a very large number (according to our earlier calculation for 1 cGy, 4×10^3 events per cell multiplied by the number of cells exposed), this causes the second term to be very small. However, as the number of *hits* increases, the exponent decreases, and the probability of the remaining events per cell not being a *hit* grows. The third term of Equation 6.1 represents the binomial coefficient for all the ways hits and misses can be assigned to $\mathcal{D}$:

$$\binom{\mathcal{D}}{h} = \frac{\mathcal{D}!}{h!(\mathcal{D}-h)!}. \tag{6.2}$$

The coefficient is read: "$\mathcal{D}$ choose h" because it indicates the number of ways to choose a subset of h elements from a fixed set of $\mathcal{D}$ elements. For example, if $\mathcal{D}$ represents the consecutive four numbers 1, 2,

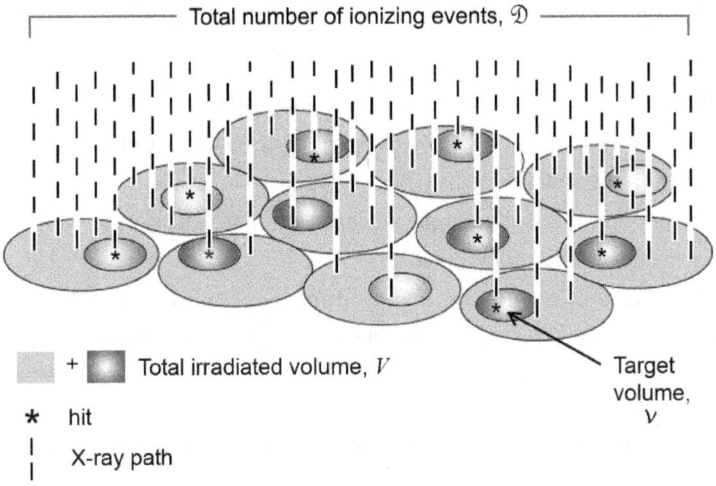

Figure 6-4 Visual representation of Target Theory components. The target volume, ν, can be represented by the volume of a cell nucleus. The total irradiated volume, V, is the total of the cellular volumes, including the volumes of the cell nuclei. The number of ionizing events, $\mathcal{D}$, includes all the interactions of all the secondary electrons set in motion by the exchange of energy from all the photons passing through the radiation field (limited to the area of the cells). A *hit* (*) is the deposition of energy in the target volume (less any instances of restitution).

3, and 4, and you were to choose two (h), you might choose 1 and 2, or 1 and 3, or 1 and 4, or 2 and 3, or 2 and 4, or 3 and 4. These are the number of ways you might choose h out of $\mathcal{D}$. For small h, the number of choices is very large. As h increases, the number of choices decreases. The mathematical solution evokes the factorial expression ($n!$) that denotes the product of all whole numbers from 1 to n. In the case of the numerator, $\mathcal{D}! = 1 \times 2 \times 3 \times 4 \times 5 \times \ldots \times \mathcal{D}$. The large total number of events seems to dominate the expression, but because the fraction $\mathcal{D}!/(\mathcal{D} - h)!$ depends on factorials; it changes rapidly with h, as does ($1/h!$). An excellent online reference is provided at the end of this chapter for any student interested in reviewing probability statistics (Lane & Osherson, n.d.).

All of this is well and good but recall that our quest was to determine the cell survival for any given exposure (a model that generates Figure 6-3). So far, we have determined the probability that a cell will receive *hits* (the processes up to and including "DNA Damage" in Figure 6-2). To determine cell survival, we need to understand how the cell responds to *hits* (the chemistry, biochemistry, and biology bracketed by "Cell Damage" and "Repopulation" in Figure 6-2). Does it survive? Recall that Lea was a physicist. He understood the physics of Figure 6-2 well. Unfortunately, because the target was unknown at the time, the chemistry and biochemistry of Figure 6-2 between "DNA Damage" and "Cell Death" or "Repopulation" represented a black box. So, Lea placed an unknown mechanism into his theory. He let $H(h)$ be a function that describes the response of a cell to h *hits*. The function is not defined, and the form is unknown. Everything in Figure 6-2 between "DNA Damage" and "Repopulation/Cell Death" is tucked into this black box. The probability that a cell will survive h *hits* (P) can then be expressed as:

$$P(\rho,h,\mathcal{D}) = \left(\rho^{h}\right)(1-\rho)^{(\mathcal{D}!-h)}\binom{\mathcal{D}}{h}\left(H(h)\right). \tag{6.3}$$

The probability of a cell surviving h *hits* differs from the probability of a cell receiving h *hits* (Equation 6.1) in two ways. The most obvious is that the undescribed function for cell survival, $H(h)$, becomes a factor. The second is that $\mathcal{D}$ in the second exponent becomes $\mathcal{D}!$ because the order of the events may influence survival. The terms of Equation 6.3 are summarized in Table 6-1.

Table 6-1 Symbol definitions for development of Lea's Target Theory.

Symbol	Definition
ν	The size of the sensitive target (we can approximate this by the size of the nucleus)
hits	A deposition of energy within the target volume, ν, capable of causing a stable molecular lesion
V	The total irradiated volume of cells
D	The number of events that occur per volume, V (proportional to dose, which is in J/kg)
ρ	The *hit probability* for a target; equal to ν / V
$\mathcal{D}$	The total number of events in the irradiated volume of cells; equal to DV
$H(h)$	An unknown function that describes cell survival as the number of *hits* increases
$p(\rho, h, \mathcal{D})$	Probability that a cell will receive *h hits*
$P(\rho, h, \mathcal{D})$	Probability that a cell will survive *h hits*

Now, $P(\rho, h, \mathcal{D})$ accounts for the probability that a cell will survive *h hits*. But we want an expression that will describe the probability that a cell will survive any number of *hits*. Figure 6-3 plots survival against exposure, and as the exposure ($\mathcal{D}$) increases, the probability of *hits* increases. In other words, we want an expression that is independent of the number of *hits*, *h*. Because a cell may have a nonzero probability of survival for any $h \leq \mathcal{D}$, the sum of survival probabilities for all values of h to $\mathcal{D}$ must be considered i.e.,

$$S(\rho,\mathcal{D}) = \sum_{h=0}^{h=\mathcal{D}} P(\rho, h, \mathcal{D}!). \tag{6.4}$$

$S(\rho, \mathcal{D})$ is the General Survival Equation for Target Theory. If we make some reasonable assumptions, we can test the model against the relationship in Figure 6-3. Assume:

- A cell will only survive 0 *hits*; $H(h) = 1$ for $h = 0$.
- One or more *hits* will result in cell death; $H(h) = 0$ for all $h \geq 1$.
- The hit probability for zero hits, ρ_0, is very small.

The third assumption is fair as ν/V, the size of the nucleus compared to the size of the irradiated field, is indeed extremely small. Although the first two assumptions seem arbitrarily restrictive, the implementation of assumptions is the means by which a model is tested. If these assumptions lead to something resembling the empirical data, then something about cell survival has been discovered. If the resulting plot is not quite right, the assumptions can be modified until the model agrees with the experimental data. Implementation of these assumptions resolves Equation 6.3 to

$$P(\rho_0, h, \mathcal{D}) = (\rho^0)(1-\rho_0)^{(\mathcal{D}-0)}\,(1)(1) = (1-\rho_0)^{(\mathcal{D})}. \tag{6.5}$$

Another way of writing this is (because of the mathematical identity),

$$P(\rho_0, h, \mathcal{D}) = e^{\mathcal{D}\ln(1-\rho_0)}. \tag{6.6}$$

Because $S(\rho, \mathcal{D})$ is the sum of all $P(\rho, h, \mathcal{D}!)$, and because we have only one case here ($h = 0$), Equation 6.6 is also the formula for survival (substitute for P in Equation 6.4). Excellent. Notice that Equation 6.6

is an exponential function that would produce linear semi-log plots such as the one in Figure 6-3. Some mathematical manipulation will clean things up a little more. For small values of ρ_0 (the third assumption earlier), $-\ln(1 - \rho_0)$ is approximately equal to ρ_0. Substitution into Equation 6.1 (for $h = 0$) indicates that ρ_0 is equivalent to p. This yields the final form of the Single Hit Model for cell survival:

$$S = e^{-p_0 \mathcal{D}} \tag{6.7}$$

The Single Hit Model tells us that cell survival depends on the probability that a cell will receive no hits (p_0) and the number of energy deposition events in the radiation field ($\mathcal{D}$); two rather difficult numbers to determine empirically. A little logic allows us to state the exponent in terms of dose, although at Lea's time, this was not much of an improvement. Recall that $\mathcal{D} = DV$, and D is proportional to dose. Let's declare a new term, D_0, which is the *mean lethal dose.* It is mathematically the dose required to reduce the surviving fraction of cells (S, the number of surviving cells/number of cells plated) one natural log (equal to 1/e). It is also the reciprocal of $\rho_0 V$. Now the model can be expressed in terms that are more intuitive (and more useful):

$$S = e^{-(\rho_0 V)D} = e^{-\frac{D}{D_0}} \tag{6.8}$$

Sometimes, clinicians prefer to use the quantity D_{10}, which is the dose required to reduce the surviving fraction one log, or 0.1. In other words, how much dose do you have to deliver to reduce the surviving fraction from 0.1 to 0.01? Both D_0 and D_{10} are measures of the slope of the survival curve at less than 0.1 fractional survival (Figure 6-5B).

The final step in modeling biological behavior is to test the model against the data collected from experiments. Dose escalation experiments were performed on the single-celled organisms that could be cultured at the time. For some cell types, certain bacteria and haploid yeast, the plotted data points on a semi-log graph of survival fraction versus dose, were perfectly fitted by the Single Hit Model (Equation 6.8) straight line. However, diploid yeast displayed a significantly different survival pattern. So, Lea's theory

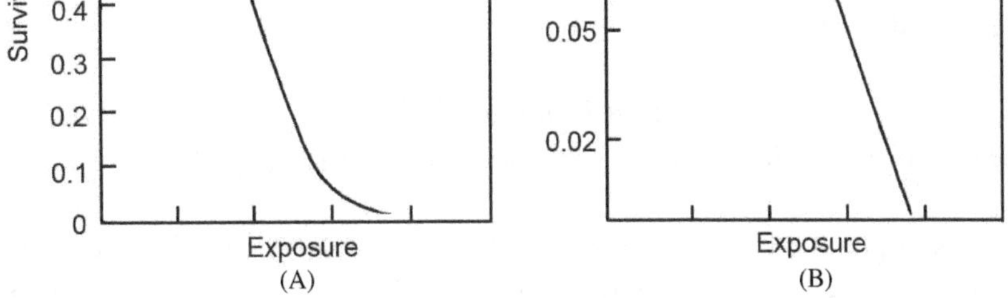

Figure 6-5 Cell survival curves representing diploid yeast, and in later experiments nearly all mammalian cell types. **(A)** The number of counted colonies per number of cells plated (survival fraction) is plotted for each radiation exposure delivered. **(B)** When plotted on a semi-logarithmic graph, the relationship between surviving fraction and exposure becomes a straight line at lower survivals (< 0.2) only.

is only true for some specific cases. Why? Well, remember the definition for a *hit*. A *hit* is an energy transfer event that occurs within the target (an *active event*), which can produce nonreversible damage. For example, say a *hit* is a fixed DNA double-strand break (dsb). If a species has a small genome and is repair deficient (e.g., if haploid it would lack HR capacity), a single DNA dsb might likely prove fatal. As we progress through cell survival theory, keep thinking back on Lea's definition of a *hit*. His definition is almost as good as anything modern radiation biophysics has proposed.

6.2.2. *Multitarget-Single Hit Model*

What is to be done about the diploid yeast response to radiation exposure? If data from a dose escalation study of diploid yeast is plotted, the survival curves looks more like Figure 6-5. The shape of the curve is sigmoidal, and on a semi-logarithmic graph, the lower exposures bend; they form a shoulder, whereas the higher exposures resemble haploid yeast. A modification of the Single Hit Model must be required. Lea reasoned that there might be multiple targets and some specified number of targets must be incapacitated to induce reproductive cell death. Again, he relied on probability. He assumed that each target has an equal probability of being hit (i.e., the volumes, ν, are equal). This assumption does not hold up to modern scrutiny, although it is not a serious deficit, and we will discuss the impact of that later. The Multitarget-Single Hit (MTSH) survival model makes the following assumptions:

1. There are n targets in each cell.
2. Each target has the same probability, q, of being hit.
3. In the general case, there will be a hit survival function, $B(b)$.

Following the same logic as for the Single Hit Model, we arrive at the probability that a cell will survive, P, following b hit targets.

$$P(q, b, n, D) = \left(1 - e^{-qD}\right)^{b} \left(e^{-qD}\right)^{n-b} \binom{n}{b} \left(B(b)\right). \tag{6.9}$$

As before, D is the number of energy transfer events per volume; q is the probability of a *hit* (the inactivation coefficient). Think of the exponent of e^{-qD} as an expression reflective of the number of successes resulting from a large number of trials. For example, 100 coin tosses ("D") with a 0.5 probability ("q") of resulting in "heads," yields 50 "heads" (0.5×100). The exponent would be −50. The first term of Equation 6.9 is the probability of a hit and the second term of the equation is the probability of a miss. The binomial coefficient accounts for n hits taken b at a time. It is analogous to the ratio of factorial expressions represented in Equation 6.2, substituting n for $\mathcal{D}$ and b for h. The hit survival function now depends on the number of targets hit (b) rather than the number of hits (h), so it is renamed B (b). Then Lea made the following testable statements:

1. For $b < n$, $B(b)$ assumes a value such that $P(q, b, n, D)$ is 1. The probability of cell survival is 1.
2. For $b \geq n$, $B(b)$ is 0 and therefore, $P(q, b, n, D) = 0$. The probability of cell survival is 0.

With these statements applied, we can develop an expression for cell survival. The survival probability of all values of b less than n is 1. The survival probability of the nth hit is 0; and also, as in Equation 6.9, it is $(1 - e^{-qD})^n$ because $b = n$ for this case. The survival probability of the nth hit (which indicates cell death) can be subtracted from the survival probability of all hits up to and including −1, which is 1. The final formulation for the MTSH model is therefore:

$$S(q, n, D) = 1 - (1 - e^{-qD})^{n} \tag{6.10}$$

Table 6-2 Symbol definitions for development of the MTSH model.

Symbol	Definition
D	The number of events that occur per volume (proportional to dose)
n	The number of targets
q	Probability of a *hit*
b	The number of targets that have been "hit" (have the potential to be inactivated)
$B(b)$	An unknown function that describes survival probability
$P(q, b, n, \mathcal{D})$	Probability that a cell will survive b hit targets

The terms used to develop the MTSH formula are summarized in Table 6-2. This expression produces the curves depicted in Figure 6-5. Moreover, considerable biology can be derived from the interrogation of the presentation on a semi-logarithmic plot. As we noted before, for surviving fractions less than about 0.1, the plot becomes a straight line. Whatever is causing the shoulder does not influence the higher dose straight line portion of the graph, or put another way, the biological effect responsible for the death of some small number of cells at low doses is different from the biological effect responsible for the majority of cell deaths. When $n = 1$, Equation 6.10 takes the form of the Single Hit Model, as it should. Also, comparing the semi-log plot to Equation 6.10 provides some easy ways to derive important variables.

1. Extrapolation of the straight line portion of the graph back to the y-axis yields the value of n: the number of targets.
2. Extrapolation of the straight line portion of the graph to 100% survival (i.e., fractional survival = 1.0) yields a value for the quasi-threshold dose (D_q).
3. $1/q$ equals the dose for 37% survival decrease, within the straight line portion of the survival plot.
4. The slope of the curve at 100% survival (0 Gy dose) is 0; the curve is flat.

The logical extension of the first relationship indicates that the shoulder increases in breadth (i.e., the shoulder encompasses a greater range of dose) as the number of targets (n) increases, and *vice versa* (Figure 6-6). The straight line portion slides left and right (the dose range changes) dependent upon n. The quasi-threshold dose (D_q) is descriptive in nature. It indicates a dose at which the plot becomes a straight line, or approximately how much dose must be delivered before the biology responsible for the shoulder diminishes to insignificance. Because the inverse of the inactivation coefficient $1/q = D_0$ (definition of D_0 presented during the derivation of the Single Hit Model; aka the *mean lethal dose*), the following relationship can be derived from Equation 6.10:

$$D_q = D_0 \ln n. \tag{6.11}$$

Because of the ease of graphical solutions, the MTSH model has been perpetuated in the literature and remains a favorite method for rapid analysis. However, the model has serious shortcomings, one at either extreme of the dose range. First, data from empirical trials indicate that initial slope of the curve as $D \to 0$ is not 0 (flat). Chromosome aberrations and cell death initiate at very low doses. Second, the low survival tail of the straight portion of the plot is not straight. Data derived from contemporary experiments using mammalian cells disclosed that the value of D_0 tends to decrease as survival decreases. The disagreement at low doses (survival fraction near 1.0) can be alleviated by multiplying the MTSH formula (Equation 6.10) by $e^{-q'D}$, where q' is an additional requirement for numbers of hits. There is no fix for the low survival deviation, however. A new model is required.

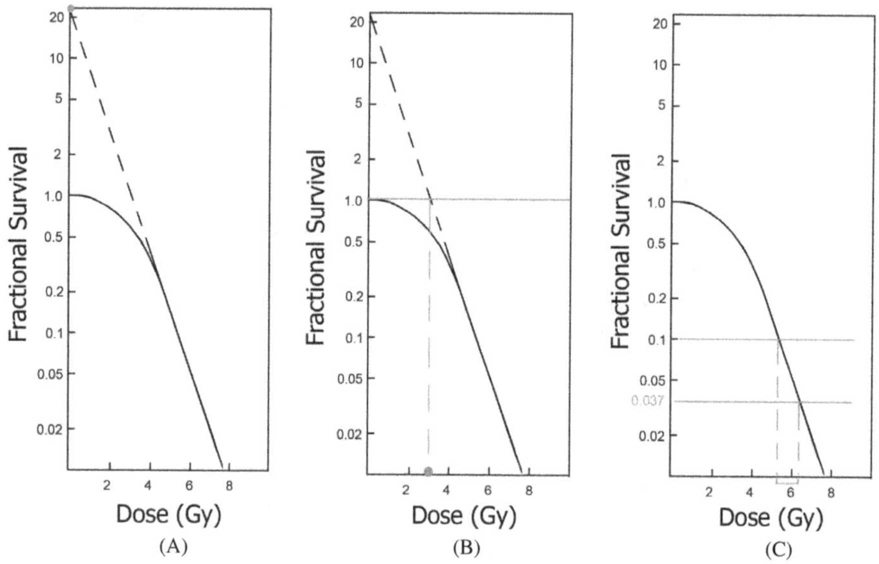

Figure 6-6 Graphical solutions to the MTSH model. (**A**) Extrapolation of the straight line portion of the survival curve to the intersection with the ordinate yields n, the number of targets (here, approximately 21). (**B**) The intersection of the extrapolation of the straight line portion of the survival curve with the 100% survival abscissa yields D_q, the quasi-threshold dose (here, approximately 3 Gy). (**C**) The difference between the dose required to reduce survival to 0.1 and the dose required to reduce survival further by $1/e$ (0.037) yields D_0 (here, approximately 1.1 Gy).

6.2.3. *Linear Quadratic Model*

In the 1950's Theodore Puck (Puck & Marcus, 1955) developed a method for culturing mammalian (human) cells. In addition, it became technically possible to measure the deposition of energy in small volumes of water, and so the concept of *dose* became well defined (1 Gy = 1 J/kg) and easily measured. Watson and Crick resolved the chemical structure of DNA in 1953 (using Rosalind Franklin's essential x-ray crystallography data) followed rapidly by the realization that the integrity of this structure must be imperative for cell function. The hypothesis that this, therefore, must be the target for cellular radiation damage (see Chapter 3, Section 3.1) logically followed. Now, all the pieces were in place to develop a physiological, hypothesis-driven model for radiation damage (i.e., not based on simple probabilities but rather on biological functionality).

We will follow Professor Ed Alpen's (Alpen, 1990) excellent example here to derive what has come to be called the Linear Quadratic (LQ) formula. We present it as a refinement of Chadwick's and Leenhouts' (Chadwick & Leenhouts, 1973) "molecular model." Here are their initial assumptions.

1. Lea's "target(s)" is (are) DNA and the critical damage is a DNA dsb.
2. The effective action of radiation (indirect or direct) is a rupture of the molecular bonds in DNA.
3. The lesions (broken bonds) are capable of being repaired, and the efficiency of repair can be expressed as radiobiological effect.
4. Repair processes include restitution, charge transfer, enzymatic repair, and recombination.

These assumptions sound totally reasonable in retrospect, but remember that this was cutting edge discovery in the 1970s. Also, notice that restitution was nonexistent to Lea; he considered only *active events* (now known to be stable DNA damage) leading to cell death a *hit*. We now must define a new set of terms to represent the players in this model motivated by biophysical realities. Call the number of chemical bonds per cell available for damage, N_0 (refer to Figure 3-9). Let N be the number of undamaged bonds following radiation exposure. D is dose, and K is the probability for damage of a single bond per unit dose (see Table 6-3). K is a constant: the probability of damaging any given bond is assumed the same for all bonds. Logically, the decrease in the number of unbroken bonds per cell, as dose increases, is simply the product of the probability of damage and the number of remaining *unbroken* bonds per cell:

$$-\frac{dN}{dD} = KN, \tag{6.12}$$

leading to the solution,

$$N = N_0\, e^{-KD}. \tag{6.13}$$

The number of *broken* bonds (per cell, per dose) would then be $N_0 - N$. Substituting for N (Equation 6.13):

Table 6-3 Symbol definitions for development of the LQ Model.

Symbol	Definition
	For ssbs
N_0	The number of vulnerable chemical bonds per cell
N	The number of undamaged chemical bonds following irradiation
K	The probability of rupturing a chemical bond per unit dose
D	Dose
f	The proportion of bond breaks not repaired
	For dsbs
n_1	The number of critical chemical bonds on strand 1
n_2	The number of critical chemical bonds on strand 2
n_0	The number of sites that can sustain a dsb ($n_0 \leq n_1$)
k	The probability constant per bond per unit dose
k_0	The DNA dsb hit probability constant
f_1	The unrestored fraction of bonds on strand 1
f_2	The unrestored fraction of bonds on strand 2
f_0	The fraction of unrestored DNA dsb
q_1	The number of strand 1 broken bonds per cell
q_2	The number of strand 2 broken bonds per cell
Δ	The fraction of dose that induces DNA dsb through a single event
E	An efficiency factor representing the likelihood that two DNA ssb occur sufficiently close in time and space to produce a DNA dsb
Q_i	The mean number of DNA dsb/cell resulting from one event
Q_{ii}	The mean number of DNA dsb/cell resulting from two independent DNA ssb
Q	The mean number of DNA dsb per cell
Q_L	The mean number of lethal DNA dsb per cell
χ	$n_0 f_0$
ϕ	$E n_1 n_2 f_1 f_2 f_0$
p	The proportionality constant between DNA dsb and cell death

$$N = N_0 - N_0\, e^{-KD} = N_0\,(1 - e^{-KD}) \tag{6.14}$$

So, this accounts for the DNA bond damage incurred. But sometimes the damage reverses (restitution) and sometimes it is enzymatically repaired. We must account for that. Let f be a constant representing the fraction of damage that is *not* resolved, the fraction of damage remaining. The number of lesions leading to a DNA single-strand break (ssb) is thus:

$$N_0 - N = fN_0\,(1 - e^{-KD}). \tag{6.15}$$

We know from Chapter 4 that reproductive cell death results from DNA dsb. And we know that dsb can be the result of a single two lesion event, or two independent lesion events close in time and space (Figure 4-8). We can use the expression for ssb events (Equation 6.15) to derive the expression for lesions leading to reproductive cell death. Analogous to the previous derivation, n_1 is the number of critical chemical bonds on strand 1 and n_2 is the number of critical chemical bonds on strand 2 ($n_1 = n_2$); k is the probability constant per bond per unit dose, and f_1 is the unrestored fraction of bonds on strand 1; f_2 is the unrestored fraction of bonds on strand 2. Now, we need to account for the two mechanisms for DNA dsb production (a single event vs. two independent events), so let Δ be the fraction of dose (D) that induces the first mechanism and $1 - \Delta$ be the fraction of dose that induces the second mechanism.

The number of ssb on each DNA strand is therefore:

$$q_i = f_i n_i[1 - e^{-k(1-\Delta)D}], \tag{6.16}$$

where "i" is 1 for the first strand and 2 for the second strand. We need to account for the two different mechanisms leading to a DNA dsb. When a DNA dsb is the result of two events, they must occur close enough together that the first is not repaired before the second occurs, and they must be within 10 bp on opposing strands. Let's include an efficiency factor, E, that accounts for the likelihood that these two conditions are satisfied. With this in mind, we can express the mean number of unrepaired DNA dsb per cell that occur as a result of *two independent* DNA ssb events (Q_{ii}) as:

$$Q_{ii} = En_1n_2f_1f_2f_0[1 - e^{-k(1-\Delta)D}]^2, \tag{6.17}$$

where f_0 is the number of unrestored DNA dsb. The mean number of unrepaired DNA dsb per cell that occur as a result of a *single* event is analogous to Equation 6.16:

$$Q_i = f_0n_0[1-e^{-k_0\Delta D}], \tag{6.18}$$

where n_0 is the number of sites that can sustain a dsb ($n_0 \le n_1$), and k_0 is the DNA dsb hit probability constant. The mean number of DNA dsb per cell resulting from both mechanisms is just be the sum of the two mechanisms:

$$Q = f_0n_0\left[1-e^{-k_0\Delta D}\right] + En_1n_2f_1f_2f_0\left[1-e^{-k(1-\Delta)D}\right]^2. \tag{6.19}$$

One more consideration: not all unrepaired DNA dsb lead to reproductive cell death (see Figure 6-2). So, although Equation 6.19 provides the number of DNA dsb per cell, it does not indicate the probability of survival — which was our quest.

Let's clean things up a bit before we develop an expression for cell survival. Let $\chi = (n_0f_0)$, and $\phi = (En_1n_2f_1f_2f_0)$. If not all DNA dsb result in "lethal" damage — damage capable of reproductive cell death — then some proportion of them do. So, let's invent a constant of proportionality, p, that represents the number of DNA dsb capable of causing cell death and that, again, is not necessarily known. The mean number of lethal DNA dsb per cell is:

$$Q_L = p\left\{\chi\left[1-e^{-k_0\Delta D}\right]+\phi\left[1-e^{-k(1-\Delta)D}\right]^2\right\}. \tag{6.20}$$

This still does not translate to cell survival because a cell may have several lethal DNA dsb, but it can die only once. And if we wish to know the number of cells killed, we must not only consider the number of lethal DNA dsb per cell, but also the number of cells available to be killed — a number that decreases as cells die. The fraction of cells that are *killed* can be expressed as a Poisson distribution of the mean number of lethal DNA dsb,

$$F_d = 1-e^{-Q_L}, \tag{6.21}$$

And likewise, the fraction that *survive* can be expressed as

$$S = e^{-Q_L} = e^{-p\left\{\chi\left[1-e^{-k_0\Delta D}\right]+\phi\left[1-e^{-k(1-\Delta)D}\right]^2\right\}}. \tag{6.22}$$

Wow; that expression is a bit cumbersome even using the substitutions of χ and ϕ. If we tidy things up by letting $\alpha = (p, f_0, n_0, k_0, \Delta)$ and $\beta = (p, f_0, E, n_1, n_2, f_1, f_2, k^2, (1-\Delta)^2)$, and assume again that k and k_0 are small, then we can express cell survival as a simple exponential with a linear term and a quadratic term:

$$S = e^{-(\alpha D + \beta D^2)}. \tag{6.23}$$

Equation 6.23 represents the final formalism of the currently most popular model for cell survival. Though not a perfect model for reasons that we shall discuss, the LQ model satisfies many predictive measures in a variety of circumstances. The form is simple, and the interpretation of the linear and quadratic terms is straightforward. It is elegant and useful. The LQ model is almost universally accepted for calculating therapeutically equivalent doses for different delivery schemes. As we begin applying this model, remember the components of the two variable terms, α and β (see Figure 6-7). The linear term, α, depends on the number of sites that can sustain a DNA dsb (n_0), the DNA dsb hit probability constant (k_0), the number of unrestored DNA dsb (f_0), the proportion of DNA dsb that are potentially lethal (p), and the fraction of dose that induces DNA dsb through a single event (Δ). This last term is critical for two reasons: (1) α accounts for cell death resulting from a single event (the relationship is *linear*), and (2) the physical characteristic determining the fraction of dose that induces single event DNA dsb is *track structure*. We introduced track structure in Chapter 2 (specifically Figures 2-19 and 2-21) and we will explore track structure in depth later, when we discuss microdosimetry. The number of unrestored DNA dsb, f_0, reflects radiation sensitivity in the form of repair competence. The quadratic term, β, depends on the same or similar factors (n_1 and n_2, k, f_0, f_1 and f_2, and p), but the two strands must be considered separately to achieve the value for two interdependent events. Similarly, the probability of a hit for each strand (k) is multiplied to account for the probability of two hits, one on each strand (k^2). The final term, $1 - \Delta$, is obviously the fraction of dose that induces DNA dsb through any combination of events other than a single event, that is, two independent events (see Figure 4-8). And as before, that fraction must be multiplied to account for two strands, $(1 - \Delta)^2$. So, β is quadratic. In other words, two events must happen before a DNA dsb is formed; a DNA dsb that has the potential to be lethal and thus has the potential to cause cell death.

The linear quadratic equation (Equation 6.23) has graphical solutions (Figure 6-8). If the initial slope, as $D \to 0$ Gy, is extrapolated to higher doses, the distance from the 100% survival abscissa (fractional

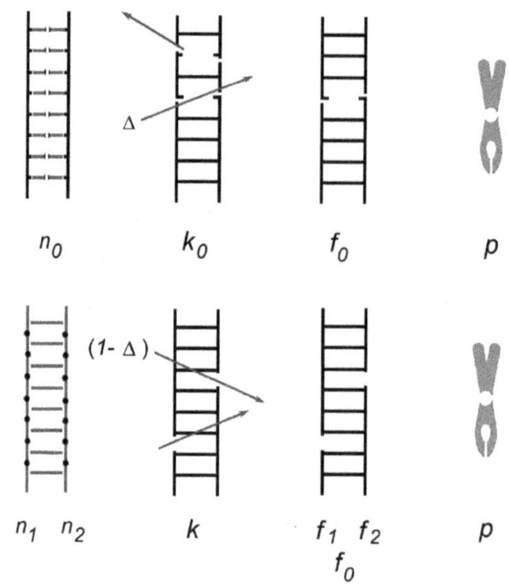

Figure 6-7 A visual depiction of the terms represented by the linear quadratic exponent. Top row: α. The number of DNA dsb susceptible sites, n_0, is represented by spaces in the base pairs, although Chapter 3 provides a more complete compilation of possible dsb chemistry. The fraction of dsb from a single event, Δ (varies with radiation quality), and the probability that a susceptible site will be hit, k_0, represent the radiation hit in the second frame of the top row. The fraction of unrestored dsb that may result in lethal damage, f_0, is represented in the third frame and potentially lethal damage, p, is represented by chromosomal damage in the fourth frame. Bottom row: β. The number of ssb susceptible sites on strand 1, n_1, and strand 2, n_2, are represented by nodes on the phosphate backbone, although a more complete compilation of possible ssb chemistry is provided in Chapter 3. The probability that a susceptible ssb site on one strand will be hit, k, must be considered twice (k^2), once for each strand; and the fraction of dsb resulting from one hit on each strand must also be considered twice, $(1 - \Delta)^2$. The third frame in the bottom row indicates the fraction of unrepaired breaks on each strand, f_1 and f_2, and the fraction of those that are sufficiently close in time and space to potentially cause lethal damage, f_0. Potentially lethal damage, p, is represented by a damaged chromosome in the final frame of the bottom row.

survival = 1) to the extrapolated line is equal to αD, and the distance from the extrapolated line to the survival curve is equal to βD^2. Given that, it is straightforward to measure the two variables, α and β, at a dose of 1 Gy. Furthermore, there exists a dose at which the distance above the extrapolated line is equal to the distance below the extrapolated line. At that dose,

$$\alpha D = \beta D^2, \tag{6.24}$$

and the dose is equal to the quantity α/β:

$$\alpha/\beta = \frac{D^2}{D} = D. \tag{6.25}$$

This quantity, α/β, serves as a gauge of the radiation sensitivity of the cells under observation (f_0) and the effectiveness of the radiation (Δ). Radiation biologists rely on the MTSH value D_0 or the LQ value α/β to describe the radiation quality systems with a single value. Although it is relatively easy to determine α/β from the survival curve (Figure 6-8A), determining α and β at 1 Gy is difficult, especially β because that value is an order of magnitude less than α. We can resolve this difficulty through a simple mathematical manipulation. If we take the natural log of both sides of Equation 6.23,

$$-\ln S = \alpha D + \beta D^2, \tag{6.26}$$

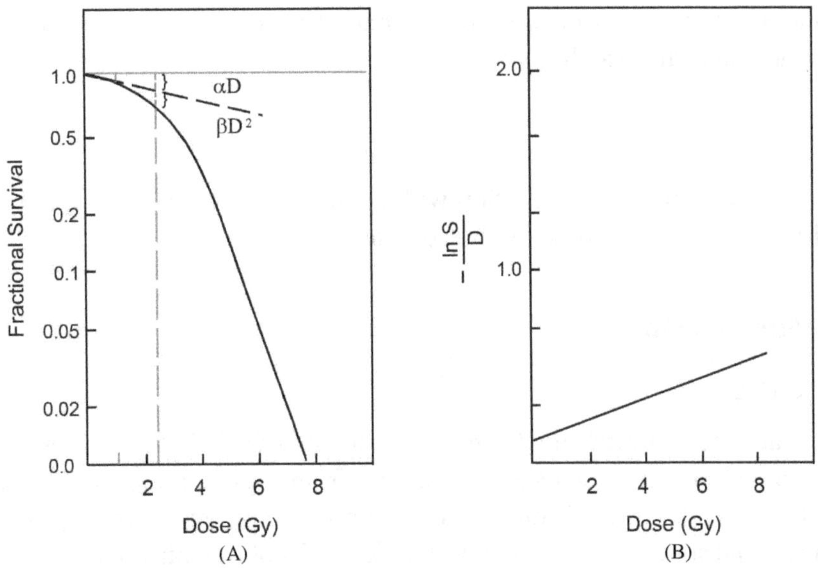

Figure 6-8 Graphical solutions to the LQ model. (**A**) A line extrapolated from the initial slope of the survival curve together with the boundaries of 100% survival and the survival curve, determine the values of the variables α and β. At the dose of 1 Gy, the fractional survival above the extrapolated line equals (α)(1) and the fractional survival below the extrapolated line equals (β)(1)^2, directly yielding the variables. Furthermore, when the extrapolated line bisects the boundaries, i.e., $\alpha D = \beta D^2$, the dose (2.4 Gy in this case) is equal to α/β. (**B**) When the linear-quadratic equation is rearranged, $y = -(\ln S/D)$ and $x = D$; the intercept equals α and the slope of the line yields β.

divide both sides by dose and rearrange,

$$-\ln S/_{D} = \beta D + \alpha . \tag{6.27}$$

Equation 6.27 characterizes the form of a straight line plot: $y = mx + b$. So, if we plot ($-\ln S/D$) vs. dose (D), the slope of the line will yield β and the intercept will yield α (Figure 6-8B). Both plot forms are popular in the literature because they provide strong visual comparisons among different systems and provide "radiation biology at a glance."

The LQ formalism is not without detractors. Decades of attempts to show a direct correlation between DNA dsb and cell death have failed to one degree or another. This is not so much a failure of the model as it is an indication that the assumptions of the LQ model development are not necessarily correct. The mathematical derivation suffers from the assumption that dsb accrue according to Poisson statistics. Modern techniques to measure DNA dsb induction and repair, such as pulsed-field gel electrophoresis and γH2AX foci phosphorylation (see Chapter 5, Section 5.5), indicate that repair is likely a Gaussian distribution, and that repair, rather than induction, likely drives cell survival. The model does not account for a time variable, whereas cell cycle dependence and repair kinetics are clearly important factors. Furthermore, when the sensitivity of cells to neutron radiation was compared to the sensitivity of cells to x-rays, the ratio of $(\alpha/\beta_{\text{neutrons}}) / (\alpha/\beta_{\text{x-rays}})$ was found to be greater at lower doses. Mathematically, that can't happen unless either the numerator or the denominator (or both) change with dose. The most concerning failure of the LQ model arises at survival fractions less than 10^{-3}, where empirical data exhibit a decreasing slope. To address this failure, Douglas and Fowler (1976) suggested a *three lambda model* with a superposition of exponential terms:

$$S = e^{-\lambda_3\left(1-e^{\lambda_1 D}\left(1-\left(1-e^{-(\lambda_2-\lambda_1)D}\right)^2\right)\right)}. \tag{6.28}$$

Almost two decades later, in 1993, Michael Joiner (2009) proposed a linear-quadratic-cubic model that added a third exponential term to address the low survival complication:

$$S = e^{-\alpha D - \beta D^2 + \gamma D^3}. \tag{6.29}$$

These modifications have gained some traction with experimentalists, but clinicians continue to favor the LQ model, which suffices to address their needs adequately.

6.2.4. *Boutique Survival Models*

6.2.4.1. *Cybernetic Model*

Around the same time that Chadwick and Leenhouts motivated the LQ model, Kappos and Pohlit (1972) published a model based on three "cell states" resulting from radiation exposure. They proposed that cells of the first state proceed through mitosis following radiation exposure to produce a discernable colony. Cells of the second state present reparable damage. Cells of the third state have sustained damage that is not repairable. Although the publication refers to an "essential molecule," DNA, for all intents and purposes, they avoided the strict mechanistic approach of the LQ model. Cell sensitivity is assumed to increase with dose and decrease with "recovery actions" including enzymatic repair and restitution. The kinetics of the latter two are considered, including influences of dose and dose rate that affect the instantaneous concentration of free radicals as well as the subsequent reduction of free radical concentration resulting from competition by biomolecules that may be upregulated by the cell.

The basic assumptions of the Cybernetic model are the following:

1. There exists an essential molecule that experiences an interaction with radiation resulting in an observed outcome. The molecule depends on the behavior under observation.
2. Because the behavior under observation is cell survival, the essential molecule is DNA.
3. There exist two pathways resulting in damage to the essential molecule: a direct pathway and an indirect pathway (the reaction rate, $\eta = \eta_{\text{dir}} + \eta_{\text{ind}}$).
4. At least two types of reactions with radiation are possible: the cell can be repaired, or the cell cannot be repaired.
5. Exposure to radiation results in three cell states. Cells of the first state may proceed through mitosis (designated A). Cells of the second state experience repairable damage (designated B); and cells of the third state receive irreparable damage (designated C).

The Cybernetic Model can be described by three rate reactions for state transitions:

$$\frac{dN_A}{dt} = -\eta_{AB}\dot{D}N_A - \eta_{AC}\dot{D}N_A + \varepsilon_{BA}N_B, \tag{6.30}$$

$$\frac{dN_B}{dt} = +\eta_{AB}\dot{D}N_A - \eta_{BC}\dot{D}N_B - \varepsilon_{BA}N_B, \tag{6.31}$$

and

$$\frac{dN_C}{dt} = +\eta_{AC}\dot{D}N_A + \eta_{BC}\dot{D}N_B. \tag{6.32}$$

$A \underset{\varepsilon_C}{\overset{\eta_C}{\rightleftharpoons}} A'$	*restoration*
$A \underset{\varepsilon_{BA}}{\overset{\eta_{AB}}{\rightleftharpoons}} B$	*repair*
$B \xrightarrow{\eta_{BC}} C$	*fixation*
$A \xrightarrow{\eta_{AC}} C$	*lethal damage*

Figure 6-9 Kinetics of the Cybernetic Model. Following radiation exposure, damaged cellular DNA (cell state A′) may chemically revert through restitution to undamaged (cell state A). Damage appears according to the rate constant η_c, and restitution proceeds according to the time constant ε_c. Exposure to radiation may also induce reparably damaged DNA (cell state B) that may be enzymatically returned to undamaged (cell state A). Repairable damage accumulates according to the rate constant η_{AB} and is repaired corresponding to the time constant ε_{BA}. A cohort of the pool of repairable damage (cell state B) evades repair and transitions to irreparable damage (cell state C) according to the rate constant η_{BC}. Finally, it is possible for radiation to directly induce lethal damage (cell state C) corresponding to the rate constant η_{AC}.

Table 6-4 Terms of the Cybernetic Model.

Term	Definition
$\eta_{AB}\dot{D}N_A$	The rate at which state A cells become state B cells
$\eta_{AC}\dot{D}N_A$	The rate at which state A cells become state C cells
$\varepsilon_{BA}N_B$	The rate at which state B cells are repaired (return to state A)
$\eta_{BC}\dot{D}N_B$	The rate at which repairable state B cell damage converts to irreparable
$\eta_{AC}\dot{D}N_A$	The rate at which state A cells receive direct irreparable damage

Consult Figure 6-9 as we consider these mathematical statements. Equation 6.30 proposes that the number of cells in state A (N_A) will decrease (dN_A/dt) according to the rate at which cells receive repairable damage, convert from state A to state B (η_{AB}), and according to the rate at which cells receive irreparable damage, convert from state A to state C (η_{AC}). Repair offsets the decrease in N_A by returning state B cells back to state A at a rate of ε_{BA}. The reduction in the number of cells in state A due to damage depends upon the dose rate ($\dot{D}$), but DNA repair in state B cells does not depend on the dose rate. Equation 6.31 proposes that the number of cells in state B (N_B) will increase (dN_B/dt) according to the rate (η_{AB}) at which cells receive damage (convert from state A to state B), but will decrease corresponding to the rate at which the damage becomes irreparable and the cells convert to state C (η_{BC}). N_B will also decrease corresponding to the rate at which the DNA is repaired (η_{BA}) returning cells to state A. Again, repair does not depend on dose or dose rate, unlike the occurrence of DNA damage ($\eta_{AB}\ \dot{D}N_A$) and the rate at which damage becomes irreparable ($\eta_{BC}\ \dot{D}N_B$). Think of it this way; a DNA dsb may form from two ssbs if the two lesions arise close together in time and space, which is more likely to happen at high dose rates. On the other hand, repair is a cellular housekeeping function. Equation 6.32 proposes that the increase in state C cells (dN_C/dt) — those with irreparable lesions — depends on the rate of direct, irreparable damage (η_{AC}) and the rate at which repairable DNA damage becomes irreparable (η_{BC}). Both transitions are dependent upon dose rate. These three equations are simple statements of kinetics. According to Figure 5-6, states A and B represent buckets that fill (+) and leak (–); state C only fills (+).

Kappos and Pohlit assumed that the indirect pathway for damage dominates x-ray radiation reactions leading to repairable damage (B) and that the direct pathway for damage dominates x-ray radiation reactions leading to irreparable damage (C) from undamaged cells (A). For review of these pathways, refer to Chapter 3 (Sections 3.4 and 3.5). The kinetics of free radical reactions depend on the local concentration of radicals. The local concentration of radicals, in turn, depends on dose absorption. Kinetics furthermore depend on the availability of free radical scavengers that compete with DNA for radical acquisition (see Equation 3.5). Therefore, η_{AB} and η_{BC} represent dose-dependent factors, whereas η_{AC} represents a rate constant. Nonetheless, because $\eta_{BC} << \eta_{AB}$ (the rate at which repairable DNA damage converts to irreparable damage is much smaller than the rate at which repairable DNA damage appears), η_{BC} may also be treated as a constant. Thus, only the rate "constant" η_{AB} varies with relative concentrations of free radicals and scavengers. We will not present the algebra here, the reader is referred to the original publication, but based on assumptions regarding the situational concentrations of free radicals and scavengers, Kappos and Pohlit presented this relationship for the conditional rate constant:

$$\eta_{AB} = \eta^*_{AB}\left(1 - e^{-\eta_c \dot{D}t}\right), \tag{6.33}$$

where η^*_{AB} is the rate of cell transition from state A to state B in the absence of scavenger molecules, and η_c is the rate at which competing scavenger molecules are removed from the system. Note that in Figure 6-9, this is the rate of restitution (for justification, refer to Chapter 3, Section 3.4). The interplay between free radical concentration and free radical scavenger concentration, and each of the reaction component's dependence on dose and dose rate, can be described simply as radiation sensitivity.

What does the model look like and how does it help us to predict survival behavior? Let's examine a couple of cases (limiting boundary conditions). In the case where the dose rate becomes diminishingly small ($\dot{D} \to 0$), the rate of transition from A to B equals the rate of transition back from B to A; restitution dominates. Equation 6.30 becomes

$$\frac{dN_A}{dt} = -\eta_{AC}\dot{D}N_A, \tag{6.34}$$

and integration provides

$$N_A = N_{A,0}e^{-\eta_{AC}\dot{D}t}, \tag{6.35}$$

which is a straight line on a semi-logarithmic plot; slope revealing η_{AC}. Notice that $N_A/N_{A,0}$ is equivalent to the survival fraction (S/S_0), as it is the number of dividing cells per the number of dividing cells plated. In the case, where the dose rate becomes extremely high ($\dot{D} \to \infty$), the rate of repair is negligibly small (by analogy, almost all DNA ssb interact to become DNA dsb rather than being repaired) and the third term of Equation 6.31 can be ignored. The change in the number of dividing cells becomes

$$\frac{dN_A}{dt} = -(\eta_{AB} + \eta_{AC})\dot{D}N_A. \tag{6.36}$$

Furthermore, under these conditions, η_{AB} is approximately η^*_{AB}, and integration of Equation 6.36 provides:

$$N_A = N_{A,0}e^{-(\eta^*_{AB} + \eta_{AC})\dot{D}t}. \tag{6.37}$$

which is a straight line on a semi-logarithmic plot; slope revealing $\eta^*_{AB} + \eta_{AC}$. Furthermore, at extremely large doses ($D \to \infty$), almost all cells transition to state B, having received damage, and then to state C as the damage becomes irreparable. Therefore, any reduction in the number of state A cells compared with lower doses results from the rate of damage conversion to irreparable, η_{BC}. In other words, fewer B cells are available for repair affecting the kinetics of the reaction from B to A. As with the MTSH model, a plot of the instantaneous number of cells in state A relative to the initial number of cells in state A (N_A/N_{A0}) vs. $\eta_{AB}D$ can be used to graphically determine η^*_{AB}/η_c. Akin to Figure 6-6B, extrapolation of the straight line portion of the curve back to $N_A/N_{A0} = 1$ provides an abscissa value of η^*_{AB}/η_c.

The Cybernetic Model was a novel approach for predicting cellular response to radiation. It considered the kinetics of cell transitions between damage states rather than molecular action. Unfortunately, Kappos' and Pohlit's assumptions and rationales did not stand up to the test of time, as the mechanisms of enzymatic repair became better understood. Even at the time of publication, the model proved to be more qualitative than quantitative. A fudge factor of 1.09 had to be used to achieve better fit to experimental data. Derived parameters generally attained the correct order of magnitude when compared to empirical results but failed to achieve accuracy. Although conclusions can be drawn from the values derived, absolute evaluations cannot be made. The model struggled also when Kappos and Pohlit attempted to use it to decipher the results from a "split dose" experiment. Nonetheless, a similar approach was taken by Stan Curtis a few years later.

6.2.4.2. *Repair–Misrepair Model*

Tobias (Tobias *et al.*, 1980) proposed a uniquely different model based on DNA repair processes. While both the MTSH model and the LQ model are premised on DNA lesion creation — that is, the physical production of DNA damage in Figure 6-2 — the Repair–Misrepair (RMR) model is based solely on the biochemistry initiating with DNA damage and resolving with cell death or continued cell cycling (repopulation). The formulation of this model begins, again, with a set of assumptions.

1. The initial deposition energy ultimately produces long-lived molecular configurations as the result of radiation chemistry.
2. Radiation chemistry is followed by biochemical processes including repair, increased damage, or damage fixation, and by subsequent cell physiological states.
3. Surviving cells may express changes in phenotype.

These assumptions closely parallel the highlighted area of interest in Figure 6-2. Two repair states were proposed: R_L, the linear yield per cell resulting from monomolecular reactions and R_Q, the quadratic yield per cell. Tobias *et al.* propositioned that there existed "uncommitted lesions" (U) that resolve to one of these two states, and that R_Q is proportional to the square of the density of U. If the unresolved lesions are homogeneously distributed in the reaction volume and the radiation dose is delivered within a time that is short compared to the rate of repair (we will deal with protracted dose rates later), then

$$\frac{dU}{dt} = -\lambda U(t) - kU^2(t) \tag{6.38}$$

where λ and k are rate constants for the linear and quadratic repair processes, respectively. The number of uncommitted lesions will decrease as they commit to either linear or quadratic repair, and so we must integrate these terms over time:

$$U(0) - U(t) = \int_0^t \lambda U(t)\,dt + \int_0^t kU^2(t)\,dt. \tag{6.39}$$

The integrals define R_L and R_Q:

$$R_L = \int_0^t \lambda U(t)\,dt \tag{6.40}$$

And

$$R_Q = \int_0^t kU^2(t)\,dt, \tag{6.41}$$

leading to the simplification

$$U(0) = U(t) + R_L(t) + R_Q(t). \tag{6.42}$$

The number of lesions undergoing linear repair plus the number of lesions undergoing quadratic repair plus the number of remaining uncommitted lesions, at any time point following irradiation, is equal to the total number of (uncommitted) lesions immediately following irradiation, U.

The rate constants, λ and k, are independent of time and dose, and the ratio, λ/k, constitutes the "repair ratio." Tobias named the repair ratio ε. Instantaneously following irradiation, the number of lesions occupying each state, R, is equal to 0, and at infinite time, the remaining uncommitted lesions equal 0 (all lesions are resolved). Given those boundary conditions, we can solve the integrals of Equations 6.40 and 6.41 and substitute the solutions into Equation 6.42 to derive expressions for the decrease in uncommitted lesions and increases in lesions committed to linear or quadratic repair. These expressions look like this:

$$U = \frac{U_0 e^{-\lambda t}}{1+\left(U_0/\varepsilon\right)\left(1-e^{-\lambda t}\right)} \tag{6.43}$$

$$R_L(t) = \varepsilon \ln\left[1+\frac{U_0}{\varepsilon}\left(1-e^{-\lambda t}\right)\right] \tag{6.44}$$

$$R_Q(t) = \frac{U_0\left(1+U_0/\varepsilon\right)\left(1-e^{-\lambda t}\right)}{1+U_0/\varepsilon\left(1-e^{-\lambda t}\right)} - \varepsilon \ln\left[1+\frac{U_0}{\varepsilon}\left(1-e^{-\lambda t}\right)\right]. \tag{6.45}$$

Notice that contained within the boundary conditions stated earlier is the assumption that all lesions are faithfully repaired. We know that they are not. So, Tobias introduced two new terms. The *probability* that linear lesions are correctly resolved is represented by ϕ and the *probability* that quadratic lesions are correctly repaired is represented by δ. The publication by Tobias (Tobias *et al.*, 1980) extrapolates the rates above (Equations 6.40, 6.41, and 6.42) using the probabilities of correct repair (as well as the probabilities of misrepair: $1 - \phi$ and $1 - \delta$) to derive survival under several circumstances wherein uncommitted lesions at time t are considered lethal. Let's look at one set of circumstances to disclose the differences between RMR and LQ models. Suppose that all linear repair results in a return to the original sequence without error ($\phi = 1$), and all quadratic repair results in mutation, that is, misrepair ($\delta = 0$). For the sake of brevity, the reader is referred to Tobias' original publication for algebraic resolution employing Poisson statistics. The resultant expression for survival, when all linear repair is accurate and all quadratic repair is faulty, is

$$S(t) = e^{-U_0}\left[1+\frac{U_0}{\varepsilon}\left(1-e^{-\lambda t}\right)\right]^{\varepsilon}. \tag{6.46}$$

Now, we need to define the implicit relationships between U and dose, and rate and time. Let's assume that U_0 is proportional to dose,

$$U_0 = \alpha D. \tag{6.47}$$

To couple rate and time, we know that there is a time when the repair process completes (Figure 4-9). Simplistically, we can think of this point as the release of the cell cycle block from G_2 to M. We can justify a duration for the repair process from irradiation to mitosis (due to the rate of repair),

$$T = \left(1-e^{-\lambda t_{max}}\right), \tag{6.48}$$

and we can substitute T into Equation 6.46 for that time,

$$S = e^{-\alpha D}\left[1+\frac{\alpha D}{\varepsilon}T\right]^{\varepsilon}. \tag{6.49}$$

Recall that in Equation 6.48, $e^{-\lambda t} = 1/e^{\lambda t}$, so when λt is large, $e^{-\lambda t} \rightarrow 0$ and $T \rightarrow 1$, yielding the final form of the survival equation for the case where $\phi = 1$ and $\delta = 0$,

$$S = e^{-\alpha D}\left[1+\frac{\alpha D}{\varepsilon}\right]^{\varepsilon}. \tag{6.50}$$

RMR differs from MTSH and LQ in that the rate of production of the initial lesions is linear with dose (implicit in U) and is time dependent (implicit in the rate). The model fits the data well, generally better than MTSH or LQ and modification of the formula during the fitting process can inform investigators regarding the accumulation of damage, the proportions of linear or quadratic damage, the rates of repair of each type of damage and the extent to which damage is repaired. Unfortunately, the beneficial flexibility of the model regrettably diminishes usability. Ultimately, the model never appealed to the scientific community that viewed it as ill-defined, complicated to use, easily misapplied, and saw the results as difficult to interpret. Nonetheless, RMR has the potential to better model cell survival than either MTSH or LQ.

6.2.4.3. *Lethal–Potentially Lethal Model*

In 1986, Curtis published a somewhat familiar-sounding model based on the concept of "potentially lethal damage." In this model, three states are possible: undamaged DNA (state of A), potentially lethal lesions (state of B), and lethal lesions (state of C). The number of potentially lethal lesions and the number of lethal lesions both are linearly related to dose (as is U_0); the constants of proportionality are η_{AB} and η_{AC}, respectively (Figure 6-10). The potentially lethal lesions are either resolved and returned to the undamaged state or the damage is unresolved, and the lesions are converted to the lethal state.

$$A \underset{\varepsilon_{BA}}{\overset{\eta_{AB}}{\rightleftarrows}} B$$

$$B \underset{\varepsilon_{BC}}{\rightarrow} C$$

$$A \underset{\eta_{AC}}{\rightarrow} C$$

Figure 6-10 Transitions between allowed states of the LPL cell survival model. The number of species in the three states are undamaged bonds (**A**), potentially lethal lesions (**B**), and lethal lesions (**C**). The rates of transition are dictated by the constants associated with the directional arrows.

Curtis (1986) began with the following assumptions:

1. The potentially lethal lesions are created in the nucleus of a cell and are repairable by a first order enzymatic reaction. If the lesions are not repaired, they interact to form irreparable lethal lesions by a second order reaction.
2. Lethal lesions may additionally be created on a very short time scale if they are created simultaneously or nearly simultaneously. This set of lesions is described by the yield per unit dose, η_{AC}.
3. The rate of formation of lesions is proportional to the dose rate, $\dot{D}$, and the constant of proportionality for potentially lethal lesions is η_{AB}; the constant of proportionality for the formation of simultaneous lethal lesions is η_{AC}. The rate constant for the restoration of potentially lethal lesions is ε_{BA}. The rate constant for the conversion of potentially lethal lesions to lethal lesions is ε_{BC}. Because two sites are required to interact for this conversion to take place, the reaction is second order.
4. The rate of repair is not dependent on the number of lesions; there is no saturation of the repair process.

The formation of lethal lesions does not depend on the spatial relationship between interacting lesions (i.e., they do not have to be within 10 bp), but rather depends on the number of potentially lethal lesions. Obviously, the more lesions created, the more likely it is that they will be close together. The Lethal–Potentially Lethal (LPL) model is the first to identify *dose rate* as an important parameter. Recall in our earlier discussion, for two DNA ssb to result in a DNA dsb, the breaks must be concurrent. If the first DNA ssb resolves prior to the appearance of the second DNA ssb, formation of a DNA dsb is not possible. Thus, although not addressed in the previous models, the rate at which energy is deposited appears to be fundamentally important.

Like the molecular models (Lea's target theory and MTSH), the LPL model begins by considering the introduction of lesions during irradiation. And as in the Cybernetic Model, LPL considers the kinetics of lesion formation and resolution, rather than the physics of energy deposition and probability statistics. The rate of change of the number of potentially lethal lesions (B) during irradiation can be represented thus:

$$\frac{dB(t)}{dt} = \eta_{AB}\,\dot{D} - \varepsilon_{BA}B(t) - \varepsilon_{BC}B^2(t). \tag{6.51}$$

Reading the equation form left to right, potentially lethal lesions will accumulate proportional to the dose rate ($\eta_{AB}\,\dot{D}$) but will disappear as they are repaired ($\varepsilon_{BA}\,B(t)$) or are converted to lethal lesions ($\varepsilon_{BC}\,B^2(t)$). When we think about kinetics, we can always use our bucket analogy (Figure 5-6). The "bucket" of potentially lethal lesions fills according to dose rate and empties through two holes: one draining into the pool of repaired DNA and one draining into the bucket of lethal lesions. The rate of change of the number of lethal lesions (C) during radiation can be represented thus:

$$\frac{dC(t)}{dt} = \eta_{AC}\dot{D} + \varepsilon_{BC}B^2(t). \tag{6.52}$$

The "simultaneous" lethal lesions occur proportional to the dose rate ($\eta_{AC}\dot{D}$) and the converted B lesions appear at the same rate at which they disappear from the bucket of B lesions in Equation 6.51 ($\varepsilon_{BC}B^2(t)$). This bucket has two sources and no holes. We can integrate these expressions using the boundary condition that initially all DNA is undamaged, that is, $B = C = 0$. Then,

$$B(t) = \frac{2\eta_{AB}\dot{D}\left(1 - e^{-\varepsilon_0 t}\right)}{\varepsilon_0 + \varepsilon_{BA} + (\varepsilon_0 - \varepsilon_{BA})e^{-\varepsilon_0 t}}, \tag{6.53}$$

where ε_0 is $\sqrt{\varepsilon_{BA}^2 + 4\varepsilon_{BC}\eta_{AB}\dot{D}}$, and

$$C(t) = \eta_{AC}\dot{D} + \varepsilon \ln\left[\frac{2\varepsilon_0}{\varepsilon_0 + \varepsilon_{BA} + (\varepsilon_0 - \varepsilon_{BA})e^{-\varepsilon_0 t}}\right] + \frac{(\varepsilon_0 - \varepsilon_{BA})^2 t}{4\varepsilon_{BC}}, \tag{6.54}$$

where ε is the ratio of the constants $\varepsilon_{BA}/\varepsilon_{BC}$. These expressions for the accumulation of lesions during irradiation are not intuitive but it is possible to gain some understanding. Examine the numerator of Equation 6.53. The number of potentially lethal lesions (B) will accrue according to $\eta_{AB}\dot{D}$, just as stated in the assumptions, but that number will be reduced by number of lesions that are repaired or converted to lethal damage ($\eta_{AB}\dot{D}e^{-\varepsilon_0 t}$). The lesions restored or converted to lethal lesions are subtracted from the number of lesions accumulated. This second term is time dependent. What remains in Equation 6.53 is the term $2/((\varepsilon_0 + \varepsilon_{BA}) + (\varepsilon_0 - \varepsilon_{BA})e^{-\varepsilon_0 t})$. Notice that this term also appears in Equation 6.54. Furthermore, the expression within that term ($\varepsilon_0 - \varepsilon_{BA}$) appears several times in the two equations. The constant ε_0 is composed of the rate constant for restitution of potentially lethal lesions (squared because two lesions must interact) plus the rate constant for the conversion of potentially lethal lesions to lethal lesions (which is dose rate dependent). If $4\varepsilon_{BC}(\eta_{AB}\dot{D})$ is very small, ε_0 approximates $\sqrt{\varepsilon_{BA}^2}$ or ε_{BA}. In that case, ($\varepsilon_0 - \varepsilon_{BA}$) is 0. So, the magnitude of $4\varepsilon_{BC}(\eta_{AB}\dot{D})$ is important, and ($\varepsilon_0 - \varepsilon_{BA}$) indicates conversion of potentially lethal to lethal (B to C). This term is always time dependent. Equation 6.54, then, is not so complicated. The number of lethal lesions at time t depends on the direct formation of lethal lesions, $\eta_{AC}\dot{D}$, plus two other terms accounting for the conversion of potentially lethal lesions to lethal lesions. First, we add the ratio of the rate constants for restitution of potentially lethal lesions to conversion of potentially lethal lesions to lethal lesions (ε). Are more lesions becoming lethal or returning to undamaged? Because $\varepsilon = \varepsilon_{BA}/\varepsilon_{BC}$, the faster lesions convert to lethal (B to C), the smaller the second term becomes. The ratio modifies the natural log of ε_0, normalized by the expression examined previously. In other words, the conversions of B to C resulting from the interactions of two independent B sites ($2/((\varepsilon_0 + \varepsilon_{BA}) + (\varepsilon_0 - \varepsilon_{BA})e^{-\varepsilon_0 t})$).

Equations 6.53 and 6.54 describe the modification of states A, B, and C during irradiation, while the number of potentially lethal lesions is increasing due to the ongoing deposition of energy ($\eta_{AB}\dot{D}$). What happens after the radiation is turned off? Clearly, the bucket of potentially lethal lesions will continue to drain, but the input has ceased. The rates of change of B and C (Equations 6.51 and 6.52) look the same, except the first term (creation of A) of each is removed. The rate of change for B is negative, because the number is decreasing (after removing $\eta_{AC}\dot{D}$, multiply Equation 6.48 through by −1). Integration of these modified equations yields:

$$B(T+t_r)=\frac{N_{PL}e^{-\varepsilon_{BA}t_r}}{1+\frac{\left(N_{PL}\left(1-e^{-\varepsilon_{BA}t_r}\right)\right)}{\varepsilon}}, \quad (6.55)$$

and

$$C(T+t_r)=\frac{N_L+[1+N_{PL}/\varepsilon](1-e^{\varepsilon_{BA}t_r})}{1+\frac{\left(N_{PL}\left(1-e^{\varepsilon_{BA}t_r}\right)\right)}{\varepsilon}}-\varepsilon\ln\left[1+\frac{N_{PL}\left(1-e^{\varepsilon_{BA}t_r}\right)}{\varepsilon}\right], \quad (6.56)$$

where T is the duration of irradiation and t_r is the time available for repair; N_{PL} is the number of potentially lethal lesions (B) at secession of radiation (T), and N_L is the number of lethal lesions (C) at secession of radiation. These numbers, N_{PL} and N_L, result from Equations 6.53 and 6.54 at time T. The numerator of Equation 6.55 simply declares that the number of potentially lethal lesions existing at the instant the radiation is turned off (N_{PL}) will *decrease* exponentially as repair continues. The same term appears in the denominator when you multiply through by N_{PL}. Nearly the same term appears several times in Equation 6.56, except that the exponent is positive, so the number *increases* mirroring the previous trend resulting from a negative exponent. As B decreases, A and C increase. However, A and C do not increase equally and are dependent upon their rate constants ε_{BA} and ε_{BC}. Dividing by ε is equivalent to multiplying by $\varepsilon_{BC}/\varepsilon_{BA}$, so each time ε appears in the dividend, the expression adjusts for the fraction of potentially lethal lesions being converted to lethal lesions. Notice that in Equation 6.55, the number of potentially lethal lesions (at T) decreases exponentially with time, but Equation 6.56 indicates that the number of lethal lesions increases additively. The increase in lethal lesions is complicated because of the requirement for lesion interaction, but perhaps it is not critical that these equations be understood in detail. Our goal is to determine cell survival.

To derive the form of the survival equation, Curtis used the Poisson distribution once again to determine the probability that a given cell has no lethal lesions. He also assumed that at time ($T + t_r$), all remaining lesions are lethal because the cell can no longer repair potentially lethal lesions; the cell cycle block has been released. Thus,

$$S=e^{-(N_{tot})(T+t_r)}. \quad (6.57)$$

where N_{tot} at time ($T + t_r$) is $[B(T + t_r) + C(T + t_r)]$. We can substitute Equations 6.55 and 6.56 for N_{tot}. After some algebra we get:

$$S=exp\left(-(N_{PL}+N_L)+\varepsilon\ln\left[1+{N_{PL}\left(1-e^{-\varepsilon_{BA}t_r}\right)}\Big/{\varepsilon}\right]\right)=exp\left(-N_{tot}\left[1+\frac{N_{PL}}{\varepsilon\left(1-e^{-\varepsilon_{BA}t_r}\right)}\right]^{\varepsilon}\right). \quad (6.58)$$

The result is not complicated, but the relationship is not intuitive, either. Curtis (1986) derived an expression for relatively high dose rates using a power series expansion of the logarithmic form of the survival equation ($-\ln S = (N_{tot})(T + t_r)$). That expression yielded a linear-quadratic formalism wherein

$$\alpha=\eta_{AC}+\eta_{AB}e^{-\varepsilon_{BA}t_r} \quad (6.59)$$

And

$$\beta = \frac{\eta_{AB}^{2}}{2\varepsilon}\left(1-e^{-\varepsilon_{BA}t_r}\right)^2. \tag{6.60}$$

Curtis tested his expression for survival using previously published data and found that the values obtained for α and β agreed with those published using the LQ method. The LPL model fits the data slightly better at survival fractions less than about 10^{-3}. Nonetheless, the model considered unrepaired and misrepaired lesions as lethal lesions at t_r. As discussed in Chapter 5, these lesions are more likely involved in genomic instability leading to cell transformation rather than resulting in cell death. Furthermore, the rate of repair is assumed to be constant in both LPL and RMR models, but DNA dsb repair kinetics indicate a continuous spectrum of repair capabilities (Figure 4-9).

Still, the concept of potentially lethal lesions proved advantageous, and we will see it again. The quadratic component of the LQ model was envisioned in Chapter 4 as DNA ssb forming cooperative lesions, a two-step process. This concept required that the DNA ssb be within 10 base pairs, but as we saw, the LPL model violated that precept without consequence. Furthermore, in the LQ model, most DNA ssb were hypothesized to be formed through the *indirect* process: energy deposition in water leading to free radicals that in turn damage the DNA. Lethal lesions ("simultaneous lesions" of the LPL) were envisioned to be *direct* damage. The concept of potentially lethal damage conveniently provides a "black box" allowing the application of various models to describe cell survival even as biophysicists continue to develop an understanding of the nature of the lesions. If we identify two categories of lesions, one that behaves linearly (lethal lesions) and one that behaves quadratically (potentially lethal lesions), we do not have to define exactly what form these lesions take. Clearly, the two types of DNA damage have something to do with strand breaks; nonetheless, DNA ssb and DNA dsb appear to be necessary but not sufficient to result in cell death.

6.2.4.4. *Threshold Energy Repair Saturation Model*

In 1985, Dudley Goodhead proposed a new model based on the assumptions that the efficiency of repair decreases with increasing dose, and that this decrease in repair efficiency is caused by DNA repair kinetics saturation. The TERS model does not require "sublesions" like the RMR and LPL models. We won't derive the survival equation here, as the model never gained much favor. The survival relationship looks like this:

$$S = exp - \left(\frac{n_0 - c_0}{1 - \left({c_0}/{n_0} \right) e^{kt(c_0 - n_0)}} \right), \tag{6.61}$$

where n_0 is the initial number of unrepaired lesions, c_0 is the initial number of repair enzyme molecules, k is a constant of proportionality and t, of course, is time. The basic tenants of the model are inconsistent with empirical evidence. Dozens of DNA dsb may be created per Gy, while the concentration of repair enzymes ranges from 10^4 to 10^8 molecules per cell. The reactions are unlikely to saturate.

6.2.4.5. *Giant Loop Binary Lesion Model*

Following WWII, from around 1945 to around 1985, a bolus of experimental techniques and theoretical models were published by newly converted biophysicists and their proteges. Hypotheses advanced from

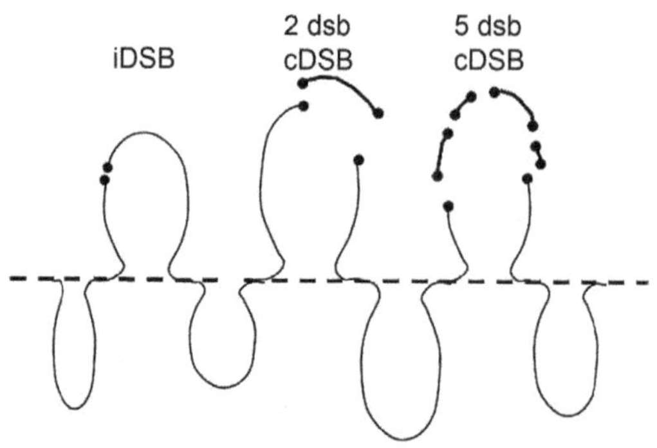

Figure 6-11 Cartoon depiction of chromatin breaks. Chromatin forms loops anchored to the nuclear matrix (dashed line). The dsbs sever the chromatin. These breaks may occur as a single event in a loop (iDSB) or several dsb may occur in a single loop (cDSB). Sections of chromatin are dissociated from the nuclear matrix when two or more dsbs occur within a single loop. Reproduced and modified with permission of the Radiation Research Society from Friedrich *et al.* (2012).

molecular models to biologically based models to models motivated by DNA repair kinetics. Several minor models have been proposed since the mid-1980's, models that tinkered around the edges of modern radiation biology but failed to produce substantially better data fitting than the LQ model. Then, in 2013, Friedrich (Friedrich *et al.*, 2013) proposed a model based on megabase-pair-length chromatin loop structures (see Figure 3-5). They asserted that two classes of damage exist: a single dsb within a single loop, or multiple DNA dsbs within a single loop (Figure 6-11). They further required that each DNA dsb occurring within different loop domains is repaired independently. The repair fidelities associated with the two damage classes are presumed to be different. As before, we begin with Friedrich's assumptions.

1. Cell inactivation is represented by the mean number of lethal events for a large population of cells.
2. The number of lethal events is Poisson distributed.
3. The survival probability can be obtained by $S(D) = e^{-\varepsilon(D)}$, where ε is the number of lethal events.
4. Chromatin domains of approximately 2 Mbp composing a *loop*, determine the cellular radiation response.
5. A single DNA dsb within in a loop is called an *isolated dsb* (iDSB). The DNA dsb leads to a break in the chromatin, the rejoining of which is facilitated by access from both sides of the DNA dsb and anchorage of both ends to the nuclear matrix.
6. Two or more DNA dsb within a single loop are referred to as a *clustered dsb* (cDSB). Excised fragments are no longer connected to the nuclear matrix and can easily diffuse away.
7. Lesions in different chromatin loops are processed independently.

Then, we let the probability an iDSB resolving to a lethal lesion be ε_i and the probability a cDSB resolving to a lethal lesion be ε_c. Friedrich claimed that all dsb, iDSBs and cDSBs, can be treated in the same way because ε_i and ε_c represent averages of lethal events. And, because iDSBs are more likely to be repaired than cDSBs, $\varepsilon_c >> \varepsilon_i$. The survival probability can then be written,

$$S = e^{-(n_i\varepsilon_i + n_c\varepsilon_c)} \tag{6.62}$$

where n_i is the average number of chromatin loops containing an iDSB and n_c is the average number of loops containing cDSB. If the genome size is S_G and the average loop size is S_L, then the genome contains

N_L loops, $N_L = S_G/S_L$. Because of assumption 2, we again use Poisson statistics to derive the number of loops containing no dsb (n_0), one dsb (i.e., iDBS; n_i), and two or more dsbs (i.e., cDSB; n_c):

$$n_0 = N_L e^{-Y_{Loop}D}, \tag{6.63}$$

$$n_i = N_L (Y_{Loop} D) e^{-Y_{Loop}D}, \tag{6.64}$$

and

$$n_c = N_L - n_0 - n_i; \tag{6.65}$$

where Y_{Loop} is the dsb induction frequency per Gy, per loop. In Chapter 4, we indicated the induction frequency was approximately 40 dsb per Gy in each cell nucleus; this value, α_{DSB}, varies with cell type. To get the number of dsb per loop, $Y_{Loop} = \alpha_{\mathrm{DSB}}/N_L$. The expressions n_0, n_i, and n_c can be substituted into Equation 6.62 to arrive at the final expression for survival.

$$S = \exp-\left[\left(N_L\left(Y_{Loop}D\right)e^{-Y_{Loop}D}\right)\varepsilon_i + \left(N_L - n_0 - n_i\right)\varepsilon_c\right]. \tag{6.66}$$

Once again, n is based on average values and the presumption is made that the number of lethal events, as well as the numbers of iDSB and cDSB, are Poisson distributed. In fact, the effect of a dsb depends on circumstance. If a dsb occurs within a single copy gene, the likelihood that it will be lethal is greater than if it occurs within a nontranscribed sequence. Likewise, the lethality of a cDSB will depend on whether it results in a complex rearrangement or to a deletion. So, is this approach justified? Does the Giant Loop Binary Lesion (GLOBLE) formalization fit empirical data as well as the LQ formalism? Using assumption 3 for the case of LQ (Equation 6.23).

$$\varepsilon = \alpha D + \beta D^2. \tag{6.67}$$

Using Equation 6.62 for the case of GLOBLE,

$$\varepsilon = n_i\varepsilon_i + n_c\varepsilon_c, \tag{6.68}$$

where n_i and n_c are nonlinear functions of dose (D). Comparing Equation 6.26 to Equation 6.67 discloses that $\varepsilon = -\ln S$. Therefore, we can define α and β by finding the limits of the first and second derivatives with respect to dose:

$$\alpha = \lim_{D\to 0}\frac{d(-\ln S)}{dD}, \tag{6.69}$$

$$\beta = \frac{1}{2}\lim_{D\to 0}\frac{d^2(-\ln S)}{dD^2}, \tag{6.70}$$

And applying Equations 6.69 and 6.70 to arrive at expressions that relate the LQ α and β to the GLOBLE ε_i and ε_c:

$$\varepsilon_i = \frac{\alpha}{\alpha_{\mathrm{DSB}}} = \frac{1}{\alpha_{\mathrm{DSB}}}\alpha, \tag{6.71}$$

$$\varepsilon_c = 2\left(\frac{N_L\beta + \alpha_{\text{DSB}}\alpha}{\alpha_{\text{DSB}}^2}\right) = \frac{2N_L}{\alpha_{\text{DSB}}^2}\beta + \frac{2}{\alpha_{\text{DSB}}}\alpha. \tag{6.72}$$

Keep in mind that α_{DSB} is simply the number of DNA dsb per Gy per cell (40 by our estimation). Thus ε_i is proportional to α, so you might expect that the low dose part of the survival curve fits comparably to LQ. Equation 6.72 indicates that ε_c likewise has a component proportional to β, but a multiple of the anticorrelated α value is added. This has the effect of lessening the slope at higher doses, in agreement with empirical data. Because all cDSBs are considered equivalently lethal, regardless of the number of dsb greater than two, the process is maximally efficient at two dsb per loop. Additional lethal DNA dsb do not increase cell death. GLOBLE addresses only DNA dsb; there is no consideration of ssb. However, in this model, it is possible for two damaged loops to interact and increase lethality, either iDSB or cDSB. GLOBLE asserts no proximity requirement, as there is in the LQ model for two ssbs to combine to form a DNA dsb. Distant dsb may be brought into proximity through chromatin compaction. Because ssbs are ephemeral, a model that addresses only DNA dsb is inherently appealing. Furthermore, distant interactions on the order of micrometers is consistent with investigational chromosomal aberration data. Kinetics data also support this model. Figure 4-10 indicated a slow and a fast component for repair, which we interpreted as ssb and dsb, but this behavior could just as easily reflect iDBS and cDSB repair. We also know that the survival curve indicates two processes ("linear" and "quadratic"). The identification of two classes, iDSB and cDSB, also responds to these empirical results.

A drawback of the GLOBLE model is that it relies heavily on empirical input parameters. The initial yield of dsb (α_{DSB}), the size of chromatin loops (S_L), and the number of loops (N_L) can be fairly accurately derived in the case of cultured cell lines, but accurate estimations for *in situ* tissue and tumors in particular, are difficult. Also, GLOBLE assumes that all DNA dsb result in a chromatin break (Figure 6-11). However, we know that the structure of chromatin includes nucleosomes and supercoiling (refer to Figure 3-5) that increase the integrity of the chromatin strand. It seems that multiple DNA dsb within the chromatin fiber would be required to release nucleosomes and coiling tension to result in a subsequent chromatin break. Nonetheless, if we drop the assumption that a DNA dsb produces a loop break and instead state that there exists a chromosome break, irrespective of the mechanism of that break, we have not substantially altered the model's assumptions. The model does not distinguish between nonhomologous end-joining (NHEJ) and homologous recombination (HR), which differ in repair fidelity. This could be addressed by attributing different average numbers of lethal events to iDSBs processed by NHEJ and HR, respectively. In conclusion, the application of GLOBLE shows promise, particularly at very low survival rates, and certainly the premises are more consistent with the recent literature.

6.3. Summary

A sampling of cell survival, aka dose–response models, have been presented herein. The sampling is most definitely not exhaustive and is intended to give the student a grasp of the processes typical in the evolution of a model that is adequate to predict the biological response. Model development follows a standard protocol: assumptions are set forth, a mathematical framework is developed, the model is tested against experimental results. Contemporary understanding of the biophysical, biochemical, and physiological radiation response influenced the assumptions upon which cell survival models have been based. All things considered, the LQ model dominates modern clinical application (Brenner, 2008). LQ survives by supplanting the original assumptions involving ssb and dsb with the less restrictive descriptions of potentially lethal damage and lethal damage. These terms encompass historical damage descriptions from Lea to Friedrich. The two state classifications of potentially lethal damage and lethal damage will likely suffice as descriptors into the foreseeable future.

- Models facilitate the prediction of events based on historical behavior.
- Most cultured cells are derived from cell lines and the lineage of most of those are cancers.
- Most cell lines are immortal, which means that their cell cycle control and life span mechanisms are not normal.
- Fibroblasts are more easily cultured than epithelial cells, however, most cancers are epithelial in origin.
- Cell survival (clonogenic survival) is defined as the ability of a cell to divide and produce a visible colony (~50 cells for most mammalian cell lines).
- Lea's target theory (1947) relied on interaction probabilities and defined a target volume as well as a "hit," which represented a potentially lethal event.
- Target theory resulted in a single hit model (i.e., a single hit was lethal) that adequately fit data for irradiated bacteria and haploid yeast.
- MTSH expanded on the single hit model to produce a model that by-and-large fit data generated from irradiated diploid cells including mammalian cells.
- The MTSH model, $S(q, n, D) = 1 - (1 - e^{-qD})^n$, must be modified to account for reduced D_0 at high doses: $S(q, q', n, D) = e^{-q'D}\,[1 - (1 - e^{-qD})^n]$.
- The mean lethal dose, D_0, is the amount of additional radiation dose required to reduce the number of surviving cells 1/e (37%). The alternative D_{10} is often used clinically; it is the amount of additional radiation dose required to reduce the number of surviving cells one $\log_{10}$ (e.g., 0.01 fractional survival to 0.001 fractional survival).
- Important parameters describing survival characteristics can be derived graphically from the MTSH curve: D_q, D_0, and n.
- MTSH interprets the breadth of the cell survival curve shoulder as indicative of cell type repair capacity.
- The LQ Model (1973), $S = e^{-(\alpha D + \beta D^2)}$, utilized newly disclosed molecular biology to formulate the basic model assumptions.
- Radiation sensitivity is described by the ratio α/β according to the LQ formalism. The variable α describes linear damage (single hit lethality). The variable β describes quadratic damage (dual hit lethality).
- The LQ variables can be determined graphically.
- LQ model remains the favored model in the clinic today.
- The Cybernetic model formalized damage repair kinetics using the two damage induction mechanisms: direct and indirect.
- The RMR model follows the rate of repair of molecular damage, but also allows for misrepair resulting in altered phenotype. The repair is assumed to be either linear or quadratic in nature.
- According to RMR model, the rate of production of initial lesions is linear with dose and is time dependent.
- The LPL model (1986) refined the Cybernetic model, formally recognizing the dependence of lesion formation on dose rate and the independence of lesion processing (to repaired or irreparable states) on dose rate.
- The LPL model introduced the concepts "potentially lethal damage" and "lethal damage."
- Graphical solutions provide rate transition constants and the number of each type of lesion.
- The foundation for the GLOBLE model (2013) reflects contemporary interpretation of the genetic damage/repair processes resulting in clonogenic survival, cell death, and mutation.

- The GLOBLE model assumes that two types of chromosomal damage determine cell survival: single chromosomal breaks and multiple chromosomal breaks within DNA loops attached to chromatin; and proposes that the survival probability can be obtained from the transitional rate constants, $S = e^{-(n_i\varepsilon_i + n_c\varepsilon_c)}$.
- Graphical solutions to GLOBLE yield the number of single chromosomal breaks (iDSB) and the number of multiple chromosomal breaks (cDSB) as a function of dose.
- Survival curves can be fitted by the GLOBLE formalization $S = \exp - \left[\left(N_L \left(Y_{Loop} D \right) e^{-Y_{Loop} D} \right) \varepsilon_i + \left(N_L - n_0 - n_i \right) \varepsilon_c \right]$ if S_G and S_L are estimated, yielding ε_i and ε_c.

6.4. Problems

The following data are provided for the clonal survival of a hamster cell line grown *in vitro* immediately after irradiation. Use these data for Questions 1–5.

Dose (Gy)	Surviving fraction
0	1.0
2	0.8
4	0.5
6	0.22
8	0.10
10	0.04
12	0.016
14	0.005

1. Plot the data by hand on four cycle semi-logarithmic paper. Graphically determine the MTSH parameters D_0, D_Q, and n.
2. From your solution to Question 1, determine n and q. Substitute for n and q in the MTSH equation for survival (Equation 6.10). Find the survival fraction resulting from 3 Gy, 5 Gy, 7 Gy, and 9 Gy doses. Plot these points on the hand-drawn graph produced for Question 1.
3. Use any data plotting software package (e.g., Excel) to plot Surviving Fraction vs. Dose from the data provided. Estimate the linear quadratic value of α/β.
4. Use any data plotting software package (e.g., Excel "Solver") to fit the plot of Question 3 using the linear quadratic equation (Equation 6.23). Find the values of α, β, and α/β.
5. Plot the data on a graph of $-\ln S/D$ *vs* D. From the plot, what is α, β and α/β?
6. Compare your solutions for Questions 3, 4, and 5. Why might the solutions differ?
7. According to the Cybernetic model, *at low dose rates* the relative fraction of dividing cells (fractional survival) will decrease with time as irreparable damage accumulates (Equation 6.35); that is, all reparable damage is repaired (ssb do not interact; all state B cells return to state A cells). Use the following data from a fictitious experiment conducted at 0.1 Gy/min. Plot the data on a semi-logarithmic graph using any software package and determine the rate constant for the accumulation of cells in state C from the slope of the plot. Provide appropriate units (always!).

N_A/N_{A0}	Time (min)
0.25	0.1
0.06	0.2
0.016	0.3
0.001	0.5

8. Plot the survival data (dose–response curve) described by Equation 6.50 if $\alpha = 0.1$ and $\lambda/\kappa = 0.7$. What attributes of the formula provide for the shape?
9. Evaluate the anticipated survival fraction at a dose of 20 Gy, according to the GLOBLE model (Equation 6.66). Assume the size of the genome is 3.235×10^9 bases and the average size of a loop is 3.7×10^5 bases. Let $\alpha = 0.1$ and $\beta = 0.02$.

6.5. Bibliography

Alpen, E. L., 1984. Theories and models for cell survival. In: *Radiation Biophysics*. San Diego: Academic Press, pp. 132–168.

Alpen, E. L., 1990. *Radiation Biophysics*. 2nd ed. San Diego: Academic Press.

Bodgi, L. *et al.*, 2016. Mathematical models of radiation action on living cells: From the target theory to the modern approaches. A historical and critical review. *Journal of Theoretical Biology*, 394, pp. 93–101.

Brenner, D. J., 2008. The linear-quadratic model is an appropriate methodology for determining isoeffective doses at large doses per fraction. *Seminars in Radiation Oncology*, 18(4), pp. 234–239.

Chadwick, K. H. & Leenhouts, H. P., 1973. A molecular theory of cell survival. *Physics in Medicine and Biology*, 18, pp. 78–87.

Curtis, S. B., 1986. Lethal and potentially lethal lesions induced by radiation — a unified repair model. *Radiation Research*, 106(2), pp. 252–270.

Douglas, B. G. & Fowler, J. F., 1976. The effect of multiple small doses of X rays on skin reactions in the mouse and a basic interpretation. *Radiation Research*, 66, pp. 401–426.

Friedrich, T., Durante, M. & Scholz, M., 2013. Modeling cell survival after photon irradiation based on double-strand break clustering in megabase pair chromatin loops. *Radiation Research*, 178, pp. 385–394.

Hall, E. J. *et al.*, 1972. Survival curves and age response functions for chinese hamster cells exposed to X-rays or high LET alpha-particles. *Radiation Research*, 52(1), pp. 88–98.

Joiner, M. C., 2009. Models of radiation cell killing. In: M. C. Joiner & A. van der Kogel, eds. *Basic Clinical Radiobiology*. London: Hodder Arnold, pp. 41–55.

Kadauke, S. & Blobel, G. A., 2009. Chromatin loops in gene regulation. *Biochimica et Biophysica Acta*, 1789(1), pp. 17–25.

Kappos, A. & Pohlit, W., 1972. A cybernetic model for radiation reactions in living cells. I. Sparsely-ionizing radiations; stationary cells. *International Journal of Radiation Biology and Related Studies in Physics, Chemistry, and Medicine*, 22, pp. 51–65.

Kellerer, A. M. & Rossi, H. H., 1978. A generalized formulation of dual radiation action. *Radiation Research*, 75(3), pp. 471–488.

Lane, D. M. & Osherson, D., n.d. *Online Statistics Education: An Interactive Multimedia Course of Study*. [Online] Available at: http://onlinestatbook.com/2/probability/probability.html [Accessed 2 October 2019].

Lea, D. E., 1947. *The Actions of Radiation of Living Cells*. London: Cambridge University Press.

Puck, T. T. & Marcus, P. I., 1955. A rapid method for viable cell titration and clone production with HeLa cells in tissue culture: The use of x-irradiated cells to supply conditioning factors. *Proceedings of the National Acadamy of Sciences*, 41(7), pp. 432–437.

Tobias, C. A., Blakeley, E. A., Ngo, F. Q. H. & Yang, T. C. H., 1980. The repair-misrepair model of cell survival. In: R. E. Meyn & H. R. Withers, eds. *Radiation Biology and Cancer Research*. New York: Raven Press, pp. 195–230.

Chapter 7

Applications of Dose Response Models

7.1. Introduction

In Chapter 6, we explored the development of several dose response models. Our next task will be to see how well these models predict biological response; in other words, we will test the models. Prior to 1950, Lea's model successfully tested against haploid bacteria and yeast, and so seemed viable. Unfortunately, a few pesky diploid yeast species required modification of the formulation to account for the sigmoid shape of their survival curve, so the universality of the model was always in doubt. Modern cell cultures include haploid single-celled species and diploid single-celled species as well as mammalian immortal cells including fibroblasts and epithelial cells, and several primary cell cultures. Various types of radiation can be delivered through an entire array of schema. If valid, a model should fit the data acquired from any experimental protocol using any cell type or radiation type and the assumptions fundamental to the model should extrapolate to hypotheses regarding the biology dictating cell survival under the protocol conditions. In this chapter, we will consider the effects of target size, genetic lineage, biochemical competence, cell cycle phase, apoptosis, dose rate, and radiation type on dose response, and we will consider the implications of various model assumptions. Finally, we will discuss the manipulation of survival curve characteristics by external influences.

7.2. Interpretation of the Dose Response Curve

Practical considerations when performing dose response experiments include a less than perfect plating efficiency (PE). Not all the cells deposited into a petri dish will attach, thrive, and divide. For this reason, a PE factor must be derived and incorporated into the calculation of survival. Suppose a researcher plates 100 cells in a few milliliters of medium into a petri dish or tissue culture flask. The cells will settle and attach to the bottom and begin dividing. If the researcher observes the dish or flask after several days of incubation, he or she will observe between 20 and 90 small colonies derived from the plated cells, not 100. The efficiency of this colony formation process is called PE and is calculated by dividing the average number of colonies counted by the number of cells originally plated in the flasks, that is, PE = [#colonies counted]/[#cells plated]. To calculate the survival fraction (S) for a particular dose of radiation, the number of colonies counted is divided by the number of cells plated, corrected for PE:

$$\frac{S}{S_0} = \mathrm{S} = \frac{\textit{colonies counted}}{(\textit{cells plated})\mathrm{PE}}. \tag{7.1}$$

This is important in a radiation dose response experiment because to determine the S we need to calculate the fraction or percentage of cells that survive following a dose of radiation compared with the number of cells that survive when not exposed to radiation.

Now, let's examine each of the models presented in Chapter 6 to see if they (1) agree with experimental data and (2) suggest reasonable justifications for dose response behavior. As depicted in Figure 7-1A, a

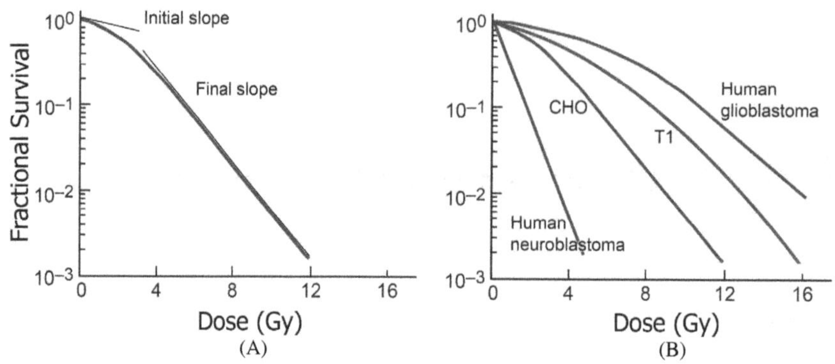

Figure 7-1 Typical *in vitro* survival curves. **(A)** The curve derived from a dose response experiment using CHO cells represents the "ideal" shape that may be explained by each of the models for diploid cells. The region between the tangents to initial slope and the final slope is called the shoulder. **(B)** Dose response curves for several different cell lines. A typical curve for human glioblastoma cell lines displays an initial slope and a final slope, but the shoulder persists to relatively high doses (more than 10 Gy, here). The shoulder encompasses the entire dose range for T1 lymphocyte cells, and human neuroblastoma cells exhibit no shoulder whatsoever.

typical semilogarithmic plot of dose response commences (at radiation doses approaching 0) with a small negative slope. Because the multi-target single hit (MTSH) formalism implies that the initial slope equals 0, MTSH struggles to interpret this portion of the plot. Contemporary advocates for the model have described an initial slope of "D_1" that represents single event cell killing. This interpretation invokes more recent models, the linear quadratic (LQ) model in particular. At increasing doses, a typical plot exhibits a curving shoulder the width of which has been the subject of much interpretive conjecture. According to the MTSH model, the width of the shoulder is determined by the number of targets. Recall that n (the target multiplicity) is derived by the width of the shoulder and the slope of the descending portion of the survival curve (Figure 6-6A). That relationship (Equation 6.11) is repeated here:

$$D_q = D_0 \ln n. \tag{7.2}$$

Also, recall that according to Lea, "the inactivation of n targets leads to loss of reproductive capacity". What does "inactivation of a target" mean? As our understanding of macromolecular biochemistry has evolved, the definition of target inactivation has progressed from a single DNA strand break (ssb), to a double DNA strand break (dsb), to a chromosomal break, and most recently to a DNA break that leads to a giant loop scission. The term "potentially lethal damage (PLD)" conveniently encompasses all these manifestations of target inactivation, with inactivation of n targets resulting in "lethal damage." Then what would "target multiplicity" imply? If a DNA ssb represents target inactivation, then the target must be any bond available for breakage. If DNA loop scission represents target inactivation, then the target must be a section of loop. Clearly, regardless of description, target multiplicity does not vary greatly among mammalian cell lines. So, how would one explain the difference in shoulder extent between the Chinese Hamster Ovary (CHO) and glioblastoma plots in Figure 7-1B? MTSH does not seem to provide a satisfactory interpretation of the shoulder. The final slope denoted in Figure 7-1A results from an exponential function similar to that seen in haploid cells reflecting the probability that a cell will receive less than the critical number of hits (too few hits to result in cell death). Thus, the slope depends on more than one event. The slope of this portion of the curve is defined by the MTSH parameters D_0 or D_{10}, the additional dose required to reduce survival $1/e$ or the additional dose required to reduce survival one $\log_{10}$, respectively. The final slope, according to MTSH, is the dose required to deliver one inactivation event per cell, on average.

According to the LQ model, the initial slope of a survival curve (at doses approaching 0 Gy) is linear; it is an exponential function characterized by the proportionality constant α. Recall that the LQ model assumed that a DNA dsb represented theoretically lethal damage, and that α incorporated: the number of sites that can sustain a DNA dsb (n_0), the DNA dsb hit probability constant (k_0), the number of unrestored DNA dsb (f_0), the proportion of DNA dsb that are potentially lethal (p), and the fraction of dose that induces lethal DNA dsb through a single event (Δ). The first two terms are constants for any given cell type (the number of sites that can sustain a DNA dsb and the DNA dsb hit probability) while f_0 and p (the number of unrestored DNA dsb and the proportion of DNA dsb that become lethal) reflect the capacity of the cell type to accurately repair damage. The final term, Δ, is a function of the radiation quality (track structure). The initial slope represents single events leading to lethal damage. The final slope is an exponential function characterized by the proportionality constant β. The constant β depends on the same or similar factors as α, but two interdependent events are required to produce a single PLD site, resulting in a quadratic relationship. Thus, the shoulder is described by the dose at which $\alpha D = \beta D^2$, or the quantity α/β. It is the ratio of the constants for single hit lethality to multiple (two) hit potential lethality. To clarify, "lethal damage" does not indicate cell inactivation; it indicates a type of damage that may possibly inactivate a cell without requiring a second event (sublethal) or misrepair (repair that converts potentially lethal to lethal).

The repair–misrepair (RMR) expression for fractional survival, Equation 6.50, is valid under the assumption that the initial number of lesions is proportional to dose ($U_0 = \alpha D$) and provided that the repair process has completed (i.e., sufficient time has passed to reach a state where all lesions are either repaired or misrepaired), leading to

$$S = e^{-\alpha D}\left[1+\frac{\alpha}{\varepsilon}D\right]^{\varepsilon}. \tag{7.3}$$

In this expression, ε represents the ratio of the rate constants for the two possible repair outcomes (λ/k). Notice that when the dose approaches 0, the second term of Equation 7.3 approaches 1. Tobias' RMR model therefore proposes an initial slope dominated by $e^{-\alpha D}$, identical to LQ. Because this formula was derived assuming that all single event damage is accurately repaired and that all interdependent damage fails to repair, it would be illuminating to examine the behavior of the model under these conditions. The rate constant for accurate repair (λ) is approximately 1.6×10^{-5} hr^{-1} and the rate constant for misrepair (k) is approximately 0.15 hr^{-1} (Brinkman *et al.*, 2018), yielding $\varepsilon = 10^{-4}$. Substituting for ε, it becomes apparent that Equation 7.3 indicates an underlying linear exponential plot ($e^{-\alpha D}$) influenced by a slowly increasing factor ($[1+\frac{\alpha}{\varepsilon}D]^{\varepsilon}$). Because the previous conditions are unrealistically restrictive, one might ask, "what happens if some percentage of the interdependent damage is accurately repaired?" In an experiment where strand breaks were artificially introduced using the endonuclease Cas9, a ratio of $\varepsilon = \lambda/(x\lambda + yk) = 0.28$ fit the empirical data (Brinkman *et al.*, 2018). In this case, the second term of Equation 7.3 is greater than 1.0 and increases more rapidly than in the first case. Figure 7-2 compares dose response plots assuming these two conditions. In the first plot (A), because the second term of Equation 7.3 approximates 1.00 from 0 Gy to 20 Gy, the response resembles the situation Lea observed with haploid cells. The exponential term of Equation 7.3 dominates resulting in plots resembling those in Figure 6-3, modeled by the single hit Equation 6.8. This stands to reason. If all single hit damage is accurately repaired, only interdependent damage results in cell death; only one type of damage is expressed. The lower curve of the inset in Figure 7-2A demonstrates that this response is nearly linear on a semilogarithmic plot. On the other hand, if we allow some of the interdependent damage to be accurately repaired, the plot begins to look more like the

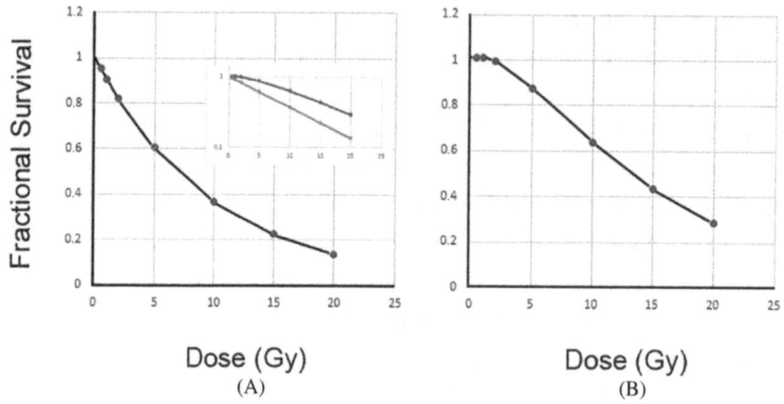

Figure 7-2 The effect of repair rate on the survival curve shape. (**A**) Plot of Equation 7.3 when $\Phi = 1$ and $\delta = 0$, using $\lambda = 1.6 \times 10^{-5}$ hr^{-1} and $k = 0.15$ hr^{-1} as the rate constants for perfect repair and misrepair, respectively. (**B**) Plot of Equation 7.3 when $0 < \delta < 1$ using the empirical value $\varepsilon = 2.8$. The inset in **A** shows a comparison of the semilogarithmic plots.

response curves demonstrated in diploid cells (Figure 6-5) as exemplified in Figure 7-2B and the upper curve of the inset in Figure 7-2A. The comparison of plots resulting from these two repair parameters clearly displays that the ratio of repair rates (ε) determines the shape of the shoulder of the survival curve.

Because the Cybernetic and lethal-potentially lethal (LPL) models share a fundamental foundation, let's examine them together. Curtis derived a linear-quadratic expression for S wherein α and β are based on transition rates between damage states (undamaged, potentially lethal, and lethal). The constants of proportionality in this case are provided by Equations 6.59 and 6.60:

$$\alpha = \eta_{AC} + \eta_{AB} e^{-\varepsilon_{BA} t_r} \tag{7.4}$$

and

$$\beta = \frac{\eta_{AB}^{2}}{2\varepsilon}\left(1 - e^{-\varepsilon_{BA} t_r}\right)^2. \tag{7.5}$$

Rather than depending on p, f_0, n_0, k_0, and Δ, in this model α is determined by the conversion of undamaged sites through single event damage resulting in lethal lesions and the conversion of undamaged sites by multiple event damage leading to potentially lethal lesions (A $\rightarrow$ B and A $\rightarrow$ C). The loss is reduced by repair of PLD (B $\rightarrow$ A). And, rather than depending on $p, f_0, E, n_1, n_2, f_1, f_2, k^2$, and $(1 - \Delta)^2$, β is determined by the reduction in undamaged sites through conversion to potentially lethal lesions via interdependent damage mechanisms, countered by restoration through repair (which in turn is reduced by the conversion of potentially lethal lesions to lethal lesions). Notice that the difference between the LQ and LPL derivations for α and β is basically philosophical. The former considers formation of DNA ssb and DNA dsb; the latter considers formation of potentially lethal lesions and lethal lesions. The LQ model incorporates the fraction of "unrestored" bonds while the LPL model accounts for repair of potentially lethal lesions. Thus, it is not surprising that the shoulder of the survival curve, as modeled by LPL, is a manifestation of the ratio of lesions resulting from direct lethal damage plus lethal lesions resulting from PLD (including unresolved lesions) to lethal lesions resulting from PLD acting in quadrature (or those that remain unresolved).

Lastly, GLOBLE proposes that S is a function of $e^{-\varepsilon}$, where ε depends on the number of lethal events (assumption 3). Model development disclosed that,

$$\varepsilon = \varepsilon_i \eta_i + \varepsilon_c \eta_c, \tag{7.6}$$

where ε_i is the probability that isolated loop scission results in lethal damage and ε_c is the probability that clustered loop scission results in lethal damage; η_i and η_c are the average numbers of loops presenting isolated breaks or clustered breaks, respectively and are dose dependent. As illustrated for LQ, where $\varepsilon = \alpha D + \beta D^2$, there exists a dose where $\varepsilon_i\eta_i = \varepsilon_c\eta_c$, and that dose describes the shoulder of the survival curve. Thus, the shoulder is described by the probability that an isolated break results in lethal damage compared with the probability that multiple breaks result in lethal damage.

In each of these models, the width of the shoulder is determined, in part, by the mechanism that describes the log linear portion of the cell survival curve. According to MTSH, the more available targets a cell line presents, the more sensitive the cells are to radiation (more cells die at lower doses). The log linear slope is described by D_0, which is dependent upon n (by rearrangement of Equation 7.2, $D_0 = D_q/\ln n$). LQ claims that the frequency of indirect damage reflects the quadratic response, that is, the tendency for DNA ssb to combine to form DNA dsb determines the survivability of cells at increasing doses (straight line portion of the curve). RMR proposes that DNA quadratic repair capacity drives the final slope: the faster and more reliably uncommitted damages are repaired through the quadratic mechanism, the more radiation resistant a cell line appears. LPL and cybernetic models examine the pool of lethal damages that are converted from PLD. This is calculated in quadrature, just as β is when using the LQ formalism. In fact, similarities between LQ and LPL should be apparent. GLOBLE predicts the final slope based on the appearance of clustered loop lesions and the probability that these lesions result in cell death.

We have not yet considered the problematic curves of Figure 7-1B: the dose dependent responses of human neuroblastoma and T1 cells. Let's consider the neuroblastoma cell line first. Following exposure to increasing doses of x-ray radiation, this cell line exhibits no shoulder when the S data are plotted. In fact, the decrease in survival is a linear exponential that looks a lot like haploid cells. What could cause this response? Well, Lea would say this is a function of hit survivability: a single hit is sufficient to kill a cell. MTSH would propose that $n = 1$; that is, there is inactivation through a single target. The LQ, RMR, LPL, and GLOBLE models interpret the linear exponential by assuming that the final curve extrapolation of Figure 7-1A *is* the curve. In the latter four cases, there is no shoulder; whatever attribute the shoulder represents must be absent from this cell line. The continuous curvature presented by T1 cells cannot be fitted by the MTSH or LQ/LPL models, and therefore these models fail. The T1 data is better fit by the RMR and GLOBLE models that more closely reproduce the survival curve shoulder, but the models also fail significantly at higher doses. Whereas the biological activity represented by the shoulder is absent in the neuroblastoma cell line, it dominates the T1 line.

The multiplicity of proposed biological rationale for the dose response shoulder must be resolved by testing the hypotheses empirically, as is always the case. The success of any model depends on how well it holds up when applied to experimental results. However, as often is the case, development of a definitive protocol design requires a certain degree of cleverness. Mortimer Elkind and Harriet Sutton designed a particularly ingenious experiment (Elkind & Sutton, 1959) to examine the shoulder of the survival curve. Elkind and Sutton irradiated two cell lines (either CHO or V79 to establish the generality of the effect) following the normal procedure of dose escalation, incubation, and colony counting (Elkind *et al.*, 1961). First, they determined the dose required to reduce S to approximately 0.1 (10% of the unirradiated survival). Then they repeated the experiment, but they greatly increased the number of plates irradiated and divided them into several groups. The cells were irradiated as before up to the dose determined to

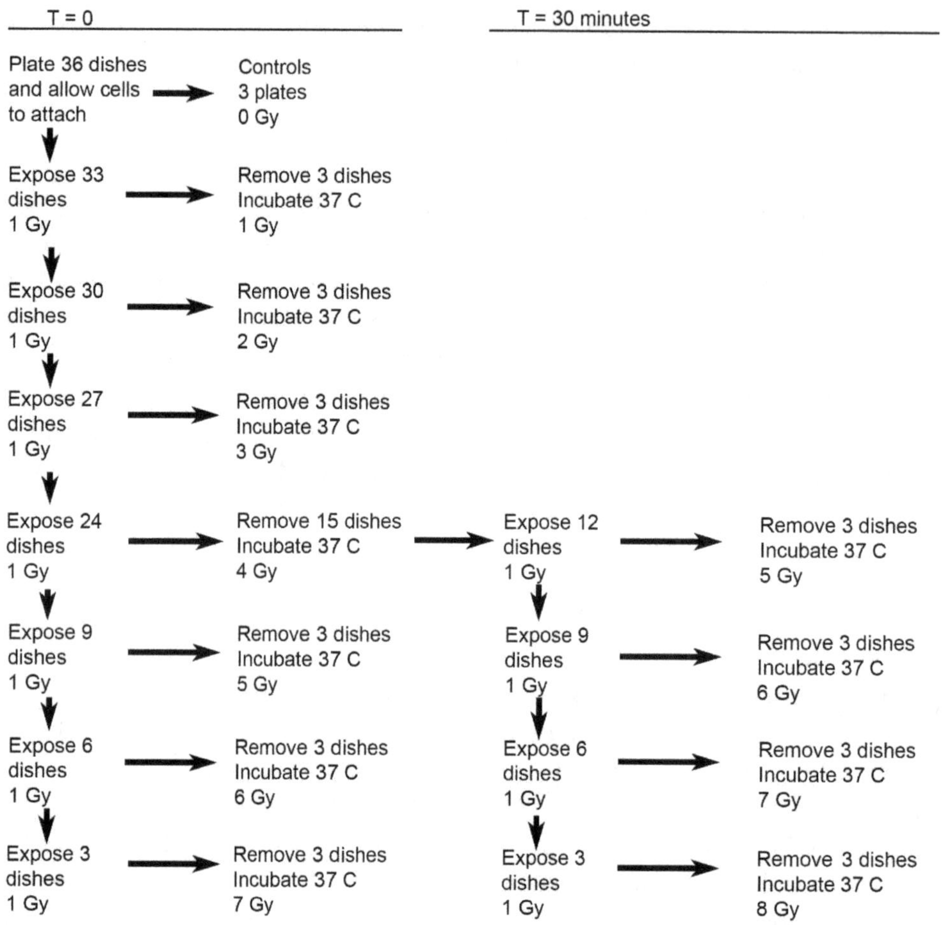

Figure 7-3 Flowchart depicting the protocol for the first two groups of an Elkind and Sutton split dose experiment (Groups A and B of Figure 7-4). Three samples are averaged at each dose and time point to assure accuracy. To reproduce the entire experiment, 24 more dishes are plated and exposed to 4 Gy x-ray radiation in column 1, removed to incubate for 2 hours (12 dishes) or 6 hours (12 dishes) at row 5, exposed as column 3, and incubated as column 4 of the flowchart.

result in 0.1 survival, but only a quarter of them were exposed to greater doses (Group A). The remainder of the cells was returned to the incubator. Following a time lag, a set of plates was removed from the incubator and exposed to further dose escalation. Following an additional time lag, a second set of plates was removed and exposed to further dose escalation. Following an additional time lag, a third set of plates was removed and exposed to further dose escalation. The flowchart (Figure 7-3) represents the procedure followed for the first two groups: Group A and Group B. We have limited the flowchart to two groups for simplification. To describe all 4 groups (the experiment portrayed in Figure 7-4), 24 additional plates are required (column 1, row 1 becomes 60 plates). These additional plates also are exposed up to 4 Gy x-ray radiation, removed to incubate for 2 hours (12 dishes) or 6 hours (12 dishes) at row 5. At the designated lag time, they are exposed to radiation (as in column 3) and incubated (as in column 4). This would create 4 more columns on the flowchart initiating at row 5, 2 under the header "T = 2 hours" and 2 under the header "T = 6 hours." These column entries would be identical to columns 3 and 4. For this reason, describing only two groups should be sufficiently illustrative.

Approximately 7 or 10 days after the irradiations had been performed, the dishes were removed from the incubator and the colonies counted. Typical resulting survival curves are presented in Figure 7-4. To

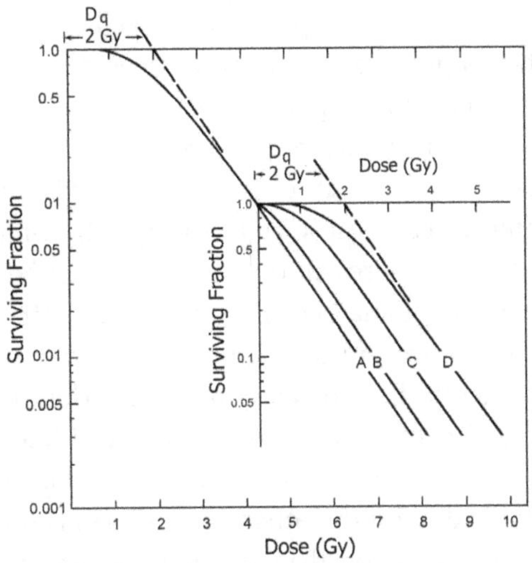

Figure 7-4 Representation of the Elkind–Sutton experiment. In this fictitious depiction, sparsely plated cells are exposed to increasing doses of x-ray radiation. All cells are irradiated to a dose known to reduce the fractional survival to approximately 10% of the unirradiated controls. The plates are then divided into groups and irradiated further, following a resting incubation period. Curve A was produced by cells further irradiated immediately, curve B cells incubated for 30 minutes, curve C cells incubated for 2 hours, and curve D cells incubated for 6 hours. Reproduced with permission of Elsevier, from "Radiation Biophysics." E. L. Alpen, Academic Press 1990.

understand this plot, ignore all the curves and internal axes, and look only at the curve designated "A." This curve represents a typical survival curve as depicted in Figure 6-5B. No surprises here; a quick check of the flowchart in Figure 7-3 column 2 shows that plates have been irradiated at 1 Gy, 2 Gy, 3 Gy, 4 Gy, 5 Gy, 6 Gy, and 7 Gy. The curve manifests an initial slope, a shoulder, and a log linear falloff. Because the Elkind and Sutton split dose experiment was conducted in 1961, when MTSH was the reigning dose response model, let's analyze the curve using MTSH terminology. An extrapolation of the final slope yields a D_q of 2 Gy. Now, ignore everything in Figure 7-4 except curve D and the internal axes. After 6 hours, the cells (plates) are removed from the incubator and irradiated with 1 Gy, 2 Gy, 3 Gy, or 4 Gy. These survival data recorded on a semilogarithmic plot (internal axes) produce a curve identical to curve A at the same doses. Notice that a line extrapolated from the final slope yields a $D_q = 2$ Gy. However, these samples were exposed to a 4 Gy dose of x-rays prior to being placed in the incubator. Therefore, instead of plotting the postincubation dose at 1 Gy (as on the internal axis), we plot it at 5 Gy (on the "true" axis). Furthermore, the initial 4 Gy dose resulted in the same cell killing in the Group D samples as in the Group A samples, so the number of live cells after 6 hours of incubation (insufficient time for the cells to divide) would be the same as the number of Group A cells after 4 Gy. Thus, our first datum point for this curve is displaced to 5 Gy (true axis) and a fractional survival resulting from that (cumulative) dose (true axis). Now, if you compare the survival of cells after 5 Gy (0.05) with the survival after 5 Gy when the final 1 Gy is delivered following a 6 hour lag (approximately 0.1), it becomes apparent that more cells survive when the dose is split. Splitting the dose, that is, providing a lag between doses, decreases cell killing. More interestingly, after the 6 hours lag the cell response to radiation appears to return to the Group A response. The survival curves are identical when curve D and the internal axes are translocated to overlay Group A and the true axes. The only apparent difference between curves A and D is that after the initial 4 Gy exposure, there are fewer cells available in each plate when the second dose escalation begins. This is interesting.

What do the other curves tell us? What happens when cells exposed to 4 Gy x-ray radiation are incubated for 30 minutes or 2 hours prior to continued dose escalation? An examination of Figure 7-4 discloses that the final slopes of all curves are parallel, that is, the D_0 for curve A is the same as the D_0 for curve B, for curve C, and for curve D. However, the width of the shoulder of curve B is not the same as the widths of the shoulders of curve C or curve D. In other words, the D_q for curves B, C, and D are different. Elkind and Sutton interpreted these results thusly. The final slope, the log linear survival portion of the curve, was understood to represent the radiation sensitivity of the cell line. Between 1950 and 1960, several mammalian cell lineages had been exposed to radiation and examined with respect to dose response. Although the sampling was quite small compared to today's experience, the general scientific consensus agreed that D_0 exemplified the amount of radiation required to kill a cell. Because all cells exposed in the split dose experiment arose from the same cell line, the D_0 for each curve in Figure 7-4 remains consistent. Regardless of protocol, all cells of a particular lineage exhibit the same inherent radiation sensitivity. The shoulder width, on the other hand, depends on the lag of the split dose. Elkind and Sutton recognized that the phenomenon driving the shoulder width must be completely absent when there is no lag because under that condition, there is no shoulder. If the radiation delivery is interrupted and the cells are incubated for even 30 minutes, a shoulder begins to reappear. At 6 hours, the shoulder is fully reconstructed, and additional incubation does not affect the shape (until cells begin to divide). Based on this behavior, Elkind hypothesized that cells recover a capacity to sustain some measure of damage without being killed. Existing radiation damage resulting from the preliminary 4 Gy exposure must be repaired during the lag. The logic goes like this. If 100 DNA ssb are induced by an initial radiation exposure, any additional DNA ssb generated by a subsequent exposure are likely to interact with the preexisting DNA ssb to form DNA dsb. However, if time passes between exposures, some original DNA ssb may be repaired. Additional DNA ssb introduced by the subsequent exposure are less likely to interact because of the reduced number of residual DNA ssb. So, Elkind and Sutton described the damage as "sublethal" and hypothesized that sublethal damage (SLD) was repaired during the lag. To clarify, SLD requires an additional insult to become lethal; PLD is either repaired or transitions to lethal damage. Based on the results of the Elkind, Sutton, and Moss experiment (Elkind *et al.*, 1961), we can now describe the three regions of the survival curve. The initial slope indicates cell killing induced by instantaneous lethal damage. The shoulder describes cellular repair capacity. The final slope describes cell type radiation sensitivity.

Do the models selected in Chapter 6 for closer examination (Table 7-1) support the split dose experiment conclusions? MTSH predicts that the initial slope, as the dose approaches zero, is flat; the initial

Table 7-1 Factors determining survival curve shoulder shape for several models

Model	Interpretation of Shoulder	Term
MTSH	Number of *targets*	n
LQ	Proportionality constants for *single (linear) hits* to *interdependent (quadratic) hits*	α/β
RMR	Rate constants for "linear" *damage repair* to "quadratic" *damage repair*	λ/κ
LPL	Conversion to *linear lethal damage* compared with conversion to *quadratic lethal damage*	$\dfrac{\eta_{AC}+\eta_{AB}e^{-\varepsilon_{BA}t_r}}{\dfrac{\eta_{AB}^2}{2\varepsilon}\left(1-e^{-\varepsilon_{BA}t_r}\right)^2}$
GLOBLE	Likelihood of *single loop breaks* resulting in cell death compared with the likelihood of *clustered loop breaks* resulting in cell death	$(\varepsilon_i)/(\varepsilon_c)$

slope is zero. This fails to agree with empirical results and would imply that there is no direct lethal damage. MTSH also claims that the shoulder reflects the number of targets whereas the experimental evidence indicates that the shoulder represents repair capacity. So, MTSH fails. Nonetheless, MTSH provides a convenient way to describe the shape of the curve (D_q, D_0, D_{10}, n) stipulating that the interpretation of the terms may be inaccurate. LQ professes that the shoulder reveals the relationship between α and β. Recall the terms represented by α and β: $\alpha = (p, f_0, n_0, k_0, \Delta)$, where f_0 is the fraction of unrestored DNA dsb, and $\beta = (p, E, n_1, n_2, f_0, f_1, f_2, k^2, (1-\Delta)^2)$, where f_0 is the fraction of unrestored DNA dsb and $f_{(n)}$ is the fraction of unrestored DNA ssb on either strand. The relationship between $kn_{(n)}$ (the average number of breaks) and $f_{(n)}$ (the number of unrestored breaks) describes the ability of a cell line to repair breaks. The relative efficiencies of DNA dsb repair (linear) versus DNA ssb repair plus interlesion DNA dsb repair (quadratic) affect the shape of the shoulder. Allowing that LQ assumes that the relevant damages are DNA ssb and DNA dsb, the model does not conflict with Elkind's results. A more accurate depiction would replace "DNA ssb" and "DNA dsb" with "SLD" and "PLD." Thus, the LQ model supports the split dose experiment conclusions. The third model of Table 7-1, the RMR model, is spot on. In RMR, the survival curve shoulder is described as a reflection of linear uncommitted damage repair compared with quadratic uncommitted damage repair. Uniquely, the effector damage is not described by this model and so the model avoids becoming trapped by disagreements regarding physiochemical evidence and evolving understanding of the specifics of DNA damage repair. RMR relies entirely on the empirical evidence that the initial portion of the survival curve can be mathematically described as exponentially linear and the high dose region can be described as exponentially quadratic.

Kappos and Pohlit (1972) claimed that the Cybernetic Model (and thus the LPL model) can explain the biology behind the split dose experiment on the basis of scavenger molecule concentration (refer to Equation 3.5). From Equation 6.33, the transition rate from state A cells (undamaged) to state B cells (reparably damaged) during the initial dose (up to 4 Gy) would look like this:

$$\eta_{AB}^{D_1} = \eta_{AB}^{*}\left(1 - e^{-\eta_c D_1}\right). \tag{7.7}$$

According to Kappos and Pohlit's hypothesis, during the rest interval (t_{int}) the reaction rate "constant," η_{AB}, decreases because competing scavenger molecules are replenished by the cell. The rate of change in the number of cells with lethal lesions (state C) can be restated (compare to Equation 6.32):

$$\frac{dN_c}{dt} = \varepsilon_c\left(N_{c,0} - \left(N_{c,0}e^{-\eta_c D_1}\right)\right), \tag{7.8}$$

where ε_c is the time constant for scavenger replenishment. Notice that this expression claims that as the scavengers are replaced, the number of cells converting to state C decreases according to $N_{c,0}e^{-\eta_c D_1}$ ($N_{c,0}$ is the number of cells in state C at $t_{int} = 0$). According to Kappos, it is not that cells in state C are being repaired; the number of cells transitioning to state C is increasingly reduced as the lag time (and thus number of scavenger molecules) increases. During the incubation lag between exposures,

$$\eta_{AB} = \eta_{AB}^{*}\left(1 - e^{-\eta_c D_1}\right)e^{-\varepsilon_c t_{int}}, \tag{7.9}$$

where $\eta_{AB}^{*}(1 - e^{-\eta_c D_1})$ reflects the rate of transition to state B (Equation 7.7) and $e^{-\varepsilon_c t_{int}}$ determines the rate of transition back to state A effected by scavenger competition. To find the rate of transition during the second exposure, assume the number of scavenger molecules is replenished during recovery and again depleted during the second dose. Through logical extension,

$$\eta_{AB} = \eta^*_{AB}\left\{1-\left[1-\left(1-e^{-\eta_c D_1}\right)e^{-\varepsilon_c t_{int}}\right]e^{-\eta_c D_2}\right\}. \tag{7.10}$$

Kappos and Pohlit considered only restitution in the formulation represented above. In other words, scavenger molecules competed with DNA for adsorption of free radicals and restored damaged DNA according to Equation 3.5 (DNA· + [R – H] → [DNA – H] + R·). The resulting theoretical curve did not fit experimental data. The time constant, ε_c, like some of the other "constants" employed in the cybernetic model, varies with t_{int}. Kappos postulated that the mismatch indicates a stepwise production of free radicals, scavengers, and/or repair molecules and derived an expression for the variable time "constant":

$$\varepsilon_c = \varepsilon^*_c\left\{\frac{\varepsilon_c}{\eta_c\dot{D}+\varepsilon_c}+\frac{\eta_c\dot{D}}{\eta_c\dot{D}+\varepsilon_c}e^{-(\eta_c\dot{D}+\varepsilon_c)t}\right\}. \tag{7.11}$$

As demonstrated in Figure 7-5, the shoulder of the survival curve cannot be described with a single variable. Although development of the Cybernetic model leads to the conclusion that restitution and repair occur at the same variable rate, more recent insight indicates that this cannot be universally true. At best, the fitted curves agree better with split dose data from diploid yeast cells than with data from mammalian cells.

GLOBLE presents a more complicated formalism than LQ (Equation 6.66 vs. Equation 6.23) and relies heavily on empirically derived constants (N_L and Y_{Loop}). Nonetheless, a careful examination of the survival equation reveals only two terms pertinent to the shape of the shoulder, ε_i and ε_c. Fully substituting for n_0, n_i, and n_c in Equation 6.66 yields,

$$S = \exp-\left[\left(N_L\left(Y_{Loop}D\right)e^{-Y_{Loop}D}\right)\varepsilon_i+\left(N_L-N_Le^{-Y_{Loop}D}-N_L\left(Y_{Loop}D\right)e^{-Y_{Loop}D}\right)\varepsilon_c\right]. \tag{7.12}$$

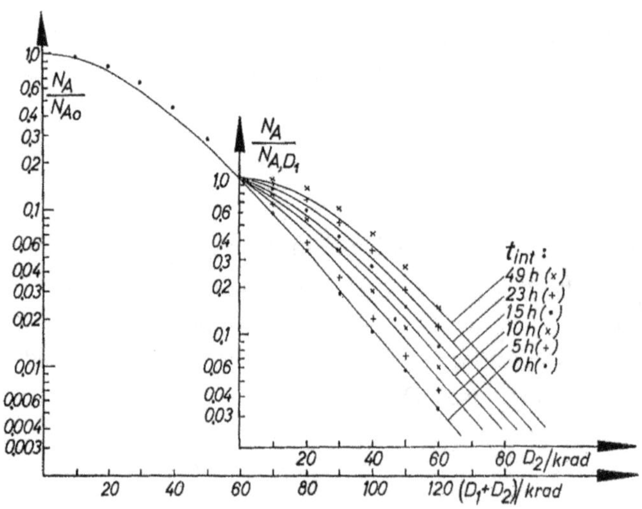

Figure 7-5 Kappos and Pohlit cybernetic model fit to a split dose experiment. The experiment was performed on diploid yeast cells (best fit conditions) and the initial dose was 6 Gy x-rays. Reproduced with permission of the Radiation Research Society from Kappos, A. & Pohlit, W., 1972. A cybernetic model for radiation reactions in living cells. I. Sparsely-ionizing radiations; stationary cells. *International Journal of Radiation Biology and Related Studies in Physics, Chemistry, and Medicine*, 22, pp. 51–65.

Notice that the exponent varies exponentially with dose (D). Recall that N_L is the number of loops per cell and Y_{Loop} is the induction frequency per Gy in each cell nucleus divided by the number of loops per nucleus. These are constants that can be derived empirically (at least theoretically). So, this expression can be simplified to yield

$$S = \exp - [((CD)e^{-BD})\, \varepsilon_i + (A - Ae^{-BD} - (CD)\, e^{-BD})\, \varepsilon_c], \tag{7.13}$$

by substituting for the constants, $N_L = A$, $Y_{\text{Loop}} = B$ and $N_L Y_{\text{Loop}} = \alpha_{DSB} = C$. Survival depends on a repeating linear exponential function of dose (e^{-BD}) dependent upon a proportionality constant comparable to the DNA dsb induction frequency of the LQ formalism. To see how ε_i and ε_c impact the shoulder, let's begin with Equations 6.71 and 6.72. We saw that the LQ interpretation of the shoulder is described as α/β. From Equation 6.71,

$$\alpha = \varepsilon_i\, \alpha_{DSB} = \varepsilon_i C. \tag{7.14}$$

From Equation 6.72,

$$\beta = \frac{\left[\varepsilon_c - 2\left(\alpha/\alpha_{DSB}\right)\right]\alpha_{DSB}^{2}}{2N_L} = \frac{1}{2}CB\left(\varepsilon_c - 2\alpha/C\right) = \frac{1}{2}CB\left(\varepsilon_c - 2\varepsilon_i C/C\right) = \frac{1}{2}CB(\varepsilon_c - 2\varepsilon_i). \tag{7.15}$$

To complete our analogy,

$$\frac{\alpha}{\beta} = \frac{\varepsilon_i C}{\frac{1}{2}CB(\varepsilon_c - \varepsilon_i)} = \frac{2}{B}\left[\frac{\varepsilon_i}{(\varepsilon_c - \varepsilon_i)}\right]. \tag{7.16}$$

If the likelihood of an iDSB resolving to a lethal lesion is small, ε_i can be ignored in the denominator. And if we can extend the LQ analogy to GLOBLE as per Equation 7.16, then the shoulder reflects the probability of iDSB resolving to lethal lesions compared with the probability of cDSB resolving to lethal lesions. What does "resolving to lethal lesions" mean? The small gap created by iDSB and the large gap created by cDSB each may be repaired by the Non-Homologous End Joining (NHEJ) or the Homologous Recombination (HR) DNA repair pathways, or the gap may convert to chromosomal damage. Thus, the efficiencies of repair and the likelihood that loop scissions will become lethal chromosomal aberrations determine the shoulder. In this case, the two damage descriptors governing the survival curve shape are not DNA ssb and DNA dsb, or direct damage and indirect damage; they are iDSB and cDSB. So, GLOBLE claims that the shoulder is defined by processes that diminish the conversion of loop damage to lethal chromosomal aberration. The GLOBLE model supports the split dose conclusions.

It is important to discriminate among cell death, mutation frequency, and chromosomal aberration frequency. Clearly, cDSB are more likely to result in mutation than iDSB, but we have not considered mutation frequency in any of the cases that we have applied to the Elkind and Sutton experiment. Surviving cells include mutated cells. Also, it is true that "cell death" is an ambiguous endpoint reflecting not only the demise of a cell but also the inability of a cell to divide six times and produce a visible colony. Therefore, the possibility that mutation may result in second- or third-generation metabolic fatality has not been considered. Let it go. Our application of models to the survival curve shape focuses only on the creation of lethal damage leading to the types of death reported in Chapter 4, Section 4.6. Within this analysis, GLOBLE considers only whether the loop gap, large or small, is likely to be converted to a lethal construction. If the two free ends are rejoined and chromosome integrity is reestablished,

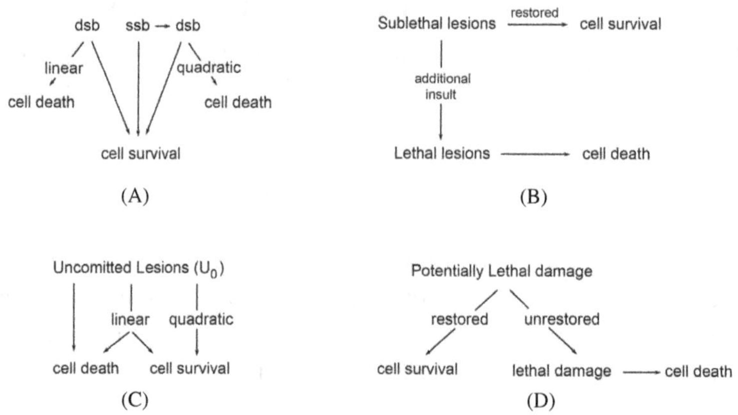

Figure 7-6 Comparison of cell survival schema. **(A)** The LQ model considers cell death resulting from direct dsb and from the interaction of ssb to form dsb. Remaining ssb are either restored or repaired. The unknown repair mechanisms are described by the mathematical description of the survival curve: linear or quadratic. **(B)** The RMR model considers the impact of uncommitted lesions on cell survival. Any unrepaired lesions result in cell death; any lesions repaired by the quadratic mechanism result in cell survival. The lesions repaired by the linear process may result in misrepair leading to cell death or eurepair leading to cell survival. **(C)** Elkind and Sutton proposed the concept of "sublethal damage" that may be resolved resulting in cell survival. If a second event converts the sublethal damage to lethal, the damage cannot be repaired, and the cell dies. **(D)** PLD is either restored to a stable state or not restored leading to cell death. The flexibility of this scheme leaves plenty of room for repair details to be inserted into the mechanisms of restoration and restoration failure (unrestored).

if cell division successfully proceeds, the cell has been repaired. Figure 7-6 provides a schematic comparison of the models tested here against the data derived from the split dose experiment.

7.3. Factors Affecting Cell Survival

Interpretation of the dose response curve relies on only two possible outcomes: a cell divides several times or it does not. Nonetheless, we can use survival curves and the proposed models to investigate radiation response mechanisms as well. If the shoulder represents repair, what mechanisms are involved in repair? And if the final slope represents radiation sensitivity, what determines sensitivity?

7.3.1. *Target Size*

The Single Hit Model proposes that cell survival depends on the probability that a cell receives no hits. MTSH refined this assumption to require that n targets must be hit to accomplish cell death and LQ assumes that survival is dependent upon the number of critical sites (n_n) and the probability of damage (k) to each site. From these assumptions, it follows that a smaller genome, that is, fewer targets, should require a higher photon fluence (dose) to establish a fatal number of hits. Logically, if higher doses are required to kill a cell, then the species must be more radiation resistant. For example, 90% of mammalian cells can be killed at x-ray doses of about 5–10 Gy. To kill 90% of yeast cells, doses greater than 500 Gy are required. The slope of the curve at low survival (D_0) indicates radiation sensitivity. Figure 7-7 presents typical survival curves for several species. It is immediately apparent that D_0 correlates with target size; the greater the dose of radiation required to kill additional cells, the greater the D_0. Apparently, target size is a factor determining radiation sensitivity. What else might be involved? Let's think about phages

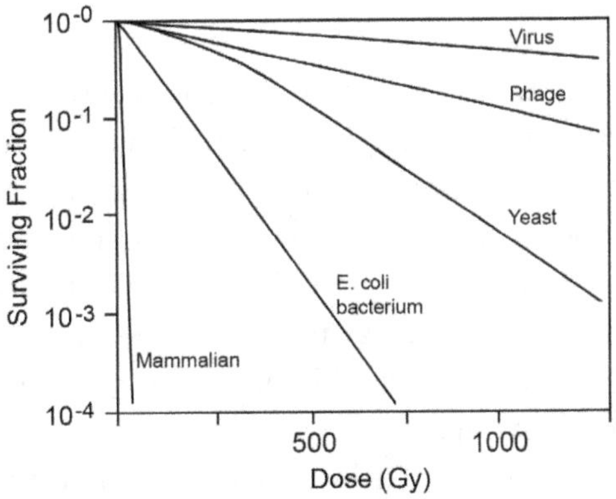

Figure 7-7 The effect of target size on survival curve shape. Shown are typical survival curves for each species. As the volume of the target (number DNA bases) increases from viruses to mammalian cells, the dose of x-ray radiation required to kill the host reduces from hundreds of grays to tens of grays. The inverse relationship between inactivation dose and target size reflects the probability of a hit. The slope of each curve reflects the radiation sensitivity of each species. Viruses and phage do not possess necessary proteins, do not repair damage, and therefore do not manifest a survival curve shoulder. The shoulder of the yeast curve is robust. The dose response shoulders for bacteria and mammalian cells are difficult to see at this scale.

and viruses. These species exist as capsules of genetic material. They do not contain enough proteins and other cellular machinery to independently grow and therefore do not divide. To determine viability, viruses and phages must be permitted to infect cells expressing the necessary DNA, RNA, or protein duplication machinery to enable proliferation. Two factors determine the survival of viruses and phages: the capacity to infect host cells and the viability of the damaged genomic material. Because the host cell has not been exposed to radiation, its machinery is fully intact. The survival curve implies that smaller fragments of viral DNA retain effectiveness once delivered to the host nucleus, and likely the delivery mechanism remains robust. Because a virus fuses with the cell membrane, a passive process, it seems likely that this may be more robust than active phage material insertion following irradiation. If so, the success of genetic material transfer also may be an important component of radiation sensitivity. Nonetheless, the extreme radiation resistance of viruses and phages is overwhelmingly due to their extremely small target size.

Now, consider yeast. Yeast shares many gene homologues with humans. Important repair functions such as replication, recombination, and cell cycle checkpoints are expressed in yeast. It is therefore not surprising that the survival curve exhibits a pronounced shoulder and resembles mammalian curves, albeit at orders of magnitude higher dose because of reduced genome size. The genes contributing to chromatin remodeling, chromosome segregation, and transcription affect both repair and by overlap, sensitivity. Other activities that contribute to cell cycle functions such as nuclear pore formation and cytokinesis, Golgi and vacuolar activities, ubiquitin-mediated protein degradation, mitochondrial activity, and cell wall maintenance are directly responsible for radiation sensitivity. In other words, the genome size determines the location of the curve on the plot and to a large extent the slope, but the D_0 is also a product of metabolic activities. Some confusion is generated by the overlap of repair with sensitivity. Logically and biologically, repair impaired cell lines are more radiation sensitive than repair proficient cell lines. The phenomenon of the quadratic exponential decline in cell survival reflects DNA repair as a component of sensitivity (e.g., SLD). The shoulder of the curve, according to Elkind, is wholly dependent upon repair capacity.

7.3.2. *Repair Capacity*

We can observe the consequences of DNA repair impairment in several diseases impacting human well-being. For example, patients with ataxia-telangiectasia (AT) are extremely radiation hypersensitive. Recall the ATM (ataxia-telangiectasia mutated) protein is critical to cell cycle control and DNA repair (Figure 4-18). Although unfortunately designated "mutated" by researchers mistakenly believing that they had discovered an altered form of the AT gene, this normal, unaltered gene is located on chromosome 11 and actual mutation can result in AT syndrome. Other diseases: Seckel syndrome, Nijmegen breakage syndrome, Fanconi's anemia, and several homologues of the RecQ-Bloom syndrome can likewise be traced to impaired DNA repair capacity.

Repair-deficient cell lines can be generated by randomly mutating cellular DNA with chemicals, ultraviolet (UV), or ionizing radiation and then selecting for repair-deficient cells. Zolan *et al.* (1988) developed a particularly clever protocol for creating and identifying repair-deficient mutants of *Coprinus cinereus*, a small edible mushroom. Zolan cultured fungal hyphae, the vegetative haploid phase of the fungal life cycle. She then exposed the cultures to UV radiation and subsequently allowed the hyphae to fuse (asexual reproduction), producing diploid cells. Under the appropriate conditions of temperature, light cycle, and humidity, these cells produce fruiting bodies — mushrooms. Mushrooms produce haploid spores that disseminate to produce new hyphal colonies (sexual reproduction). These spores cover the gills on mushrooms. Perhaps you have experienced spores, a brown powder on the inside of a package purchased at a grocery store after the mushrooms have opened. Zolan recognized that *Coprinus* mushrooms incapable of producing spores are white rather than brown. Because meiosis requires crossover — DNA dsb formation, transfer of chromosomal fragments and DNA repair — the white mushrooms are likely repair-deficient. Thus, Zolan selected repair-deficient mutants by exposing the white mushroom hyphal cultures to ionizing radiation and observing the resultant survival curves. Figure 7-8 shows the survival curve for a *C. cinereus* mutant designated rad 9-1, compared with the unmutated, parental "wild type." Notice that when the only difference between two cell lines is a single mutated DNA repair gene (confirmed in subsequent analyses) the D_0 is not affected greatly. The

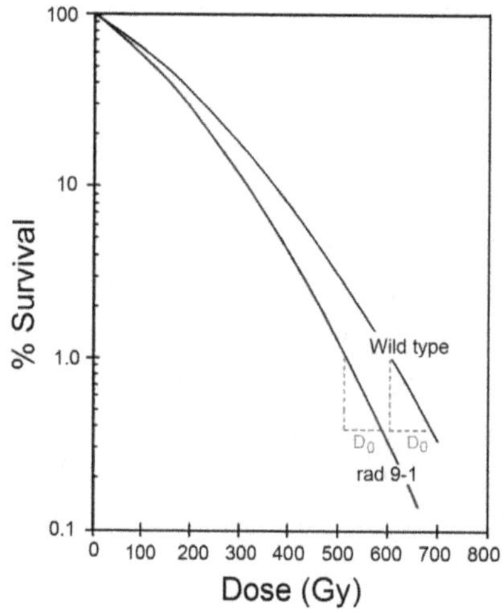

Figure 7-8 The effect of repair impairment on survival curve shape. Wild type fungus, *Coprinus cinereus* and a repair-deficient mutant, rad 9-1, were each exposed to escalating doses of x-ray radiation. Although the quadratic slopes are similar exhibiting only about a 5 Gy difference in D_0, the shoulder of the curve fitted to the mutant data is reduced by approximately 50 Gy when compared with the wild type.

difference between the rad 9-1 D_0 and the wild-type D_0 is only about 5 Gy (about 1% of the dose). On the other hand, the extent of the shoulder portion of the rad 9-1 curve is approximately 50 Gy smaller than the wild-type shoulder. It is the reduced shoulder that is displacing the rad 9-1 curve toward the left. The effect of repair deficiency as reflected in shoulder extent is not unique to fungal species, or haploid species. It also can be observed clearly in mammalian cell lines.

The impact of mechanisms that supersede repair also is clearly reflected in the extent of the shoulder for each cell line. Section 7.3.4 discusses the effect of apoptosis on the survival curve. Scientific literature is resplendent with examples of dose response curves derived from repair-deficient species. A reduction in the width of the shoulder accompanied by a small decrease in D_0 is a universal characteristic without regard for the specific, defective repair gene or repair compromising process.

7.3.3. *Cell Cycle and Synchronization*

We have previously discussed the relationship between radiation sensitivity and cell cycle stage, also referred to as cell age (Chapter 5, Section 5.2.2). Now, we can examine the relationships between repair and sensitivity within each stage of the cell cycle. Cultures can be synchronized primarily by one of two techniques: chemical block or mitotic harvest. After plating and cell attachment at 37°C, the synchronized cycling cells are exposed to radiation at delays equal to the times required to reach each cell cycle phase. In other words, cells exposed within an hour after plating remain in mitosis; cells exposed after that (the exact timing depends on the cell line — between 1 and 11 hours), are in G_1. After conclusion of G_1, 12–15 hours for HeLa cells, the cells are in S phase, and for 3 hours after that, the cells are in G_2. Figure 7-9 illustrates that the survival curve produced by cells undergoing mitosis reveals a small D_0 — indicating relative radiation sensitivity — and manifests no shoulder — indicating no repair activity. During mitosis (prophase, metaphase, anaphase, and telophase), DNA condenses into chromosomes, spindle fibers attach to the kinetochores, the nuclear membrane dissolves, chromosomes align along the equatorial plate, and then migrate along the spindle fibers to opposite poles.

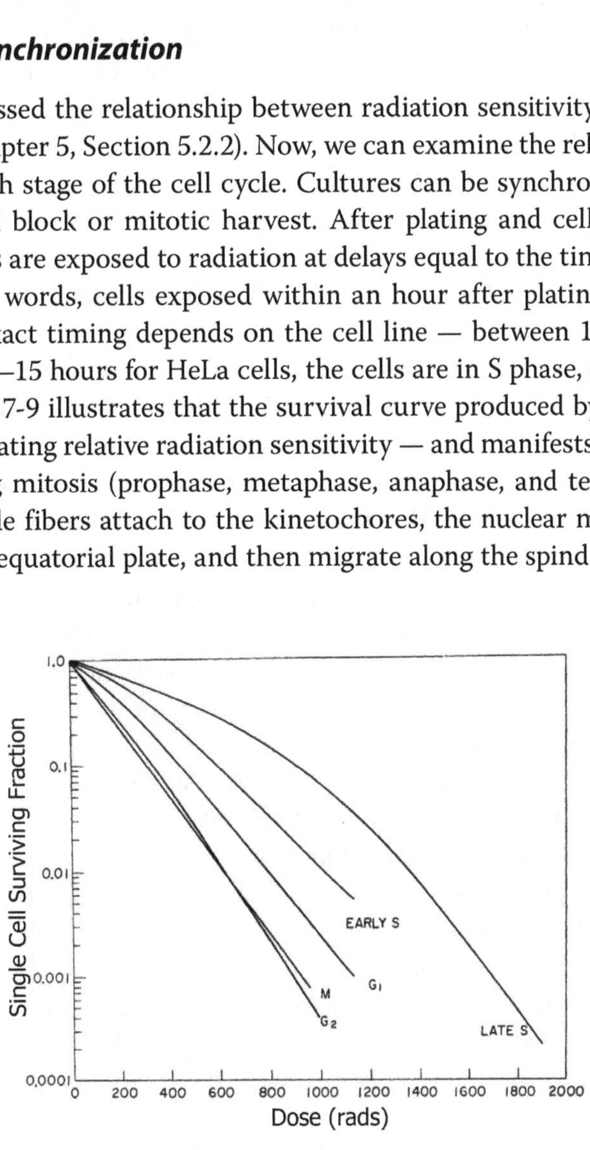

Figure 7-9 The effect of cell cycle age on the shape of the survival curve. CHO cells were synchronized by mitotic shake-off and exposed to increasing doses of x-ray radiation at appropriate times following plating. The first dose escalation occurred during G_1, the second and third during S, the fourth during G_2, and the fifth during M. Visible colonies were counted 6 days after exposure. Reproduced with permission of the Radiation Research Society from Sinclair, W. K. & Morton, R. A., 1966. X-ray sensitivity during the cell generation cycle of cultured Chinese hamster cells. *Radiation Research*, 29, pp. 450–474.

Finally, the cell constricts along the equatorial plane and forms two independent cells. Structural damage to the spindle fibers, kinetochore attachment, and membrane remodeling mechanisms might be expected to severely interfere with the carefully choreographed dance that is mitosis. Furthermore, imagine the effect of cDSB during chromosome condensation. Cells that fail to successfully complete mitosis die. Thus, cells are predictably radiation sensitive during mitosis. Why are cells repair-deficient during mitosis? The usual suspects deserve interrogation. Structural biophysicists would say that the supercondensed chromosome structure prevents access to damaged sites. If heterochromatin suppresses transcription, then surely condensed chromosomes would be inaccessible to repairasomes. Biochemists would argue that additional DNA dsb induced by radiation during mitosis overwhelm the available enzyme concentration at a time when enzyme recruitment through transcription is unavailable. In general, mitotic chromatin is transcriptionally silent to ensure the integrity of the separating chromosomes. Although it has been suggested that transcription remains active for a few specific sites during early M phase prior to condensation (Liu *et al.*, 2017), mRNA screens indicate that most genes, including repair enzymes, are not transcribed at any time during mitosis. Both hypotheses, structural and biochemical, indicate that repair is inactive during M phase in agreement with the dose response curve that presents no shoulder. Here is another surprising but logical fact. All mammalian cells exhibit the same M phase dose response curve, that is, all the curves superimpose. During cell division, all mammalian cells, regardless of their behavior during other phases of the cell cycle or the behavior of asynchronous populations of a particular cell type, exhibit the same radiation sensitivity and lack of repair during mitosis.

At the other extreme, survival data from cells irradiated during late S phase exhibit an extensive shoulder encompassing approximately 11 Gy (1132 rad). Although the D_0 is only slightly greater for late S phase than M phase, late S phase is commonly described as the most radiation-resistant cell cycle phase because of the combined effects of repair and sensitivity. Resources that declare late S phase to be radiation resistant are quick to describe that the resistance is likely due to HR repair following replication. This muddies the waters when interpreting the dose response curve. DNA replication not only enables HR, but also (as we learned in Chapter 4) facilitates proofreading, unwinding, and repairasome binding. These characteristics explain the shoulder of the dose response curve. Cells irradiated in late S phase are likely resistant, that is exhibit a slightly larger D_0, because damage incurred by structures and processes peripherally overlapping repair or independent of repair have time to recover during G_2, prior to cell division. Both DNA and metabolic repair occur during the cell cycle checkpoint in G_2. Why would cells irradiated during early S phase exhibit a smaller shoulder than cells irradiated during late S phase? Evidence indicates that radiation interferes more effectively with the initiation of replication at the origin and less effectively during strand elongation. ^{3}H-thymidine incorporation is impaired to a greater extent during early S phase. Thus, we might conclude that during early S phase, when replication initiates, radiation reduces DNA repair efficiency coincident with replication proofreading and repairasome binding. Nonetheless, the quadratic slope is similar, only slightly shallower compared with the slope exhibited by cells irradiated in late S phase (by about 40 rads). Processes and structures less connected to DNA repair experience similar kinetics to those impacted in late S phase and so early and late S phase cells are nearly equally radiation sensitive (or resistant, depending on whether your cup is half empty or half full).

Cells irradiated during G_1 appear to exhibit a further reduction in the extent of the repair shoulder but the G_1 curve is in Figure 7-9 shifted to the left of the early S curve primarily because the linear α component of damage is more pronounced, that is, the initial slope of the dose response curve is more negative. The linear DNA dsb damage that has been described as "direct" or "instantaneous" is complex (has multiple locally damaged sites) and requires sophisticated repair. G_1 repair is achieved by NHEJ because homologous chromosomes are not available prereplication. Although this repair is efficient, it leads to serious mutations and chromosomal aberrations, increasing cell death. While we have not been considering the initial slopes of the curves in Figure 7-9, because our discussion has been limited to the conclusions of Elkind and Sutton, notice that as we move from the leftmost curve (M) to the one furthest right (late S), α decreases. This physical damage parameter that is likely driven by chromatin configuration, more compact to the left of G_1 and less compact to the right of G_1, clearly determines the placement of the curves on the plot.

If late S phase irradiated cells fare so well, why do cells irradiated during G_2 exhibit such poor repair capacity? Replication completes during S phase. Cells in G_2 contain a replicated genome available for HR, but damage incurred following the G_2 cell cycle checkpoint has little opportunity for repair. Cells enter mitosis with critically fragmented DNA, and most do not survive division. Although the curves for M and G_2 cross, the curve for G_2 has a small shoulder resulting from repair of DNA exposed prior to and during the cell cycle block.

Overall, the curves for M, G_1, and late S exhibit very similar D_0, with the curve for G_2 indicating a smaller D_0 (slightly more sensitive) and the curve for early S indicating larger D_0 (slightly more resistant). But by-and-large, Elkind and Sutton would claim that the radiation sensitivity does not vary throughout the cell cycle. This seems logical; these cells are all CHO cells and the metabolic functions supporting radiation hardness would be consistent within a single cell line. On the other hand, the shoulder of the late S curve is by far the most pronounced, followed by the shoulders produced by exposure during G_1 and early S, which are similar. For cells exposed during mitosis (M), no visible shoulder is present. Repair, then, is most successful during late S, moderately successful during G_1 and early S, and unsuccessful during G_2 and M. Nonetheless, superficial assessment of the plot of Figure 7-9, would indicate that cells are most "radiation sensitive" in G_2/M and least "radiation sensitive" in late S. The temptation to interpret the data in this way is overwhelming, and this is what you will read in the literature and hear in conversation. Our more pedantic examination uses empirical evidence (Elkind & Sutton, 1959) to test the assumptions that the shoulder of the cell survival curve represents repair while the quadratic slope reflects radiation sensitivity, which is influenced by repair as well as the percentage of α-type damage and nonrepair metabolic processes. There are additional factors that influence the variation in radiation sensitivity through the cell cycle. For example, Elenore Blakely has shown that the level of cellular natural antioxidants (free radical scavengers) such as glutathione increases as cells enter S phase and maximizes in mid- to late-S phase. Free radical scavenging could help explain late S phase radiation resistance to x-rays (see Section 7.7.2).

7.3.4. *Apoptosis*

Let's try another gedanken experiment (thought experiment). Given that a radiation response curve displays the number of cells that survive at increasing doses of radiation, how would you expect two survival curves for two cell lines to compare if one expresses apoptosis and one does not? Apoptosis, remember, is a programmed form of cell death that triggers off DNA dsb damage. Our logic might proceed something like this. As radiation dose increases, two things happen: individual cells experience higher DNA dsb frequency, and a greater number of cells experience DNA dsb. Relatively few unresolved DNA dsb trigger apoptosis in a given cell, so an increased number of DNA dsb per cell has little effect on apoptotic death. If 10 DNA dsb trigger apoptosis, an additional 100, or 1000 are just "overkill." However, as the number of cells receiving DNA dsb damage increases, more cells experience apoptosis

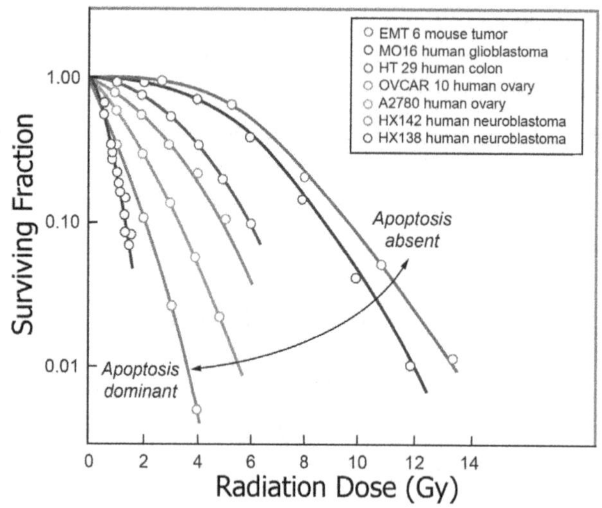

Figure 7-10 The effect of apoptosis on survival curve shape. Seven dose response curves are shown, each for a different cell line, ranging from the least radiation sensitive (EMT 6 mouse tumor; uppermost curve) to the most radiation sensitive (HX 138 human neuroblastoma; leftmost curve). The legend identifies the cell lines that produced each curve, from least sensitive to most sensitive (top to bottom). Sensitivity (D_0) correlates with apoptotic activity. DNA repair activity (shoulder) anticorrelates with apoptotic activity. Reproduced and modified with permission of Elsevier from Hall, E. & Giaccia, A., 2012. Radiobiology for the Radiobiologist.

and succumb. A cell line that does not express apoptosis is more tolerant of DNA damage; here the amount of damage per cell becomes important. Chromosomal damage must be relatively severe to result in nonapoptotic death. Furthermore, cells lacking apoptotic mechanisms have greater opportunity for repair during cell cycle arrest. Once DNA dsb damage is sensed, CHEK2 is activated triggering cell cycle arrest as well as phosphorylation of E2F1 and PML (see Figure 4-17). A cell line not expressing apoptosis might be expected to respond to increasing doses of radiation with cell cycle arrest, repairing damage, or experiencing senescence or mitotic death through any of the nonapoptotic mechanisms. We would anticipate that the nonapoptotic cell line survival curve would exhibit a repair shoulder and a decreasing higher dose survival resembling the CHO cell survival curve of Figure 7-1. On the other hand, we might expect that the apoptotic competent cell line would exhibit a diminished shoulder reflecting the reduced opportunity for repair (which competes with apoptosis; Figure 4-18), and a steeper final slope (decreased D_0) because cells are not provided an opportunity to arrest and repair nongenomic components. Our intuition does not fail us. Figure 7-10 compares the dose response curves for several cell lines expressing different apoptotic competencies from the extremely apoptotic human neuroblastoma HX 138 line to the EMT 6 mouse tumor line that completely lacks apoptotic activity. Apoptosis makes cells more radiation sensitive (decreased D_0) and impedes DNA repair (reduced shoulder).

7.3.5. *Dose Rate*

Imagine you are caught in a heavy rain storm and you are getting very wet. Now imagine a drizzle where just a few drops fall every few minutes. In the second case, it takes much longer for you to get equally wet not only because it takes longer for the same volume of water to fall on you, but also because you will dry a bit between drops. Now suppose that a cell is being exposed to x-ray radiation via similar scenarios. If radiation hits occur in rapid succession, the number of complex DNA dsb (or cDSB) will be relatively

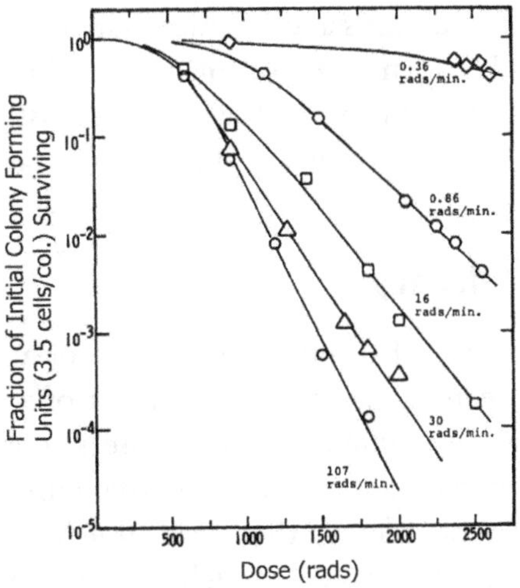

Figure 7-11 The effect of dose rate on the survival curve shape. Chinese hamster CHL-F cells in G_1 were exposed to ^{60}Co gamma rays using various dose rates. One rad is equal to 0.01 Gy. Reproduced with permission of the Radiation Research Society from Bedford, J. S & Mitchell, J. B., 1973. Dose-rate effects in synchronous mammalian cells in culture, *Radiation Research,* 54, pp. 316–327.

high (SLD converts to lethal). This represents a high dose rate; multiple grays of dose are delivered per minute. On the other hand, if only a few hits occur over a long period of time, the percentage of DNA ssb and simple DNA dsb (or iDSB) increases and the percentage of complex damages decreases (SLD is repaired). The dose response curves for these two situations would be expected to present very different repair shoulders and the cells would be expected to have very different survival probabilities. Be mindful of the difference between dose rate and dose. One gray of radiation can be delivered in 1 second or 1 minute. In these two schemes, the dose is the same (1 Gy) but the rate is different (1 Gy/minute or 1 Gy/second). Thus, you can increase the dose while holding the dose rate constant by increasing the exposure duration.

Figure 7-11 illustrates changes in repair capacity as well as changes in radiation sensitivity resulting from decreasing dose rates. The repair shoulder indicates similar repair capability at the relatively high dose rates of 1.07 Gy/min and 0.30 Gy/min. These rates are typical during radiation therapy and should not be considered particularly high dose rates. The shoulder broadens by about 0.5 Gy indicating improved repair functionality at 0.16 Gy/min. At 0.0086 Gy/min repair improves again, the shoulder broadening an additional ~ 0.5 Gy. What do you suppose it is happening at the extremely low dose rate of 0.0036 Gy/min? The plot indicates virtually no cell killing up to 20 Gy (2000 rads), where the repair shoulder begins. Do you think that repair is so efficient that only about 1% of cells are killed at the relatively high dose of 20 Gy? Statistics would suggest that lethal damage would occur more often. So, what else could account for the high survival? Remember that at a fixed dose rate, the time required to deliver a given dose increases as the dose increases. For example, at 1.07 Gy/min, it takes 18.69 minutes to deliver 20 Gy. At 0.0086 Gy/min, it takes 2326 minutes to deliver 20 Gy, or 38.76 hours. Recall that mammalian cells complete the cell cycle in about 24 hours (see Chapter 4, Section 4.2); they double in number every 24 hours. So, at very low dose rates, new cells replace killed cells and the population apparently remains constant. For this reason, it is important to select an appropriate dose rate when conducting dose response experiments. Proliferation must not be permitted to impact the data. Now let's see what the curves tell us about radiation sensitivity. Clearly the D_0 for each curve is different. As the dose rate is reduced, all cellular structures and functions fare better. Cells have time to generate free radical scavengers, replace damaged proteins, repair damaged membranes, reorganize mitotic structures, and so on. The slower the onslaught of antagonists, the more successful the biological survival strategies become. Therefore, cells are more radiation resistant at lower dose rates.

7.4. Fractionated Dose Delivery

There are two ways to reduce dose rate. If you have an accelerator at your disposal, you can adjust the fluence of electrons at the target, thereby adjusting the number of x-ray photons produced per second (see Chapter 1, Section 1.3.3). Under these conditions, dose rate is continuously variable. However, if you have a solid source, say ^{60}Co, you could periodically uncover the source so as to deliver a partial dose, put the shield in place for a period of time, remove the shield to deliver another partial dose, replace the shield, and so on. The second method is referred to as delivering a "fractionated" dose. It delivers dose at a higher dose rate for short periods of time, and you have just seen that has consequences (Figure 7-4). One option would be to wait for the ^{60}Co activity to decline. Obviously, that solution is not practical for many reasons. Additionally, the characteristics of fractionation are beneficial for therapy. So, a careful examination of the radiobiological effects of fractionated dose delivery is in order.

The Elkind and Sutton split dose experiment represents a two stage dose fractionation. In their experiment, an initial set of doses was delivered, a delay was imposed, and a second set of doses was delivered. Depending on the lag period between the sets of exposures, the breadth of repair shoulder ranged from totally absent (for no delay) to completely reconstructed. The reconstructed shoulder is identical to the original shoulder of the initial dose escalation exposure. Remember that Elkind and Sutton believed that the shoulder reflected the repair of SLD, in other words the removal of a type of damage that requires a second event to become lethal. For this reason, much of the literature refers to SLD when describing fractionation. Remember also that exposure to radiation triggers cell cycle checkpoints, particularly the checkpoint in early G_2. This cell cycle block collects cycling cells at G_2. Repeated periodic exposures (fractionation) reinforce synchronization. Synchronized cells respond differentially to radiation depending on the cycle phase at the time of exposure (Figure 7-9). If the delay between exposures matches the cell cycle periodicity, the shoulder of each fraction will be affected according to the repair competence of the repeatedly targeted stage. If the lag between fractions is sufficiently long, cells will "resort" and become asynchronous again. When designing fractionated treatments, medical physicists, radiation biologists, and radiation oncologists must take these characteristics into consideration.

Fractionated survival curves can be constructed according to Figure 7-12. Begin by considering the Elkind–Sutton split dose experiment depicted in Figure 7-4, but compare only the curves resulting from continuous dose delivery (no lag) and those resulting from an optimal lag sufficient to complete repair without allowing repopulation (Figure 7-12A). Then suppose that we extend the experiment such that at 99% cell killing (1% survival, or 0.01 fractional survival), we continue the dose escalation, but also allow a set of plates to incubate for the optimal recovery time followed by a dose escalation exposure series as we did at 90% survival (Figure 7-12B). The identical shoulder will reappear at 1% survival because, although there are fewer cells in each incubated plate, each cell has repaired the DNA damage and has returned to its initial genetic state. Next, we remove the portion of the curves resulting from continued dose escalation lacking recovery periods and retain only the data from the cells that have been allowed to rest and recover from DNA damage prior to continued radiation exposure (Figure 7-12C). These cells have experienced fractionated dose delivery. The cells are irradiated, allowed to recover, and are irradiated again; the cycle is repeated several times. In practice, the cycle periodicity is determined by external

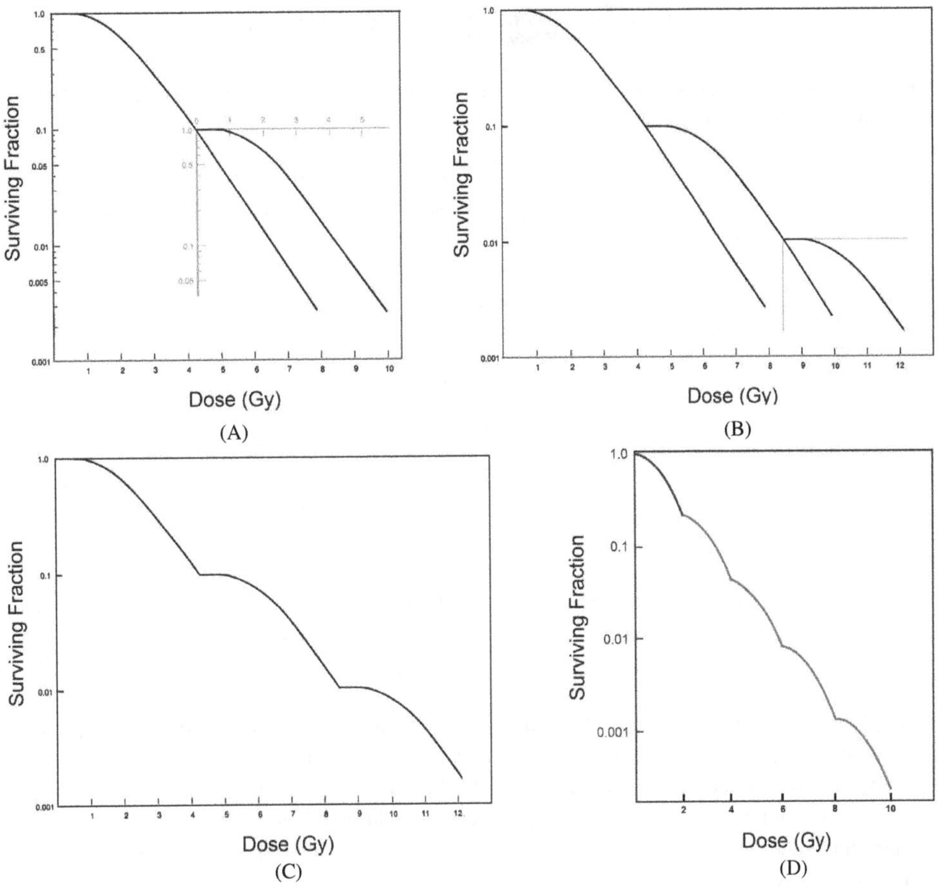

Figure 7-12 Construction of a fractionated survival curve. (**A**) The split dose experiment establishes the foundation of a fractionation survival curve, if only the curves for no lag (T = 0) and the minimal lag allowing for maximal repair ($T = t_{max}$) are included. (**B**) If the split dose protocol at 10% survival is repeated at the total dose resulting in 1% survival, the repair shoulder manifests again for surviving cells. (**C**) If the "no lag" curves are removed, only cells exposed to increasing doses of radiation followed by incubation and repair, followed by increasing doses of radiation, remain. This is fractionation: irradiate, rest, irradiate, rest, repeat. (**D**) Representation of a survival curve for 5 fractionated rounds of 2 Gy radiation exposure including recovery periods.

considerations. For example, during clinical therapy, patients' tumors are commonly exposed to 2 Gy x-ray radiation at 24-hour intervals. We shall reveal the rationale behind the choice of 2 Gy per dose a little later. The interval of 24 hours is convenient for clinical personnel and patients. It is long enough to permit recovery and short enough to not permit cell division (repopulation). In Figure 7-12D, the cycles are not constrained to logs of survival, but rather determined by 2 Gy fractional doses, as in the clinical example.

Both SLD and PLD benefit from fractionated dose delivery. In the case of SLD, the initial lesion has an opportunity to repair prior to creation of a second lesion. PLD also can be repaired (it is *potentially* lethal) when provided the opportunity between exposures. Thus, obviously, fractionated dose delivery must result in a different cell killing efficiency. The S for cells exposed to a single exposure of 8 Gy is anticipated to be less than the S for cells exposed to four 2 Gy fractions of radiation, with a 6-hour recovery period between each fraction (Figure 7-13). The periodically repaired cells (those exposed to

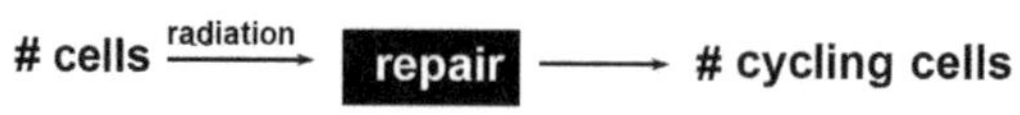

Figure 7-13 Protracted radiation delivery and cell survival. The number of surviving cells that produce visible colonies depends on the extent of repair achieved.

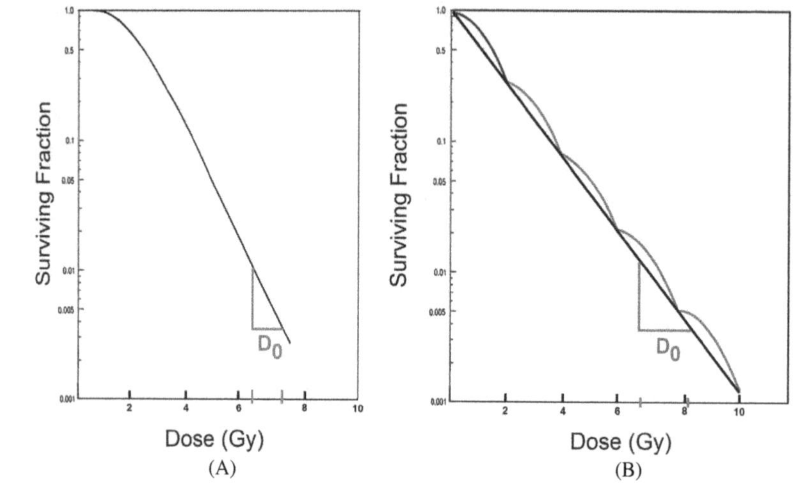

Figure 7-14 A comparison of single, acute dose vs. fractionated dose radiation sensitivity. **(A)** Dose response curve for single doses delivered to cells in culture. **(B)** Fractionated dose response curve for cultured cells permitted to recover between partial doses. A line is extrapolated to connect the fractional survival at the initiation of each recovery phase. Radiation sensitivity is indicated by the dose of radiation required to reduce the survival an additional 37% (1/*e*) from 0.01 survival. A comparison of the acute dose D_0 and the fractionated dose D_0 reveals the requirement for approximately 50% more dose to achieve the same reduction during fractionated delivery.

fractionated dose delivery) should appear more radiation resistant than cells exposed "acutely" (to a single dose). How can we quantify this?

Recall the LQ equation (Equation 6.23):

$$S = e^{-(\alpha D + \beta D^2)}. \tag{7.17}$$

In the case of fractionated delivery, the total dose (D) delivered is equal to the number of fractions (n) times the dose per fraction (d), or

$$S = e^{-(\alpha n d + \beta n d^2)}. \tag{7.18}$$

Notice that this is also equivalent to

$$S = e^{-(\alpha n d + \beta d D)} \tag{7.19}$$

because $D = nd$ (for example, 4 fractions × 2 Gy/fraction = 8 Gy). Equation 7.19 shows clearly that cell survival depends both on the total dose (D) and the dose per fraction (d), and that this relationship occurs in the quadratic portion of the survival curve.

Figure 7-14 compares the continuous exposure of Figure 7-4 with the fractionated exposure of Figure 7-12D. To delineate a survival slope for the fractionated scheme (Figure 7-14B), we have connected the origins of each shoulder — the S at the initiation of each recovery period. The D_0 of the fractionated protocol can be determined from the extrapolated line. Suppose a clinic delivers a fractionated dose to a patient and the medical physicist wants to know how much damage healthy tissue will suffer if incidentally exposed. The D_0 indicates the radiation sensitivity of the healthy tissue for any proposed fractionation scheme. Note that we just said, "healthy tissue." Cancers are composed of rapidly dividing cells that frequently bypass cell cycle checkpoints in favor of quantity over quality. Fractionation will not result in repair of these rapidly dividing cells because they have discarded or have impaired G_1 and/or G_2 cell cycle blocks. Because of deficient repair, the split dose survival of cancer cells is represented by the intermediate curves (A, B, C) of Figure 7-4 — the survival achieved under conditions of incomplete repair and recovery. This difference between healthy cell cycling and cancer cell cycling provides the basis for the fractionated dose delivery therapeutic advantage. Much more on this in Chapter 10.

By comparing Figure 7-14 with Figure 7-11, it becomes apparent that fractionated dose delivery and reduced dose rate each increase the apparent radiation resistance of a particular cell line (as exhibited by the derived D_0). In radiation biology, these schemes are referred to as "protracted" and the process is called "dose protraction." The effects of dose protraction depend on the specific dose rate and fractionation recovery period. Thus, we need to account for time — the duration of the recovery period and the rate of dose delivery. Fortunately, the LQ formalism is sufficient for our purposes. The argument supporting the use of the LQ formalism goes as follows. The saturable repair model, which has never completely been discarded, relies on the hypothesis that the rate of repair decreases as dose increases (because the kinetics of repair depend on the concentration of repairasomes). David Brenner showed that the saturable repair model resolves to a form with the same time factor as the generalized LQ model. If Brenner's proof is correct, the LQ formalism applies to the saturable model as well as other nonsaturation models. The latter is transparent from Brenner's assumptions (Brenner, 2008), listed below.

1. Radiation produces DNA dsb with a yield proportionate to the dose.
2. These DNA dsb can be repaired, with first-order rate constant λ (equal to $\ln 2/T_{1/2}$, where $T_{1/2}$ is the repair half-time). In practice, there may be more than one class of DNA dsb, which may be repaired with different rate constants.
3. In competition with DNA dsb repair, binary misrepair of pairs of DNA dsb produced from different radiation tracks can produce lethal lesions, the yield being proportional to the square of the dose. The two independent radiation tracks can occur at different times during the overall regimen, allowing repair of the first DNA dsb to take place before it can undergo pairwise misrepair with the second.
4. Single radiation tracks can produce various lethal lesions, possibly by a variety of mechanisms, the yield being proportional to dose.

We subsequently assume that the LQ formalism is universally valid. To allow for dose protraction using the LQ formalism, it is necessary only to include a time factor:

$$S = e^{-(\alpha D + G\beta D^2)} \tag{7.20}$$

where G is the Lea-Catcheside time factor and is assumed to equal 1.0 in Equation 7.17. Inclusion results in a more general form of the LQ survival equation. Notice that G acts only on the quadratic component, as assumption 3 states clearly that binary misrepair leading to lethal lesions occurs only in the case of interacting independent DNA dsb. In other words, G affects the quadratic slope and not the initial slope (as $D \to 0$). The time-dependent yield (Y) of lethal lesions is proportional to the exponent of the generalized LQ survival equation,

$$Y \propto \alpha D + G\beta D^2. \tag{7.21}$$

The Lea-Catcheside Time Factor, G

The Lea-Catcheside time factor accounts for the removal of dsb through repair and the subsequent reduction in misrepair of interacting, independent dsb that results from protracted dose delivery. The generalized form of G is,

$$G=(2/D^2)\int_{-\infty}^{\infty}\dot{D}(t)dt\int_{-\infty}^{t}e^{-\lambda(t-t')}\dot{D}(t')dt'.$$

where λ is $\ln 2/T_{1/2}$, and D is dose. The second integral term represents the initial DNA dsb of the pair required to produce a lethal lesion. It describes the reduction in the number of initial dsb with time. The first integral term represents the second dsb that can interact with the remaining dsb produced earlier. G can take any value between 0 and 1, with 1 representing a single acute dose.

Protracted dose may be either fractionated or delivered at a continuous low dose rate. For short-lag fractionated delivery, G takes the form,

$$G = [2\theta/(1-\theta)][n-(1-\theta^n)/(1-\theta)],$$

where θ is $e^{-(\lambda t)}$ and n is the number of fractions. When the time between fractions is large, the time factor becomes,

$$G = 1/n.$$

For continuous low dose rate, G takes the form

$$G = [2/(\lambda T)^2][\theta - 1 + \lambda T]$$

where T is the duration of the radiation delivery.

7.5. Radiation Quality

The previous drivers that modify the nominal, cell type-specific survival curve have depended on biochemical response. The capacity for repair differentiates among cell states. Residual quadratic damage determines dose protraction effects. The size of the genome determines radiation sensitivity. Cell cycle stage establishes repair opportunity. The dominance of apoptosis within a cell line affects both the shoulder and the D_0. Nonetheless, when we consider radiobiological effects, we always must remember that there are two factors: biological and physical determinants of outcome. Once again, recall the factors that determine α according to our mechanistic assumptions: the number of sites that can sustain a DNA dsb (n_0), the DNA dsb hit probability constant (k_0), the number of unrestored DNA dsb (f_0), the proportion of DNA dsb that are potentially lethal (p), and the fraction of dose that induces DNA dsb through a single event (Δ). The final term, Δ, is a physical function of the radiation quality. Cellular response to various types of radiation: x- or γ-ray, electron, proton, neutron, or heavy charged particle, reveals the radiation's quality. Thus far, we have limited our discussion to photon radiation (x-rays and γ-rays). However, in Figures 2-17, 2-19 and 2-21, we pointed out that various radiations interact with matter differently. Because photons ionize atoms to produce secondary electrons, and because secondary electrons are responsible for subsequent biochemical endpoints (review Chapter 2), a description of electron radiation quality represents photon radiation quality well. Thus, the Δ factor and specifically the linear component (α) of the LQ formalism are dependent upon charged particle interactions with matter.

In 1977, the International Commission on Radiological Protection (ICRP) released *Publication 26*, a document establishing radiation quality factors to be used to determine human risk from radiation exposure. The quality factors (Q) were derived from physical interactions of radiation with matter;

Table 7-2 Quality factors for principle radiations

Radiation	Quality Factor (Q)	LET (keV/μm)
γ-rays and x-rays	1	(measured in path length, $1/\mu$)
Electrons	1	≤3.5
High-energy protons	10	53
Neutrons of unknown energy	10	(from the weighted average)
α-particles	20	173

primarily the linear energy transfer (LET) derived from microdosimetry of the particle track structure.[1] Table 7-2 lists Q for several particles of interest. Notice the Q for both photons and electrons equal 1.0 because, as pointed out, photons ionize secondary electrons and therefore the radiobiological effect is the same. At the other end of the spectrum, α-particles are 20 times more damaging than electrons because their mass (four times the mass of a proton) is large resulting in deposition of energy over a short range. Proton mass is one-fourth of an α-particle but still orders of magnitude larger than electron mass. This results in a dense, linear deposition of energy and a Q value of 10. Neutrons are a special case. Like photons, neutrons have no charge. Recall that depending on energy, neutrons interact with matter through several different mechanisms, each producing charged particle products. Depending on the products, Q will vary. In most cases of human exposure, the mixture of neutron energies is unknown so, the ICRP derived a weighted, most probable mixture value. We briefly discussed particle track structure in Chapter 2. Now, let's closely examine LET.

7.5.1. *Linear Energy Transfer*

LET reflects the pattern of energy deposition along a particle path; it reports how much energy is transferred from the particle to the medium per distance traveled. In general, LET is defined as the restricted stopping power; we examined stopping power in Chapter 2 (Equation 2.38). The expression for the *restricted* stopping power of electrons looks like this:

$$\frac{1}{\rho}\left(\frac{dE}{dl}\right)_{\Delta}=2\pi r_0^2 N_e \frac{\mu_0}{\beta^2}\left[\ln\frac{2(E+2\mu_0)(E-\Delta)\Delta}{\mu_0 I^2}+\frac{E}{E-\Delta}+\frac{\Delta^2/2+\mu_0(2E+\mu_0)\ln\left(\frac{E-\Delta}{E}\right)}{(E+\mu_0)^2}-1-\beta^2-\delta\right], \quad (7.22)$$

where Δ indicates the energy transfer limit to be considered as the restriction. As before, r_0 is the electron radius, N_e is the number of electrons per gram of the medium, E is the kinetic energy of the electron, I is the mean excitation energy, μ_0 is the mass energy equivalent ($m_0c^2 = 0.511$ MeV), β is v/c, and δ is a density correction factor. Notice that dl replaces dx as a reminder that the measured distance is along the particle path and is not the linear distance measured in the medium. This distinction becomes important in the case of light particles that experience relatively large deflections resulting from collisions with electrons (see Figure 2-19). Equation 7.22 provides an example of restricted stopping power; the expression for S_Δ varies with the particle under consideration. Nonetheless, the key term in all $(dE/dl)_\Delta$ formulas is the restrictive factor, Δ. The value of Δ is chosen to eliminate energy deposition beyond

[1] In 1991, the application of Q for determination of risk was abandoned and replaced by weighting factors, w_R, that were derived based on biological response rather than radiation physics.

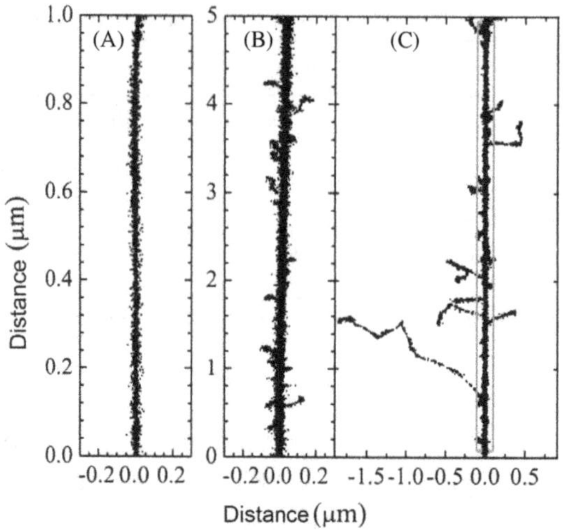

Figure 7-15 Track structure and LET. Three particle tracks in water, **(A)** 0.15 MeV proton, **(B)** 1.75 MeV/nucleon alpha, and **(C)** 25.5 MeV/nucleon C^{+6}, illustrate the transfer of energy (small dots) by secondary electrons and delta electrons. The choice of Δ determines the cylindrical boundary of the LET calculation represented by the gray cylinder in (c). Reproduced and modified with permission of the Radiation Research Society from Yusa Muroya *et al.*, 2006. High-LET ion radiolysis of water: visualization of the formation and evolution of ion tracks and relevance to the radiation-induced bystander effect. Radiation Research, 165, pp. 485–491.

a specified range. Energetic charged particles may knock off bound electrons imparting sufficient energy to generate delta, ionizing electrons. The energy transfers from delta electrons happen at some distance perpendicular to the primary electron track. Because LET is the energy transferred to the medium along the primary path, we want to eliminate the delta electron energy deposition. Figure 7-15 presents heavy particle tracks demonstrating the deposition of energy in water. Each fine black dot indicates an energy transfer event. Branching delta electrons can be seen in panels (B) and (C). If we want to know the energy transferred per distance traveled, the LET, we want to eliminate these branching energy transfers. The limiting factor, Δ, sets the diameter of a cylinder encompassing the path beyond which energy transfer events are not included. A gray cylinder in panel (C) of Figure 7-15 represents the limit; the surface expands or contracts with choice of Δ. An alternative way of thinking about Δ goes like this: Δ sets a limit on the amount of energy transferred in a single event. For any event above that limit, the energy is not included in the calculation. Why? Because large energy transfers are likely to generate delta electrons. To provide a tangible example of the function of Δ, the LET for a 1 MeV electron in water is 0.1256 keV/μm if the value of Δ is chosen to be 0.001, and 0.1477 keV/μm if Δ is chosen to be 0.01. For comparison, the stopping power for a 1 MeV electron in water is 0.1852 keV/μm.

Consider this: as the particle of interest travels through the medium, it continuously loses kinetic energy until it reaches the end of range and comes to rest ($T = 0$). If LET is energy dependent, which it is, then LET must also change as the particle loses energy. For this reason, a particle cannot be described by LET. We can speak only of "residual energy" (or "residual range"), or LET at a given depth that is coincident with the particle energy at that depth ($dE/dx = S$ and $(dE/dl)_\Delta = \text{LET}$). Furthermore, the greater the kinetic energy of a particle, the less the LET. This can be understood by examining Equation 7.22, but the logic is the same as before for stopping power. A particle possessing elevated kinetic energy travels at a high velocity ($E = ½mv^2$). As the particle slows, the potential for electronic interaction improves — the "stickiness" increases ($\beta = v/c$). As a particle approaches the end of range, energy

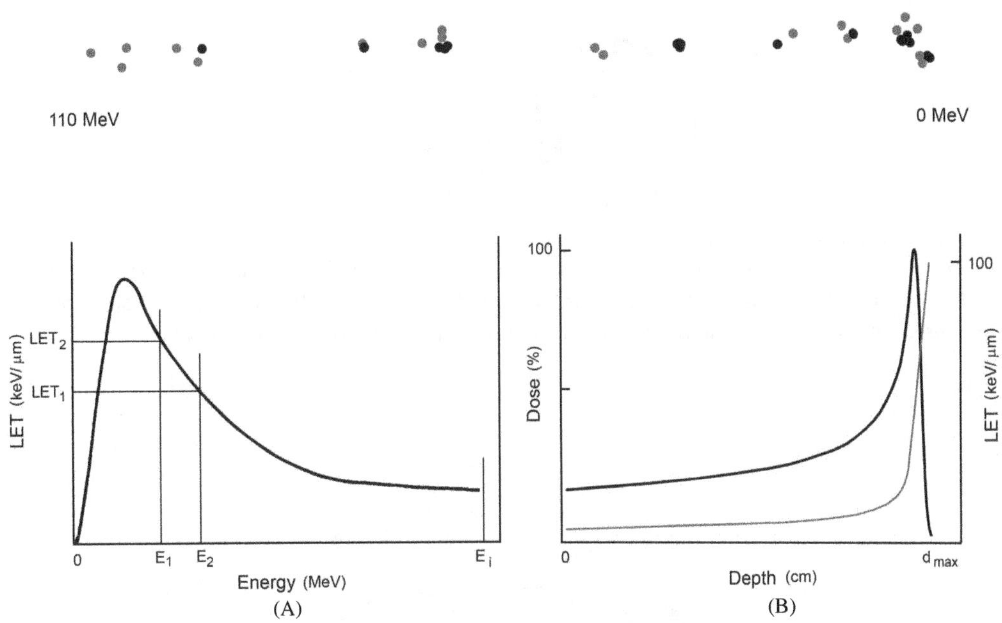

Figure 7-16 Comparisons of energy, dose, and LET. At top: an idealized proton track in water. The proton enters the medium at left with maximum energy (110 MeV). As the particle transfers energy (dots), the proton energy decreases. As the proton reaches maximum range, energy transfer events cluster. Gray dots represent excitation events, black dots represent ionization events. Panel A portrays the relationship between particle energy and LET. The x-axis increases from 0 MeV at the origin and therefore the particle enters the medium (E_i = initial energy) at the right side of the plot. Panel B examines the relationship between delivered dose and LET. Dose (black line) increases rapidly near the end of range (maximum depth) because of transfer event clustering. No dose is delivered beyond the range, $\mathfrak{R}$. The LET (gray line) increases rapidly near the end of range due to transfer event clustering.

deposition events occur closer and closer together. In other words, the LET increases. Recall how close together the events mapped near the end of range in Figure 2-19.

Figure 7-16 compares energy, dose, and LET. The graphic at the top shows an idealized proton track in water — the gray dots represent excitation events and the black dots represent ionization events. In this representation, the proton enters the medium at the left where it possesses maximum energy. As the particle penetrates deeper into the water (as it travels from left to right), it transfers energy through interactions with bound electrons (review Chapter 2, Section 2.7). Stopping power (S) rapidly increases as the proton nears its range ($\mathfrak{R}$) resulting from energy transfer event clustering. Figure 7-16 panel A demonstrates the relationship between particle energy and LET. Recalling that LET is simply the restricted stopping power, it follows that LET increases coincident with S as particle kinetic energy decreases, and increases rapidly approaching $\mathfrak{R}$. Because LET is energy dependent, each value along the x-axis has a corresponding value on the y-axis. LET is determined by particle energy. The x-axis in Figure 7-16 increases from 0 MeV at the origin, as customary. The particle enters the medium at a maximum initial energy, E_i, at the extreme right of this plot, and the particle can be envisioned as advancing toward the left as it loses energy. Panel B examines the relationship between delivered dose and LET. The transfer of energy to water results in the deposition of dose (review Chapter 2, Section 2.6.3). Therefore, the dose deposition as the particle travels deeper into the water phantom reflects stopping power and LET. Because S (dE/dx) and LET ($[dE/dl]_\Delta$) increase rapidly approaching $\mathfrak{R}$, so does the dose. Notice that LET (gray line) is small at higher energies (lower dose) and increases near $\mathfrak{R}$. No dose

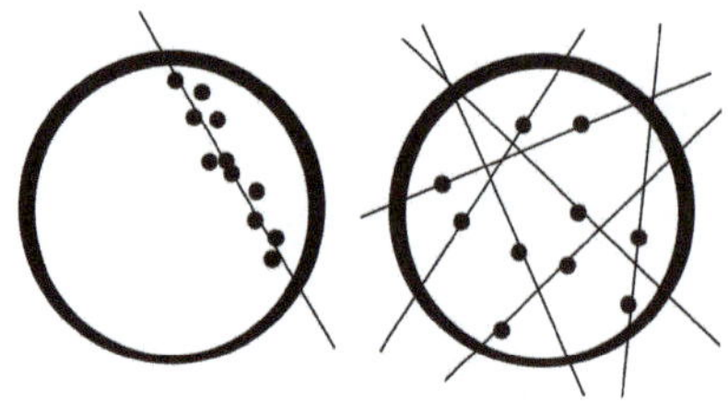

Figure 7-17 Particle tracks in microvolumes. The ~1 nm radius microvolume on the left represents exposure to heavy charged particles while the ~1 nm radius microvolume on the right represents exposure to photon radiation. Equal doses are delivered to the two volumes, each experiencing energy transfer events (dots) summing to equal intensity (in kJ).

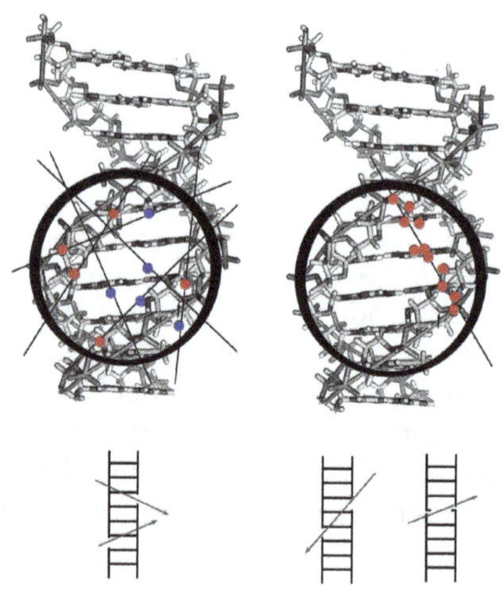

Figure 7-18 Treating DNA as an irradiated microvolume. The tracks of Figure 7-17 are superimposed on the DNA double helix. The image at left represents photon-induced secondary electrons. The blue excitation events release free radicals from water; the red ionization events result in direct damage to DNA. The image at right represents a heavy charged ion (e.g., proton) track. Heavy ion events are direct and clustered, causing multiple local damages. The schematics below illustrate the resultant types of damage, as per Figure 4-8.

is delivered beyond $\mathfrak{R}$; LET increases to a maximum coincident with $\mathfrak{R}$. The precipitous fall off of LET to zero at $\mathfrak{R}$ is difficult to model.

Now we can begin to understand why particles have different radiation qualities and why Q is based on LET. It seems that dose is not the whole story. Microdosimetry examines the pattern of dose distribution in small volumes (nm), as opposed to standard dosimetry which examines averaged dose in comparably large volumes (mm). The microvolumes in Figure 7-17 receive equal dose. A heavy charged particle penetrates the microvolume on the left while the microvolume on the right depicts electrons released by photon radiation. Recall that the probability of photon interaction (the path length, μ^{-1}) is on the order of cm, so it is unlikely that individual photons will release multiple electron tracks per microvolume. The patterns of energy deposition in the two microvolumes are very different, even though the dose is the same.

Recalling that the basis for radiation response is cell death, and that the relevant target for radiation within the cell is DNA, the impact of microdosimetry becomes clear. Figure 7-18 demonstrates the influence of track structure on DNA damage. Assume the DNA double helix constitutes a microvolume of approximately 2 nm diameter. Typical photon tracks superimposed on the double helix (left) might result in the energy transfer events indicated. Damage occurs primarily (~70%) through interaction of electrons with water molecules in the DNA hydration layer (blue dots). The probability of a direct hit (red dots) is low because energy transfer events occur infrequently. Individual events tend to cause DNA ssb and are separated in time, which links lethal damage to dose rate. On the other hand, superposition of the heavy particle track on a double helix microvolume might induce the energy transfer events indicated on the right. In this case, the probability of direct damage increases significantly. Energy deposition events frequently fall directly within the DNA strands and the interactions result in regions of multiple damages. This damage is not time dependent as events occur on a femtosecond scale, and therefore the damage is not dose rate dependent. Furthermore, although secondary electrons set in motion by heavy charged particles interact with water as readily as those set in motion by photons, because the free radicals are clustered in proximity, they tend to recombine to reform water with greater frequency. Figure 7-18 is somewhat naïve, but the comparison is easily extrapolated to DNA loops or chromosomal damage resulting in lethal aberrations.

We stated above that LET is generally referred to as the restricted stopping power. The International Commission on Radiation Units and Measurements (ICRU) defined LET as "the average energy, dE_L, locally imparted to the medium by a charged particle of specific energy traversing a distance dx." Therefore, LET cannot be — strictly speaking — the restricted stopping power for several reasons including

- Stopping power is valid for photons and neutrons (which are not charged particles).
- Implicit in the ICRU definition is a continuous slowing down approximation (CSDA).
- Each track structure is unique.
- LET is an average rate of local deposition of energy.

Consider Figure 7-16A. LET only can be stated precisely if the energy loss over a distance, $\Delta E\ (x_2 - x_1)$, and the energy transfer over a distance, $\Delta LET\ (y_2 - y_1)$, are extremely small. These restrictions have given rise to two definitions of *LET* (see Figure 7-19): track averaged (LET_{av}) and energy averaged ($(LET_{av})_E$).

The formalism for track averaged LET makes use of stopping power and looks like this:

$$(LET_{av}) = \frac{\int_0^\infty \varphi_r(x) S(r)\, dr}{\int_0^\infty \varphi_r(x)\, dr}, \tag{7.23}$$

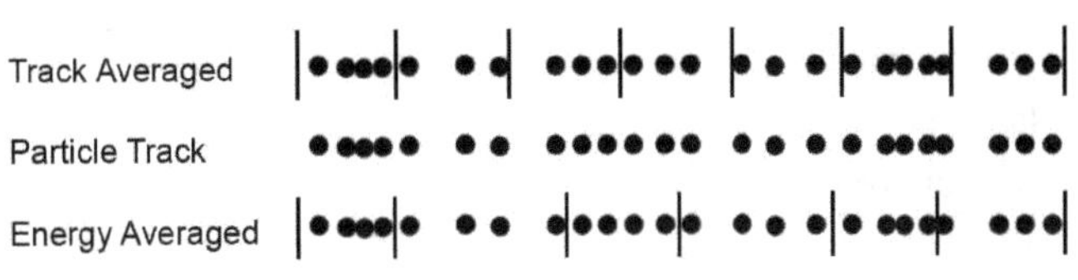

Figure 7-19 A schematic comparison of LET averaging methodologies. The track averaged LET (top) separates events into equally sized bins determined by path length. In this case, Δx is constant, and the energy varies. The energy averaged LET (bottom) separates events into equally sized energy bins. In this case, E is constant, and Δx varies.

where $\varphi(x)$ is the fluence of particles at x with residual ranges between r and dr, and $S(r)$ is the stopping power of particles with residual range r. We use residual range because the initial energy of particles will vary, but at a specified distance (r) short of $\Re(E_R = 0)$, the energy is precisely derivable.

To simplify calculation of LET, track averaged LET is often specified as track *segment* averaged $(LET_{av})_S$ when data is collected as a particle passes through a thin target, thus capturing only a short segment of the track (dx). This way, $(x_2 - x_1)$ is limited. In this case,

$$(LET_{av})_S = \frac{1}{(E_2 - E_1)} \int_{E_1}^{E_2} [LET(E)]\, dE. \tag{7.24}$$

In other words, $(LET_{av})_S$ is the sum of all LET values (along the y-axis of Figure 7-16A) defined by two energies (along the x-axis of Figure 7-16A), divided by the difference between the two energies — if the difference between the two energies is extremely small. This is true if the distance (dx) is small.

The energy averaged *LET* $(LET_{av})_E$ sums all the initial energies of all the particles set in motion by the primary particle and divides by the track segment length (Δl):

$$(LET_{av})_E = \sum_{i=1}^{n} E_i \left(\frac{1}{l_2 - l_1} \right). \tag{7.25}$$

The particles set in motion are generally secondary electrons in the case of proton radiation but various in the case of neutron radiation. It is also important that the distance over which the energy is measured is along the track (dl), not linear in the medium (dx). The differences between these two averages are significant, and it is important that caution be observed when quoting these averages.

Relating track structure to LET seems relatively straightforward when we are examining heavy charged particles (and neutrons, which emit heavy charged particles upon nuclear interaction). Low LET radiations, such as x-rays and gamma rays — that can only be considered as having LET values because they interact via light charged particles (electrons) — present a more complicated analysis. Table 7-3 indicates a 10 fold difference between the track averaged LET for ^{60}Co gamma rays (~1 MeV) and 250 kVp x-rays. Electron tracks are tortuous, manifesting large differences between dx (linear distance or depth in the medium) and dl (distance along the track). For that reason, we rely on CSDA to evaluate stopping power. Therefore, the general definition of LET — the restricted stopping power — is always used when discussing low LET radiation.

Table 7-3 A comparison of track averaged LET and energy averaged LET for several radiations.

Radiation	Track Average LET (keV/μm)	Energy Average LET (keV/μm)
^{60}Co gamma rays	0.27	19.6
250 kVp x-rays	2.6	25.8
3 MeV neutrons	31	44
Radon alpha rays	118	83
14 MeV neutrons	11.8	125
Recoil protons	8.5	25
Heavy recoils	142	362

Source: Reprinted with permission from Alpen (1990).

Rossi's Method for Determining Energy Averaged LET

Harald Rossi (1959) developed an ingenious method for determining the energy averaged LET. Because each heavy particle track is unique and therefore a generalized average energy is impossible, Rossi reasoned that tracks could be envisioned as energy deposited in spheres of uniform mass (such as water). An energetic charged particle originating inside or outside the sphere deposits energy (primary and secondary interactions) within the sphere. Let

E_γ be the energy deposited in the sphere by a heavy charged particle,
Y be the size of each energy deposition event,
d be the diameter of the sphere, and therefore
$E_\gamma/d = Y$ (for constant LET, Y is a constant).

It is now possible to examine the probability ($P(Y_d)$) that an event of size Y will occur in a volume of diameter d. The distributions of Y for various values of d can be plotted and compared. The advantage of this approach arises from a comparison of spheres of various diameters with energy deposition events of size Y, eliminating the need to know the precise track structures of each particle under consideration. The Rossi formalism looks like this:

$$\overline{Y}_D = \int_0^{\infty} Y \frac{\int_0^{\infty} Y^2 P(Y_d)\,dy}{\int_0^{\infty} YP(Y_d)\,dy}. \tag{7.26}$$

Rossi compared spheres of diameter 1.5 μm, 3 μm, and 6 μm for 500 keV neutrons in 1968. He found good agreement with experimental values of maximum LET (at the Bragg peak) for the two smaller spheres. The 6 μm diameter sphere failed to predict LET because many particle tracks were contained within the sphere; they did not completely traverse the sphere thus weighting the energy events toward lower Y (energy events at high energy, upstream of the narrow Bragg peak).

To relate Rossi's technique to radiation quality, let

Z be the locally deposited energy, and
ΔZ be the increment of locally deposited energy.

Now, if $Y = 80$ keV/μm and $d = 1.5$ μm, then ΔZ will be 10.9 Gy. This scenario is described by a single track intersecting a sphere of $d = 1.5$ μm as in Figure 7-17. Examination of the terms defined here discloses that Z predicts the frequency of events of size Y (i.e., LET). Furthermore, the probability ($P(Z)$) of a local energy density of Z in the irradiated medium defines radiation quality.

Mozumder and Magee (1966) described electron track structure in water (inducing hydrolysis) as three patterns: spurs, blobs, and tracks. Short tracks transferring 500 eV to 5 keV are rare in low LET radiation, but spurs transferring 6 eV to 100 eV, and blobs transferring 100 eV to 500 eV are ubiquitous. In fact, the clustered pattern of energy transfer events at the end of an electron track, the physics that requires an increase in energy transfer frequency as the energy approaches zero (the Bragg peak consequence of the Bethe–Bloch equation, Equation 2.38), results in the deposition of a blob. This higher LET volume with a diameter of approximately 7 nm provides the local multiple damages leading to the initial "α-component" (irreparable damage) slope of the low LET radiation dose response curve (Figure 7-1A). The representative electron track in Figure 7-20 has large, black dots signifying primary electron ionization events (connected by the black track). Small black dots indicate secondary electron ionizations and excitations. The difference between the linear distance (dx) and the path length distance (dl) is indicated by the double-headed arrows. A typical spur releasing three ion pairs resulting in six water free radicals

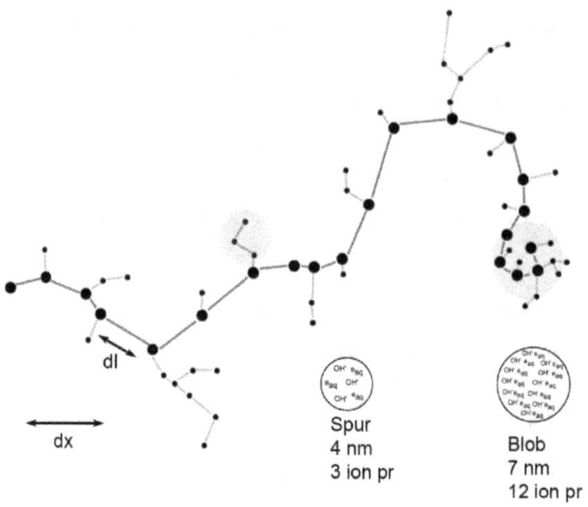

Figure 7-20 Depiction of an electron track in water. Large, black dots represent primary electron ionization. Small black dots represent secondary electron interactions. The difference between the linear distance (*dx*) and the path length distance (*dl*) is indicated. The primary electron possesses maximum energy at left and minimum energy at right. A typical spur releasing 3 ion pairs resulting in 6 water free radicals is indicated by the 4 nm gray disk. The low energy end of range exemplifies a typical blob releasing 12 ion pairs resulting in 24 water free radicals.

(see Figure 3-3) is indicated by a 4 nm gray disk. The low energy end of range exemplifies a typical blob releasing 12 ion pairs.

7.5.2. *Radiation Quality and the Survival Curve*

Again, our intuition tells us that if high LET radiation induces local multiply damaged sites and cDSB, high LET radiation should be repair resistant. In other words, the direct energy transfer from high LET charged particle interactions with DNA would be expected to cause complex irreparable, lethal damage. How would irreparable, lethal damage manifest in the shape of the survival curve? Well, if the shoulder reflects repair, then we might expect the shoulder to be diminished or absent. Figure 7-21 compares three common therapeutic high LET particles: neutrons, carbon ions, and protons, with typical x-rays. The curves represent data from independent experiments, but the collective presentation correctly portrays typical survival curve shoulders under similar conditions.

If we are to continue using the LQ formalism to model cell survival, the determination of α, β, and α/β becomes problematic for high LET radiations. Are the neutron and carbon slopes of Figure 7-21 proportional to α or β? How can the initial proton survival slope be accurately determined? Scholz *et al.* (1997) provided a resolution for this dilemma. In their publication, they modified the LQ relationship for x-rays (Equation 6.23) in such a way as to permit modeling survival for a specific average energy within the nucleus. They assumed that the dose dependence of cell inactivation by x-rays expresses as two parts: a shoulder region and a log linear region evident at doses greater than some threshold dose, D_t. Under these conditions, the LQ formalism looks like this:

$$-\ln S = \alpha_x D + \beta_x D^2 \tag{7.27a}$$

for $D < D_t$ (inclusive of the shoulder), and

$$-\ln S = \alpha_x D + \beta_x D^2 + s_{\max}(D - D_t) \tag{7.27b}$$

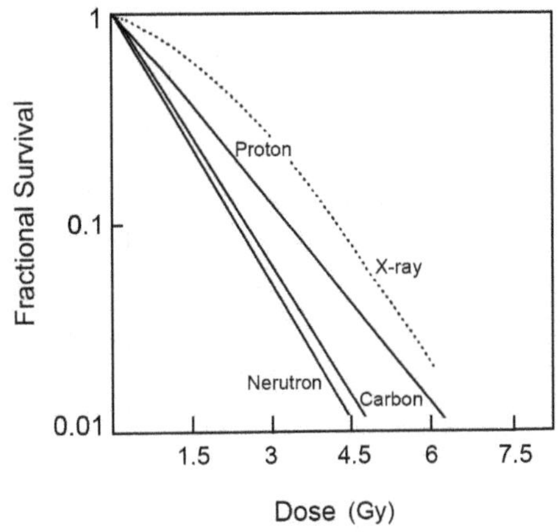

Figure 7-21 The effect of radiation quality on survival curve shape. Neutrons and carbon ions with initial energies of 8 MeV/u and protons with initial energy of 8 MeV are compared with typical standard x-rays. Cell survival following neutron radiation exhibits no perceptible shoulder while the carbon shoulder is barely discernable. The proton survival curve shoulder can be analyzed only if the ordinate is greatly expanded. The x-ray dose response curve is provided for comparison.

for $D \geq D_t$, where $s_{max} = \alpha_x + 2\,\beta_x D_t$ (the log linear slope) and the subscript x indicates x-ray radiation. To get from x-ray radiation to high LET radiation, it is necessary to pick a point in the nucleus rather than choosing an average energy over the entire nucleus (because the energy decreases relative to the residual range of each particle). Scholz designated the local dose at a given point, *d*, as the sum of the contributions of all ion tracks at the given point. The survival probability for a cell containing a nucleus that is experiencing tracks leading to the dose *d* at the selected point can be found by integrating *ln (S/d)*. The quantity $ln\ (S_x/d)$ corresponds to the x-ray survival curve over a nucleus of volume *V*. So, in general,

$$\ln S = \int (S_x(d))dV/V. \tag{7.28}$$

The survival probability, *S(D)*, at a given dose *D* can be derived as the average value of *S* from a large number of irradiated cells. Given *ln S* calculated for a few doses, *D*, you already know how to find α and β from Equation 6.27. Simply plot the linear form,–*(ln S)/D*, to get the intercept, α, and the slope, β (as in Figure 6-8).

Scholz's method, which has come to be known as the local effect model (LEM), fails at doses greater than about 10 Gy, but holds for most clinical predictions in particle therapy because few fractional doses exceed single digits. Friedrich (Friedrich, *et al.*, 2013) endorsed LEM and proposed that because the GLOBLE model predicts a continuously increasing negative slope from extremely low doses to the log linear ($D \geq D_t$) region, it should be possible to use low x-ray doses to predict survival resulting from high LET exposures. He proposed that after determining the number of iDSB and cDSB for a given particle at a given energy, the photon parameters ε_i and ε_c can be used to ascertain survival for cells exposed to charged particle radiation. Thus, Equation 6.62 can be considered universal, applying to both photon and high LET radiations. Many radiation biophysicists rebuff attempts to extrapolate x-ray survival methodology to particle radiation because of the ubiquitous assumption that inactivation can be expressed as a Poisson distribution. Whereas this assumption has always been troublesome but justifiable in the case of photons, it is considerably more egregious in the case of particles exhibiting unique, situationally dependent track structure. For this reason, Monte Carlo modeling remains the gold standard for particle radiation analyses.

Here is something else to consider. What radiobiological effect would fractionation (or protracted dose) have when using a high LET radiation source rather than a low LET radiation source? Ed Alpen

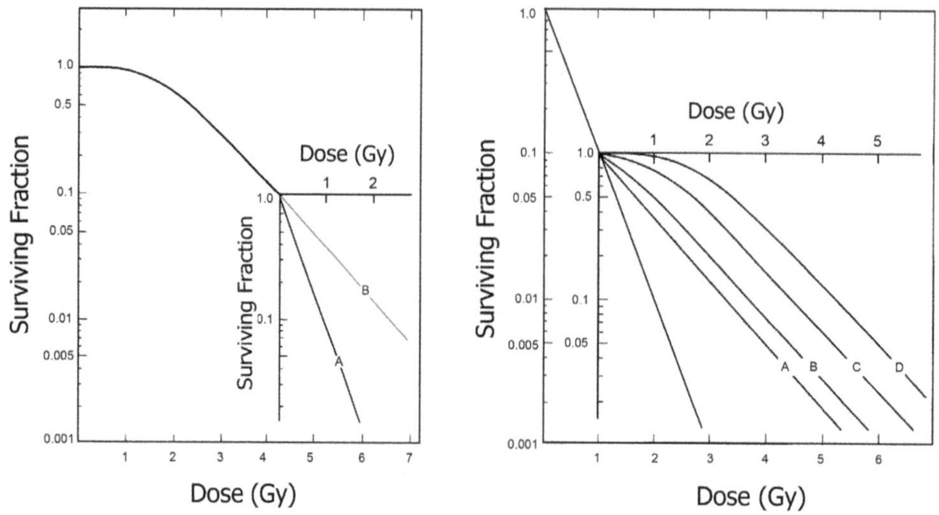

Figure 7-22 The Elkind–Sutton experiment hypothesized with high and low LET exposures. Left panel: cells are exposed to escalating doses of x-ray radiation up to the dose producing 1% survival (~4.2 Gy). The cells are permitted to recover for 30 minutes, 2 hours, or 6 hours followed by exposure to escalating doses of high LET radiation (**A**). Also shown is the predicted survival for a continued x-ray dose escalation with no recovery lag (**B**). Right panel: cells are exposed to escalating doses of high LET radiation. After 0 minutes (**A**), 30 minutes (**B**), 2 hours (**C**), or 6 hours (**D**) of recovery incubation, selected cells are exposed to escalating doses of x-ray radiation. Reproduced with permission of Elsevier, from "Radiation Biophysics." E.L. Alpen, Academic Press 1990.

(1990) considered this question and produced a representation of the Elkind experiment based on published data from several sources. Here is what they might have found (Figure 7-22). If first they exposed cultured cells to escalating doses of x-ray radiation, as described in Figure 7-3, but following the incubation increments they used escalating doses of high LET radiation (column 3 of the flowchart), the cell survival data plots as in the left panel of Figure 7-22. The x-ray survival plot is unremarkable, exhibiting the characteristic shoulder and log linear tail (continuous dose escalation is shown in gray at less than 90% survival). However, regardless of the repair incubation period, all high LET dose escalation data superimpose on a single survival plot (A). The results can be explained thusly. At 1% survival, 90% of the cells have been killed (or at least are incapable of forming a colony). The remaining 10% repair to the extent permitted by the incubation period. Regardless of their state of repair, exposure to high LET radiation introduces novel lethal damage. Inactivation of cells, repaired or not, results from the damage induced by the high LET radiation. Then Alpen considered cells exposed first to high LET radiation (Figure 7-22, right panel). At 1% survival, following 1 Gy exposure, cells are either exposed to a dose escalation using high LET radiation or they continue the dose escalation using low LET radiation (A). Plates of cells are incubated at 1% survival as we saw in Figures 7-3 and 7-4. Curves A, B, C, and D appear identical to Figure 7-4. Why? The surviving cells have experienced nonlethal damage, else they would not have survived. There are 90% fewer cells, but they are all viable. The 1% surviving cell chromosomes following high LET radiation exposure appear identical to the 1% surviving cell chromosomes exposed to low LET radiation. Subsequent low LET radiation exposure induces SLD that either compounds existing damage to become lethal or, in the case of repaired cells, introduces novel sublethal lesions. PLD repairs (or not) analogous to *de novo* x-ray exposure. The inactivated cells exposed under the two experimental protocols acquire lethal damage through different mechanisms, and so the cell sensitivity (D_0) differs between the two radiations. The D_0 for curve B in the left panel is the same as the D_0 for curve A in the right panel. The slope of A in the left panel is the same as the high LET slope of the right panel.

The mechanisms for producing damage differ, but the endpoint — the expression of lethal chromosomal damage or the absence of lethal chromosomal damage — is identical regardless of the radiation quality. Elkind would claim that fractionation has little effect on high LET radiation survival response. For decades, dose rate independence was particle therapy dogma. Nonetheless, a dose rate dependence during proton irradiation of haploid cells was first reported in 2004.[2] The dose rate effect during proton irradiation is now well documented.

High LET radiation does not exhibit the cell cycle dependence observed in Figure 7-9. Presumably, cell cycle dependence is suppressed because, regardless of the cell cycle stage at which irradiation occurs, the resulting damage is largely complex and irreparable. The reader is referred to work by John Ward (1994) and his colleagues for in-depth examination of radiation-induced complex, multiply damage sites in DNA.

7.6. Relative Biological Effectiveness

Suppose you are a clinician, and you would like to treat a child's brain tumor with protons rather than x-rays. You prefer this radiation because protons have a finite range, so it is possible to better conform the radiation to the tumor while protecting healthy brain tissue. Unfortunately, as we have just seen, 1 Gy of high LET radiation does not have the same biological effect as 1 Gy of low LET radiation (Figure 7-21), so you are uncertain what dose you should prescribe. The radiation quality is different. If you know the effective dose to cure the tumor using x-rays, can you determine the effective dose using protons? The necessary comparator is the relative biological effectiveness (RBE),

$$RBE = \frac{\text{dose for a given end point (reference radiation)}}{\text{dose for a given end point (test radiation)}}. \tag{7.29}$$

For example, if experience has shown that 6.75 Gy x-ray radiation is sufficient to reduce cell survival 99% and proton radiation requires 5.67 Gy to reduce cell survival 99% (as indicated in Figure 7-21), then the RBE of proton radiation compared to x-ray radiation is 1.2. The equivalent dose for a 60 Gy x-ray treatment would be 60 Gy/1.2 = 50 Gy of protons. Similar calculations can be performed for carbon ions and neutrons. It seems straightforward. Unfortunately, RBE ratios are based on assumptions that dose response curves for biological effect are described by identical functions. In fact, endpoint selection determines RBE. Figure 7-23 illustrates some of the disparities arising from end point choice in the case of calculating RBE for neutrons. Presented are commonly used end points: 10% survival, 1% survival, D_0, and α/β ratios. For example, in panel A, RBE = [x-ray dose resulting in 0.1 survival]/[neutron dose resulting in 0.1 survival]. In panel B, RBE = $D_{0,\text{ x-rays}}/D_{0,\text{ neutrons}}$. Panel C highlights another problem. Historically, ^{60}Co gamma radiation represented the universal standard reference. But gamma dose rates decline with the age of the source, and that can influence the dose response curve (Figure 7-12). Additionally, x-ray generators became more readily available and researchers began to favor 250 kVp radiation for the standard reference. Clinical linear accelerators generally operate above 6 MV accelerating potential (6 MV, 18 MV, and 24 MV are particularly popular accelerating voltages). So, RBE determination in a clinical setting should probably rely on high voltage machines to establish the appropriate reference. When we use each of these references, we arrive at different RBE values for neutron radiation. Table 7-4 lists the range of RBE values acquired under the circumstances presented in Figure 7-23.

The parameters discussed in this section represent only a few of the possible causes for variability in reported RBE values. Any physical, chemical, or biological factor responsible for altering the shape of the

[2] Split dose results for *C. cinereus* presented at the 2004 Particle Therapy Cooperative Group (PTCOG) workshop by S. B. Klein.

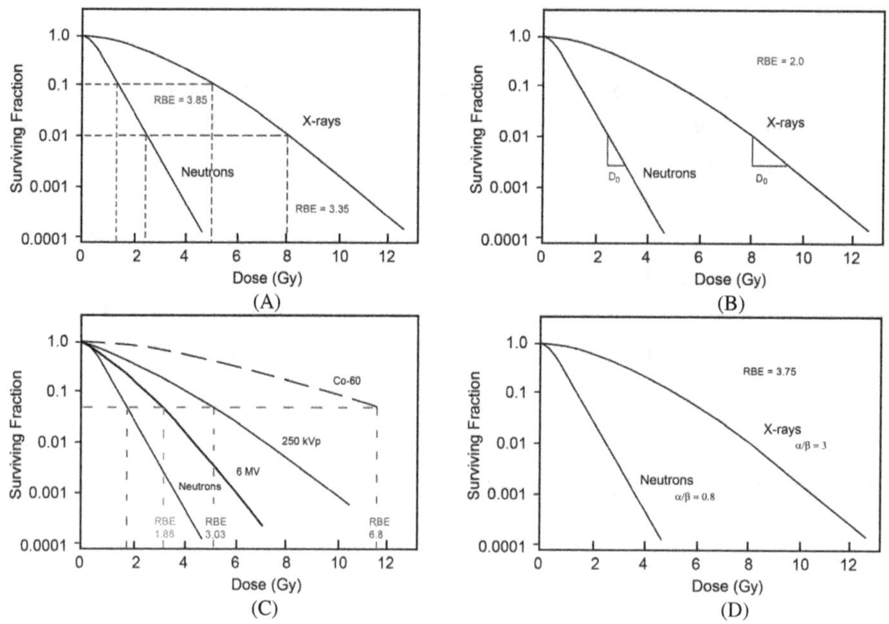

Figure 7-23 Alternative methods for determining the RBE for neutrons. (**A**) RBE is calculated at either the 10% survival (red) or the 1% survival (blue) doses. (**B**) RBE is calculated as a ratio of D_0 values. (**C**) Here, again, RBE is calculated at a given survival fraction; the numerator of the RBE ratio — the reference — employs photons of different energies. (**D**) RBE is calculated as the ratio of α/β values for the reference and test radiations.

Table 7-4 A comparison of RBE values for neutron radiation arrived at using endpoints and radiations illustrated in Figure 7-23

End Point	Reference Radiation	RBE
10% Survival	Diagnostic x-rays	3.85
1% Survival	Diagnostic x-rays	3.35
D_0	Diagnostic x-rays	2.0
5% Survival	6 MV x-rays	1.86
5% Survival	250 kVp x-rays	3.03
5% Survival	Co-60-γ-rays	6.8
α/β	Diagnostic x-rays	3.75

reference dose response curve (the numerator) also alters the RBE. For example, dose rate, repair competency, cell cycle stage, apoptosis capacity, fractionation scheme, cell line, or target size all affect the shape of the reference radiation survival curve and therefore the RBE. If we consider the effects of scale, RBE also varies with tissue type, tumor type, subject species, and individual patient. The most optimistic use of RBE as a comparator of radiation effect must be limited to the specifics of the situation. Most importantly, the circumstances under which an RBE value was derived must be thoroughly appreciated. Researchers and clinicians must never, never, never compare apples with oranges.

One final interesting point regarding RBE: increasing LET diminishes the repair shoulder and reduces D_0 indicating more effective cell killing. Does this imply that greater LET corresponds with increasing RBE? It does, to a limit. Figure 7-24 reveals the relationship between LET and RBE. Each point in Figure

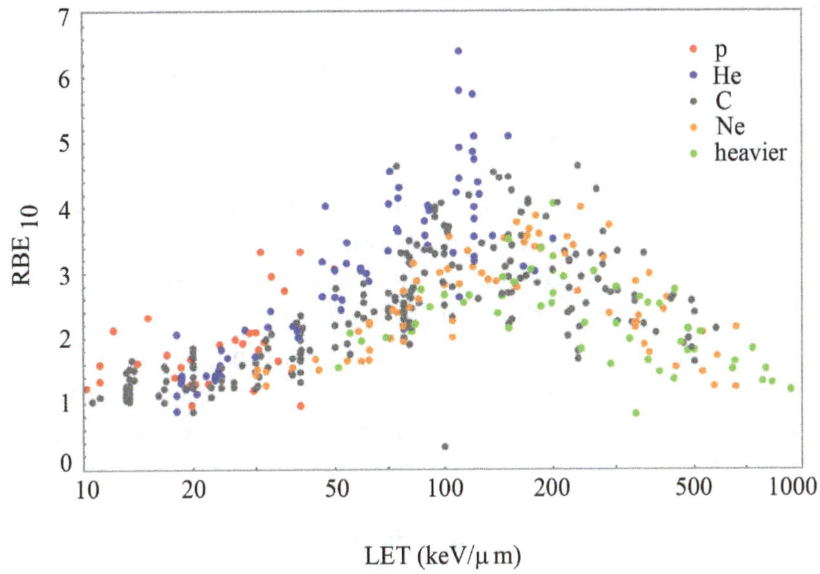

Figure 7-24 RBE vs. LET. Each point represents an ion evaluated at a specific energy, $(dE/dx)_\Delta$= LET. Legend describes ions measured. RBE reaches a maximum just above 100 keV/μm for all ions, and then declines. Data from the PIDE database, https://www.gsi.de/bio-pide, plotted by Thomas Friedrich. Reproduced with permission by CC BY-NC-ND 4.0 from Durante, *et al.*, 2019. Applied nuclear physics at the new high-energy particle accelerator facilities. *Physics Reports*.

7-24 represents a dose response experiment conducted using particle radiation, as in the studies of Figure 7-21. The initial particle energy dictates the radiation LET reported for each survival curve. For example, the proton curve of Figure 7-21 is typical of 60 MeV proton acceleration. To calculate the corresponding RBE, the test endpoint is the 10% S (RBE_{10}) resulting from particle irradiation and the reference is the 10% S resulting from 250 kVp x-rays, as in Figure 7-23. A clear trend can be seen in Figure 7-24. All particles exhibit maximum RBE at just above 100 keV/μm. Beyond that, the RBE declines. Why? Think about track structure. Sufficiently high LET has an elevated probability of inducing DNA dsb (Figure 7-18) leading to lethal chromosomal aberrations. If the LET increases — if the track structure is more densely ionizing (i.e., if there are more dots per distance along the track in Figure 7-18) — the same number of DNA dsb result, even though each individual lesion may be more severe. So, a LET increase beyond some point has no effect on individual cell survival. The excess energy transfer events are wasted. Additionally, if more energy is deposited along an individual track, and the dose is held constant, fewer tracks are responsible for the deposition of dose. Therefore, fewer cells experience tracks resulting in lethal damage. The survival curve begins to shift up; the D_0 increases. Therefore, the ratio of reference/test radiations decreases. The RBE versus LET curve slopes downward at LET greater than approximately 200 keV/μm.

7.7. Radiation Response Modification

Ideally, clinicians would like to make tumors more responsive to radiation and healthy tissue more resistant. We have examined radiation biophysics, biology, and biochemistry; and we have introduced several models that might prove useful for guiding manipulation of radiation response. A phenomenon that we have not discussed is the "oxygen enhancement ratio (OER)."

7.7.1. *Oxygen Enhancement Ratio*

Equation 3.6 described the fixation of DNA damage through the addition of molecular oxygen:

$$DNA\cdot + O_2 \rightarrow [DNA\cdot - \text{peroxide}]. \tag{7.30}$$

Oxygen fixation initiates the process that results in SLD and PLD induced by water free radicals. Obviously therefore, oxygen plays an important role in radiation DNA damage. Presumably, if oxygen is unavailable damage fixation will not occur. To test this hypothesis, researchers perform a dose response experiment in the traditional manner with cells held in ambient air, then repeat the experiment using cells held under nitrogen gas (N_2). The survival curves look something like those in Figure 7-25. The cells irradiated under nitrogen are more resistant because the chemical reaction responsible for fixation of damage is suppressed. The comparison of dose response curves is called the OER and it is calculated as

$$OER = \frac{dose\ in\ N_2}{dose\ in\ O_2}, \tag{7.31}$$

at a designated S (0.1 in Figure 7-25). As might be expected, OER is ubiquitous. Because the effect results from basic chemistry, OER can be detected in bioactive molecules, haploid cells, diploid yeast, and mammalian cells. But is it pertinent at larger scales?

OER experiments have been conducted on spheroids — balls of cells grown in culture — and several attempts have been made to alter the oxygenation of tissue *in situ*, but it is difficult to imagine that this technique could be useful for altering radiation sensitivity at the scale of an organism. Nonetheless, some physicians will place patients in a hyperbaric chamber prior to radiation therapy to increase systemic oxygenation. Radiation resistance resulting from reduced oxygenation becomes extremely important in the context of tumor necrosis and anoxia. Tumor biology is fascinating and complex. We will examine it in some detail later in this text. For now, we note that oxygen diffuses out of capillaries where hemoglobin within red blood cells exchanges O_2 for CO_2. Diffusing O_2 is absorbed by cells; it does not effectively permeate tissue more than 150 μm. So, once a tumor grows beyond about 30 cells in diameter, the center cells begin to suffer anoxia. Tumors respond by secreting vascular endothelial growth factor (VEGF) and other cytokines that encourage nearby capillaries to expand into the tumor. This "neovasculature" inevitably fails to provide sufficient oxygen at some point during tumor growth;

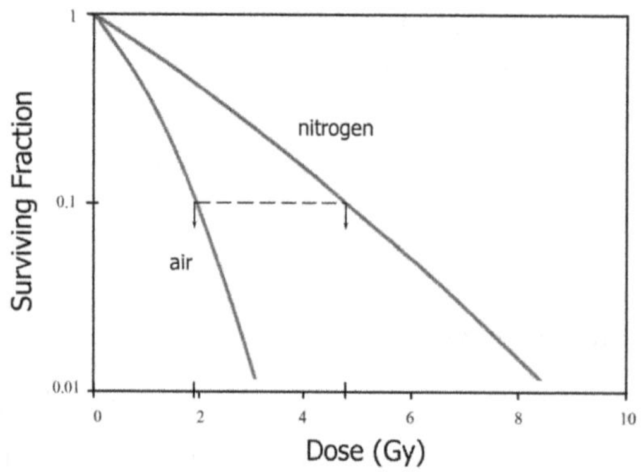

Figure 7-25 Oxygen enhancement ratio. Two typical x-ray dose response experiments are represented, one performed in air and one performed under nitrogen gas. Here, the OER is determined at 0.1 survival. Ticks on the abscissa indicate the doses required in each experiment to attain 90% cell inactivation. The OER = 2.5.

tumors always contain areas of hypoxia (reduced oxygen levels), anoxia (severely reduced oxygen levels), and necrosis (oxygen deprived dead cells). These regions of the tumor are radiation resistant reflecting the OER phenomenon. Radiation delivery techniques have been developed to promote reoxygenation of these cells and return them to a similar sensitivity state as the remainder of the tumor. Although the radiation sensitivity differences between oxygenated tumor cells and anoxic tumor cells can be explained by Figure 7-25, it is inappropriate to refer to "OER" on this scale. The definition of OER (Equation 7.31) holds only within the laboratory tissue culture setting.

High LET radiation diminishes OER. Because high LET radiation acts primarily through direct DNA damage, and because the basis of OER is fixation of DNA radicals resulting from radiolysis (indirect damage), oxygenation status is less relevant for high LET radiation cell inactivation. Heavy ion radiation thusly becomes advantageous for the treatment of large tumors.

7.7.2. *Chemical Modification [Radioprotectors and Sensitizers]*

In Chapter 3, Section 3.4, we described several chemical reactions resulting from the interaction of hydrolysis products with DNA resulting in radicals within the DNA macromolecule. The restoration of native structure involves exchange of hydrogen (Equation 3.5). Might it be possible to bias the reaction kinetics in favor of the restitution reaction? Shortly before 1950, it was discovered that high concentrations of the amino acid cysteine could impart radiation protection to irradiated mice. Cysteine (Figure 7-26) belongs to a chemical family referred to as "thiols" that contain a sulfhydryl (SH) group with a relatively loosely bound hydrogen atom. Equation 7.32 depicts a generalized restoration reaction involving a DNA radical and a thiol.

$$DNA\cdot + [R - SH] \rightarrow [DNA - H] + RS\cdot \tag{7.32}$$

Thiols prove effective when added to the medium of cultured mammalian cells, and mice injected with about 150 mg/kg of cysteine or cysteamine (an analogue of cysteine) are twice as resistant to radiation as control animals. In addition to the reaction represented in Equation 7.32, thiols scavenge water free radicals. However, the local concentrations of free radicals induced by low LET radiation would require extremely high concentrations of thiols to have significant impact. Furthermore, the radioprotectant would need to be delivered locally and immediately prior to irradiation; the rate of production of hydrolysis products overwhelms diffusion constants for thiols. In general, radioprotectants do not function through the removal of hydrolysis free radicals for these reasons.

Following World War II and the deployment of nuclear devices with horrific post-detonation consequences, the U.S. army synthesized and tested more than 4,000 potentially radioprotective compounds. One thiol, WR2721 (S-(2-(3-animo propylamino))ethyl-phosphorothioic acid) proved to be relatively effective *in vitro*, but was at best minimally protective (1.5–3 fold) in most tissues and not at all protective for the central nervous system (CNS). Even worse, at protective doses WR2721 was toxic, inducing nausea, vomiting, hypertension, and depression. Nonetheless, because it was not protective of the CNS,

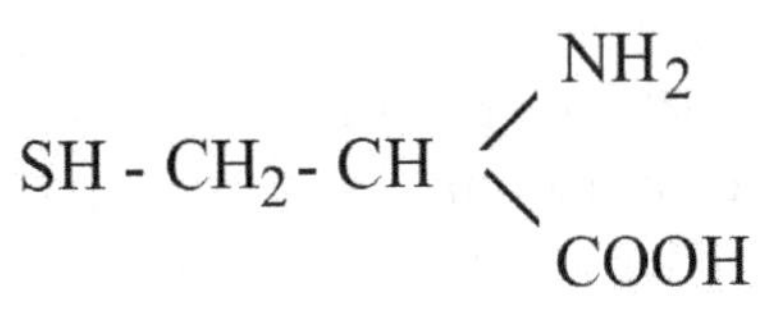

Figure 7-26 Structure of cysteine.

Reversal of OER by NO·

Cells become resistant to radiation-induced damage in the absence of oxygen because Equation 7.30 leading to DNA · fixation becomes impossible.

$$\text{DNA}\cdot + \text{O}_2 \nrightarrow [\text{DNA}\cdot - \text{peroxide}]$$

Restitution of nascent DNA becomes the favored reaction given sufficient opportunity. Adding O_2 to the anoxic gas used to generate the results of Figure 7-25 resurrects cell sensitivity to radiation by reinstating the fixation chemistry of Equation 7.30, leading to stable base damage that requires base excision repair (BER).

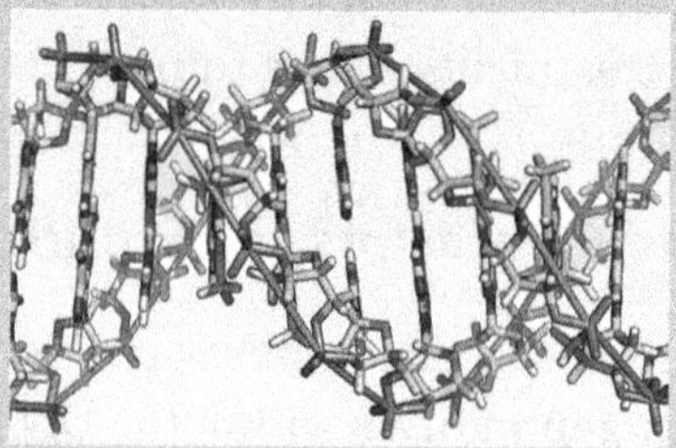

Nitric oxide (NO·) likewise reacts with DNA· but in this case forms stable, atypical bases (e.g., xanthine) that must be subsequently removed through nucleotide excision repair (NER).

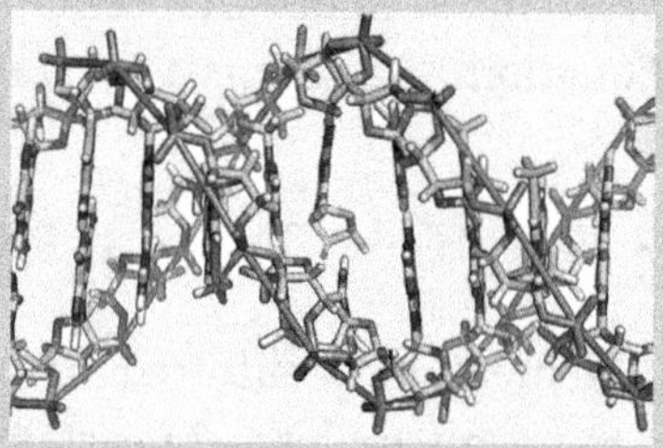

The addition of NO· to oxygen-deprived cultures diminishes OER (restores cell sensitivity to radiation) without resupply of O_2 because NO· establishes base damage independently through reaction of DNA radicals with nitric oxide radicals. Evidence suggests that potentially lethal damage established by supplemental NO· is more resistant to repair than O_2-induced damage, and subsequent cell death is highly replication-dependent (because it likely results from stalled DNA replication forks; see Figure 4-4).

the Food and Drug Administration (FDA) approved the drug (known as amifostine or Ethyol®) for clinical use, specifically for cancers of the head and neck. During radiation treatments, WR2721 protects the skin, and preserves salvation and speech while maintaining radiation sensitivity in the CNS tumor. The effect may prove unexpectedly universal; it has been shown that WR2721 penetrates tumor tissue more slowly than normal tissue. If the drug is administered subcutaneously approximately 30 minutes prior to treatment, a (non-CNS) tumor may fail to be protected while surrounding tissue is protected. Treatment with WR2721 remains unpopular because there is always a possibility that the drug might provide protection to the tumor, in which case the prescribed dose will fail to be effective.

A class of radiosensitizers called nitroaromatics became popular in the 1970s. Nitroaromatics perform the same function as oxygen during fixation of DNA damage and successfully reversed OER.

There was some enthusiasm for the use of these compounds in large tumors, anticipating that hypoxic cells could be returned to nominal sensitivity. Unfortunately, in practice, standard fractionation protocols known to be effective had to be altered to allow for the administration of the drug. Physicians voiced concern about abandoning tried and true methodology. Furthermore, when the drug was used coincident with standard fraction protocols, no effect was detected. Because tumors reoxygenate between radiation fractions (see Chapter 8, Section 8.5.3), this also may help explain the lack of efficacy for hypoxic sensitizers.

When 5-bromouracil incorporates into the DNA of a test sample, the genetic material becomes super-sensitive to radiation. *In vitro,* it is sufficient to supply culture medium supplemented with 5-bromouracil. Cells take up the thymidine substitute and incorporate it onto the newly synthesized DNA strand during replication. The bromine atom in the halogen-substituted pyrimidine is uniquely radiation sensitive and renders multiple DNA ssb upon exposure to any type of radiation. Delivery of 5-bromouracil specifically to tumors remains elusive, but it is of interest because the mechanism of action is unique and independent of radiation LET.

7.8. Summary

- When calculating survival, it is necessary to account for PE. Not all plated cells attach, metabolize, and divide to form a colony.
- A typical dose response curve exhibits an initial slope, a final slope, and a shoulder.
- MTSH fails to accurately depict the initial slope exhibited by most mammalian cell survival curves and does not provide a justifiable explanation for the variability in shoulder width. However, the MTSH parameter D_0 is useful for describing the low survival, log-linear slope; and D_q is useful for describing the initial portion of the curve.
- LQ adequately describes the initial slope, the form of the shoulder, and the final log-linear slope.
- LQ model assumptions are based on DNA dsb. More recent evidence suggests that these assumptions need to be modified to provide accurate interpretation of the mechanics determining cell inactivation.
- RMR formalism indicates that repair rates determine the dose response.
- Cybernetic and LPL models replace the LQ assumptions based on DNA dsb with assumptions based on lethal damage and PLD. Lethal damage is assumed to be derived from either (1) direct events or (2) interdependent events (PLD or damage in quadrature).
- GLOBLE describes the shape of the survival curve shoulder as the ratio of the probability that cell death results from a single loop scission to the probability that cell death results from a cluster of loop scissions (two or more).
- The log linear, final slope of the survival curve is presumed to be determined, in all models, by two interdependent events. This explains the quadratic nature of the slope.
- Elkind and Sutton empirically determined the mechanism responsible for the survival curve shoulder: DNA repair.
- According to empirical evidence, the initial slope of the survival curve reflects cell inactivation by direct, single event, irreparable DNA damage. The shoulder represents DNA damage repair. The final slope specifies cell type radiation sensitivity.
- The following traits affect the shape of the survival curve:
  - Cell type
  - Target size

  - Repair competence
  - Cell age (cell cycle phase)
  - Apoptotic activity
  - Dose protraction (rate or fractionation schema)
  - Radiation quality
- Fractionation creates a therapeutic advantage because cancer cells respond to radiation differently from healthy cells. Often, cancer cells do not stop to repair before dividing.
- The general form of the LQ equation allowing for dose protraction is $S = e^{-(\alpha D + G\beta D^2)}$ where G is the Lea-Catcheside time factor.
- The ICRP established radiation quality factors to facilitate determination of human risk from radiation exposure. The quality factors (Q) were derived from radiation physics, primarily LET.
- LET is a mathematical description of track structure.
- LET is energy dependent. Thus, LET must be referred to as LET at energy E, or LET at residual range $\mathfrak{R}$, or LET at depth d.
- The most common definition of LET, but the least precise, is the restrictive stopping power: $(dE/dx)_{\Delta}$.
- Definitions of LET include "track averaged" and "energy averaged."
- Hydrolysis events along an electron track can be described by spurs (6 - 100 eV), blobs (100–500 eV), or tracks (500–5000 eV).
- Dose response curves resulting from high LET radiations exhibit diminished shoulders. Repair of local multiply damaged sites is ineffective.
- According to LEM theory, the LQ parameters: α, β, and α/β, can be determined for high LET radiation by plotting $-lnS/D$ for x-rays. LEM loses validity at doses above 10 Gy.
- The dose rate effect is suppressed during high LET irradiation because most DNA damage presents as local multiply damaged sites that are irreparable.
- $\text{RBE} = \frac{\text{dose for a given end point (reference radiation)}}{\text{dose for a given end point (test radiation)}}$.
- The RBE value is influenced by reference radiation selection and the specifics of the dose response protocol.
- A plot of RBE versus LET reaches a maximum at approximately 100 to 200 keV/μm.
- $\text{OER} = \frac{\text{dose in } N_2}{\text{dose in } O_2}$ at a designated S.
- The OER > 1.0 is demonstrable in biomolecules and cell culture.
- Although radiation resistance due to anoxia is documented in tumors, the oxygenation status of tumor cells is transient and complicated. It should not be referred to as OER.
- Thiols function as radioprotectants through the donation of hydrogen that resolves DNA radicals.
- 5-bromouracil sensitizes DNA to radiation; however, delivery *in vivo* is problematic.
- Analysis of dose response curves provides us with "The 4 R's of Radiobiology":
  - Repair
  - Reassortment
  - Repopulation
  - Reoxygenation

7.9. Problems

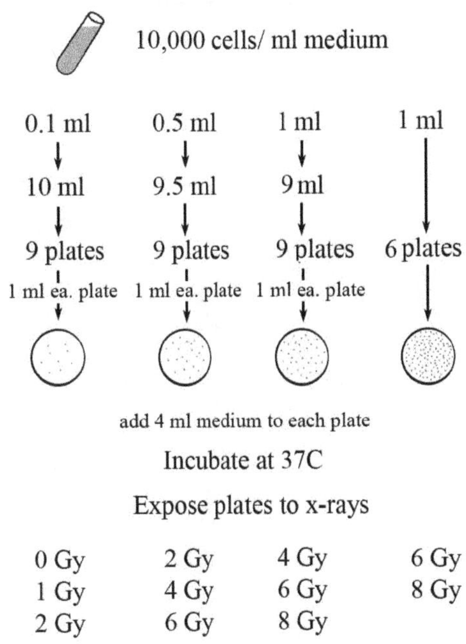

1. Suppose you are performing a dose response experiment. You harvest several plates of mammalian cells, suspend them in tissue culture medium and count them in a hemocytometer. You adjust the concentration to 10,000 cells per mL in 20 mL of medium. The illustration to the right represents the experiment setup. You draw out 100 μL and add these cells to 10 mL medium. Of this, you place 1 mL into each of 9 petri dishes and then add 4 mL TC medium to each dish to provide 5 mL of nutrient. Draw out 500 μL of the 20 mL suspension and add that to 9.5 mL medium (to get 10 mL). Add 1 mL of this suspension to each of 9 plates and top off each with 4 mL medium. Then draw out 1 mL of the 20 mL suspension and add it to 9 mL medium. Place 1 mL into each of 9 plates. Top off with 4 mL medium. Finally add 1 mL of the 20 mL cell suspension to each of 6 plates. Top off with 4 mL medium. Incubate plates at 37°C and allow cells to attach. Irradiate 3 plates from the first group to 1 Gy and return to incubator. Irradiate 3 plates of the first group to 2 Gy and return to incubator. Irradiate 3 plates of the second group to 2 Gy and return to incubator, and so on. After 6 days, remove the plates and count the colonies.
   a. Why is it necessary to plate different densities of cells?
   b. How many cells are delivered to each plate in each of the four groups?
   c. Why did we expose three plates to each dose?
   d. What is the plating efficiency?
   e. Complete the spreadsheet.
   f. Plot the survival curve.

Cells plated	Dose (Gy)	# Colonies	<survival>	S/So
	0	93		
	0	87		
	0	91		
	1	56		
	1	53		
	1	61		
	2	43		
	2	40		
	2	38		
	2	209		
	2	195		
	2	206		
	4	58		
	4	45		

Cells plated	Dose (Gy)	# Colonies	<survival>	S/So
	4	43		
	6	11		
	6	3		
	6	8		
	4	107		
	4	94		
	4	97		
	6	14		
	6	17		
	6	12		
	8	0		
	8	0		
	8	4		
	6	150		
	6	146		
	6	134		
	8	15		
	8	10		
	8	13		

2. Suppose you performed the experiment in Question 1 and added a split dose subset after 6 hours of incubation (maximum recovery). Using the MTSH parameters D_0, D_q and n, sketch the resulting plot that displays the original, continuous dose escalation and the repaired/recovered dose escalation (similar to Figure 7-12A).
3. If you delivered a fractionated dose regimen of 2 Gy per fraction to cells of the same cell line used in Question 1, with a lag of 6–12 hours, what would your dose response curve look like (see Figure 7-12D for an example)? Provide a sketch. Compare the acute dose D_0 with the fractionated dose D_0.
4. Construct the flowchart (correct and complete Figure 7-3) for the experiment that would generate the results illustrated in Figure 7-4.
5. According to Kappos and Pohlit, the average reaction rate constant for water free radical with DNA (Equations 3.1–3.4) decreases because competing scavenger molecules (Equations 3.6 and 7.32) are replenished by the cell during the incubation lag. In other words, the concentration of scavenger molecules increases during the lag. Assume that the reaction is first order with respect to the concentration of reactive species,

$$r_1 = \Delta[DNA]\Delta[OH\bullet] \quad and \quad r_2 = \Delta[SH]\Delta[OH\bullet],$$

   and that the instantaneous production of hydrolysis product (represented here as OH•) upon irradiation remains constant. The nominal reaction rates for both water free radicals with DNA and sulfhydryl are on the order of 10^8 mol/s (see Chapter 3, Section 3.4). What would this situation imply with respect to competition between DNA and scavenger (*SH*) for free radical capture? Compare the theoretical change in reaction rate for scavenger (r_2) when the concentration is reduced during radiation exposure, that is, without lag versus with recovery during the lag. Are there other competing reactions? If so, name one.
6. Explain Figure 7-9 using only the degree of chromosome condensation at each cycle phase. Assume G_1 phase chromatin is the baseline.
7. Determine the D_0 and shoulder breadth for each curve of Figure 7-10.

8. The standard clinical dose rate is 2 Gy/min. Calculate the treatment duration if a 60 Gy treatment is delivered in a single fraction (as it might be for stereotactic radiosurgery). Calculate the treatment duration at 2 Gy per fraction each weekday (as it might be for prostate therapy). Would the two treatments be equally effective at killing cancer cells? Why or why not? If not, how would you adjust them?
9. Approximately what is the minimum stopping power, in keV/μm, of an electron that can produce a DNA dsb?
10. The stopping power (dE/dx, and thus LET, $(dE/dx)_\Delta$) for 250 kVp photons (Table 7-3) differs from the stopping power of 250 keV electrons. Why might you expect this to be true?
11. A proton enters a water tank at 175 MeV. What is the energy at 16 cm depth (z = 16 cm)? Use the tables provided by NIST at https://physics. nist. gov/PhysRefData/Star/Text/intro. html (select p star for protons). Show your calculations.
12. Use the following table (LET given in keV/μm) to determine the track averaged LET of the proton in Question #11 (Data from Fada Guan *et al.*, 2015. Analysis of the track- and dose-averaged LET and LET spectra in proton therapy using the geant 4 Monte Carlo code. *Medical Physics*, 42(11)).

	$z = 0$ cm	$z = 1.9$ cm	$z = 3.95$ cm	$z = 4.75$ cm	$z = 4.85$ cm
LET_t	0.89	1.1	1.9	6.2	10.4

13. Equation 7.27a can be used to find α and β by graphing $-lnS/D$ versus dose (D). Show that Equation 7.27b resolves to a similar form for $-lnS$. Evaluate your expression at a dose of 8 Gy, using the threshold dose at S = 0.01 of Figure 7-14A ($D = 6.2$ Gy).
14. Find the RBE for proton radiation if the α/β for protons is 6.2 and the α/β for 250 kVp x-rays is 9.3.
15. Palcic and Skarsgard presented some noteworthy results while investigating the graph OER (reproduced with permission of the Radiation Research Society from Palcic and Skarsgard, 1984. Reduced oxygen enhancement ratio at low doses of ionizing radiation. *Radiation Research*, 100, pp. 328–339). Why was it interesting that they found a reduced OER at low doses (see the title of their paper)? To answer this question, consider these logical steps.
    a) What is the chemical rationale explaining the OER effect?
    b) Is this pertinent to linear or quadratic damage? Why?
    c) Examine the graph carefully. Do you suppose that they used one of the models discussed in Chapters 6 and 7 (MTSH, LQ, RMR, LPL, GLOBLE)? Why or why not?
    d) Where would this data plot on Figure 7-25? Is this region dominated by linear or quadratic events?
    e) Why was it interesting that they found a reduced OER at low doses?

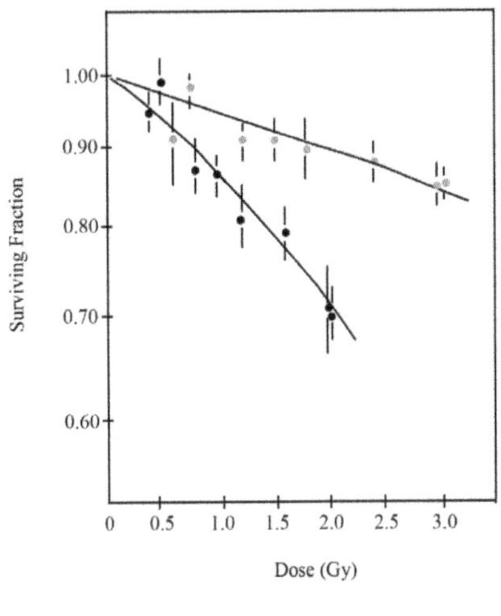

7.10. Bibliography

Alpen, E. L., 1990. *Radiation Biophysics.* 2nd ed. San Diego: Academic Press.

Brenner, D. J., 2008. Point: The linear-quadratic model is an appropriate methodology for determining isoeffective doses at large doses per fraction. *Seminars in Radiation Oncology,* 18(4), pp. 234–230.

Brinkman, E. K. *et al.*, 2018. Kinetics and fidelity of the repair of Cas9-induced double-strand DNA breaks. *Molecular Cell,* 70(5), pp. 801–813.

Elkind, M. M. & Sutton, H., 1959. X-ray damage and recovery in mammalian cells in culture. *Nature,* 184(4695), pp. 1293–1295.

Elkind, M. M., Sutton, H. & Moses, W. B., 1961. Postirradiation survival kinetics of mammalian cells grown in culture. *Journal of Cellular and Comparative Physiology,* 58(3), pp. 113–134.

Friedrich, T., Durante, M. & Scholz, M., 2013. Modeling cell survival after photon irradiation based on double-strand break clustering in megabase pair chromatin loops. *Radiation Research,* 178, pp. 385–394.

Goodhead, D. T., 1994. Initial events in the cellular effects of ionizing radiations: clustered damage in DNA. *International Journal of Radiation Biology,* 65(1), pp. 7–17.

Kappos, A. & Pohlit, W., 1972. A cybernetic model for radiation reactions in living cells. I. Sparsely-ionizing radiations; stationary cells. *International Journal of Radiation Biology and Related Studies in Physics, Chemistry, and Medicine,* 22, pp. 51–65.

Liu, Y. *et al.*, 2017. Transcriptional landscape of the human cell cycle. *Proceedings of the National Academy of Sciences,* 114(13), pp. 3473–3478.

Mozumder, A. & Magee, J. L., 1966. Model of tracks of ionizing radiations for radical reaction mechanisms. *Radiation Research,* 28(2), pp. 203–214.

Rossi, H. H., 1959. Specification of radiation quality. *Radiation Research,* 10(5), pp. 522–531.

Scholz, M., Kellerer, A. M. & Kraft-Weyrather, W., 1997. Computation of cell survival in heavy ion beams for therapy. *Radiation and Environmental Biophysics,* 36, pp. 59–66.

Ward, J. F., 1994. The complexity of DNA damage: relevance to biological consequences. *International Journal of Radiation Biology,* 66, pp. 427–432.

Winterbourn, C. C., 2008. Reconciling the chemistry and biology of reactive oxygen species. *Nature Chemical Biology,* 4(3), pp. 278–286.

Zolan, M. E., Tremel, C. J. & Pukkila, P. J., 1988. Production and characterization of radiation-sensitive meiotic mutants of Coprinus cinereus. *Genetics,* 120(2), pp. 379–387.

Chapter 8

In Vivo Radiation Response

8.1. Introduction

With a firm foundation of radiation dose response in hand, we now address the critical question: "So what?" What relevance does this research, performed at the cellular or molecular scale, have to organisms? In the comparatively tightly controlled environment of tissue culture, isolation and manipulation of experimental factors is reasonably straight forward. Pertinent variables include DNA damage characteristics, damage repair, cell cycle progression, and processes resulting in cell inactivation. At the scale of organisms, however, hormones and immunology may impact the responses of the irradiated target(s). These responses can also be influenced by internal or external environmental variables such as an organ's differentiation state or internal organization that can alter the responses and reactions to stresses such as radiation. Individuals vary in gender and age with biochemical ramifications to those states. So . . . Is it possible that data obtained from cultured cells can be extrapolated to whole organisms?

The only way to answer this question with certainty is to perform observations at the scale of organisms and compare the results to those obtained at the scale of cell culture. We can form hypotheses based on our understanding of cell survival and test for success or failure in more complex systems. Fortunately, many intrepid researchers have gone before us and provided empirical evidence that affords some confidence that in many cases cellular radiation dose response characteristics extrapolate to higher scale.

8.2. Scaling Up from *In Vitro* Dose Response

Unlike physics and engineering, analyses in biology are challenging because component parts cannot *a priori* be examined in isolation and then be expected to behave comparably when studied in an organism. Biology is complex. For example, DNA radiation damage can be analyzed in a test tube and biochemical actors can be added and tested, but DNA in a cell is organized and compacted into chromatin, heterochromatin, and chromosomes; cellular DNA is modified during cell cycling and its epigenetic state can be altered, and it may be fragmented through apoptosis. Each level of biological organization, each scale, adds complexity to the function under examination. Thus, it is necessary to reexamine the phenomena at each scale to establish the impacts of an increasingly elaborate and complex environment.

In this text, we casually have been referring to three experimental systems that we shall now define for the sake of clarity. The least complex system is *in vitro*. The Latin phrase refers to biological experiments conducted in test tubes, culture plates, flasks, or other apparatus allowing the examination of material external to an organism. When studying something *in situ*, the phenomenon or feature under examination resides "in place" within an organism. The organism supports the attribute under examination in a manner analogous to tissue culture (supplying nutrients, controlling gas partial pressure, maintaining temperature, etc.). However, the most inclusive study needs to consider all components of a living animal. Often, these animal models are genetically identical and are held under restrictive conditions to minimize the number of environmental factors, but when studying humans, isolation is seldom ethical. These studies are identified as conducted *in vivo*, "in life," as part of the living organism. The three

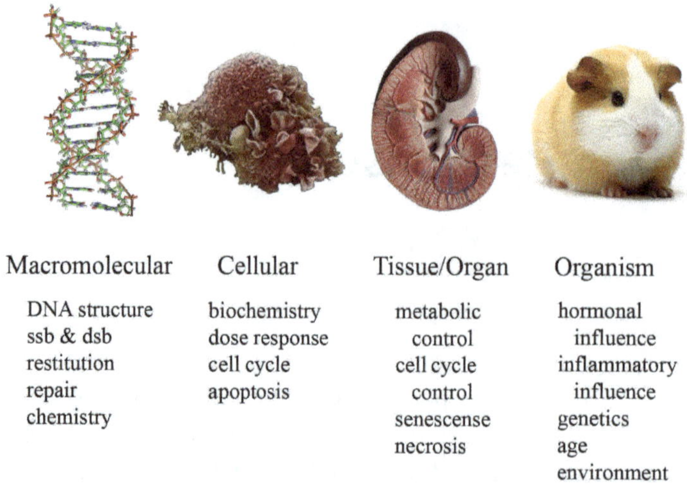

Figure 8-1 Matters of scale. As the system under examination includes increasing phenomena that influence radiation response, characterizing the interplay of factors becomes increasingly difficult. Extrapolation of hypotheses from one scale to the next requires verification. These four scales represent convenient levels of complexity.

investigative systems might be considered increasing scales of complexity, although the overlap between *in situ* and *in vivo* can sometimes be ambiguous. Physiological scales, such as those presented in Figure 8-1, present sharper boundaries. Additionally, we can address radiation sensitivity at each scale and interrogate the extension of *in vitro* dose responses (at molecular and cellular scales) to the two more inclusive systems (*in situ* and *in vivo*) and the two higher scales (tissues and organisms). Best practice identifies both the system and the level of scale when designing and analyzing experiments.

8.2.1. *Colony Forming Unit Assays*

Rod Withers developed an *in situ*, dose response protocol exploiting mouse skin (Withers, 1967), a conveniently accessible organ. Using shielding, he sterilized an annulus of skin with sufficient radiation to inactivate the skin stem cells and permanently prevent repopulation (Figure 8-2). The annulus exposure was performed on many, many mice. He then removed the shielding from each mouse and exposed the previously protected inner disk of skin to radiation. Delivering the dose was tricky; it had to be high enough to damage the dividing cells (stem cells) but low enough to not sterilize the test patch. After recovery, the disks of skin that were exposed to prescribed doses of radiation exhibited small patches of recovered growth. Withers assumed that each recovered skin growth was the product of a single surviving stem cell that produced a colony of differentiated skin cells. The colonies were counted as in a traditional *in vitro* experiment. Wither's mouse skin experiment represents a clever but straightforward extension of a second scale (cellular) experiment to a third scale (organ) experiment. It also exemplifies the expansion of radiation biology studies from *in vitro* to *in situ*.

Two years after he published his *in situ* work using mouse skin, Withers published a paper with Elkind (Withers & Elkind, 1969) reporting on the radiation sensitivity of mouse gastrointestinal crypt cells. Jejunal crypt cells of the gastrointestinal lining provide an excellent, isolated, *in vivo* system for the investigation of radiation sensitivity. The small intestine is lined with villi — fat, soft bristles that increase the surface area of the intestinal wall. The villi improve absorption of nutrients from digested food. As food passes through the gut, cells forming the thin epithelial surface of the villi are shed at the tips of the villi and need to be constantly replenished. So, at the base of each villus is a pit — a crypt — where stem

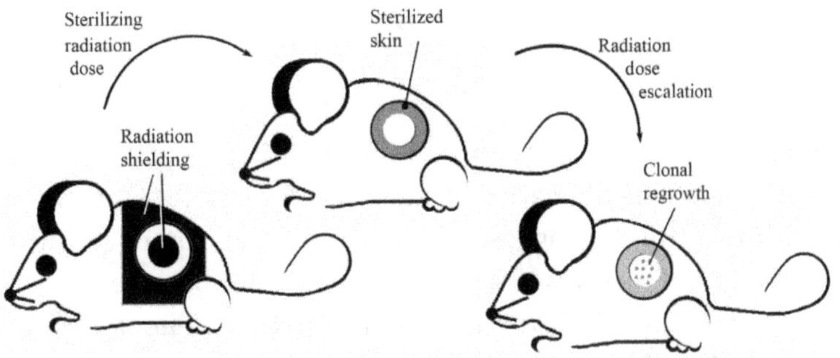

Figure 8-2 Withers' mouse skin CFU assay. An annulus of skin was sterilized using high doses of radiation. Then, a patch of skin interior to the annulus was exposed to a dose sufficient to damage cycling stem cells and halt cell division. After sufficient time, cells recovered and resumed cycling, creating clones of surviving cells. The number of recovered clonal colonies could then be scored to reveal the percent survival similar to *in vitro* clonal assays.

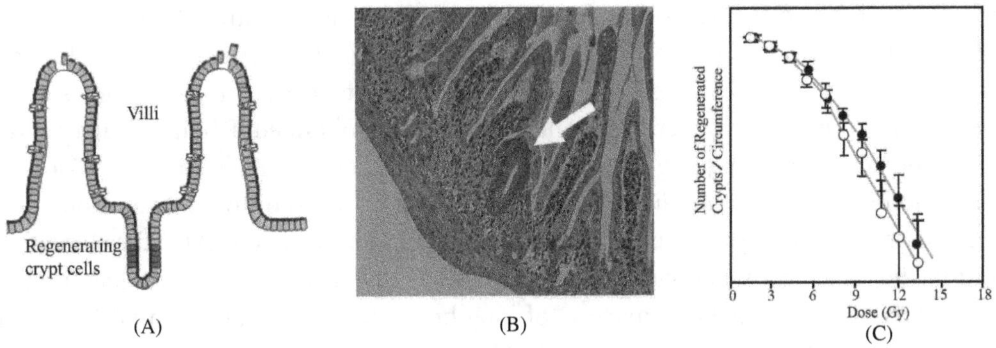

Figure 8-3 *In vivo* radiation response in organs. Jejunal crypt cells rapidly divide to replenish the epithelial lining of the small intestine. (**A**) Schematic cross section of the organ lining illustrating two villi and the associated crypt of Lieberkühn. This schematic is modified from Boumphreyfr — https://commons. wikimedia. org/wiki/File:Epithelial_shedding. png. (**B**) Histology of intestine cross section showing the intestinal lumen, a crypt (arrow), and several villi twisting in and out of the slice plane. The villi contain vascular tissue and RBC. (**C**) Experimental results obtained from irradiation of mouse gut using mixed energy (1 MeV–70 MeV) protons (•) or low energy (> 1 MeV) protons (o).

cells divide, continuously producing one new stem cell and one progenitor cell that migrates up the villus and differentiates into an epithelial villus cell. Irradiating the gut with sufficient dose kills cycling stem cells in the crypt without affecting the noncycling intestinal cells that migrate up the villi and shed off, so the villi will quickly shorten over time. Radiation response can be determined by simply counting the number of regenerating crypts visible in the cross sections of mouse small intestine. The regenerating crypts are akin to cell colonies *in vitro*, appearing a few days postirradiation. John Gueulette, Blanche De Coster, and S.B. Klein used this system to interrogate the relative biological effectiveness (RBE) of wide bandwidth, mixed energy, therapeutic proton radiation versus low energy, narrow bandwidth, proton radiation in the Bragg peak (Figure 8-3). If x represents the number of regenerating crypts and N is the total number of crypts in control animals divided by the circumference of the intestinal cross section, the fraction of crypts sterilized (f) can be determined by:

$$f = \frac{N - x}{N}. \tag{8.1}$$

The average fraction of surviving clonogens can be found by assuming Poisson statistics ($f' = N\,(-\ln f)$), yielding,

$$f' = N\left(-\ln\frac{N-x}{N}\right). \tag{8.2}$$

Plots of the number of clonogens (or the number of regenerating crypts) versus dose resemble *in vitro* dose response curves with shoulders indicative of sublethal damage (SLD) repair and D_0 reflecting sensitivity. The data plotted in Figure 8-3 suggest that crypt cells are more sensitive to the lower energy (higher LET) proton radiation. These results are not surprising, but the shape of the dose response curves justifies the extrapolation of LQ formalism to higher scales. The *in vivo* system under examination requires no synthetic isolation; the scale is at the level of an organ.

Spermatogenesis presents another convenient *in vivo* system for radiation sensitivity studies. Spermatogonial stem cells divide to produce a stem cell and a type B cell that differentiates into a primary spermatocyte. The primary spermatocyte divides meiotically into two haploid spermatids, and the spermatids are transformed into spermatozoa by the process of spermiogenesis. Withers took the advantage of this process as well (Withers *et al.*, 1974). He irradiated the testes of mice to 16 increasing ^{137}Cs doses, allowed the testes to recover for 42 days, removed, fixed, and examined the cross sections using standard histology. The results are shown in Figure 8-4. Although it is difficult to assess low-dose effects and establish the shoulder of the curve, the sensitivity of spermatogonia can readily be resolved from the exponential killing observed at higher doses and the D_0 can be obtained. Repair capacity (the shoulder) can be ascertained by conducting a split dose experiment and deriving D_q as in Figure 7-4.

A combination of *in situ* and *in vivo* techniques also can be used to investigate colony forming unit (CFU) dose response at the organ/tissue scale. In 1961, Till and McCulloch (1961) developed a technique for examining the radiation sensitivity of bone marrow cells. The skeleton provides more than just structural support. The spongy bone marrow of long bones harbors multipotential hemopoietic stem cells. These cells divide to produce either lymphoid progenitors or myeloid progenitors. The lymphoid progenitors differentiate into B cells, T cells, and NK cells, all responsible for infection clearance. The myeloid progenitors differentiate into mast cells (histamine secretion), erythrocytes (red blood cell [RBC]), thrombocytes (clotting), or myeloblasts. The myeloblasts differentiate into basophils, eosinophils, neutrophils, or monocytes. The monocytes differentiate into either dendritic cells or macrophages (clearance of cellular debris). What an extraordinary organ! The multipotent stem cells cycle

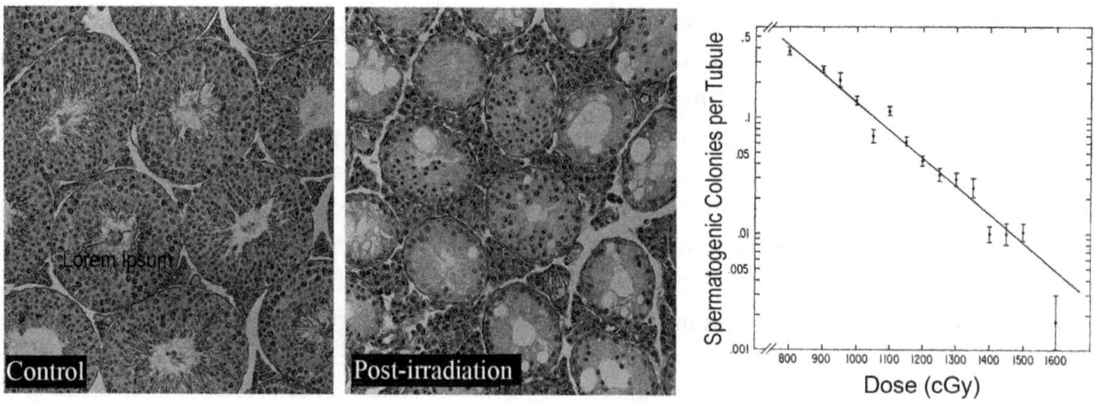

Figure 8-4 Radiation sensitivity of spermatogonial stem cells. The fraction of surviving cells is determined from a comparison of the number of cells in the tubules of unirradiated mouse testes (left) to the number of cells post-irradiation (center). Radiation sensitivity is determined by plotting the number of surviving cells per tubule per control cells vs. dose (right). Reproduced and modified with permission of the Radiation Research Society from Withers *et al.* (1974).

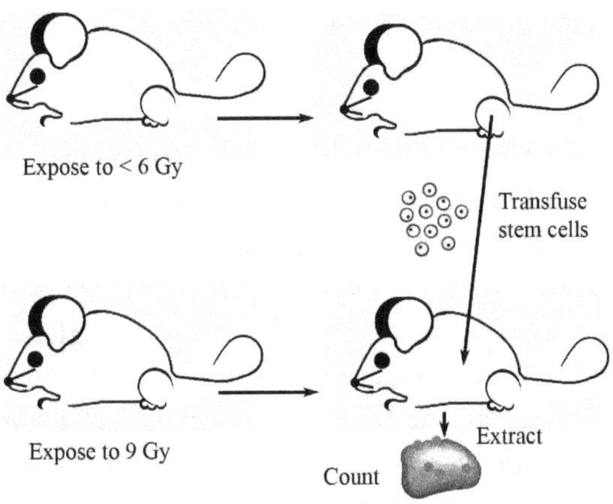

Figure 8-5 Clonogen transplant assay for bone marrow stem cell radiation sensitivity. A cohort of recipient mice are exposed to 9 Gy x-ray doses to sterilize the spleen. Cohorts of test mice are exposed to doses ranging from 0 to 6 Gy. Following exposure, nucleated marrow cells are withdrawn from the test mice and injected intravenously into the recipient mice where they form nodules on the spleen. The spleens are removed, and the nodules counted.

continuously, producing short-lived progeny fulfilling this multitude of functions. As continuously cycling cells, the multipotent stem cells would be expected to be extremely radiation sensitive. But how is one to measure this, when the progeny repopulates at different rates? Till and McCulloch approached the problem this way (Figure 8-5). Cohorts of mice were irradiated with increasing x-ray doses, according to standard dose escalation protocol. A second set of recipient mice were exposed to 9 Gy total body dose — a dose sufficient to sterilize the blood stem cells in their bone marrow and spleens. Following exposure, nucleated, isologous bone marrow cells were withdrawn from the test mice blood and injected intravenously into the recipient mice. Some of the injected cells find their way into the spleen and produce nodules of new growth. Each nodule requires tens of thousands of injected cells because many of the donor cells have differentiated (are not stem cells) and only a small fraction of injected cells migrate to the spleen and embed. The fraction of cells that succeed in producing nodules is akin to a plating efficiency and is assessed in a parallel manner using unirradiated donor mice. The surviving fraction of multipotent stem cells at each dose is then determined by,

$$\mathrm{SF} = \frac{\textit{nodules counted}}{(\textit{cells innoculated})\mathrm{PE}}. \tag{8.3}$$

This technique irradiates the stem cells *in vivo* and grows the nodules *in situ*. The methodology is called a "clonogen transplant" assay. Survival curves constructed using data from these assays agree well with those constructed from similar experiments where extracted bone marrow cells are irradiated *in vitro* and then injected into recipient mice, as well as monoclonal cultures of bone marrow cells subjected to standard *in vitro* dose escalation protocols. The results agree well, but not exactly. It has been hypothesized that small differences in D_0 arise from quiescent cells recruited from G_0 following *in vivo* irradiation.

8.2.2. *Functional Assays and Fractionation Studies*

Jack Fowler, a prolific radiation biologist and infamous character, began publishing in the 1960s about the effects of fractionation on radiation dose response expressed in pig skin. He did not perform CFU assays, however. Instead, he used skin reaction as a measure of radiation response. Skin reaction is a complicated interplay of cell death, immune infiltration, inflammation, and wound repair, but nonetheless follows a predictable and consistent pattern of development. Fowler created an arbitrary scoring system for the severity of response provided in Table 8-1. Erythema is a reddening of the skin caused by dilation of irritated capillaries. It is familiar to those who have become sunburned during a beach vacation. Desquamation is blistering and loss of skin resembling a heat-induced burn. In his publication (Fowler *et al.*, 1963), Fowler exposed patches of pig skin to increasing doses of radiation. Shown in Table 8-1 are three doses: 17.5 Gy, 20 Gy, and 22 Gy, delivered in a single, acute exposure. Also shown are three doses delivered in 21 fractions over 28 days: 45 Gy, 60 Gy, and 72 Gy. The reactions to the highest doses appear rapidly and are severe, diminishing with time. Lower doses tend to produce less severe reactions that increase slightly and plateau over time. Fractionation clearly increases skin tolerance; a single dose of 17.5 Gy induces the same response as 45 Gy of fractionated dose at 28 days. Fowler refined his scale using mouse hind leg and mouse foot skin, but the system remained qualitative at best.

Tissue kinetics are an important aspect of investigations at this scale. Notice that Fowler observed irradiated skin over a period of 100 days. The relationships between dose responses varied over time unlike clonogenic assays wherein an optimal sampling time point can be determined. In *in vivo* systems such as Fowler's — inclusive of vascular tissue, fibroblasts, infiltrating monocytes, skin stem cells, and so on, each responding to radiation doses at different rates — the reaction must be observed for a sufficient time. Results must always be reported at a specific dose/dose rate and a specific time point postirradiation. For this reason, functional radiation biology assays are generally presented as graphs of response versus time rather than survival versus dose. Nonetheless, the ratio of α/β, aka LQ analyses, can be obtained from these experiments. Fractionation is again the key. If Jack Fowler's data is evaluated for biologically equivalent dose — that is, the dose that delivers identical biological effect ("isodose" in the language of therapy) — we arrive at the values presented in Table 8-2 where the fractionation schemes are presented as # fractions delivered over # days (f/d). The table tells us that an acute dose of 17 Gy will produce visible erythema (score = 1.5), as will 35 Gy delivered in 5 fractions over 28 days, 45 Gy delivered in 9 fractions over 28 days, and 55 Gy delivered in 21 fractions over 28 days. These fractionation schemes all produce the same biological effect.

A plot of the reciprocal dose versus the dose per fraction for biologically equivalent endpoints provides α/β as per Equation 6.27 and Figure 6-8B. Fowler presented his data in this format in a subsequent publication (Figure 8-6). At the time of publication, the old unit of rad was still in use. Recall that 1 rad = 1 cGy. To compensate for the large total doses delivered, the units are provided in krads (1 krad = 10^3 rads = 10^3cGy = 10 Gy). The abscissa is equivalent to reciprocal total dose (1/2.5 krads = 4 krads^{-1} or, 1/25 Gy = 0.04 Gy^{-1}). Thus, α/β may be derived by either of two methods:

Table 8-1 Jack Fowler's system for scoring skin response to radiation exposure. Three doses were delivered and observed daily for 100 days. Scores are shown at day 28 and day 60 postirradiation.

Score	Skin Reaction	Single Dose (Gy)		Dose Over 21 Fractions (Gy)	
		28 Days	60 Days	28 Days	60 Days
0	No visible reaction				
1	Faint erythema	17.5		45	45
2	Erythema	20	17.5, 20		60
3	Marked erythema		22		
4	Moist desquamation < 50% of the irradiated area			60	72
5	Moist desquamation > 50% of the irradiated area	22		72	

Table 8-2 Jack Fowler's pig skin irradiation data evaluated on the basis of biological equivalency.

	Dose in Gy			
Score	**Acute**	**5f/28d**	**9f/28d**	**21f/28d**
1.5	17	35	45	55
2.0	19	40	50	66
2.5	24	47	55	72

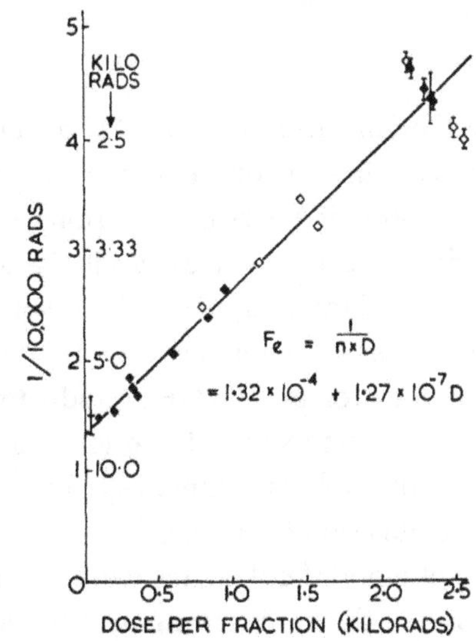

Figure 8-6 Derivation of α/β from fractionation protocols. Douglas and Fowler used mouse skin reactions to find α/β. The abscissa is equivalent to the reciprocal of total dose (provided to the right of the tick). $\alpha/\beta = 0.013/0.00127 = 10$. Reproduced with permission of the Radiation Research Society from The Effect of Multiple Small Doses of X Rays on Skin Reactions in the Mouse and a Basic Interpretation. B. G. Douglas and J. F. Fowler, Rad Res 1976; 66:401–426.

$$\frac{\text{intercept}}{\text{slope}} = \frac{\alpha / \ln S}{\beta / \ln S} = \frac{\alpha}{\beta}, \quad \text{or} \tag{8.4}$$

$$F_e = \frac{\alpha(n)D + \beta(n)D^2}{nD} = \alpha + \beta D \rightarrow \frac{\alpha}{\beta}. \tag{8.5}$$

Justification for extrapolation of LQ formalism to *in situ* and *in vivo* studies appears fairly solid based on the empirical evidence presented thus far. However, our discussion has been centered on cell sensitivity, not cell repair capacity. For that analysis, it is necessary to present the initial slope and shoulder, or at least a value for D_q, on a graph of clonogen survival versus dose. Good statistics at low doses require unacceptably large numbers of animals and targeted doses below about 10 Gy fail to reliably elicit scorable regrowth. High doses, on the other hand, exceed lethality and many of the animals perish before clonal growth appears. These practical difficulties are resolved by fractionation, that is, breaking up large total radiation doses into smaller fractions. For example, let's reexamine Withers' and Elkind's crypt cell regeneration experiment. The data presented in Figure 8-3C display the number of regenerating crypts per circumference versus dose. Equation 8.2 may be applied to derive the number of clonogens. To develop a smooth low dose curve and extended log-linear slope, Withers and colleagues (Thames *et al.*, 1981) performed many exposures using 1–16 fractions and plotted the data as clonogen survival versus dose per fraction. Acute exposure data were used to construct the log-linear slope (maximum doses per fraction, 10–16 Gy), a small number of fractions provided the shoulder (2–10 Gy/fraction) and the greatest number of fractions (≤2 Gy/fraction) revealed the initial slope. The plotted data can be precisely fitted using LQ formalism. Withers also published results for spermatogonia using the same technique. As always, this analysis relies on total single and fractionated radiation doses that result in the same biological endpoint such as $LD_{50/30}$[1] of mice, 50 crypts per intestinal cross section, or a pig or mouse skin score of 2.5.

8.2.3. ***Addressing Tissue Kinetics***

LQ formalism often performs less well on *in situ* or *in vivo* CFU data than *in vitro* survival data. Inferior fit may reflect a failure to address tissue kinetics; organ systems are less isolated than cells in culture. Occasionally, insufficient repair time is provided between fractions, or recruitment from the G_0 population is neglected. In functional studies, the relevant external tissue may not have been considered. Although cellular infiltration and vascular damage appear relatively rapidly, delayed effects such as scarring in the skin results from an inflammatory process that kills off functional epidermal cells and induces cytokines such as transforming growth factor beta (TGF-β) and tumor necrosis factor alpha (TNF-α) that support fibroblast amplification and connective tissue formation (see Chapter 10, Section 10.2). Perhaps the best way to illustrate the impact of tissue kinetics is a discussion of the effect of radiation on mature bone marrow–derived cells. Acquisition of these cells is straightforward and not detrimental to the host. One simply draws a sample of blood. Leukocytes (white blood cells), platelets (thrombocytes), and erythrocytes (RBC) are postmitotic — they are terminally differentiated and do not divide. As such, they are radiation resistant. Nonetheless, the precursors to each of these are differentiating stem cells. The stem cell population is mitotically active and consequently radiation sensitive. Whole body radiation exposure generates a dose-dependent reduction in bone marrow stem cells. Thus, the influx of the various blood cell types from the bone marrow into the blood is reduced over time. The number of radiation-resistant mature blood cells decreases through normal aging and cell death. Figure 8-7 shows the impact of kinetics on radiation sensitivity assays of mature bone marrow cell populations.

For humans and dogs, the $LD_{50/30}$ for whole body exposure is 2.5 Gy to 4.5 Gy of x-ray radiation. Death is driven almost entirely by the irreversible loss of blood cell types no longer being supplied by the irradiated bone marrow, infection due to loss of white blood cells and lymphocytes, internal bleeding due to loss of platelets, and anemia due to RBC loss. So, when dogs are exposed to approximately 3 Gy of x-rays, and blood is drawn periodically for 28 days, the decreases in leukocytes, platelets, and RBC over time reflect the depletion of the stem cell pool and the coincident normal mortality of the differentiated population in circulation. Note that the scales of the y-axes in Figure 8-7 have been adjusted to compensate for different population numbers in circulation. The differentiated stem cell pool is critically depleted relatively quickly following irradiation; the nonsurviving ~50% of animals (solid black line) cannot recover. The reserve of leukocyte precursor cells — the myeloblast population — continues to morph into leukocytes. As the white blood cell population decreases because of normal circulation life expectancy (draining out the hole in the bucket), the bucket is replenished by myeloblast morphogenesis, until the myeloblast bucket runs dry. The immediate initiation of white blood cell loss in the top left panel reflects the nominal duration in circulation: only a few hours. The slope of the curve, however, is not precipitous

[1]Exposure resulting in the death of 50% of the population in 30 days.

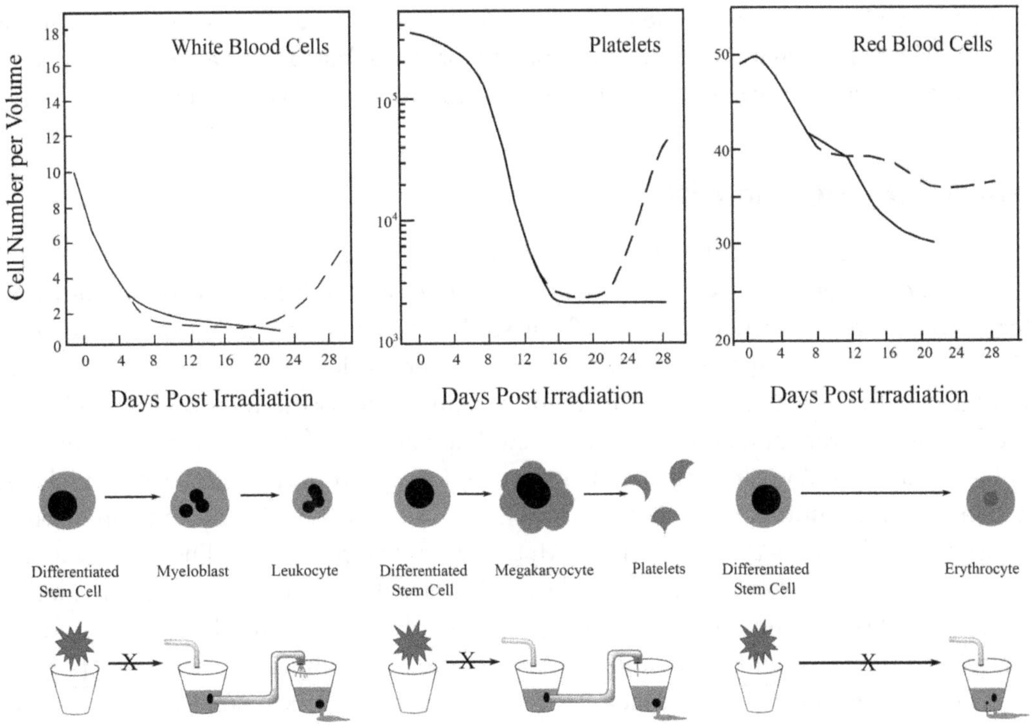

Figure 8-7 Mature bone marrow cell depopulation kinetics. Panels at the top indicate the loss of postmitotic cells from blood. Volumes vary according to protocol and the y-axis scales are set to best display the data for each cell type. The third panel is a hematocrit; the axis reads percent of total blood volume. Radiation is delivered on day 0 and blood is drawn periodically. Data for 2.6 Gy, 2.8 Gy, 3.0 Gy, and 3.2 Gy have been pooled and smoothed. Adapted with permission from Alpen (1990). The lower set of panels indicates the production of the postmitotic cells. In each case, the mature bone marrow cell is derived from a mitotically active, differentiated stem cell that divides to generate the indicated postmitotic cell. The bottom set of panels represents the analogy of Figure 5-1 as it pertains to the panel above. Exposure to radiation empties the stem cell bucket, cutting off the flow to the precursor bucket (first two panels) or final bucket (RBC). When the precursor bucket empties, the flow is disrupted into the final bucket. The drain holes in the buckets represent the flow of postmitotic cells out of the blood through normal deterioration. RBCs are lost both through cell deterioration and bleeding.

as might be expected. The population does not deplete in a matter of hours. For several days, the continued replenishment by myeloblast metamorphosis compensates somewhat for the loss. In the ~50% surviving animals, sufficient stem cell DNA repair occurs to reinstitute the flow into the precursor population. The dashed lines of Figure 8-7 indicate cell replenishment in those animals. With respect to the second panel, there are a lot of platelets in circulation (note the log scale). Just as with white blood cells, platelets have a precursor cell population that provides a buffer between the depleted stem cell population and the life cycle turnover of platelets. Because the peripheral survival of a platelet averages 9 days, the initial loss is less immediate. After a week, once the precursor bucket runs dry, the slope for platelet loss becomes precipitous (day 7 to day 16). Recovery in surviving animals is driven by replenishment of the megakaryocyte pool. Erythrocytes (RBCs; third panel) have no precursor cell; the differentiated stem cell produces a nucleated proerythrocyte in which the nucleus rapidly disintegrates. Nonetheless, the erythrocyte bucket drains very slowly as the circulating life span of an RBC is 100 days. The initial loss of erythrocytes from the blood is a direct reflection of normal peripheral survival. Something interesting happens at about 8 days, however. In nonsurviving animals, extensive bleeding occurs because of vascular damage and the impact of platelet loss (failure to clot). Bleeding removes additional RBC (and other cells), represented by a second hole in the bucket. In all cases, to measure the sensitivity of the stem cell population, the flow of surviving cells into the precursor (myoblast or megakaryocyte) pool must be measured. The precursor pool can be estimated only by counting the postmitotic cells.

8.3. Organ Radiation Sensitivity

Fowler's investigation of the impact of fractionation on irradiated pig skin deviated from *in vitro* cell survival experiments because the skin stem cells could not be isolated from the other tissues and the cell types integral to the organ referred to as "skin." His study represented an early example of organ radiation sensitivity evaluation. X-ray radiation inactivated or killed a dose-dependent number of skin stem cells, resulting in a time-dependent replenishment of skin tissue as the pool of stem cells recovered. However, other tissues were simultaneously exposed. Radiation-sensitive blood vessels and capillaries manifest pores as dying endothelial cells fail. Consequent leakage from the bloodstream includes leukocytes that mount an inflammatory response impacting additional vessels and inducing fibroblasts to extrude connective substances that result in scarring as described previously. The visualization of these external tissues provides Fowler's evaluation scale: erythema is due to capillary irritation, moist desquamation results from leakage and cellular infiltrate, and fibroblast activity resolves the wound. Although his study disclosed the power of fractionation to spare tissue — which was the intended purpose of the study — it would be inappropriate to conclude that it also revealed skin stem cell sensitivity. Fowler investigated the radiation sensitivity of skin as an organ composed of several tissues.

Another example of organ radiation sensitivity evaluated spinal cord myelopathy by observing hind limb paralysis in rats. Hind limb paralysis cannot be evaluated shortly after exposure because spinal cord injury results from cellular responses external to the nerve tissue. The long nerves of the spinal cord are not mitotically active, and so exposure to radiation has little effect on them. Nonetheless, radiation will either damage or activate neuronal supporting cells: glia, infiltrating cells, fibroblasts, and vascular endothelial cells. These cells mount a response to radiation insult over time and therefore the damage expresses several months postirradiation. Again, the process is complicated, but nevertheless it is quantifiable and consistent. In rats, 4–12 months after sufficient doses of radiation, hind limb atrophy is followed by increasing paralysis. If the number of rats experiencing hind limb paralysis is plotted against the total delivered dose, the resulting, positively sloped curve is sigmoidal — the number of rats experiencing paralysis increases with dose. An acute dose produces an almost linear curve that is extremely steep. As the number of fractions is increased, tolerance increases and the slope of the curve decreases, becoming increasingly sigmoidal. Table 8-3 describes the effect of fractionation on hind limb paralysis. In that table, the dose that induces paralysis in 50% of animals indicates spinal cord radiation sensitivity and the 10%–90% range indicates the slope of the response curve. LQ parameters can be extrapolated in the same manner as illustrated earlier for mouse skin by using the dose and the number of fractions that result in an equivalent response — for example, 50% paralysis. This experiment examines the radiation response of the spinal cord as an organ composed of several tissues, just as the Fowler experiment examined skin as an organ. Now, we have seen that organ radiation response can be modeled by LQ formalism and thus, organ radiation response can be discussed in the context of cell survival. Organs exhibit both repair and sensitivity just as cells do *in vitro*.

The spinal cord is a serial organ; that is, function is obstructed below any segment of the cord that is impaired or damaged. Serial organs may be thought of as chains in that disruption of any link breaks the functionality. Other organs, such as the liver, are parallel organs. Parallel organ function diminishes as functional sub units (FSU) are incapacitated, and subunits are incapacitated when sufficient task-critical cells are killed by radiation (Withers *et al.*, 1988). A parallel organ may be thought of as a cable in that

Table 8-3 The effect of fractionation on rat spinal cord as indicated by hind-limb paralysis following x-ray irradiation.

	Percentage of Rats w/Paralysis		
# Fractions	**50%**	**10%**	**90%**
Acute	18 Gy	17 Gy	20 Gy
2	27 Gy	24 Gy	30 Gy
3	32 Gy	30 Gy	34 Gy
5	37 Gy	35 Gy	39 Gy
10	48 Gy	43 Gy	52 Gy
15	61 Gy	56 Gy	66 Gy
20	67 Gy	63 Gy	74 Gy

Source: Data taken from A. J. van der Kogel, *Late Effects of Radiation on the Spinal Cord*, 1979.

disruption of a single strand may reduce the strength but does not greatly impact functionality. In general, an element of an organ that can be regenerated from a single surviving cell can be considered as an FSU. Organ damage becomes irreversible when the mitotically active cell pool cannot recover, as demonstrated in mature bone marrow cell survival (Figure 8-7). Regenerating crypts represent FSU because without crypts, the villi diminish in size and the functionality of the organ (absorption of nutrients) suffers. Likely you are familiar with this symptom; the loss of crypts causes nausea and diarrhea experienced by patients undergoing chemotherapy, and sometimes radiation therapy if the gut was not excluded from the treatment volume. Michael Goitein presented a persuasive argument that organ radiation sensitivity can be determined using FSU survival and LQ formalism (Goitein, 2008).

In parallel organs, there exists a critical volume — the number of FSU that can be destroyed before the organ begins to suffer a loss of function. The evaluation of organ radiation sensitivity is called the normal tissue complication probability (NTCP) and the mathematical expression of this probability should look familiar:

$$\mathrm{NTCP}(m,n,P_{\mathrm{FSU}}) = \sum_{i=(m+1)}^{n} B_n^j P_{\mathrm{FSU}}^i (1-P_{\mathrm{FSU}})^{n-1}, \tag{8.6}$$

where m is the critical number of subunits, n is the total number of subunits, and P_{FSU} is the probability of damaging a subunit. B_n are the binomial coefficients. It is assumed that all subunits are identical, and the probability of damaging FSU is identical for all. Conveniently, when $m = 0$, Equation 8.6 describes serial organs such as spinal cord as well. In this case, the formula reduces to,

$$\mathrm{NTCP}(m = 0, n, P_{\mathrm{FSU}}) = 1 - (1 - P_{\mathrm{FSU}})^{n-1}. \tag{8.7}$$

Suppose you conduct an experiment wherein you irradiate animal kidneys at several increasing doses: a dose escalation experiment. If you exposed 100 animal kidneys to a certain dose, you would acquire the number of damaged kidneys, say 3 out of 100. That would provide you with a probability of damaging a kidney at that dose. A plot of those probabilities (NTCP) versus dose derived from your study would look like Figure 8-8. The plot can be fitted using Equation 8.6, and the parameter m can be determined. If we assume that each subunit is composed of N_0 identical cells and that the subunit can be rescued by a single surviving cell, the probability of FSU collapse depends on the likelihood that all N_0 cells are killed,

$$P_{\mathrm{FSU}}\left(\overrightarrow{\mathrm{par}}_{\mathrm{FSU}}, D\right) = \left[p_{\mathrm{cell}}\left(\overrightarrow{\mathrm{par}}_{\mathrm{cell}}, D\right)\right]^{N_0}. \tag{8.8}$$

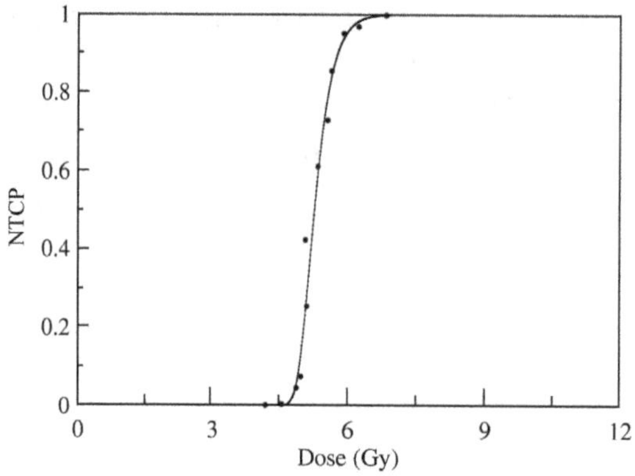

Figure 8-8 The probability of organ damage. Typical experimental results from a dose escalation experiment that exposed a parallel organ (e.g., canine kidneys) to x-rays (•) are fitted by Equation 8.6. NTCP curves are universally sigmoidal. The dose at NTCP = 0.5 indicates organ radiation sensitivity, and the linear slope segment reflects P_{FSU}, and thereby p_{cell} and N_0.

P_{FSU} is the probability of inactivating an FSU, p_{cell} is the probability of killing a cell, and the set of parameters $\overrightarrow{\text{par}}_{cell}$ characterizes the cell radiation sensitivity. According to LQ formalism, $\overrightarrow{\text{par}}_{cell} = \alpha$ and β, yielding:

$$p_{cell}(\alpha, \beta, D) = 1 - e^{-(\alpha D + \beta D^2)} \tag{8.9}$$

and thus, the set of parameters $\overrightarrow{\text{par}}_{FSU}$ is N_0, α and β. If α and β have been determined *in vitro*, the validity of the *in vitro* experiment with respect to *in situ* results can be verified.

8.4. The Brain and Radiation Response

Arguably our most cherished organ, our brain undoubtedly presents the most complicated response to radiation. FSU structures are not all anatomically distinguishable; higher order functions are unique, and functionality can be allotted to a novel physical location under some circumstances, such as injury. A cursory awareness of brain anatomy, physiology, and neuronal function will be useful for developing insights into brain radiation response.

First, let's introduce the players. The brain is composed of gelatinous tissue requiring several protective mechanisms: a bony skull, shock absorbing cerebrospinal fluid (CSF), and isolation from the bloodstream known as the blood–brain barrier. Structurally, the brain is composed of neurons, about 8.6×10^{10} of them, and an approximately equal number of non-neuronal transient, structural, or supportive cells. The neurons connect and communicate with one another by way of synapses — gaps between axonal buttons containing vacuoles of chemical transmitters and postsynaptic receptors presenting on either a dendrite or a cell body. Dendrites present postsynaptic spines and it is believed that these spines morph from "stubby" spines into "long" spines, passing through other forms: "mushroom" spines and filopodia, during the transition. One type of non-neuronal cell is a glial cell. These cells maintain homeostasis and form myelin, a fatty sheath that coats and insulates neurons. Not all neurons are myelinated; myelinated neurons comprise the white matter of the brain. Nonmyelinated neurons conduct impulses more slowly. Glial cells express as one of three types: astrocytes, oligodendrocytes, or microglia. The oligodendrocytes generate myelin. Astrocytes maintain water and ion homeostasis — critical to the axonal, ion-induced electrical pulse, the blood–brain barrier and synaptic functionality. Microglia are immunocompetent, phagocytic cells that do not originate from the same embryonic tissue

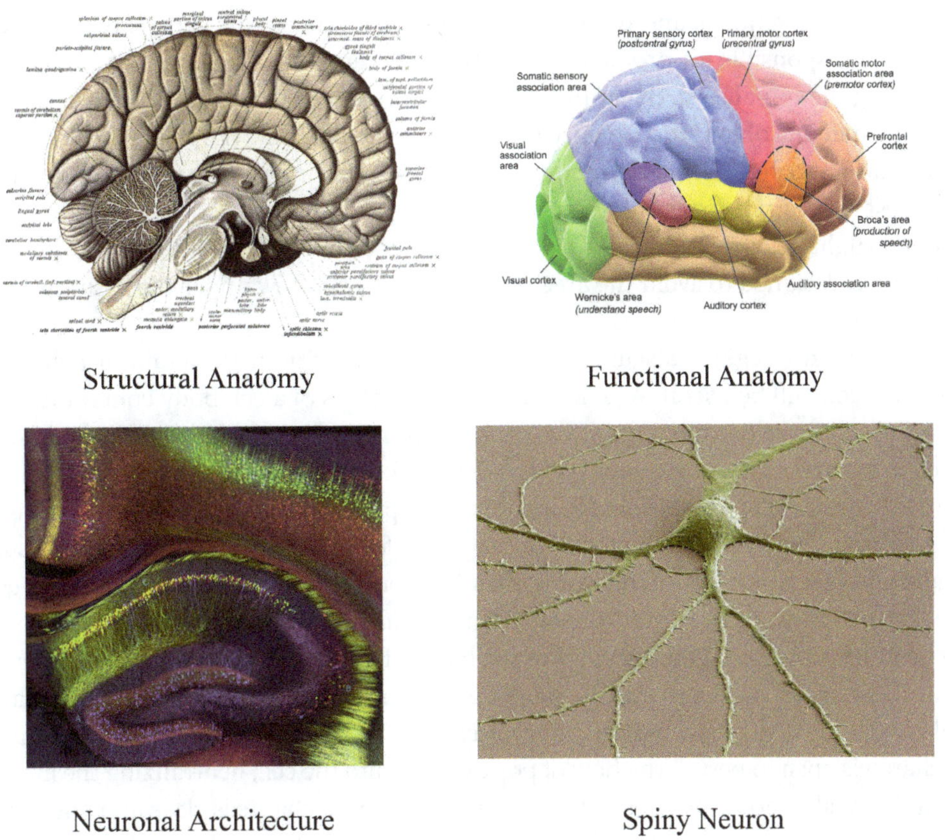

Figure 8-9 Understanding the human brain. Upper left panel reproduced from *Atlas and Text-Book of Human Anatomy Volume III Vascular System, Lymphatic System, Nervous System and Sense Organs*. The small print nomenclature is not necessary for our purposes. The right top panel shows a generalized mapping of sensory and motor centers of the cortex. This image is reproduced with permission of Blausen.com from "Medical gallery of Blausen Medical 2014; *Wiki Journal of Medicine* 1 (2). The lower left image by Tamily Weissman, Jeff Lichtman, and Joshua Sanes (2005) shows the neuronal structure of the hippocampus. Cell bodies are labeled in yellow; dendrites are labeled in green. The color-enhanced scanning electron micrograph of a neuron in the lower right panel displays a cell body, axon (extending up and then to the right) and spiney dendrites. This image reproduced with permission of Thomas Deerinck and Mark Ellisman, 2009.

as other glial cells; they have the same origin as macrophages. Microglial filopodia sense the environment inducing a response to any type of insult, including radiation. Other non-neuronal cells, pericytes, surround and seal capillaries in the brain to prevent leakage; they form the blood–brain barrier. Figure 8-9 presents three images of brain structure including brain gross anatomy, neuronal architecture of the hippocampus, and a pyramidal neuron displaying dendritic spines.

Second, brain anatomy is described either structurally or functionally, but often the boundaries between the two descriptors become confused. It is important for the radiation biologist to distinguish between them. The brain is divided into two cerebral hemispheres; the cerebral cortex comprises the outer covering. The cortex controls higher brain function, including consciousness and cognition. Structurally, the cortex is composed of gray matter (unmyelinated) arranged in six neuronal layers (neocortex) plus three or four additional neuronal layers (allocortex). The functional areas of the right hemisphere are depicted in the upper right panel of Figure 8-9. Tucked beneath the cortex, several anatomical structures — each performing specific functions — can be identified: the thalamus, subthalamus, epithalamus, and hypothalamus; the pineal gland, the pituitary gland, the amygdala, the hippocampus, the claustrum, the basal ganglia, and the three circumventricular organs. The cerebellum contains the ventricular system responsible for producing CSF; the four ventricles are structural. Finally, the base of the brain adjoining the spinal cord, the brain stem, consists of three functional constructs: the midbrain, the pons, and the medulla oblongata. Because brain functional subunits (FSU) are frequently not defined by structural anatomy, physicians and medical physicists often rely on functional imaging to define regions of interest (ROI) prior to planning radiation delivery. For example, to avoid irradiating the speech center, a patient might be asked to recite a poem while blood flow is visualized by magnetic resonance. A treatment is then designed to avoid the area of enhanced blood flow under the assumption that active cells demand oxygenation.

Third, before we can begin a discussion of radiation sensitivity in the brain, a review of neuronal anatomy and function will be useful. A generic nerve cell consists of a cell body containing a nucleus and typical organelles, an axon terminating in a synaptic button, and several dendrites expressing receptors when engaged in synapses. These receptor patches develop on dendritic spines. The spines are fascinating structures under considerable contemporary scrutiny. It is possible to visualize several forms of spines in static micrographs, but we can only hypothesize that these forms represent developmental stages. Nerves can be thought of as electrical wires and myelin can be thought of as insulation. Electrical currents pass down individual nerves from dendrite to axon terminus and impulses transit the synapse through the release of neurotransmitter from the vacuoles of the button followed by absorption in the dendritic receptors of the successive neuron. The electrical current is generated by ion channels in the axonal membrane. A nerve cell maintains a slightly negative internal charge. When stimulated by neurotransmitter, local sodium ion channels open to permit the flow of positive ions into the cell, neutralizing the internal negative charge. When the local internal charge swings positive, the sodium channels close and potassium channels open to release positive charge from inside the neuron into the extracellular environment. This flow of charges creates tiny, local currents through the membrane. Because there is a charge differential across the cell membrane, there is an electrical potential difference. The change in local potential stimulates proximal sodium channels to open, and so the process travels down the axon. It is thus an "action potential." These action potentials can be read by electrodes (as they are in an electroencephalogram, EEG) and are useful for investigating brain function. Now, nerves connect through synapses. Signaling in the brain is not, in general, one nerve to the next. One neuron can activate several ensuing neurons; the pattern of activation and quenching hides the secrets of cortical function.

Lastly, nerves form networks. Just as the gross anatomy of the brain is described by structural anatomy (e.g., ventricles and glands) and functional anatomy (e.g., visual cortex and Wernicke's area), neural networks can be described by structure and function. Consider the U.S. electrical infrastructure. The infrastructure initiates at a generator (nuclear power plant or hydroelectric dam). The cables and power lines, and poles and insulators (and transformer stations) form the physical power grid. The grid can be represented on a map. On the other hand, a diagram of power usage would look very different. It would be temporal. If one selected an instant in time and characterized the flow of electricity from the power station to each individual user structure, this would reflect the grid function. Electricity travels along the physical structure, but not all the physical structure is in use all the time, and each load varies. The same is true for neural networks. In theory, every nerve could be mapped (structural) and the connectivity of the network could be analyzed (functional). In fact, several ongoing projects are attempting to do so. Notably, the Human Connectome Project out of the National Institutes of Health (NIH) aims to build a network map of the human brain including both structure and functional elements. A Swiss research initiative, the Blue Brain Project, has been reverse-engineering human brain circuitry since 2005 to create a digital reconstruction of brain structure and function. The Human Brain Project, a European Union collaboration, is attempting to simulate human brain functionality using supercomputers and artificial intelligence (AI).

So, now we can address brain radiation sensitivity. Radiation can affect brain structure, or function, or both. There are parts of the brain that resemble the spinal cord in that they can be modeled by Equation 8.6. The cerebral cortex presents a challenge. Recall that noncycling cells are radiation resistant (i.e., the cell survival D_0 is large) and that neurons are nonmitotic. Therefore, the radiation-sensitive targets within the brain are not neurons. There is a biological caveat: evidence exists indicating low levels of neuronal cell division within the hippocampus resulting in those neurons being considered *relatively* radiation sensitive. But in general, we must look elsewhere for sources of radiation-induced brain damage. Since neurons comprise only half of the cells in the brain, potential targets are plentiful. Obviously, the vascular endothelium is actively mitotic and radiation sensitive. So are the pericytes that seal capillaries to prevent leakage across the blood–brain barrier. Glial cells *in vitro* exhibited a nearly 10-fold increase in apoptosis following exposure to 50-Gy x-rays, whereas the number of cultured neurons was unaffected by the same dose. The sensitivity of glial cells is cell-type dependent (Barbarese & Barry, 1989). Oligodendrocytes, for example, exhibit sensitivity at the differentiated stem cell stage ($D_0 \sim 1.4$ Gy), but mature oligodendrocytes are more resistant ($D_0 \sim 4.4$ Gy). Oligodendrocyte inactivation reduces the myelination of white matter. Microglia have been implicated in the remodeling of dendritic spines. Although microglia are migratory and therefore an irradiated area can be expected to recover relatively rapidly following exposure because microglia re-infiltrate, the redistribution of spine shapes may be affected longer term. Radiation-induced brain injury follows a temporal pattern — as does all radiation-induced organ damage — that can be described by a sequence of structural and functional events. The acute response, over a period of days, includes edema — an influx of fluids resulting from the microglial inflammatory response and blood vessel leakage. Weeks later, transient demyelination appears. The late response, expressing months to years following the exposure event, includes glial cell loss ("gliosis"), permanent demyelination, vascular abnormalities, and white matter necrosis. How does this structural damage manifest functionally? Brain edema expresses as headaches and drowsiness, just as it does in the case of concussion. Delayed and late structural effects correlate with speech and hearing dysfunction, seizures, double vision, loss of coordination, hormonal deregulation, and so on, depending on the area of the brain exposed. Late functional damage manifests as altered higher brain function, for example, changes in mood, cognition, intelligence, personality, memory, and attention. Interestingly, these cognitive disorders do not correlate with necrotic progression. Fractionated radiation doses can be delivered to rats in such a way as to prevent necrosis and yet the rats, nonetheless, demonstrate cognitive impairment. Something else is going on.

Decades of whole brain irradiation therapy — for example, performed prophylactically to control brain metastases during the treatment of pediatric blood-borne cancers — has provided a convincing dataset indicating decline in intelligence, short-term memory loss, attention deficit, and cognitive impairment. In adults, 50%–90% of patients receiving treatment for brain tumors experience significant cognitive impairment within six months of treatment. Furthermore, the impairment increases in severity with time. An interesting body of work in the field of extraterrestrial exploration has provided some insights. The National Aeronautics and Space Administration (NASA) is justifiably concerned that astronauts traveling beyond the Earth's protective magnetosphere may suffer mission critical cognitive impairment resulting from ubiquitous cosmic radiation in space and sporadic proton radiation emanating from solar flare activity. NASA-funded studies have indicated that brain irradiation

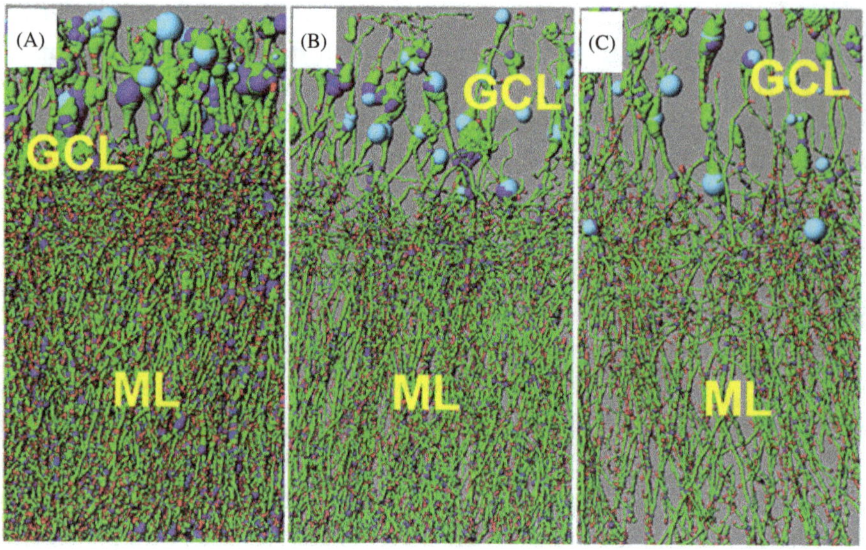

Figure 8-10 Reduced dendritic complexity following x-ray exposure. Digitally reconstructed images of hippocampus granule cell layer (GCL) and molecular layer (ML) exemplify the loss of dendrites (green) and spines (red) 30 days after whole brain exposure to (**B**) 1 Gy and (**C**) 10 Gy x-ray radiation. (**A**) shows normal, unirradiated hippocampus. Reproduced with permission of the National Academy of Science from Parihar, and Limoli (2013).

causes several behavioral changes in mice: memory is diminished, anxiety increases, and learning is impaired. Examination of mouse hippocampus neuronal structure points to a dose-dependent reduction in dendritic complexity and spine number. Figure 8-10 shows reconstructed confocal microscopy of the granule cell layer (GCL) and molecular layer (ML) of the hippocampus (the yellow and green, fluorescent structures in Figure 8-9). In this figure, transgenic mice expressing a green fluorescent protein (Thy1-EGFP) in some but not all neurons make it easier to visualize the dendrites. Spines appear red. The first image shows unexposed, control hippocampus; the second is from a mouse exposed to 1-Gy x-rays, and the third image was taken of mouse hippocampus exposed to 10 Gy. Images were acquired 30 days after exposure. These images strongly suggest that dendritic complexity is severely reduced by radiation. Nerves extend fewer dendrites and dendrites undergo reduced branching. Furthermore, long spines and filopodia seem particularly sensitive to radiation, whereas the number of stubby and mushroom spines appear unaffected (not shown in these images). The number of neurons remains unchanged from the unirradiated condition, confirming that nonmitotic neurons are not killed by radiation.

Neurons are not static; they continuously remodel their dendrites forming new synaptic connections and retracting from others. The dendritic spines morph, develop, and dissipate. Experimental evidence indicates that long spines form particularly strong synaptic connections. One hypothesis suggests that strongly connected neurons form communications hubs that route electrical impulses, receiving signals from several neurons and stimulating numerous others. Imagine an airport hub where travelers are sorted from one incoming route to another outgoing destination. The radiation-induced structural changes — the reduction in dendritic number and complexity, and alterations in spine distribution — are coincident with the behavioral modification documented in brain irradiated mice. This research suggests that radiogenic cognitive loss may be related to neural network damage and reorganization, without a loss in neuron cell number.

8.5. Another Organ-Like System: Tumors

This text, thus far, has restricted discussions to healthy, normal cells and tissues. While Roentgen enjoyed demonstrating the radiological potential of x-rays in 1896 by visualizing the bones in his wife's hand, Emil Grubbe was already treating neoplastic growths with x-rays. We have learned a lot of radiation biology over 125 years of neoplastic tissue irradiation. First, a little nomenclature would be useful. A neoplasm (adjective: neoplastic) is a recent, abnormal growth of tissue. The term neoplasm includes the broadest range of growths, including, by way of example, warts, arteriovenous malformations, polyps, and tumors. A tumor is also a growth of abnormal tissue; it can present as benign (noninvasive) or malignant (invasive). The term cancer is generally used to refer to only malignant tumors: invasive collections of cells exhibiting uncontrolled growth. Regardless, you have likely heard someone say, "she has a benign cancer." The distinctions between terms are vague and usage often relies on connotation rather than denotation. In a clinical setting, informing a patient that he or she has a tumor can be less alarming than saying he or she has cancer. In general, cancer refers to a condition or disease, whereas tumors are the manifestation of that disease. It is also useful to refer to the several tumors arising from a metastatic cancer.

Perhaps it makes sense to regard these growths as organs. There are solid tumors and blood-borne or fluid tumors. In some ways, tumors act like healthy tissue and in some ways they are unique. Different neoplasms with different origins behave differently, retaining some of their original organ characteristics. We will discuss cancer biology and the evolution from a healthy cell to a cancer cell later in this book. For now, let's consider a tumor to be a poorly organized organ, the growth being determined by the rate at which cells divide reduced by the rate at which cells are shed or die (see Section 8.2.3). To incapacitate a tumor, every cell must be killed or mitotically incapacitated. If tumors are organs, what functions do they perform? Tumors are primarily concerned with self-preservation. The radiation sensitivity of a specific tumor type establishes how much radiation must be delivered to control the tumor. Our task, as radiation biologists, is to find methods that enable determination of tumor radiation sensitivity. Fortunately, we find there is good agreement between cultured cell lines derived from tumor cells (Figure 7-10) and the radiation sensitivity of tumors *in situ*.

8.5.1. *Tumor Kinetics*

Tumor kinetics are best represented by our type C motif (Figure 5-6.) wherein there is a source of new cells entering the system, but cells do leave the system (there is a hole in the bucket). The cell loss from a tumor is generally hypothesized to result from spontaneous apoptosis but, considering the ratty condition of the ubiquitous unrepaired chromosomes expressed in tumor cells, mitotic death, autophagy, anoikis, and immunogenic cell death are all likely. All tumors experience cell loss, even small neoplasms albeit at a slower rate than larger tumors. Invasive tumors additionally lose cells through shedding; there are two holes in the bucket representing metastatic cancers. Tumor cell numbers increase by cell division. If the rate of mitosis matches the rate of cell loss, the tumor size is static. If the cells cycle rapidly, and the loss is minimal, the tumor volume increases. If cell loss exceeds mitosis, the tumor shrinks. In general, tumors grow but much more slowly than would be predicted from the cell cycle kinetics of individual tumor cells. Tumor cell dose response is determined by cell cycle kinetics and tumor radiation sensitivity is determined by tumor kinetics. In this way, the entire tumor resembles a single FSU (see Equation 8.8).

Determining tumor cell cycle kinetics is laborious and imprecise but not impossible. Here is one way it can be accomplished. Tritiated thymidine can be injected intraperitoneally into a cohort of mice with identical tumors. After a short time (<20 minutes), the ^{3}H-thymidine is flushed out with a second injection of nonradioactive thymidine. At 1-hour intervals, tumor samples are removed, fixed, and auto radiographed. Cells that have taken up ^{3}H-thymidine and incorporated it into replicated chromosomes will expose film placed beneath the tissue slice. Each preparation can be evaluated for synthesis — one strand of each chromosome is labeled. The percentage of labeled cells signifies the percent of cycling cells in the tumor. There are several sources of uncertainty inherent in this experiment. First, the cold thymidine cannot remove all the unbound ^{3}H. Second, only about 1% of the cycling cells take up ^{3}H-thymidine and incorporate it into DNA during synthesis. Third, the cells are not cycling in synchrony; only those in S-phase will incorporate ^{3}H. The statistics of small numbers demand that a lot of mouse tumors must be sampled. Nonetheless, determination of cell cycle kinetics is straightforward by verifying the number of hours passed between the first appearance of synthesized DNA and the appearance of postmitotic DNA that has passed through a second synthesis (so half of the chromosomes are labeled). Experimental tumor models have disclosed cell cycle times ranging from 10 hours to 120 hours and actively cycling percentages between 30% and 80%. For comparison, jejunal crypt cells complete a cycle in about 19 hours.

Tumors that arise from epithelial tissue (carcinomas) experience a high rate of cell loss, likely because epithelial tissue turns over continuously. For example, normal skin cells are programmed to degrade organelles (including the nucleus), flatten, and become empty bags of lipids that protect the epidermis from dehydration. Therefore, it is not surprising that over 70% of a carcinoma may be undergoing some type of cell death at any given time. The hole in that bucket is very large, indeed. On the other hand, tumors arising from fibroblasts (sarcomas) experience less than 50% cell death. Total cell loss is difficult to determine, particularly in humans where ethical concerns prohibit most experimental evaluations. Cell loss is usually determined by simply comparing the observed tumor volume doubling time with the theoretical doubling time, given the cell cycle kinetics of the tumor under examination. On average, the human tumor doubling time is about 2 months.

8.5.2. *Tumor Radiation Sensitivity Assays*

A straightforward method for assessing tumor radiation sensitivity makes use of standard, *in vitro* analysis. The tumor is excised, individual cells are isolated by pulverizing the mass and treating the clumps with trypsin. The cells are plated and then exposed using a dose escalation protocol. It does not appear that this process interferes with reliable radiation survival determination despite cell damage incurred along the way. Nonetheless, the technique has low plating efficiency and is successful for only a relatively small number of tumor lines, working best if immortalized tumor cells are first injected into an animal to create the tumor that is subsequently removed. The technique was modified by Robert Hill and Ray Bush (Hill & Bush, 1969) who investigated the radiosensitivity of KHT sarcoma. In this case, the tumors were irradiated *in situ*, excised, minced, isolated, suspended, and injected intravenously into recipient mice. Sarcoma cells migrate to the lungs and form nodules. The number of nodules arising per fixed number of injected cells reflects the tumor sensitivity because the percentage of injected live cells decreases with dose.

Tumor growth provides an obvious end point for radiation sensitivity analysis. It answers the question, "Does exposure to radiation suppress tumor growth?" To perform these experiments, the tumor under examination is removed from an animal, usually a rodent. A single cell suspension is prepared as abovementioned and 10^4–10^6 cells are injected subcutaneously into recipient animals of the same species. When the induced tumors have reached sufficient size, the rodents are subjected to a standard dose escalation protocol — several mice are exposed to each dose. The tumors are measured, usually with calipers, along two diameters at 90° separation, and the average is recorded. The y-axis of the resultant

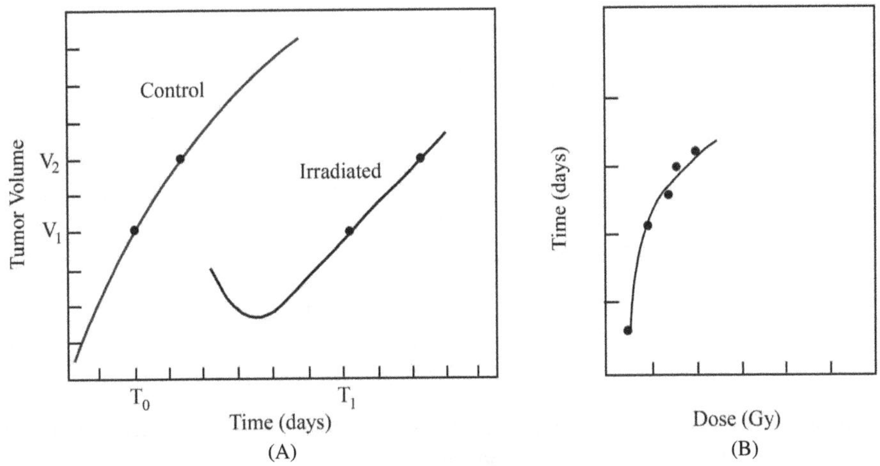

Figure 8-11 Radiation sensitivity assessment by tumor growth. **(A)** Typical untreated tumor volumes increase as portrayed by the curve on the left. If the tumor is exposed to radiation dose D at time T_0, the tumor volume decreases and then begins to increase again, as indicated by the curve on the right. Treated tumors regain the original volume, V_1, at time T_1. Growth rates for untreated tumors and treated tumors can be determined by the time required for each to increase from V_1 to V_2. **(B)** A plot of the time required for tumors to increase in size from V_1 to V_2 vs. dose.

plotted data is the average size, so α/β must be derived as before using the linear form of the LQ equation. The clever reader has already recognized that growth implies another kinetics problem. One might anticipate that a hefty dose of radiation would decrease tumor size and induce a lag prior to return to the preirradiation growth status, and this lag might be dose dependent. Indeed, this is the case. Figure 8-11A shows typical curves for tumor growth without and following exposure to x-rays of a given dose. An induced tumor exhibits exponential growth for a few weeks (until oxygen deprivation suppresses mitosis). Several days following exposure (T_0), the tumor decreases in size (diameter or volume), recovers, and reestablishes growth, albeit at a reduced rate. For this reason, growth rate provides a better endpoint than tumor size. A specified tumor size can be preselected (V_2 in Figure 8-11), and the time required to reattain that size is called the tumor growth delay, which can be recorded and plotted for each radiation dose (Figure 8-11B).

Perhaps the most obvious inquiry is this: "What dose is required to cure a given tumor?" An absolute answer to this question is difficult because of individual biological variation. The 50% tumor cure dose (TCD_{50}) — the dose that controls the tumor in half of the treated animals — provides a better statistic. Sampled tumors must be the same type and similar size, but these experiments can be performed on naturally occurring "fortuitous" cancers. For example, thousands of dogs suffering from melanomas or sarcomas might be exposed at various veterinary clinics to reasonable doses of x-rays and observed for several months. The number of dogs achieving local control (cured tumor) is calculated as a percentage of total treated animals at each dose (Figure 8-12). A plot of the data reveals the TCD_{50} for melanomas, treated with x-ray radiation, in dogs. For ethical reasons, the dose range tends to be narrow and is limited to doses that are anticipated to cure the cancer. Rigorous laboratory experiments employing control groups and identical animals presenting identical tumors extend the dose range and verify clinical results. It is important to point out that tumor radiation sensitivity is not equivalent to tumor curability. Tumor eradication depends on the probability that every single clonogenic cell of the tumor can be eliminated; that no viable cell remains to divide and regenerate the tumor. In addition, this tumor eradication must be able to be accomplished without inducing unacceptable normal tissue damage and complications in the surrounding tissues.

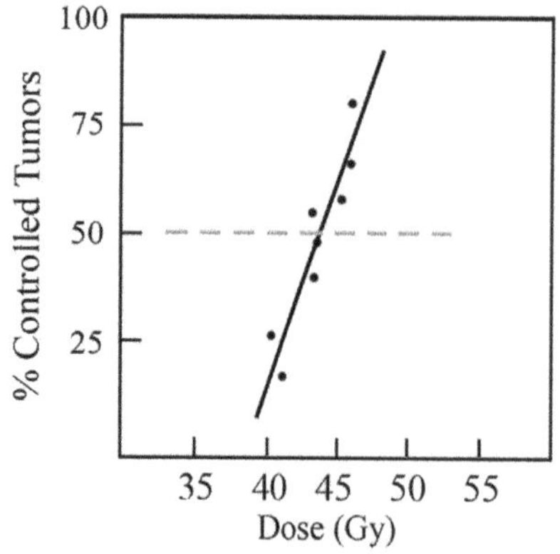

Figure 8-12 TCD_{50} analysis. The percentage of tumors locally controlled at 120 days is plotted against dose delivered to fortuitous tumors. These assays reflect patient outcome analyses.

Techniques have been developed to provide cell survival statistics in tumors and direct LQ analyses. What has come to be known as the Hewitt dilution assay was first performed in 1959 (Hewitt & Wilson, 1959). This *in vivo* system removes the liver of a mouse presenting with advanced lymphocytic leukemia and suspends extracted tumorigenic cells in ice cold Tyrode's solution. The cells are counted and diluted, and recipient mice are injected intraperitoneally with known numbers of tumorigenic leukemia cells. On average, only three injected cells will induce tumors in half of the mice. However, if the donor mouse is irradiated, some of the extracted cells will be killed by the delivered dose, and therefore more cells will be required to induce tumors in the recipient mice. For example, suppose the tumor-bearing mouse is exposed to 10 Gy of x-rays prior to liver removal. Upon recipient mouse follow-up examination, none of the mice receiving 10 cells presented any tumors but the mice receiving 65 cells all presented tumors. One of the mice receiving 15 cells developed a tumor and half of the mice receiving 30 cells developed tumors. The recipient mice have served as clonal growth incubators and provided a 50% tumor dose (TD_{50}) of 30 cells. The surviving fraction of cells, following irradiation, can be calculated as:

$$S = \frac{TD_{50,\,\text{controls}}}{TD_{50,\,\text{irradiated}}} = \frac{3}{30} = 0.1. \tag{8.10}$$

By repeating the procedure at several exposure doses, a survival curve analogous to *in vitro* experiments can be constructed. The technique can be used to assess any tumor, blood-born or solid, provided the tumor can be successfully resolved to a single cell suspension and the suspended cells are tumorigenic once injected into recipient mice. Some immortalized tumor cells are amenable to being cultured, and in those cases the suspended cells can be plated, and clonal growth can proceed *in vitro.*

The protocols discussed earlier for tumor model systems also prove useful for dose protraction studies. For example, Figure 8-11 provides a TCD_{50} analysis for an acute dose, but it is useful to also know what fractionation schema produce similar tumor control while providing radiation protection for the surrounding healthy tissues (Figure 8-13). Fractionation suppresses β — the survival curve is dominated by the shape of the acute dose shoulder. Thus, the most reliable predictor of tumor fractionated radiation sensitivity is α. As always, fractionation can be used when acute doses might prove lethal to the animal under study.

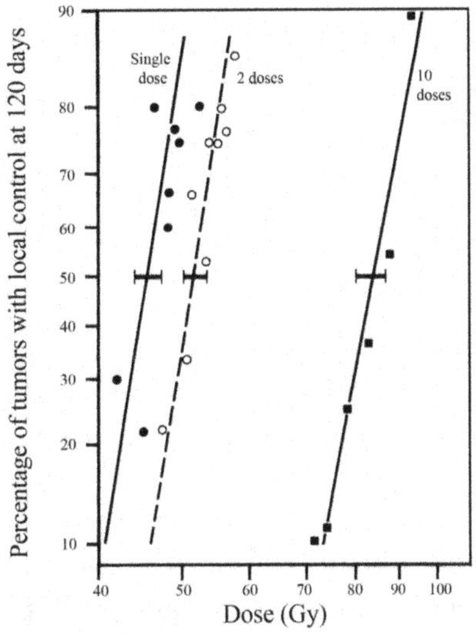

Figure 8-13 Percentage of mouse mammary tumors locally controlled by acute radiation exposure and fractionated regimens. The TCD_{50} is indicated by bars encompassing the 95% confidence interval. Reproduced with permission of the Radiation Research Society from "Radiation dose fractionation and tumor control probability." H.D. Suit, and R. Wette, *Radiation Research,* 1966; 29: 267–281.

8.5.3. *Hypoxia and Anoxia*

The oxygen enhancement ratio (OER), as described in Chapter 7, Section 7.1, is a mathematical expression representing a reduction in radiation sensitivity occurring in oxygen-deprived, cultured cells. The *in vitro* effect is attributed to altered reactive oxygen species (ROS) chemical reaction kinetics diminishing indirect DNA damage fixation. Does this reduction in persistent indirect damage express at higher scale? Do we have evidence of reduced damage fixation in tumors *in vivo*? Perhaps. High LET radiation drastically reduces or eliminates OER in cultured cells and it has been shown to be more effective than low LET radiation in the treatment of oxygen-starved tumors. Presumably, high LET radiation introduces direct damage and more complex DNA damage that does not rely on chemical processing post-hydrolysis. However, heavy ions also perturb the cellular and extracellular environment. Resulting physiological effects may enhance the tumor response. More convincing evidence exists. Certain chemicals (e.g., metronidazole, misonidazole, nimorazole, and pimonidazole) substitute for oxygen in damage fixation reactions (Equation 7.30) and these have been shown to sensitize oxygen-deprived tumor cells *in situ* but these oxygen memetics have not been widely adapted clinically due to peripheral neural toxicity. Nonetheless, radiation resistance observed in tumor cells calls more factors into play *in vivo* than *in vitro.*

The difference between the effects of OER and hypoxia is one of target as well as scale. OER is molecular and results from the fixation of DNA radicals by oxygen. Hypoxia is cellular, triggering anaerobic metabolism, activating biochemical pathways, and producing compensatory proteins. Both states imply a reduction in local oxygen partial pressure and may indicate certain cellular and extracellular consequences including reduced pH and diminished nutrients. The resultant impact on cell physiology includes altered glucose metabolism (anaerobic glycolysis results in lactic acid buildup), unstable redox reaction rates, mitochondrial disfunction, alteration of DNA repair pathway kinetics, and

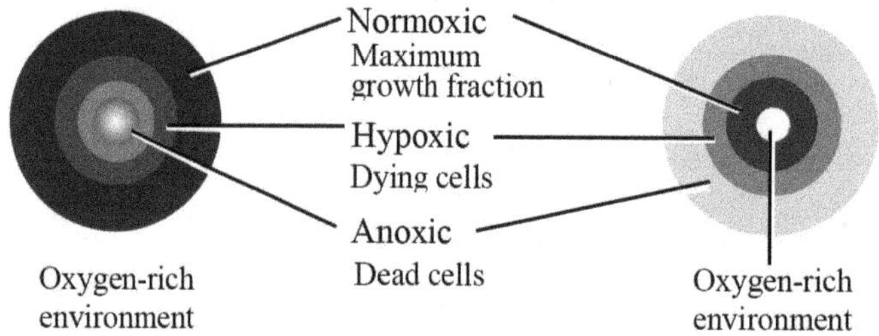

Figure 8-14 Models of hypoxia. The sphere on the left represents clonal growth of a tumor surrounded by oxygen carrying blood vessels. The cylinder on the right represents tumor surrounding a blood vessel. In each case, dying cells begin to appear at 70 μm to 100 μm beyond the oxygen source, and dead cells (necrosis) appear beyond 100 μm.

inhibited mitosis. A reduction in the number of cycling cells could easily explain the increased radiation resistance of hypoxic tumors exclusive of indirect damage inhibition. Hypoxia-induced mitotic block also could compensate for absent tumor cell cycle checkpoints, providing opportunities for structural and chromosomal repair. Repair in noncycling/quiescent, hypoxic cells may favor more robust (less error-prone) repair pathways. The radiation biology of anoxic tissue is best considered to be a complex, tissue-scale phenomenon. Figure 8-14 shows two conventional concepts for tumor anoxia. The left-hand sketch illustrates the concept that tumors grow somewhat spherically, by exponential cell division. If the original tumor cell divides into 2, and the 2 divide into 4, and the 4 divide into 16, and so on, it seems logical that these cells would form a sphere. And if we imagine that this happens between oxygen-rich capillaries, then the cells in the center of the mass would be deprived of oxygen once the radius surpasses the diffusion limit for O_2. They would become hypoxic. However, mitosis is favored near vessels and tumors recruit new vasculature, so a tumor develops wrapped around blood vessels. The image on the right is a more realistic way to envision hypoxic architecture.

Reduced O_2 partial pressure stimulates hypoxia-inducible factors (HIF) — transcription initiating cofactors that target genes responsible for oxygen deprivation coping mechanisms such as angiogenesis or apoptosis, HIF exists as a heterodimer composed of an α-subunit that is generated in response to low O_2 partial pressure, and a constitutively transcribed β-subunit. There are at least three forms of HIF, each expressed in different cell populations and each targeting different genes. For example, in Hep3B cells HIF-1 stimulates the production of erythropoietin, a hormone that increases the production of RBC and prolongs the lifetime of RBC in circulation. Anoxia also stimulates an extranuclear activity known as the unfolded protein response. Here, ribosomal translation of proteins is inhibited to prevent inappropriate post-translational folding resulting in aberrant tertiary protein structures. Thus, anoxic regions of tumors present specialized cells that are behaving differently from the normoxic regions of the tumor. This condition is not limited to large tumors suffering from vascular insufficiency; hypoxic cells have been identified in small neoplasms and micrometastases as well. The distribution of anoxic microenvironments within tumors becomes vital to the estimation of tumor radiosensitivity, as nondividing, anoxic cells are radiation resistant. And here is another measure of complication: hypoxia can be transient. In other words, an unpredictable and time-dependent factor impacts the percentage of radiation resistant cells within the tumor under consideration, *in vivo*. Cells suffering from brief (minutes to hours), mild hypoxia respond differently from cells experiencing long-term anoxia. Moreover, reoxygenated hypoxic cells do not return to their prehypoxic state immediately. Recovery of gene expression, translation, and proliferation exhibit various lags.

Anoxic microenvironments can be evaluated *in situ* and in real time by several techniques including oxygen microelectrodes, magnetic resonance imaging (MRI), positron emission tomography (PET), electron paramagnetic resonance (EPR), Doppler ultrasound, photoacoustic imaging, and molecular marker probes. These techniques bear witness to the logical expectation that fraction size, interfraction interval, dose rate, and radiation quality influence radiation-induced changes in tumor hypoxia impacting tumor control. Early on, hypoxic tumors were created in laboratory animals using simple techniques such as vascular clamping or animal asphyxiation preirradiation. These tumors were exposed using various protocols (fractionation, dose rate, etc.) and assessed as described earlier (tumor growth, TCD_{50}, etc.). Engineered, anoxic tumors were compared with identical tumors provided with unfettered oxygen supply. Anoxic tumors exhibited significantly reduced radiation sensitivity; the dose response curves emulated OER *in vitro* (Figure 7-25), with normoxic tumors producing curves resembling cultured cells exposed in air and anoxic tumors producing curves resembling cells exposed under nitrogen. Experimental design prevented radiation-induced tumor reoxygenation, so the effects of scale were minimized. *In vivo* tumors suffer vascular insufficiency chronically or acutely. Chronic hypoxia occurs in cells that are beyond the perfusion range of O_2, that is, the anoxic cells are more than 70 μm from an oxygen-rich blood vessel. In this case, an appropriate dose kills the normoxic cycling tumor cells near vessels. Following that, the dead cells are removed by white blood cells and the noncycling hypoxic cells are brought into perfusion range and reoxygenated. This process of reoxygenation allows these previously oxygen-starved cells to reenter the cell cycle and return to normoxic radiation sensitivity. A well-timed exposure kills the newly oxygenated cells; they are removed by white blood cells, a new layer of anoxic cells become oxygenated, and so on. The tumor is being peeled away in layers. Unlike the contrived laboratory experience wherein anoxic cells could not be re-oxygenated, dose protraction protocols become critical with respect to tumor cure not because of cellular function but because of tissue function. There is a complication (of course). The vascular endothelium represents some of the most radiation-sensitive tissue in the tumor microenvironment. Fractionation partially protects this tissue by providing DNA repair opportunities, but nonetheless, irradiation can introduce endothelial cell damage and vascular collapse. Two effects should be considered: radiogenic inhibition of angiogenesis and radiation damage to existing vasculature (Denekamp, 1993). The former not only deprives tumor cells of oxygen and nutrients but also reduces the concentration of tumor angiogenesis factor (TAF), diminishes endothelial cell division and migration, and prevents tumor basement membrane penetration through the impairment of neovascular budding and anastomosis. On the other hand, radiation damage to existing vasculature kills endothelial cells, increases platelet adhesion, stimulates blood coagulation, decreases blood flow, and alters intravascular fluid dynamics. The two radiogenic effects also have different physiological outcomes. Inhibition of angiogenesis obstructs tumor growth. Vascular damage induces hypoxia and necrosis thereby causing tumor regression; however, these endpoints are accompanied by local thrombosis, hemorrhaging, blood flow stasis, and collapse of the vessels. The stromal tumor environment reflects the two processes (see Section 8.6) impacting tumor response to radiation as well as determining tumor aggressiveness.

Acute hypoxia develops when the blood supply is transiently interrupted. In addition to radiation-induced vascular damage, capillaries open and close, flow and ebb under normal conditions. Tumor remodeling can pinch vessels closed briefly as can stromal tissue modification, however, inflammation may introduce nitric oxide radicals that restore tumor sensitivity to radiation despite persistent hypoxia. Neovasculature arising in response to HIF signaling tends to be malformed and fragile. These vessels frequently deteriorate producing transiently hypoxic regions. Figure 8-15 reveals the two types of hypoxia. Here, antibodies are raised against proteins unique to tumor, blood vessel, or hypoxic tissue. The antibodies are labeled with fluorescent tags and hybridized to a slice of mouse, HER2-positive breast tumor. Antibodies to the HER2 protein are tagged with a blue label, and thus the tumor tissue appears blue. A red marker is attached to an antibody to CD31, a protein expressed by vascular endothelial cells.

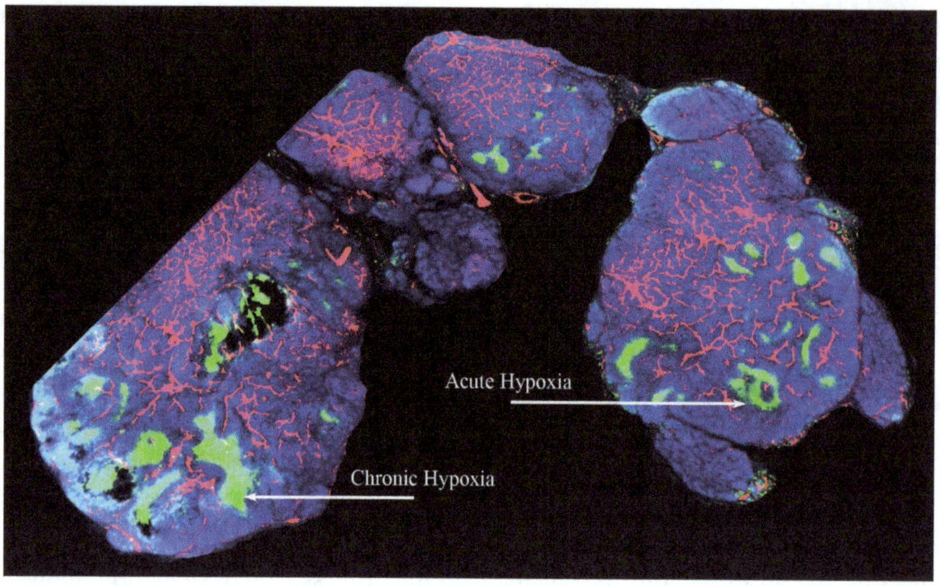

Figure 8-15 Hypoxia in mouse mammary tumor model. In this transparent tumor tomography image, hypoxic areas fluoresce green. Tumor tissue appears in blue and blood vessels are red. In the lobe on the left, hypoxic tissue resides in areas devoid of vessels and is therefore chronically hypoxic. In the lobe to the right, much of the tissue is well vascularized, but hypoxic tissue persists. Likely, these vessels lack blood flow. Image created by Steve Seung-Young Lee as part of the NCI Cancer Closeup 2016 collection.

Note that this color designates blood vessels and not blood. The vessels may be empty. An antibody to hypoxia-induced protein labels hypoxic tissue green. In the micrograph, the left tumor lobe expresses chronically anoxic tissue in areas devoid of blood vessels. This tissue is beyond the perfusion limit for O_2 carried by the blood vessels. In the tumor lobe to the right, several acutely hypoxic tissues reside in well-vascularized regions of tumor. These vessels likely fail to deliver oxygen to the tissue because no blood flows in the lumen of the vessel for one reason or another. Because acute hypoxia determines the number of cycling cells within a tumor, radiation sensitivity can vary from minute to minute and day to day as cells become hypoxic and cease cycling or become reoxygenated and reinitiate mitosis.

8.6. Radiation Biology of the Organism

In this chapter, we scaled up from studies conducted on cells in culture to examine the radiation response of organs and tissues *in situ*. We demonstrated that LQ formalism could be extended to this scale and that hypotheses developed *in vitro* provide an acceptable springboard for understanding *in vivo* dose response. However, we have only mentioned radiation sensitivity at the scale of the whole organism in passing. Let's address that now. Organisms are composed of several organ systems, just as organs are composed of several cellular systems. Within an organism, there are critical radiation-sensitive organs that impact the health of the organism just as FSU sensitivity impacts the health of the organ.

When humans are exposed to whole body doses of radiation, they experience "acute radiation syndrome." At exposures between 1 Gy and 2 Gy, few symptoms appear for several weeks (three or four). Depending on the dose, dose protraction, dose uniformity, and individual susceptibility, a small percentage of the exposed population may experience nausea and vomiting. Within four weeks, most will begin to experience hair loss, paleness, inflammation of mucous membranes, and diarrhea. They will feel weak and listless. If exposed to a 5-Gy whole body dose, all will experience nausea, vomiting, and diarrhea; the onset is almost immediate and may be persistent. By the third week postexposure, bleeding, pallor, mucous membrane inflammation, and infection commence. Higher does are increasingly fatal and cannot be rescued by medical intervention. Exposures of 20 Gy or more cause brain edema, blood–brain barrier breakdown, and collapse of the central nervous system (CNS). The Effects at 20 Gy or much higher are rapid, hours to days. These syndromes result from injury to three organ systems: neurological, gastrointestinal, and hematological. This acute effect is related to the depletion of regenerating FSU cell pools: glial cells, crypt cells, and bone marrow cells.

In addition to acute radiation syndrome, radiation injury introduces delayed and "late effects" in surviving individuals. We referred briefly to these effects when discussing functional assays in skin and the CNS. Late damage refers to any untoward changes in tissues and organs resulting from radiation that express months to years after exposure. Although the effects incidentally may be confined to an organ, the expression tracks with whole body exposure. In the case of targeted radiation delivery, late effects express coincident with the delivered dose volume.

George Casarett (Creasey *et al.*, 1976) is attributed with describing the vascular model of late damage in normal tissue. According to this model, radiation induces damage to the vascular endothelium of the microvasculature — the capillaries and small vessels. The model does not invoke DNA damage leading to cell death, nor is it concerned with the loss of a stem cell pool and replacement kinetics, but his model is consistent with these concepts. The detachment of endothelial cells causes porosity leading to vessel leakage. Mononuclear cells and cellular debris from the circulatory system induce inflammation and activate the local fibroblast population that responds by proliferating and extruding extracellular fibrotic material. The vessel walls thicken as endothelial cell proliferation overcompensates for losses and become less permeable. Radiation-induced endothelial cell proliferation and platelet aggregation cause vessel blockage and reduced circulation. Together, the vessel blockage and wall thickening reduce osmotic flow into the bloodstream, which exacerbates edema and the inflammatory response. The reduction in permeability and blood flow lowers interstitial oxygen concentration, causing organ stress and vascular regression. Figure 8-16 presents a flowchart illustrating Casarett's model leading to radiation-induced late organ damage. Notice that the lower four boxes express a self-exacerbating loop eventually resulting in organ damage. Although this model for late tissue damage does not explain all pathology expressed months to years after radiation exposure, it certainly suffices in the majority of cases.

Because different organs express late damage at different rates, it seems likely that vascular damage cannot represent the only factor responsible for late radiation effects. If so, the kinetics of vascular damage would be consistent. However, if the vascular insufficiency and low oxygen pressure resulting from endothelial damage should impact one or more FSU, then the loss of organ function would reflect the kinetics of the particular organ. In other words, acute organ damage results from FSU cell death stemming from radiation-induced DNA damage, and late organ damage results from FSU cell death caused by oxygen deprivation. Casarett's vascular response explains both stromal (structural) and parenchymal (functional) organ damage if one allows for the inclusion of FSU kinetics. The organism experiences radiation late effects according to the exposure volume. If the entire organism is uniformly exposed, then the earliest responding systems will dominate (neurological, hematopoietic, and gastrointestinal). If a multi-organ volume is exposed, tissue kinetics, including vascular tissue, will dominate. Radiation protection and the field of "health physics" are primarily concerned with preventing acute and late effects in healthy tissue. We will delve deeply into the prevention of damage through radiation delivery techniques, structural barriers, and administrative actions, and we will differentiate between the deterministic (nonstochastic) effects and stochastic effects in Chapter 11.

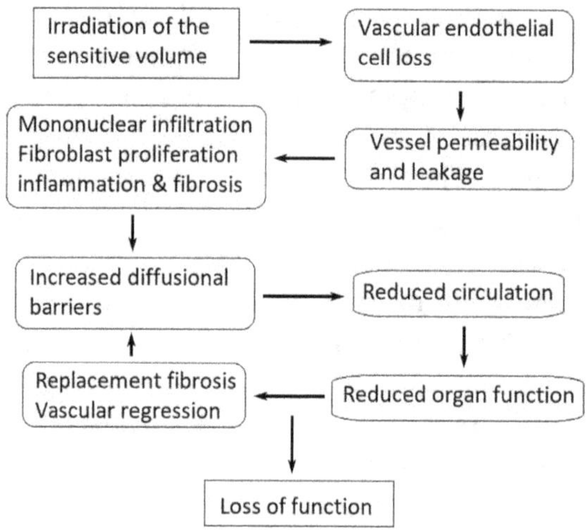

Figure 8-16 Casarett's model of radiation-induced late tissue damage. The runaway response begins with radiation damage to the vascular endothelium and progresses to a cyclic obstruction of vascular functionality leading to organ failure in part or whole.

Table 8-4 The molecular basis for radiation-induced late tissue damage based on Casarett's model.

Cell Type	Cytokine Production	Function
Endothelial Cell	IFNγ	Pro-inflammatory
	TNFα	Pro-inflammatory growth factor
	TGFβ	Pro-inflammatory growth factor
	IL-1, IL-6, IL-8	Pro-inflammatory
	CCL2	Pro-inflammatory
	ICAM-	Cell adhesion molecule, pro-inflammatory
	VCAM-1	Identifies vascular openings, assists transmigration
	E-selectin	Cell adhesion molecule
Fibroblast (radiation activation)	MCP-1	Monocyte chemoattractant
	SDF-1 (CXCL 12)	Pro-inflammatory
	Fractalkine (CX3CL1)	Stimulates monocytes, NKC, T cells, smooth muscle cells
	Rantes	Pro-inflammatory
	VEGF	Vascular endothelial growth factor; neovascularization
	ET-1	Endothelin; vasoconstrictor
	OPN	Osteopontin; inhibits vascular calcification
Fibroblast (hypoxic response)	VCAM-1	Identifies vascular openings, assists transmigration
	ICAM-1	Cell adhesion molecule, pro-inflammatory
	CX3CR1	Fractalkine receptor
	TGFβ	Pro-inflammatory growth factor
Monocyte	CCR2	Chemoattractant, pro-inflammatory
	CCR5	Chemoattractant, pro-inflammatory
	CXCR4	Chemokine receptor
	VEGFR-1	VEGF receptor
	ET_AR	Endothelin receptor; Vasoconstriction

8.7. Summary

- There are four biological scales: macromolecular, cellular, tissues and organs, and organisms.
- Because behaviors become increasingly complex, hypotheses must be verified at each biological scale.
- There are three experimental systems: *in vitro*, *in situ*, and *in vivo*.
- CFU experiments mimic *in vitro* dose response experiments in that individual, clonogenic cells form growths or structures that are counted *in situ*.
- The results from CFU experiments can be plotted in the dose response format — dose vs. fractional survival. MTSH and LQ formalisms can be applied directly.
- Clonogen transplant studies remove cells from a donor animal, create a single cell suspension, and inject the cells into recipient animals. MTSH and LQ formalisms can be applied directly.
- Functional assays generally involve multiple tissues within an organ and often the tissue under study is not the driving force of the observed endpoint.
- Functional assays invoke tissue kinetics. Results take the form of an expressed behavior over time. MTSH and LQ formalisms cannot be applied directly.
- LQ variables can be derived from functional assays by determining biologically equivalent conditions (e.g., fractionation protocols producing the same level of effect) and plotting the reciprocal of dose versus biologically equivalent condition (e.g., dose per fraction).

$$F_e = \frac{\alpha(n)D + \beta(n)D^2}{nD} = \alpha + \beta D \rightarrow \frac{\alpha}{\beta}$$

- Tissue kinetics require deconvolution of all sources and sinks of the cell type under observation: influx of new cells, cell cycle kinetics, and cell loss.
- Organs are composed of several stromal and parenchymal tissues.
- Organs can be classified as serial or parallel based on radiation response. Serial organ function is disrupted by a single point of failure (chain); parallel organs experience additive partial failure dependent on the number of single points impacted (cable).
- Organ radiation response depends on the number of FSU impacted and FSU cell survival.
- An FSU is an element of an organ that can be regenerated from a single surviving cell.
- Organ radiation sensitivity is evaluated by NTCP.
- The LQ variables, α, β and α/β can determine NTCP.
- Brain radiation sensitivity must be evaluated for either cognitive or noncognitive functions.
- Neurons compose about half of the cells in the brain. They are not radiation sensitive; they are post-mitotic.
- The radiation-sensitive cells of the brain are primarily vascular cells and glial cells.
- Radiation damage to the brain includes vascular damage causing edema and inflammation followed by transient demyelination, followed months later by gliosis, permanent demyelination, vascular malformation, and white matter necrosis. This damage correlates with the loss of functions such as speech and hearing, sight, coordination, and autonomic regulation.
- The higher functions of the cortex express radiation damage independent of demyelination and necrosis.
- Cognitive impairment may reflect damage to and reorganization of the neural network.
- Tumors may be considered a single FSU organ — every cell must be killed to prevent regeneration.

- Tumors grow more slowly than would be predicted from tumor cell kinetics. Whereas tumor cells double every day or so, tumor volume doubles in about 2 months. Tumor cell loss is due to apoptosis and other forms of inducible cell death.
- Tumor radiation sensitivity can be assessed by CFU and clonogen transplant analysis, tumor size, tumor growth, and TCD_{50}.
- The best predictor of fractionated delivery tumor radiation sensitivity is the LQ term α.
- *In vivo* radiation sensitivity of tumors agrees well with *in vitro* studies of cell lines derived from the same tumors.
- Tumors suffer hypoxia, anoxia, and necrosis resulting from masses exceeding the O_2 tissue diffusion limit of about 70 μm. Low O_2 partial pressure suppresses cell cycling, alters redox reactions, reduces pH, and affects DNA repair pathway kinetics.
- Cells distant from microvasculature are also deprived of nutrients, particularly glucose, suppressing cell cycling.
- Hypoxia can be transient and thus tumor radiation sensitivity may fluctuate in an unpredictable way.
- Acute whole body exposures can result in acute prodromal syndromes (nausea and vomiting) followed by hematopoietic, gastrointestinal, and cerebrovascular radiation syndromes due to the loss of vulnerable cells in the hematological, gastrointestinal, and neurological organ systems.
- Symptoms and timing of onset of prodromal syndromes are dose dependent, and consist of vomiting, nausea, pallor, lethargy, diarrhea, and fever.
- Late radiation effects result from vascular damage and subsequent loss in tissue permeability that debilitates organ FSU.
- Acute and late effects are not limited to a tissue or organ, they develop within the irradiated volume according to dose, dose rate, tissue sensitivity, and organism resilience.

8.8. Problems

1. Find α, β, and α/β for the following gastrointestinal crypt cell assay if the average number of jejunal clonogens per circumference is 620 and the average number of crypts per circumference is 127 in unirradiated mice. Three mice were exposed at each dose.

Dose (Gy)	# Crypts	Dose (Gy)	# crypts
0	126	10	93
	130		84
	121		81
2	120	15	32
	127		31
	129		28
5	114	20	5
	130		3
	125		13

2. Suppose the experiment by Withers *et al.* portrayed in Figure 8-4 was repeated using a split dose protocol. The results might resemble those below. Using this plot, find D_q, D_0, and, and plot the initial slope and shoulder.

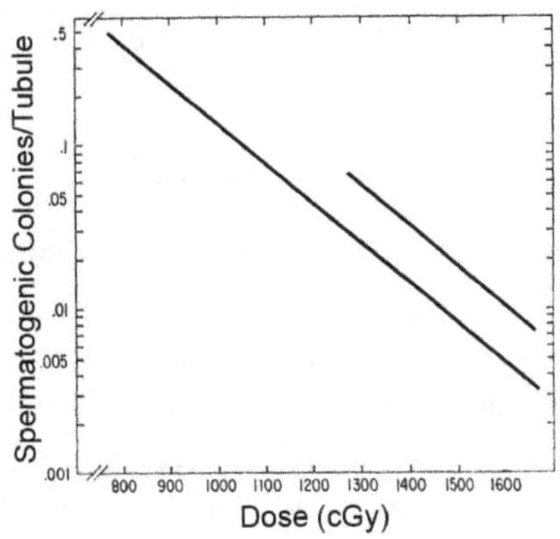

3. Table 8-2 illustrates isoeffect or biologically equivalent doses for irradiated skin response. Plot these data as reciprocal total dose vs. dose per fraction to enable LQ analysis. Find α/β.
4. The Hewitt dilution assay was carried out with a mouse lymphoma line that grows well as a single cell suspension. The following table lists the mortality rate (% dead) in groups of 10 mice per data point that received the injected number of lymphoma cells shown. Given these data, construct the curves for animal lethality versus dose of injected cells for each of the irradiation dose levels given. Construct the derived survival versus dose curve for the irradiation of this line of lymphoma cells. Estimate the D_0 and the extrapolation number, n, for this cell line.

0 Gy		1 Gy		2 Gy	
Cells injected	**% mortality**	**Cells injected**	**% mortality**	**Cells injected**	**% mortality**
2	0	6	6	8	12
4	14	8	8	12	32
6	42	10	10	14	52
8	75	12	12	16	60
10	100	14	14	18	70
12	100	18	18	—	—

4 Gy		5 Gy		6 Gy	
Cells injected	**% mortality**	**Cells injected**	**% mortality**	**Cells injected**	**% mortality**
12	0	50	10	130	0
14	6	60	25	140	10
18	40	65	50	150	42
20	49	70	52	160	51
222	70	75	69	170	80
26	100	80	95	180	100

8 Gy	
Cells injected	**% mortality**
400	0
450	5
500	17

550	30
600	51
650	70

5. Derive the expression for NTCP using the linear quadratic assumptions for cell survival (Equations 8.6, 8.8, and 8.9).
6. The statement was made, "Fractionation suppresses beta, so the most reliable predictor of tumor radiation sensitivity is alpha." Show that this is a true statement.
7. Withers and colleagues constructed a "single dose" survival curve using data collected from fractionated delivery of x-rays to jejunal crypt cells in mice. The curve can be fitted using LQ formalism and provides reliable low dose survival data as well as nonlethal high doses. Given the data provided below, plot a curve of survival vs. dose per fraction, and find α and β. Why is α/β so large?

D	*f*	*S*
13	10	0.7
14	7	0.6
16	16	0.75
26	2	0.0034
30	12	0.51
32	10	0.4
40	5	0.06
49	5	0.02
60	12	0.207
67	10	0.108
70	5	0.0021
77	7	0.0093
96	16	0.16

8. Suppose a tumor is periodically evaluated noninvasively for percent normoxic, hypoxic, and anoxic cells. The examination is repeated as the tumor grows from microscopic to palpable. A graphical representation of the tumor life history might look something like the following image. If the tumor is exposed to an acute dose of lethal radiation at time *T*, describe the biology responsible for producing the slope of the curve at A, B, C, D, E, and F. For example, at A the tumor is less than 50 μm in diameter and all cells are oxygenated; thus 0% anoxic cells.

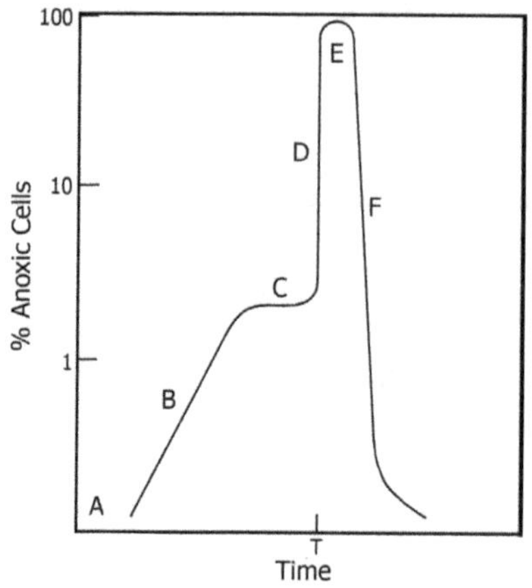

9. What is the D_0 for the initial part of the survival curve below, composed from mature (1 mm diameter) lymphosarcoma tumors in mice? What is D_0 for the final slope of the survival curve? What do the two components of the curve describe? What happens at 9 Gy? What percentage of the tumor cells are hypoxic?

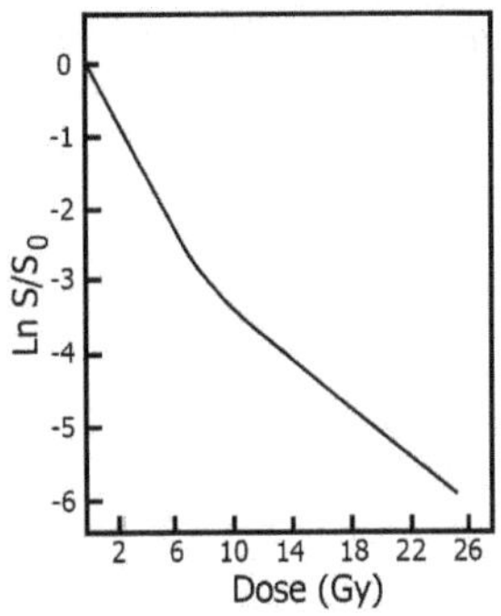

10. Within four weeks of 2 Gy whole body radiation exposure, most humans will begin to experience hair loss, paleness, inflammation of mucous membranes, and diarrhea. They will feel weak and listless. Each of these symptoms can be traced to an actively mitotic cell population within a sensitive tissue. What is the organ, tissue, and FSU for each of these symptoms?

8.9. Bibliography

Alpen, E. L., 1990. *Radiation Biophysics.* 2 ed. San Diego: Academic Press.

Barbarese, E. & Barry, C., 1989. Radiation sensitivity of glial cells in primary culture. *Neurological Science,* 91(1), pp. 97–107.

Creasey, W., Withers, H. R. & Casarett, G., 1976. Basic mechanisms of permanent and delayed radiation pathology. *Cancer,* Volume 37, pp. 999–1013.

Denekamp, J., 1993. Angiogenesis, neovascular proliferation and vascular pathophysiology as targets for cancer therapy. *British Journal of Radiology,* 66(783), pp. 181–196.

Fowler, J. F. *et al.*, 1963. Experiments with fractionated x-ray treatment of the skin of pigs. *British Journal of Radiology,* 36, pp. 188–196.

Goitein, M., 2008. *Radiation Oncology: A Physicist's-Eye View.* New York: Springer.

Greene-Schloesser, D., Robbins, M. E., Peiffer, A. M. & Shaw, E. G., 2012. Radiation-induced brain injury: a review. *Frontiers in Oncology,* 2, pp. 1–18.

Hewitt, H. B. & Wilson, C. W., 1959. A survival curve for mammalian leukaemia cells irradiated in vivo (implications for the treatment of mouse leukaemia by whole-body irradiation). *British Journal of Cancer,* 13, pp. 69–75.

Hill, R. P. & Bush, R. S., 1969. A lung colony assay to determine the radiosensitivity of cells of a solid tumor. *International Journal of Radiation Biology,* 15, pp. 435–444.

Hubel, D. H., 1988. *Eye, Brain and Vision.* New York: Scientific American Library.

Parihar, V. K. & Limoli, C. L., 2013. Cranial irradiation compromises neuronal architecture in the hippocampus. *Proceedings of the National Academy of Science,* 110(31), pp. 12822–12827.

Rockwell, S. *et al.*, 2009. Hypoxia and radiation therapy: Past history, ongoing research, and future promise. *Current Molecular Medicine,* 9(4), pp. 442–258.

Thames Jr, H. D., Withers, R., Mason, K. A. & Reid, B. O., 1981. Dose-survival characteristics of mouse jejunal crypt cells. *International Journal of Radiation Oncology, Biology, Physics,* 7(11), pp. 1591–1597.

Till, J. E. & McCulloch, E. A., 1961. A direct measurement of the radiation sensitivity of normal mouse bone marrow cells. *Radiation Research,* 14(2), pp. 213–222.

Withers, H. R., 1967. The dose-survival relationship for irradiation of epithelial cells of mouse skin. *British Journal of Radiology,* 40, pp. 187–194.

Withers, H. R. & Elkind, M. M., 1969. Radiosensitivity and fractionation response of crypt cells of mouse jejunum. *Radiation Research,* 38(3), p. 598–613.

Withers, H. R., Hunter, N., Barkley Jr, H. T. & Reid, B. O., 1974. Radiation survival and regeneration characteristics of spermatogenic stem cells of mouse testis. *Radiation Research,* 57(1), pp. 88–103.

Withers, H. R., Taylor, J. M. G. & Maciejewski, B., 1988. Treatment volume and tissue tolerance. *International Journal of Radiation Oncology Biology and Physics,* 14, pp. 751–759.

Chapter 9

Radiation and Cancer Biology

9.1. Introduction

Chapters 3, 4, and 6 followed radiation insult from physical interactions resulting in DNA sequence changes, through chemical and biochemical remodeling, to cellular responses leading to cell death. Chapters 5, 7, and 8 examined the systemic ramifications of cell death resulting from radiation damage. However, alternative radiation exposure outcomes that avoid cell death are possible. Suppose despite DNA damage, the cell survives and keeps replicating. Following DNA damage, the cell's restitution and enzymatic repair mechanisms either return the preirradiated DNA sequence and genomic structure, or else the DNA sequence and/or chromosomal structure is somehow stably altered (mutated). In the latter case, progeny surviving subsequent cell division may retain a silent mutation — one that has occurred in nontranslated DNA or one that results in a physiologically harmless state — or they may express genomic characteristics that transform the cell in ways potentially leading to carcinogenesis. This chapter examines the radiation processes leading to cancer, the path thus far not taken (with apologies to Robert Frost).

9.2. Transformation

Recall that in Chapter 4, we examined point and frameshift mutations (Figures 4-13 and 4-14) and described chromosomal translocations (Figure 4-16) and aberrations (Figure 4-15), not all of which are lethal. In Chapter 5, we discussed heritable epigenetic mutations that may be responsible for silencing some genes and activating others, and again, this does not always result in a lethal outcome. *In vitro*, these nonlethal alterations in sequence or structure may express as altered cellular anatomy and physiology referred to as cellular transformation.[1] A definitive phenotype of transformed cells exhibits a loss of normal cell contact inhibition and growth control. Normally, cells in culture stop dividing when the plate is confluent (full) because any cell in contact with other cells stops cycling. Additionally, instead of forming an organized sheet of interconnected cells *in vitro*, the progeny of transformed cells lose adhesion to the culture plate, round up, and pile one on top of the other (Figure 9-1). Transformed cells thus evade G_0 when structurally disorganized resulting from the loss of surface adhesion foci. Transformed cells exhibit the following characteristics:

- Loss of cell-cell contact inhibition — piling up into dense aggregates
- Loss of anchorage-dependent growth characteristics
- Loss of normal dependence on exogenous growth factors
- Chromosomal abnormalities
- Expression of unique surface antigens
- Escape from normal regulation of cell growth, division, differentiation

[1] Caveat: the term "transformation" also describes the insertion of foreign DNA into the genome of a cell either through viral or experimental manipulation. This procedure alters genomic content and thereby phenotypic expression.

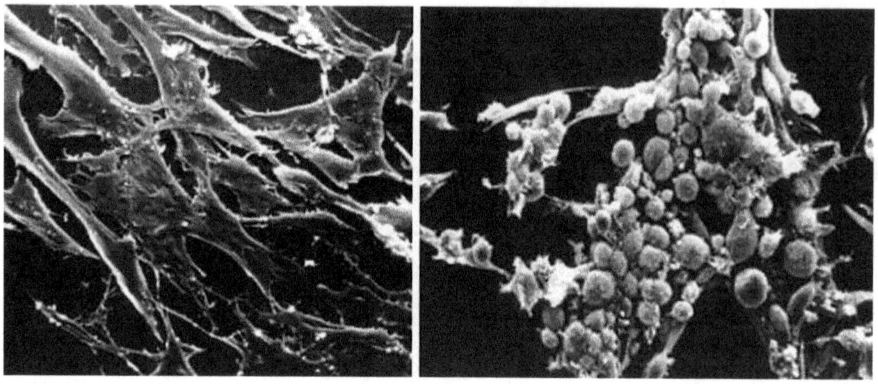

Figure 9-1 Cultured mouse fibroblasts. (**Left**) Normal, contact-inhibited fibroblasts growing as a single sheet *in vitro*. (**Right**) Transformed fibroblasts lacking normal contact responses. Photographs provided by G. Steve Martin, with permission through CC-BY 3.0, at https://www.biology-pages.info/C/CancerCellsInCulture.html.

- "Immortalization" — transformed cells may be passaged indefinitely
- Tumorigenic growth in recipient animals under appropriate conditions.

The expression of transformation is a "stochastic" event — i.e., the probability of the event is a function of mutagen dose and reflects the specific functionality disrupted by mutation. So, when we analyze transformation, we are forced to speak in terms of the *probability* of transformation at a particular dose, under specific circumstances. Also keep in mind that when performing transformation studies, the transformation efficiency excludes cells killed by the mutagen (see Figure 9-2) and thus must be normalized to dose-dependent cell survival (modeled by LQ). In transformation studies, we are examining only the surviving cell population; we calculate the dose-dependent probability of a *surviving* cell expressing transformation (Figure 9-3).

Because cancer cells and transformed cells exhibit similar characteristics, especially considering the final bullet in the earlier list, cancer biologists believe that tumors *in situ* likely result from *in vivo* cell transformation progressing to carcinogenesis, but the extrapolation is largely one of faith. The initiation of a cancer cannot be observed *in situ* in humans; systematically inducing cancer would be unethical, so we must rely on epidemiology, tissue culture, and animal studies. When extrapolating from one scale to another, several caveats must be appreciated. Fibroblast cells culture more readily than endothelial or epithelial cells, so historically *in vitro* studies have been conducted largely on fibroblasts, however, most cancers do not arise from this cell type, they arise from epithelial cells. Furthermore, progression from transformation to tumorigenesis is confirmed by implanting transformed cells into animal models, primarily rodents (Figure 8-5). Be wary when translating these results to humans. Rodents and humans have different life spans — rodents live approximately 3 years and humans live approximately 75 years — thus tumor development differs in inadequately understood ways. Rodent cells transform in culture more readily that human cells. To create successful human xenografts, human cells require at least five distinct nexus cell signaling pathways be disrupted through mutation: Ras, pRb, p53,[2] PP2A, and telomere shortening. Furthermore, rodent cells undergo spontaneous immortalization relatively easily whereas human cells are resistant to conversion. And lastly, we need to step cautiously when drawing conclusions from experiments using radiation to induce neoplastic transformation because evidence

[2] Regarding gene and protein notation: human symbols for genes are capitalized and italicized whereas symbols for proteins are capitalized but not italicized, and the specific mutation of a protein is noted through use of a superscript. Ubiquitous proteins, such as pRb, p53 and p21, are excepted through historical, common use.

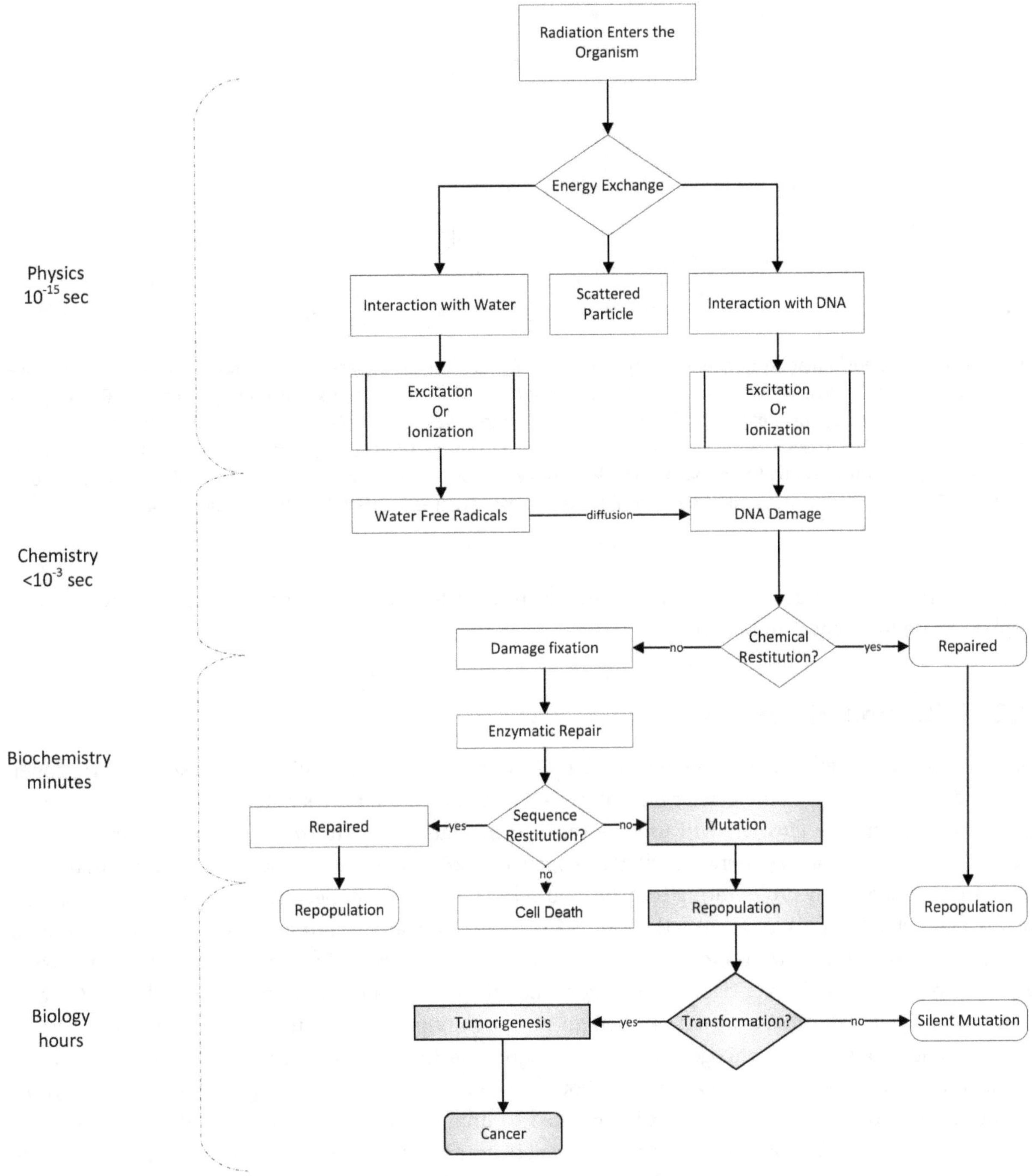

Figure 9-2 The carcinogenesis pathway. DNA radiation damage repair either returns the preirradiated sequence and genomic structure, or the chromosomal product is somehow altered. Following cell division, the surviving progeny may retain a silent mutation or may express genomic characteristics that transform the cell in ways leading to tumorigenesis. This chapter examines the radiation biochemistry pathway leading to cancer.

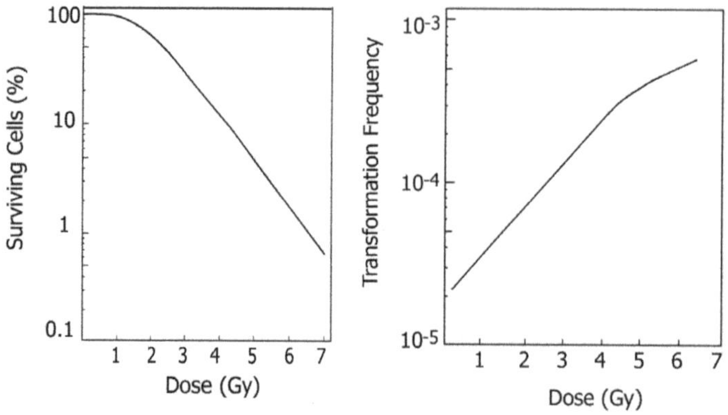

Figure 9-3 Cell transformation frequency compared with cell survival. Left: Classical cell survival curve fit by the linear quadratic equation. As dose delivered to the cultured cells increases, the fraction of surviving cells decreases. Right: Typical transformation frequency curve fit by eye. As the dose delivered to the cultured cells increases, the fraction of surviving cells that are transformed increases. At 4 Gy, 2 cells out of 10,000 are transformed and at 6 Gy, 5 cells out of 10,000 are transformed. However, note that the fraction of cells that survive decreases with dose such that at 4 Gy 1 out of 10 cells survive and at 6 Gy, 1 out of 100 cells survive; i.e., the sample size available for transformation at 6 Gy is 1/10th that available at 4 Gy.

indicates radiation-induced tumors may initiate differently from spontaneous tumors, and we shall explore these differences in Section 9.6.

9.3. Initiation of Cancer

Before proceeding, let's clarify some terminology. Cell transformation *in vitro* manifests cellular properties held in common with the cancer phenotype (see list of characteristics noted above). Transformation is less well defined *in vivo*; we will use the term to indicate that a normal cell has acquired a set of capabilities enabling it to generate a clonal neoplasm *in situ*. Thus, a neoplasm is an abnormal tissue mass formed of actively proliferating transformed cells. The neoplasm becomes a tumor when explanted neoplastic cells form a clonal growth after inoculation into an immunocompromised experimental animal such as a nude (nu/nu) mouse. Again, the term must be defined rather loosely because we assert a neoplasm is a tumor despite rigorous testing by inoculation into a test animal. As described in Chapter 8, many early stage tumor growths are benign or "carcinoma in situ"; since they do not metastasize. Carcinogenesis, also called oncogenesis, is the progression to a late stage aggressive cancer – a tumor capable of local invasion and metastasis. Although the evolution of terminology from "transformation" to "neoplasm" to "tumor" is squishy, it is intended to imply a linear accumulation of aggressive traits. Most cancer initiation hypotheses assume that genetic mutations represent an early event leading to the expression of characteristics that drive carcinogenesis. These mutations present in many forms in cancers including single nucleotide polymorphism (SNP; point mutations), sequence mutation (e.g., frameshift), microsatellite instability (changes in the number of repeating oligonucleotide sequences), loss of chromosomal heterozygosity (LOH), epigenetically altered DNA and chromatin states, and other karyotypic variations — most notably aneuploidy. Not all mutations are carcinogenic, or even tumorigenic; mutations naturally occur spontaneously in humans at a rate of something like 1×10^{-8} mutations per site per generation, though the actual mutation rate is organ and organism specific. Alexandru Das and Juliana Denekamp (Das and Denekamp 2000) present an elegant discussion of natural mutations mapped in several tissues.

9.3.1. *Origin Theories*

For now, let's assume that DNA mutation leads to cell transformation potentiation *in vivo*. In this scenario, initiation refers to the first transforming genetic event within a progenitor cell that perpetuates as a mutation detectable in the cells of the consequent tumor. This is not to say that the cell possesses only one mutation, but rather that a particular mutation motivates an alteration in cellular physiology that enables (possibly in concert with other mutations) carcinogenic progression. To differentiate the *in vitro* transformation process from the *in vivo* carcinogenesis process, allowing that they may be different, let's refer to the initial *in vivo* phenotypic changes expressed by the transformed cell as neoplasia. The primordial neoplastic phenotype is characterized most often by the loss of proliferation controls producing a clonal population with a distinct proliferation advantage. This *monoclonal origin hypothesis* can be validated through identification of at least one genetic mutation that expresses in every cell of a neoplasia and the hypothesis is supported by considerable clinical evidence. Proponents of the hypothesis note that tumor physiology can be simulated by introducing specific chromosomal damage *in vivo* and point out that both experimental and fortuitous cancers are stochastic in nature. Additionally, benign tumors, origin-specific cancers, and metastases exhibit predictable enzyme patterns — the cells within these neoplasms inappropriately express identical atopic isozymes. Karyotypic alterations tend to be consistent within the cells of individual tumors. And, because unique cell surface antigens result from specific mutations, immunoglobulins raised against a single cell recognize other cells within the same tumor. But notice that karyotypic alterations *tend to be* consistent... Tumors express incredible genetic heterogeneity. About 20,000 point mutations have been identified in breast cancers and up to 333,000 point mutations have been identified in melanomas. Lung cancers contain up to 50,000 point mutations plus as many as 54,000 additions and deletions of genetic material. Some of these are "driver" mutations that contribute to carcinogenesis, some are "passenger" mutations that do not contribute to carcinogenesis. It is possible then, that a tumor expresses clusters of cells each presenting unique subsets of mutations and chromosomal rearrangements. In that case, the *clonal evolution model* seems more likely (Figure 9-4). According to that model, two or more cells experience independent mutations that together provide symbiotic survival advantages, and these characteristics may be synergistic within the community of mutated cells thereby promoting a proliferation advantage. Analyzing the genetic fidelity of mature cancers is extremely challenging — simply extracting the DNA and performing restriction fragment

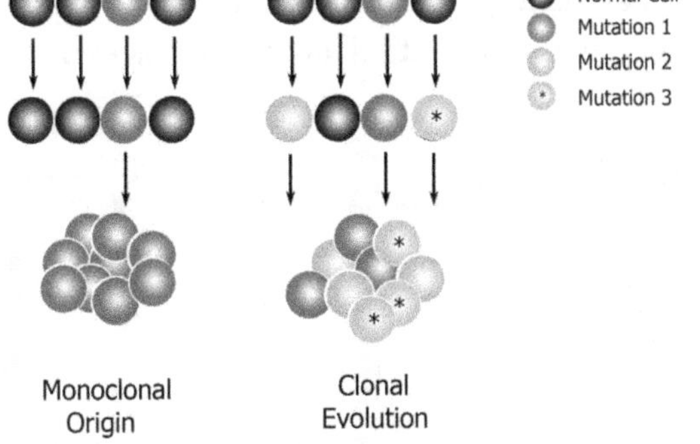

Figure 9-4 Initiation of carcinogenesis. The monoclonal origin model supposes that an initial mutation promotes cell division and thus, produces a population of clones all expressing an identically mutated gene. The clonal evolution model proposes that several cells (3 in this example) each experience a unique initial mutation resulting in several (3) clonal populations each exhibiting the mutation of their origin cell. In both models, a single initiating mutation has occurred within each of the clonal origins.

length polymorphism (RFLP) analyses to identify the points of mutation can provide a ranking of mutation frequencies, and next generation DNA sequencing can identify mutations expressed in as few as 1% of the tumor cells, but highly sophisticated techniques such as duplex sequencing are required to interrogate the clonal composition of a tumor. Because cancers can be composed of cells expressing hundreds of thousands of mutations, clonal clusters are bound to develop.

A third origin hypothesis, the *cancer stem cell hypothesis*, resembles clonal evolution but limits the tumor origin population — those cells possessing a potential to initiate a tumor — to one subset of progenitor clones. The cancer stem cell model proposes that only certain cells within a population of transformed cells possess the capacity to initiate a tumor *de novo*, just as only certain cells within a tissue divide to produce one cycling stem cell and one differentiated quiescent cell. Unlike the clonal evolution model, this model claims that the other subpopulations of cells do not derive independently via clonal mutation but are nonpotent progeny of the cancer stem cell (cell division producing one cancer stem cell and one nonpotent tumor cell). According to the hypothesis, the non-stem cells support tumor parenchyma but are incapable of metastasizing or inducing tumor regrowth. The cancer stem cells purportedly acquire stem cell characteristics through mutation-driven dysregulation of normal stem cells, somatic cell fusion, or horizontal gene transfer processes. The cancer stem cell hypothesis has interesting implications for therapeutic design for tumor treatment.

9.3.2. *Genetic Mutation*

Any of the several DNA or chromosomal reconfigurations can initiate the carcinogenesis pathway including DNA sequence mutation (SNP or frame shift), gene amplification (via replication slippage, chromosomal polyploidy, or ectopic recombination during meiosis), gene inversion, gene deletion, or chromosomal translocation (Figure 9-5). Thus far, we have focused primarily on radiation-induced mutation, but mutations also manifest during routine mitotic replication and housekeeping processes as well or may be induced by chemical mutagens and carcinogens. Furthermore, radiation-induced mutation includes non-ionizing radiation initiators; for example, the most common cancer — skin carcinoma — develops following ultraviolet (UV) radiation–induced thymine dimer formation. Although base excision repair (BER) efficiently repairs this damage, the presence of excessive thymine dimers can overwhelm the enzymatic capacity and promote misrepair. Mutated genes that induce neoplasia generally belong to one of three metabolic classes: caretaker, gatekeeper, or landscaper. Caretaker genes promote genomic stability; this class includes DNA repair genes, for example. Gatekeeper genes control mitosis and cell death, and landscaper genes encode microenvironmental factors such as extracellular matrix (ECM) structure

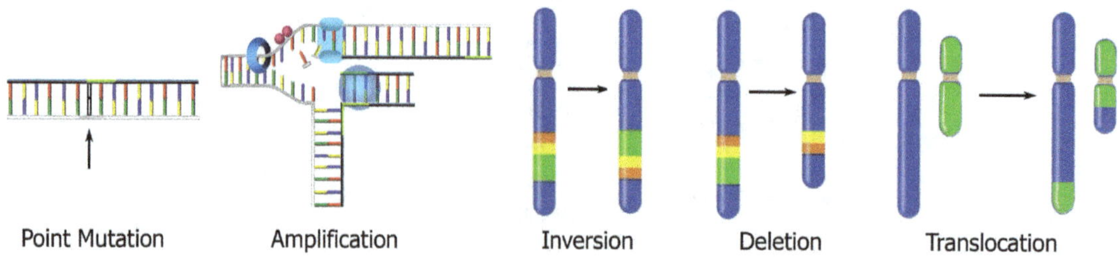

Figure 9-5 The origins of genetic mutation. A point mutation or other DNA sequence misrepair may result in a defective gene product or altered transcription signaling. Gene amplification can increase gene product; amplification can result from slippage at the replication fork as shown, through chromosomal polyploidy, or aberrant crossover during meiosis (not shown). Gene inversion, and translocation disrupt proper transcription. The Philadelphia chromosome, an exchange between chromosomes 9 and 22, is commonly found in cancers. Gene deletion knocks out gene functionality.

and components. The extracellular environment influences cell functions several ways including adhesion, migration, proliferation, and differentiation. It is important to note that mutations need not alter coding sequences; some chromosomal rearrangements can alter transcriptional control of genes. For example, untranslated promotor regions may be split leaving part associated with the original chromosome and part translocated to the new chromosome. Loss of transcription control is a common theme in cancer presentation. In addition to sequence disruption within the promotor regions of genes through chromosomal rearrangement, control is also lost through promotor scission by nucleotide substitution or loss, chromosomal deletion, gene inversion, or translocation. Mutation or deletion of genes coding for transcription factors may also disrupt gene transcriptional control. Furthermore, translocation of a gene linking it to an active promotor, say splicing it with a gene that is frequently transcribed, stimulates the overexpression of that gene. Translocation of the *myc* oncogene — a gene coding for a nuclear phosphoprotein controlling cell cycle, cell growth, apoptosis, cellular metabolism, and adhesion — downstream of the immunoglobulin enhancer leads to excessive cell proliferation. Obviously, gene amplification can also increase the expression of proteins involved in carcinogenesis and *MYC* amplification is frequently observed in cancer cells.

Of course, the disruption of gene sequences can also produce neoplastic phenotypes. Instead of being overexpressed in several cancers, the *MYC* genetic sequence is mutated disrupting any, several, or all of the phosphoprotein's critical functions. In another example, *EGFR* encodes the epidermal growth factor receptor (EGFR). Normally, the mitogen, epidermal growth factor (EGF), binds to this receptor activating a signaling pathway leading to cell division. This gene is targeted in several cancers; two forms of mutation dominate. In the first, the region of the gene encoding the extracellular portion of the receptor is lost, so the transcribed protein spans the cellular membrane but lacks the cell surface binding site for EGF. In this condition, the tail of the protein constitutively signals the cell to divide. In the second common mutation, the inhibitory region of the receptor tail is lost or deformed in such a way that the mitotic signal can no longer be extinguished by intracellular controls. Thus, once EGF binds to the receptor, the stimulus to divide persists constitutively. These mutations in *EGFR* increase the activity of EGFR protein without increasing gene transcription; in other words, the cellular concentration of EGFR protein does not increase. There are literally hundreds of other examples of gene coding mutations that influence carcinogenic progression. Nonetheless, DNA sequences that code for protein products or enzymatically active RNA molecules account for only about 1.5% of the genome and the subset of those genes capable of fostering cancer progression constitutes an even smaller cohort. Carcinogenesis is enabled through two mutational events: activation of oncogenes and loss of tumor suppressor genes. Oncogenes induce tumorigenesis through one or more of the following mechanisms: their products encourage escape from normal cell growth regulation, division, and differentiation; they prevent cell death/senescence; or they thwart tumor suppressor genes. Tumor suppressor genes promote healthy metabolic behaviors such as cell cycle arrest, apoptosis, and DNA repair. Tumor suppressor gene products additionally may inhibit oncogenes or their products. The current list of known oncogenes and tumor suppressor genes includes 70 genes associated with germline cell DNA mutations and nearly 600 genes strongly associated with somatic cell DNA mutations in human cancers. The nomenclature is a bit tricky. An "oncogene" is the mutated form of a nascent gene; the nonmutated, unaltered gene that may prompt carcinogenesis if mutated is called a "proto-oncogene." Tumor suppressor genes are nonmutated guardians of healthy cell behavior. When mutated, tumor suppressor genes cease proper function leading to tumorigenic potentiation (Figure 9-6).

With respect to cancer initiation, remember also that somatic cells contain two copies of each gene: maternal and paternal. Often, a single, undamaged copy of a gene is sufficient to provide adequate transcription of mRNA and thus protein product for normal cell metabolism; the mutation is recessive. In

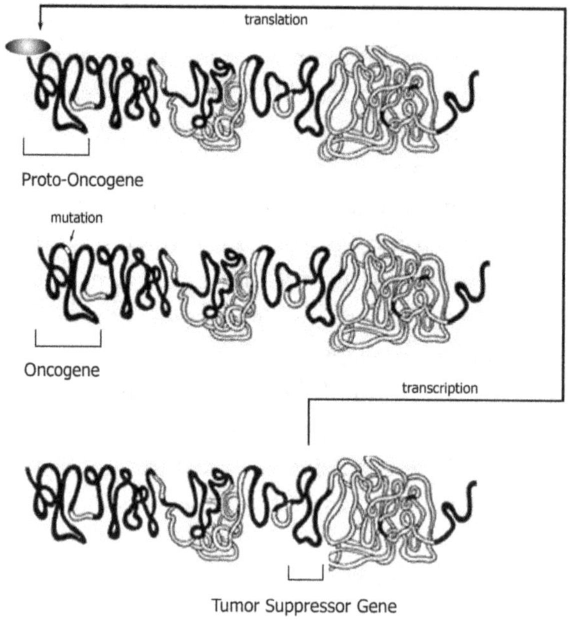

Figure 9-6 Oncogenes and tumor suppressor genes. Oncogenes express DNA sequence or epigenetic mutation that disrupts normal control of translation causing the gene to act in a way that disrupts caretaker, gatekeeper, or landscaper functionality. Tumor suppressor genes generate products that regulate proto-oncogenes. Mutations in tumor suppressor genes disrupt normal control of caretaker, gatekeeper, or landscaper genes.

these cases, both alleles must be affected before the mutation manifests, and consequently initiation requires two mutations — the first mutation does not initiate cancer, rather it creates a condition that predisposes the cell to neoplastic transformation. Duplicated genetic mutation by happenstance would seem unlikely, however, an interesting mechanism has been identified affecting several tumor suppressor mutations. In these cases, the tumor suppressor gene is deleted in one allele (first mutation) and then the other chromosome of the pair is lost (second mutation). To compensate, the cell replicates the existing chromosome (which is defective for the tumor suppressor gene) through "somatic homozygosity." Almost all tumor suppressor genes require that both alleles be knocked out for the phenotype to be expressed. In other cases, loss of one allele may be sufficient to impact cell function; these gene mutations are dominant. In general, loss-of-function defects are recessive, whereas gain-in-function defects tend to be dominant, as the presence of any mutant protein drives the untoward behavior.

9.3.3. *Epigenetic Mutation*

Clonal epigenetic pattern modification is associated with tumors at least as frequently as clonal genetic mutation. The inhibition of gene transcription through heritable hypermethylation of gene promotor CpG islands expresses in tumors up to 10 times more frequently than direct gene mutation. In colon cancer, for instance, O6-Methylguanine-DNA Methyltransferase (MGMT) gene mutation presents in about 4% of the tumors, whereas nearly all colon cancers express hypermethylated *MGMT.* In addition to *MGMT,* other DNA repair genes — *BRACA1, WRN, MLH1, MSH2, ERCC1, PMS2, XPF, P53, PCNA,* and *OGG1* — have been shown to display epigenetic hypermethylation in various cancer cells. Because DNA repair safeguards the genome against mutation, these genes represent tumor suppressor genes, and they fall into the caretaker category. Clearly, obstruction of high-fidelity DNA repair could provoke an abundance of subsequent mutation events. Precisely

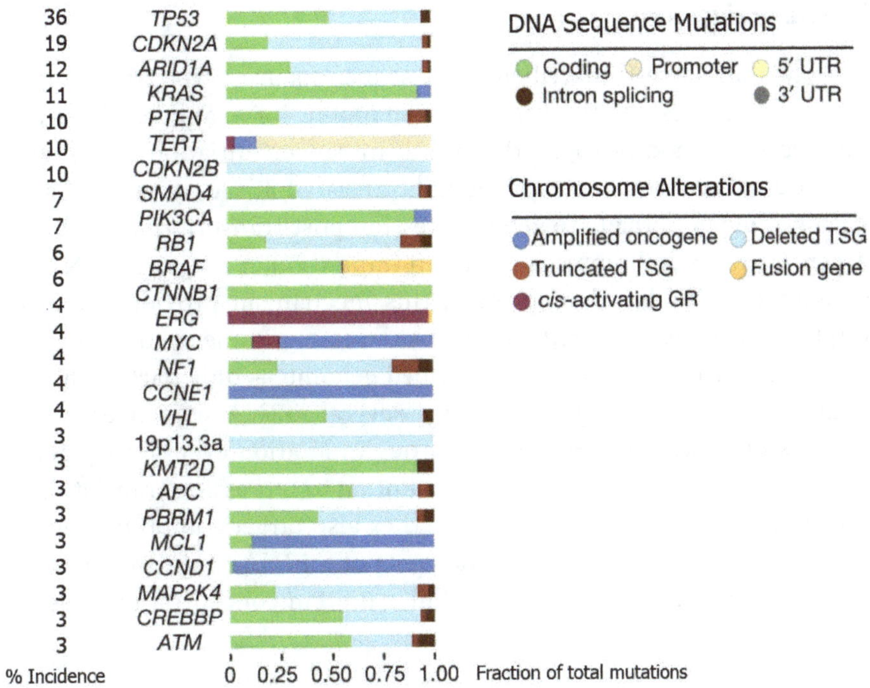

Figure 9-7 Mapping genomic alterations in cancer genes. Analysis of more than 65 tumors revealed driver mutations in the targeted genes listed (center) according to the percent of patients presenting a mutation in each gene (left).
UTR — untranslated region, TSG — tumor suppressor gene, GR — genomic rearrangement.
Reproduced and modified from ICGC/TCGA Pan-Cancer Analysis of Whole Genomes Consortium, 2020. Permission to reuse and modify granted through https://creativecommons.org/licenses/by/4.0/.

Chromatin Immunoprecipitation Sequencing (ChIP-seq)

Identification of epigenetically modified DNA and alterations in that modification can be determined by immunoprecipitating chromatin using antibodies toward various histone modification states and then sequencing the precipitated DNA. The procedure goes like this. Formaldehyde covalently bonds DNA to interacting proteins allowing immunoglobulins specific against the protein of interest to isolate the DNA/protein complex. Because the methylation status of gene promotor sites determines accessibility of transcription complex binding, if the methylation state changes, the protein binding status changes. Thus, a promotor region that is bound to a transcription complex will be isolated, whereas hypermethylated regions that are protein-free will not. The target DNA that is isolated is sequenced using any standard library preparation method to identify the epigenetically modified gene. The method reveals histone modification states and provides information regarding which regions of the DNA are in active or open chromosome states versus which regions of the DNA are transcriptionally silent.

An alternative to ChIP-seq, Methyl-seq, isolates and fragments the DNA and then separates the material into two samples. One sample is treated with bisulfite and the other is left untreated. Bisulfite treatment changes cytosine nucleotides to uracil nucleotides, leaving methylated-cytosine nucleotides unchanged. The two samples are compared to determine the extent of methylation.

because deficiencies in DNA repair can lead to a large number of subsequent mutations, identification of initiating mutations remains ambiguous in the context of a sea of multiple alternative DNA modifications. Nonetheless, evidence and logic indicate that DNA repair suppression through epigenetic mutation is likely an early event.

9.3.4. *miRNA Gene Activation*

In Chapter 4, we saw that some transcribed DNA sequences are never translated into protein but rather the RNA product represents the functional endpoint, for example, tRNA and rRNA. When considering transcriptional modifications that result in cancer initiation, we should include mutations in genes encoding a particular type of nontranslated RNA: microRNAs (miRNA). These relatively short RNA sequences regulate gene expression post-transcriptionally by binding to the 3′ region of the targeted mRNA and repressing protein translation or altering mRNA stability. Because RNA is considerably more friable than protein, this mechanism provides rapid response, short-interval, transcription and translation regulation. The human genome encodes more than a thousand miRNAs that target more than half of the transcribed genome — only part of the sequence need be complimentary, allowing a single miRNA to target several mRNA sequences. When miRNAs targeting tumor suppressor genes are upregulated through mutation events, they act as oncogenes by suppressing tumor suppressor gene expression. In addition to repressing translation, miRNAs can be exported from the nucleus to the cytoplasm where they also target complimentary mRNA inducing degradation. Mutations blocking the efficient transport of miRNA to the cytoplasm will cause aberrantly long mRNA lifetimes that can increase cellular protein concentrations without upregulating transcription.

9.3.5. *Carcinogenesis by Viral Infection*

During the early days of cancer research, over a century ago (circa 1911), much excitement surrounded the discovery that cancers in domestic chickens could be caused by a virus. That discovery of Rous sarcoma virus eventually led to the identification of many oncogenes and much of the current understanding of cancer molecular biology. As things go in research, following disappointing decades of failed attempts to discover a significant pool of additional carcinogenic viruses, the medical scientific community turned its attention to other possible drivers. Nonetheless, interest in virally induced human cancers has resurged over the last 30 years yielding the discovery of seven viruses responsible for 12% of all cancers: Epstein–Barr virus (EBV), hepatitis B virus, human papillomavirus (HPV), human T-cell lymphotropic virus, hepatitis C virus, Kaposi's sarcoma herpesvirus, and Merkel cell[3] polyomavirus. Therefore, no discussion of carcinogenic initiation would be complete without at least a cursory survey of this important mechanism.

Viruses are not efficient inducers of cancer but can stimulate oncogenesis either through genetic mutation or by directly modulating signal transduction pathways that control cell cycling and cell death, immortality, or cancer progression to metastasis. The mutational events often occur during insertion of viral DNA into the human genome — in general, incorporating randomly wherein the viral transcript may disrupt a gene promotor, gene sequence, or stop codon. Most viral DNA (or RNA) persists episomally — as small circular, free-floating nuclear genomes — occasionally inserting into the host genome after a latent period. Viruses augment DNA damage through any number of mechanisms including DNA repair impairment, production of free radicals, and cytokine upregulation. Although we have not discussed oncogenic processes subsequent to mutagenesis, viral proteins can activate these mechanisms directly without relying on genetic mutation so they must be noted here a bit prematurely. Be patient, the ramifications of these viral behaviors will become clear in the following sections. Specifically, viral

[3] A cell type below the epidermis lying close to the nerve endings that sense touch.

proteins produced by the host cell during the course of infection can deregulate cellular functions, contributing to oncogenesis. Furthermore, viruses have developed mechanisms to avoid immunological clearance. These mechanisms incidentally help virally initiated cancers evade detection. And some viral proteins degrade extracellular tissues, such as basement membrane, promoting cancer cell migration. Let's take a slightly closer look at some examples to elucidate these fascinating mechanisms of oncogenesis.

HPV infects basal epithelial cells; the viral genome initially exists episomally but two of the transcribed viral proteins appear to generate genomic instability (see Section 9.4.1) facilitating viral DNA insertion into the host genome. Cells with integrated viral DNA proliferate more rapidly, producing wart-like lesions. Some of these cells may become immortal via a different viral protein (that activates *hTERT*; see Section 9.5.1) and this event provides the opportunity for the immortalized cells to accumulate additional mutations through increased number of cell divisions over time. The protein that induces immortalization also interferes with p53 and p21 (see discussions below). A fourth viral protein interferes with an Rb protein restriction point pathway blocking cell proliferation inhibition.

EBV exists in the host lymphocyte nucleus as a circularized episome that is maintained at a constant post-mitotic copy number by the cellular DNA polymerase. EBV increases the concentration of free radicals through nicotinamide adenine dinucleotide phosphate (NADPH) activation, thereby increasing DNA single-strand break (ssb) and base damage. Nonetheless, this virus promotes carcinogenesis primarily through activating signal transduction pathways, either by producing an imposter constitutively active growth factor receptor or by switching on several native pathways. Additionally, the virus stimulates interleukin-10, an autocrine growth factor. EBV also promotes lymphocyte immortalization and basement membrane disruption. The latter enables metastasis — a definitive advanced cancer characteristic.

Hepatitis B viral DNA eventually inserts randomly into the host genome, although there are several preferential sites, or hot spots. These hot spots occur at genes transcribing cyclin E1 (controlling the G_1/S checkpoint), a methyltransferase (effecting epigenomic methylation), and *hTERT* among other oncogenic potentiators. Nonetheless, hepatitis B insertional mutagenesis takes about 20–50 years to manifest as neoplasia, so although mutation may participate in the process, it is unlikely to be the primary initiator of carcinogenesis. More likely, a single hepatitis B protein responsible for signal transduction activation, protein kinase pathway disruption, and inflammatory masking represents the primary driver of carcinogenesis. The hepatitis C core protein interacts with several transcription factors, either directly or indirectly, possibly influencing the infamous Ras/Raf/MAPK pathway (disrupted in approximately 30% of all cancer types and in about 10% of all patients with cancer). This core protein also suppresses p21 and another hepatitis C protein appears to bind to p53. The influences of these viral proteins are clearly more directly suspect than random mutation events.

The most recently discovered oncogenic virus, Merkel cell polyomavirus (MCV), commonly causes infection but is rarely oncogenic. In these rare cases, MCV DNA integrates into a specific site within the host genome and is thus clonally represented; tumor metastases all exhibit the identical viral DNA integration pattern. Thus, integration is likely the tumor initiation event. Furthermore, the process of viral DNA integration consistently truncates the MCV large T antigen codon at a specific region preventing viral replication but the resultant protein binds to pRb. With such a restrictive requirement for transformation, it is not surprising that MCV is an extremely rare cancer with long latency.

DNA transfer works both ways: viral DNA inserts into the host genome and viruses incorporate human DNA as well. As the virus is replicated by the host cell, the newly minted viruses released into

Table 9-1 Cellular genes acquired by acutely transforming viruses

Oncogene	Protein Product
myc	Transcription factor
ras	Signal transduction
raf	Signal transduction
erbB	Epidermal growth factor receptor
fms	CSF-1 receptor
kKit	Steel receptor
sis	Platelet-derived growth factor
akt	Protein kinase B
crk	SH2/SH3 adapter protein

the organism may contain integrated human DNA. The newly acquired DNA may include a proto-oncogene — or worse, the incorporation process may mutate the proto-oncogene such that the replicated virus contains a fully competent oncogene. Table 9-1 presents several oncogenes acquired by viruses from host organisms. Viral induction of oncogenesis therefore proceeds cooperatively through several mechanisms including mutation, oncogene presentation, signaling pathway manipulation of growth factors (and receptors), p53, p21, and pRb protein deregulation, persistent infection (inflammatory response), and cell proliferation induction. Viruses also support the later stages of carcinogenesis through host immune response disruption and the promotion of metastasis. Let's examine these carcinogenic drivers in some detail now, to see what our jigsaw puzzle reveals about the initiation, promotion, and perpetuation of the cancerous state.

9.4. Progression to Neoplasia

Obviously, DNA mutation in and of itself cannot induce cancer; the nucleic and proteinaceous products of mutated genes promote neoplasia through cell transformation. As we saw in the case of viruses, the viral genome may produce oncogenic proteins without a requirement for mutation of the host genome. In general (excluding viral infection), mutation of a proto-oncogene or tumor suppressor gene is necessary but not sufficient to create a neoplastic lesion. Sequencing of hundreds of human cancers has determined that around 90% of human cancers have on average four to five driver mutations per tumor. Driver mutations are a combination of oncogene activation/overexpression and tumor suppressor gene deletion or silencing that promotes and perpetuates carcinogenesis through transformed cell behaviors. Frequently, the initiation process expresses as a survival advantage via uncontrolled proliferation through aberrant growth signaling, abrogation of replicative senescence, inhibition of cell death, activation of telomerase, or some combination of these activities. A reminder of the cell biochemistry impacted by mutation of a single gene can be provided by referring to Figure 4-18, which represents a much oversimplified biochemical pathway. According to Figure 4-18, the loss of p53 functionality due to a mutation in the gene encoding p53 or a mutation in the genes that code for CHK2 or ATM (that phosphorylate and activate p53), affords escape from cell cycle arrest and allows cells to commence replication despite persistent genomic defects. This behavior would be expected to destabilize DNA repair, generating mutations and chromosomal defects as cell progeny accumulate unrepaired damage. Thus, *p53* represents a tumor suppressor gene (gatekeeper) that also may act as an oncogene (caretaker). The behaviors related to *p53* mutation provide a proliferation advantage over undamaged cells and also allow persistence of damaged chromosomes in the cells. What becomes apparent here is that the mutation of certain genes not only provides a clonal growth advantage, but also promotes further mutation of random additional genes. As we indicated earlier, these additional mutations must be necessary to promote tumorigenesis because on average five driver mutations can be identified in most cancer cells. Recall that in Section 9.2, we warned that human xenographs differed from mouse xenographs because successful human xenografts required that five distinct pathways be disrupted through mutation. Turns out that this requirement for tumorigenesis is fundamental.

Several human tumors expressing *p53* mutations support the hypothesis that coincidental promotion of additional DNA defects fosters progression to neoplasia. Instances of "clonal expansion" have been identified surrounding tumors of the colon, skin, esophagus, and kidney. In these cases, cells comprising an extensive area of tissue nearly all possess identical *p53* mutations, strongly suggesting that an original clone was responsible for the entire population. These expansions typically are discovered because cancerous or precancerous lesions presenting a *p53* mutation develop within them, inferring that the *p53* mutation was an early transformation event enabling later development of tumors. Notice that in Figure 9-8, the survival advantage of an early *p53* mutation creates a patch of clonal growth (all cells exhibiting identical *p53* mutations) but not in abnormal phenotype or discernible physiology. The tissue composed of cells carrying the mutant *p53* gene cannot be differentiated compared with the normal tissue unless molecular probes are employed. Polyps (neoplasms or tumors) are presumably formed only after subsequent genomic modification, as these growths express not only the initial *p53* mutation but also additional mutations, some expressing in all the cells of the polyp, and some expressing in a subset of cells. Apparently, the *p53* mutation provides a proliferation advantage (see Figure 9-9) that enables this genotype to dominate the local tissue, and a subsequent mutation (or mutations) enable the phenotypic evolution that presents as a neoplasm. Clinical experience has indicated that neoplastic polyps progress to a cancerous (malignant and potentially metastatic) phenotype sufficiently frequently to warrant removal and histological examination. The cancerous tumors express novel genetic mutations not identified in the polyps, so it is not unreasonable to hypothesize that one or more of these mutations advance the cancer phenotype. In addition to *p53*,

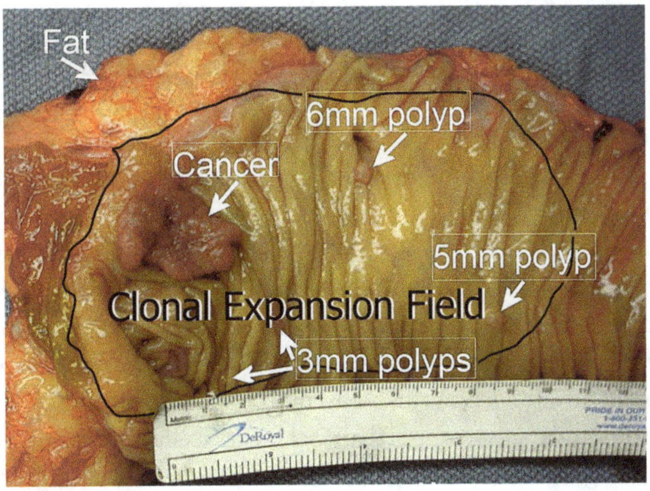

Figure 9-8 Clonal expansion field of cells expressing mutated *p53* gene in the wall of the colon. The cells containing the mutation are not transformed and cannot be distinguished from normal cells except by genomic analysis; however, the predisposition of these cells to progress to neoplasia is obvious. Reproduced and modified from Bernstein0275, with permission through CC BY-SA 3.0, at https://creativecommons.org/licenses/by-sa/3.0, via Wikimedia Commons.

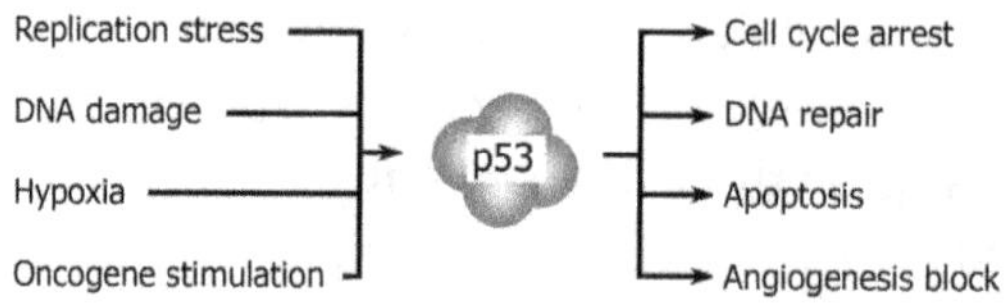

Figure 9-9 Guardian of the genome. The tumor suppressor-oncogene protein, p53, is activated through phosphorylation at several sites when initiated by replication stress, DNA damage, or hypoxia. Oncogenes can activate p53 signaling pathways via component modification. Activation of p53 supports DNA repair through cell cycle arrest and stabilization of the replication fork and promotes apoptosis via transcription upregulation of pro-apoptotic genes. In response to hypoxia, the protein inhibits production of proangiogenic factors and increases production of endogenous angiogenesis inhibitors.

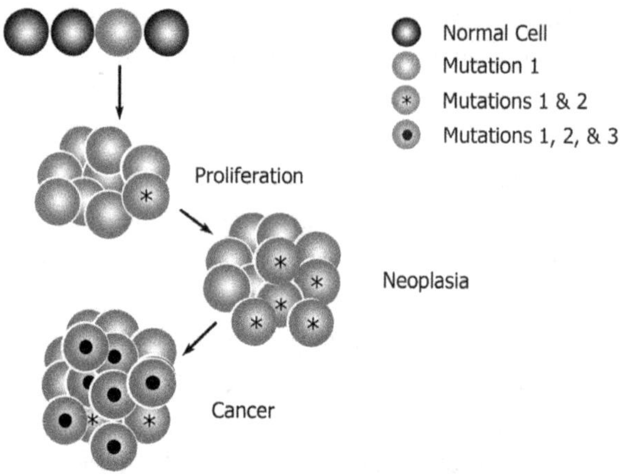

Figure 9-10 Mutator phenotype hypothesis: progression from initiation to cancer. In this scenario, an initial mutation occurs (e.g., in a gatekeeper gene with subsequent loss of normal mutation suppression through cycle checkpoints or cell mortality encouraging a proliferation advantage with coincident defective gene perpetuation). A persistent mutation destabilizes the genome expanding the number of mutations and expressing as a neoplastic phenotype. Subsequent mutations induced by defective DNA repair enable the hallmarks of cancer including immortality, angiogenesis, and metastasis.

p16 mutations have been identified in other clonal expansions in bladder, oral cavity, and esophagus. Clonal field expansion is not limited to these two examples, of course. The point is that a growth advantage provided by the loss of a gatekeeper may be an early transformation event leading to an enormous number of mutated cells increasing the probability of neoplastic progression of at least one clonal cell.

Intuitively then, should a caretaker (DNA repair) oncogene emerge, a propagation of unrepaired and erroneously repaired genes would likely flare up. Cell mortality would increase with the frequency of genomic disruption, but the initiating proliferation advantage would ensure the survival of a surplus of cells expressing various mutations. Among these surviving mutants, all three categories of oncogenes and defective tumor suppressor genes would inevitably arise. Cells expressing landscaper gene mutations would have not only a proliferative advantage, but also an environmental advantage. Cells expressing aberrant landscaping activities promote a supersupportive local extracellular milieu enabling relaxed contact requirements thus promoting cell transformation, dedifferentiation, and migration. Genes that encode excreted vascular growth signals are also landscaper genes; they promote neovascularization required for tumor maintenance. These observations that growth advantage leads to subsequent propagation of mutations generated the *mutator phenotype hypothesis* (Figure 9-10) that has dominated for several decades.

Although the pathway we have described presents a smooth, linear progression from initiation to carcinogenesis, it is important to point out now that the order of mutation events is likely irrelevant. In all cases, a mutation cascade eventually produces carcinogenesis at the point when the requisite drivers are in place.

Restriction Point Transition Oncogenesis

Cell cycling controls include biochemical pathways that variously trigger the cycle, block the cycle, arrest the cycle, or exit the cycle into G_0 or senescence. The pathways leading to cell cycle arrest (i.e., checkpoint activation) are sometimes called "cell cycle checkpoints." Occasionally, several restriction pathways converge at a single point — the transition point to cycle stimulus. Two infamous tumor suppressor genes, *p53* and *Rb*, exemplify the vulnerability of these transition points.

The *Rb* gene was discovered during investigation of retinoblastoma and so carries in its name a reference to that disease although it has been shown to have a broad impact on normal mitotic function. Examination of the Rb transition provides insight into a recognizable mechanism of oncogenesis. Briefly, growth factor signaling (see also Figure 9-13) in the presence of β-catenin and NF-κb promotes cyclin D expression, activating the cyclin D-CDK4 and cyclin D-CDK6 complexes. CDKs are cyclin-dependent kinases that bind with their appropriate cyclins to phosphorylate their target protein, e.g., Rb. This growth signal pathway is restricted (abrogated) through TGF-β activation of p15, a protein that specifically inhibits CDK4 and CDK6. Growth factors alternatively trigger cyclin E-CDK2 by inactivating p27 and p21, proteins that constitutively prohibit the kinase's activity. Thus, in the absence of mitogen the cyclinE-CDK2 pathway is restricted. The Cyclin-CDK (2, 4, and 6) complexes activate through phosphorylation of the Rb protein leading to progression through the G1/S checkpoint by release of the E2F transcription factors that promote synthesis of proteins that drive and maintain DNA replication — unless restricted. Thus, the transition from three growth factor signaling pathway restriction points flows through the Rb transition point.

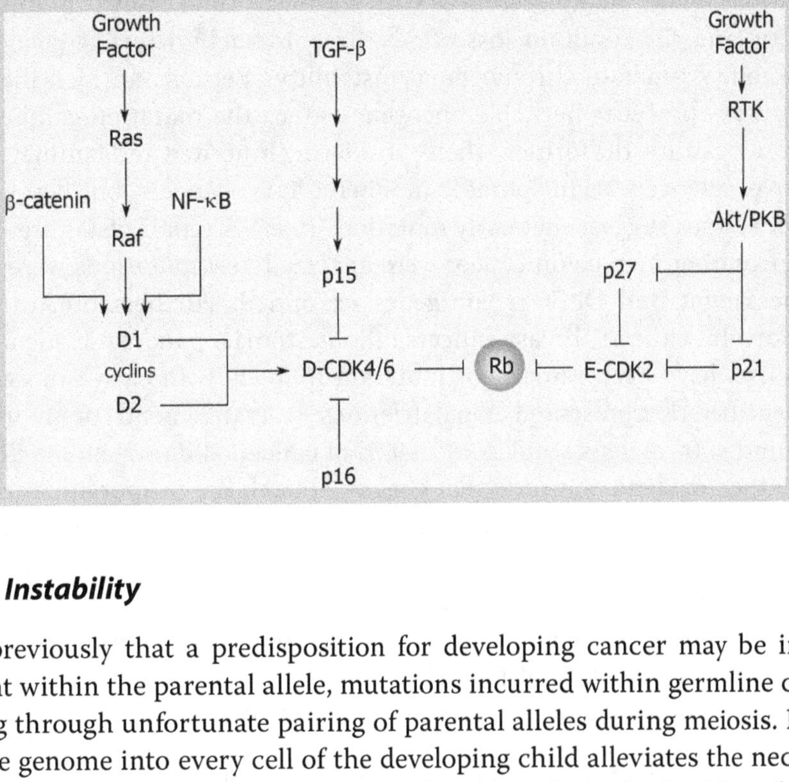

9.4.1. *Genomic Instability*

We mentioned previously that a predisposition for developing cancer may be inherited through mutations present within the parental allele, mutations incurred within germline cells, or dominant mutations arising through unfortunate pairing of parental alleles during meiosis. Replication of the mutated germline genome into every cell of the developing child alleviates the necessity for the initiating proliferation advantage of the mutator phenotype hypothesis. Nonetheless, reports of ubiquitous DNA repair gene mutations strongly support the hypothesis in hereditary cancers. For example, mutations in DNA mismatch repair genes correlate with microsatellite instability (MSI) in hereditary nonpolyposis colon cancer (or Lynch syndrome), whereas bi-allelic germline mutations in the DNA BER gene, *myh*, correlate with increased G·C to T·A conversion frequencies found in MYH-associated polyposis. Because UV radiation inflicts thymine dimerization, germline mutations in nucleotide excision repair (NER) genes predispose individuals to skin cancer. But something else additionally becomes apparent early in the progression of these cancers — the destabilization of genomic integrity. *Genomic instability* is characterized by an extremely high mutation frequency throughout the entire genome, not just within the 1.5% comprising the protein-encoding gene subset. Although we are uncertain, the onset of genomic instability may possibly be initiated by mutation of genes involved in DNA replication, DNA repair, telomere stability, or chromosome segregation. Analysis of genomic instability that requires examination of over 3.2 billion base pairs discloses a distinctive feature: MSI — the amplification of oligonucleotide sequences within untranslated regions (introns between genes) of the genome. An easily visualized karyotypic feature of genomic instability is chromosomal instability (CIN; changes in chromosome structure or number associated with metaphase checkpoint failure). CIN germline mutations associated with breast cancer include the defective DNA repair genes *BRCA1* and *BRCA2*, and their cofactors PALB2 and BRIP1. And CIN-induced mutations in genes that facilitate the resolution of interstrand crosslinks as well as DNA double-strand breaks (that including *RAD50*, *NBS1*, *WRN*, *BLM*, and *RECQL4*) promote mutations in DNA mismatch repair genes correlate with MSI in hereditary nonpolyposis colon cancer (or development of breast cancer, ovarian cancer, leukemias, and lymphomas). Genomic instability differs from chromosomal instability in frequency and specificity of chromosomal rearrangement and loci; genomic instability generates more frequent chromosomal abnormalities at more random locations. Now, we are presented with a typical "chicken or egg" dilemma: does chromosomal instability (and the resultant loss of DNA repair fidelity) invoke genomic instability or does genomic instability generate chromosomal instability? Because an identifiable germline cell genetic mutation always precedes heritable oncogenesis (i.e., the mutation can be identified in the parental germ cell), we assume the former, when considering inherited and familial cancers. Genomic instability also drives oncogenesis in sporadic (nonhereditary) cancers, but here our dilemma may be resolved. Recent studies suggest that early mutations in DNA repair genes are infrequent. When early passage cultures of human colon cancer were analyzed, few mutations were identified among 100 cell cycle checkpoint and DNA repair genes examined. Furthermore, thousands of genes obtained from colorectal cancers, breast cancers, glioblastomas, pancreatic adenocarcinomas, and lung adenocarcinomas have been ranked for mutation frequency. In each cancer type, only about four genes were identified that presented clonal deletions, rearrangements, or amplifications in more than 20% of the tumors. In various studies, 69%–97% of cancers did not exhibit clonal mutations in caretaker genes. Rather, evidence suggests that genomic instability in sporadic cancers results from oncogene-induced collapse of DNA replication forks (Figure 9-11). Remember that in somatic cells, the nontranscribed regions of the genome tend to form condensed heterochromatin that evades proofreading functions, particularly, those coincident with transcription. Nonetheless, heterochromatin relaxes during replication suggesting that genomic instability might develop during mitotic synthesis.[4] Replication is a complicated process under the control of numerous enzymes, each of which is encoded by a proto-oncogene. First, the duplexed DNA must be unwound, and the

[4]In the developing embryo, cells are pluripotent, so heterochromatin condensation is reduced and transitory and therefore DNA repair is active throughout the genome.

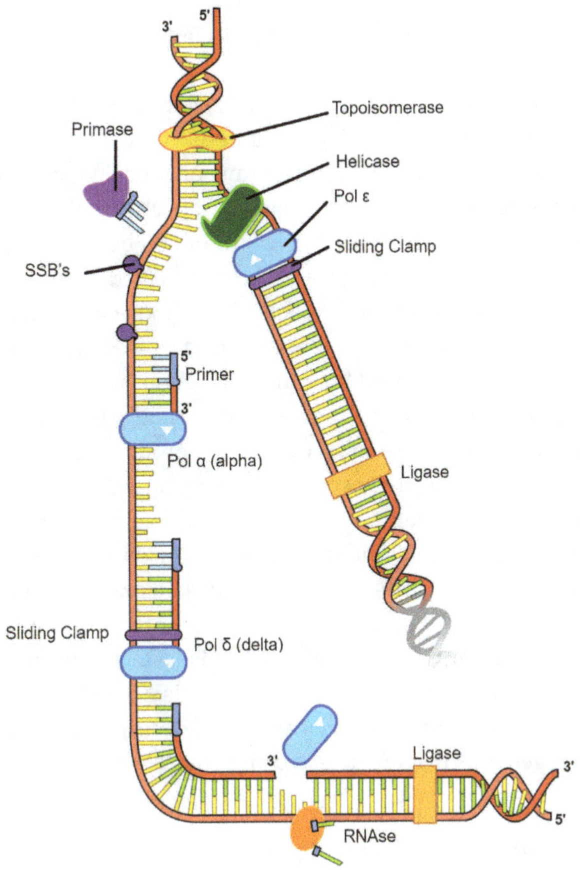

Figure 9-11 DNA replication fork proto-oncogene products. Defects in any of the enzymes required for DNA replication may promote replication stress leading to genomic instability and oncogenesis; those directly interacting with the DNA are shown here. Reproduced and modified with permission via CC0 1.0; https://en.wikipedia.org/wiki/DNA_replication.

hydrogen-bonded bases must be separated to accommodate the replication fork enzymes. These functions are performed by a topoisomerase (unwinding) and a helicase (unzipping). DNA polymerase (Pol ε) wraps around the leading strand (3′–5′) immediately adjacent to a sliding clamp protein, PCNA, that facilitates the advance of Pol ε along the lead strand. On the lagging strand, Primase adds primer RNA nucleosides initiating Okazaki fragments, also progressing 3′–5′, and Pol α adds the first several DNA nucleosides with Pol δ and a second sliding clamp completing the fragment. RNase removes the RNA primers and ligases anneal the phosphate backbone. Single-strand binding proteins (SSBs) stabilize the lagging strand at the fork. If any one of these enzymes or cofactors suffers efficiency reduction resulting from a random encoding gene mutation or dysregulation, the cell experiences replicative stress that can result in replication fork collapse. Replication malfunctions logically produce unresolved DNA dsb and single base pair mismatches. These "single nucleotide polymorphisms (SNP)" are characteristic of nonhereditary cancers, supporting the oncogenic initiation hypothesis.

High Throughput Sequencing Technology

Historically, the Sanger sequencing technique (Frederick Sanger, 1980 Nobel Prize in Chemistry) used a fluorescent labeled complementary strand of DNA that randomly incorporated nucleotides lacking the 3′-OH group required for cDNA elongation (Figure 3-4). The resulting oligonucleotide fragments were separated using single base pair resolution capillary electrophoresis, yielding an electropheragram of the original template molecule. These Sanger sequences had an average read length of 800 base pairs. However, the human genome consists of approximately 3×10^9 bases, containing approximately 20,000 genes that span 45×10^6 bases. Newer sequencing technologies read the sequence of multiple DNA molecules in parallel which increased the throughput to as much as 6×10^6 bases of DNA sequence per day. However, parallel capillary-based systems are still limited by the number of capillary columns. Using these techniques, the Human Genome Project took over a decade (and billions of U.S. dollars) to reveal the sequence of the entire human genome.

Several proprietary corporate technologies now provide high throughput sequencing (HTS) capabilities. The bridge amplification method (Illumina technology) generates small DNA clusters with identical sequences. Clusters enable multiple primer hybridization steps allowing multiple sequencing start points and paired end sequencing. Paired end sequencing information indeed increases output from a sequencing run, but also identifies splice variants through RNA-seq, and eliminates duplicate reads originating from the same template molecule. Paired end reads are also important for identifying large structural variants such as inversions, which are not apparent in short sequencing techniques. An alternative HTS technique uses polymerase chain reactions (PCR) to amplify DNA sequences suspended within an emulsified droplet (Torrent and 454 technologies). Nucleotides are sequentially passed by droplet wells where, if incorporation occurs, a series of enzymatic reactions releases light or hydrogen ions that are detected and correlated with the situational nucleotide.

Regardless of the initial transforming event or pathway taken to genomic instability, the aftermath profoundly unsettles cellular functionality through massive genetic disruption. Once genomic integrity fails, the steadfastness of the DNA code becomes unreliable, chromosomes can resort and recombine freely, gene transcription upregulates or downregulates unpredictably, and the number of chromosomes within the nucleus can change. Clearly this unstable environment fosters the elaboration of cells expressing a sufficient number of drivers of carcinogenesis, if the cell survives. The extent of the resultant genetic disruption can be appreciated in the descriptions of aberrations identified in human tumor cells. For example, Kataegis noted a cluster of nucleotide substitutions (>100 bp) in 60% of all cancers — this base replacement is always biased toward one of the two strands and the localized clusters are frequently associated with large deletions of DNA segments, chromosomal rearrangements, or gene duplications. Also documented, reciprocal translocations between chromosomes presumably resulting from the resolution of simultaneous DNA dsb on different chromosomes. The process is known as chromoplexy, and it has been identified in 18% of human cancers. Chromoplexy translocations produce shuffled chains of rearranged genes. Chromothripsis, on the other hand, presents as hundreds of clustered chromosomal rearrangements, presumably arising from a single localized event in one or a few neighboring chromosomes. Unlike chromoplexy, the motivating event appears to be tens to hundreds of simultaneous, localized DNA dsb. The DNA fragments seem stitched back together randomly, resulting in amplifications (58%), homozygous deletions (34%), and structural variants (8%) within genes and promoters. Chromothripsis was identified in 22% of all cancers.

9.4.2. *Latency*

A delay always exists between initiation and the discernable presentation of a tumor in most animals and humans. For humans, the "rule of thumb" latency period for solid cancers is very long, assumed to be 20–30 years or more, however blood-borne cancers such as leukemia can arise rather quickly (2–12 years) following an identifiable carcinogenic event. Experimentally, transformation *in vitro* and neoplasia *in situ* can be accelerated through the application of carcinogenic chemicals, such as those found in cigarette smoke, asbestos, dichlorodiphenyltrichloroethane (DDT), and gasoline fumes. Thus, these chemicals are thought to be "promotors" of cancer, and they are. Nonetheless, the historical, orderly two-step, *initiation-promotion hypothesis* has given way to an expanding spectrum of random mutations, some of which foster survival advantage, reduction in growth control requirements, vascular invasion, immortalization, or cellular migration. Additionally, the requisite initiating event may be followed by several secondary factors, some carcinogenic and some not. For example, proteases have been implicated in the process of malignant transformation and carcinogenesis. TPA (12-O-tetradecanoylphorbol 13-acetate), a known carcinogenic promotor that stimulates cell proliferation through rapid activation of protein kinase C (PKC), fails to stimulate transformation when Antipain, a protease inhibitor, is added to the medium. Neither PKC nor Antipain are carcinogenic. Secondary factors affect the latency between initiation and neoplastic presentation, reducing latency and often determining whether the drivers of carcinogenesis evolve. Application of promotors or addition of secondary factors — for example, metabolic stress from obesity or cigarette smoke — may determine cancer risk and the timing of cancer presentation.

Commencement of the latency period cannot be precisely determined for two reasons: first, it is difficult to pinpoint exactly when the initial mutation occurred, and second, multiple events — genetic, epigenetic, and extragenomic — may act synergistically to induce cellular transformation. Determination of the endpoint, a visible carcinoma, also presents uncertainty. Famously, early detection of prostate, breast, and colon cancers through modern screening technologies has significantly reduced the assumed latency of those diseases. Some tumors induce unpleasant symptoms that are difficult to ignore and send patients to the doctor early in the disease progression, others thrive virtually asymptomatically for years. But more than a century of cancer research has provided some pretty good ballpark estimates. Epidemiological studies indicate that solid tumor diagnoses begin to increase from rare to substantial when the population reaches about 45 years of age. After that age, the risk of sporadic cancer occurrence increases approximately linearly. So, a "back of the envelope" approximation can assume that if a nominal, evolutionary mutation rate is one point mutation per 10^9 base pairs per year, and if 600 genes are implicated in oncogenesis (1.6% of the 30,000 genes that comprise 1.5% of the total genome), then we derive a period of approximately 20 years between two mutation events. In support of this purely statistical analysis, studies of cancer incidence following exposure to radiation in Nagasaki and Hiroshima at the end of World War II (see Chapter 11) provided statistically significant latency estimates assuming initiation immediately after exposure and presentation of tumor as of diagnosis — survivors are examined annually, so the maximum error is 1 year.

We are left with two questions: why is there a latency period during the progression of cancer and why is the latency shorter for some cancers and longer for others? If we begin with the following assumptions:

- cancer arises from normal cells,
- carcinogenesis requires multiple driver mutations or alterations, and
- each carcinogenic trait can be attributed to one (or more) genetic, epigenetic, or extragenetic event,

then we can propose the hypothesis that the intervals between carcinogenic driver acquisition steps account for the latency period. For instructive purposes, we begin by presenting a well-documented and generally agreed upon progression from initiation to colorectal cancer presentation. The initial step in the pathway occurs when the *APC* gene (adenomatous polyposis coli; located on chromosome 5) suffers a mutation, generally through mismatch repair. The product of this tumor suppressor gene helps control cell cycling and organizes intracellular structures by managing cell-cell adhesion. The protein also ensures each mitotic daughter cell receives the correct chromosome number. The resultant increased cell cycling and reduced adhesion requirements provide a proliferation advantage creating the clinical manifestation, "stage I adenoma," or a clonal expansion field. The second mutation arises in a ras gene encoding a small GTPase that activates proteins involved in cell growth, differentiation, and apoptosis. Single point mutations leading to constitutively activated RAS GTPase upregulate signaling even in the absence of extracellular growth factors — that is, release from normal proliferation control requirements. As a result of uncontrolled mitosis, polyps appear presenting "stage II adenoma." CIN-associated chromosomal aberrations commence — likely due to the combination of APC mutation and insufficient cell cycle control — followed by an LOH of the long arm of chromosome 18 that deletes the *DCC*[5] gene, a poorly described tumor suppressor gene. The emergence of chromosomal aberrations and chromosome 18 LOH define "stage III adenoma," an alarming precancerous histology. Finally, chromosome 17 aneuploidy (deletion) disrupts p53 functionality advancing the lesion to a cancerous state and stimulating vascularization. Additional lesions are required to enable metastasis. This progression tells a nice story. Replication fork stress introduces the initial DNA point mutations followed by translated mutant protein-induced chromosomal lesions ultimately producing genomic instability leading to the oncogenic phenotype, because chromosomal rearrangement causes global effects that are not confined to a particular gene. Slowly accumulating point mutations arising from mismatch repair of DNA dsb inflicted through replication fork collapse would predict a long gap between the first and second mutations. Chromosomal instability increases the likelihood of impactful defects but random events targeting the loss of *DCC* from chromosome 18 or chromosome 17 aneuploidy remain rare. The average lag between stage II adenoma (confirmed by colonoscopy and histology examination) and malignancy is 5–10 years. However, once tumor suppression is lost and p53 protein activity begins to run amok, the latency between stage III adenoma and cancer can be quite short. The scenario presented insinuates an orderly pathway and the proposed rationale related to molecular biology motivate this progression, although the randomness of mutation targeting does not, nor does the capacity for APC defect to trigger LOH and aneuploidy. So, do the mutations necessarily occur in the order described in this commonly proposed progression? No, indeed not. Notice that in Figure 9-8, the expression of clonal expansion (stage I adenoma) is determined by identical mutations of p53 in nearly every cell of the field. In this case, the mutation of p53 prompted a proliferation advantage and the mutation occurred early in the progression to carcinogenesis, although our linear progression scenario for colorectal cancer establishes that the p53 mutation occurs late, sometime after stage III adenoma. Carcinogenesis requires each of the mutations and their constitutive stochastic lags to present, but the order is not restrictive. Here's an analogy: the components of a machine must be present, and subunits can be assembled as the components arrive, but it is only when all the pieces are available and assembled that the machine can produce the anticipated product, or in our case a cancer.

Epidemiology indicates that the polyps of nonhereditary colon cancer generally become detectable through colonoscopy examination at about 45 years of age or older. Chronic myelogenous leukemia (CML) epitomizes a rapidly developing cancer. Initiation by radiation exposure (atomic bomb fallout) and congenital genetic defect (assessed by umbilical cord genetic analysis following presentation of the disease) indicate a latency of only 2–5 years. For CML, the mutational rearrangement of a single gene through translocation between chromosomes 9 and 22, *BCRABL1*, appears necessary and sufficient to

[5] Deleted in Colorectal Cancer (DCC).

promote the disease — a small percentage of cancers apparently require only one or two driver mutations. Nonetheless, if this blood-borne cancer pathway is "one and done," why is there any latency at all? Perhaps, genetic mutation may not be sufficient to support cancer, there may be extragenetic drivers. Because cancer is malignant and metastatic, extracellular environmental factors must cooperate to sustain uncontrolled growth and enable migratory behavior. Cancer cells release angiogenic growth factors to encourage vascularization that provides glucose, glutamine, oxygen, and presents a metastatic conveyance. Chemotaxis requires fibroblast activity that is stimulated by cancer cell signaling. Furthermore, tumors must suppress immunosurveillance lest they be gobbled up by macrophages, and metastases must generate a tumor bed conducive to implantation efficiency. Each of these factors takes time to develop. Remember that tumors are organ-like systems that thrive only within an organism that is supportive.

We are all familiar with familial cancer. The doctor's office requisite health history inquiry always asks us about our family history of common cancers (as well as diabetes, heart failure, stroke, etc.). Familial cancer is not the same as inherited cancer. Inherited cancer presents a mutant gene that originates in the parental germ cell. The inherited gene can be dominant or recessive with dominant genes reducing latency and recessive genes increasing the probability of a cancer diagnosis because dual allele mutation becomes slightly more probable. Familial cancer generally does not present a common genetic mutation traceable to every occurrence; rather, it expresses as an inherited predisposition to develop cancer, sometimes specific cancers. The trait does not follow Mendelian genetics, it clusters — it often skips generations and pops up almost randomly but with significantly higher frequency than in the general population. Familial cancers exhibit the same latencies as sporadic cancer and are assumed to be the result of multiple physiological and environmental factors, such as lifestyle (diet, exercise, sun exposure, smoking...), generational exposure to known carcinogens, ancestral origin, and so on. The genetics are complex and subtle, involving altered regulation of several genes without identifiable marker mutations.

9.5. Cancer Biology

In 1971, President Richard Nixon declared a "war on cancer," legislation eliciting nonpartisan enthusiasm throughout every administration from then until now. The National Cancer Institute (NCI; authorized by President Franklin D. Roosevelt in 1937), a division of the National Institutes of Health (NIH), has spent over $90 billion on cancer research and treatment since its inception. An additional 260+ nonprofit organizations, such as the American Cancer Society (ACS), have driven the total spending to over $158 billion. Nonetheless, cancer remains the second leading cause of death in the United States (the third leading cause worldwide behind heart disease and infectious and parasitic disease). For captivating documentation on the history of cancer research, the reader is referred to *Cancer, the Emperor of all Maladies.*[6] Although incredible advances have resulted in understanding, successful treatment strategies, and improved survival, the accumulation of data has been so extensive that the information itself can become unmanageable. It is as if on a rainy day, you pull an old jigsaw puzzle out of the attic. It turns out — much later in the day — that several of the pieces are missing and many pieces belonging to other puzzles have gotten mixed in. Some of the erroneous pieces fortuitously fit into the puzzle although the images on those pieces are nonmatching, and you are left with several holes. Cancer research is currently troubleshooting the mismatched pieces and searching for the missing ones. This metaphor explains how

[6] Available as a book (Mukherjee, 2010) and as a documentary film by Ken Burns (https://www.pbs.org/kenburns/cancer-emperor-of-all-maladies/).

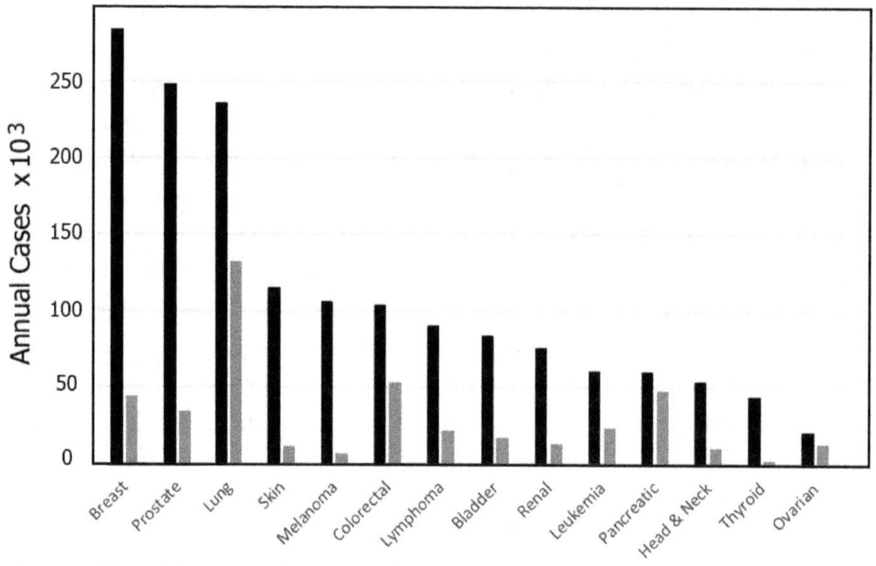

Figure 9-12 Cancer incidence and mortality rates for common cancers. Black bars represent the number of cancers diagnosed annually; gray bars indicate the number of annual deaths. Early detection has greatly reduced the mortality of breast and prostate cancers. Lung cancer early detection is more difficult, and metastasis occurs early during progression. Colorectal cancers are asymptomatic and metastasize readily through the intimately available vascular system; detection through colonoscopy has significantly reduced the mortality rate of this cancer. Early detection of pancreatic cancer is difficult, the cancer is resistant to chemotherapy, and the delayed diagnosis limits surgical options. Data from the American Cancer Society, Cancer Facts and Figures, 2021.

innocuous housekeeping proteins ended up with names like transforming growth factor (TGF-α and β) and retinoblastoma (Rb) because they were isolated and characterized within the context of a quest for cancer genes. Within this perspective, our description of cancer biology — including the previous sections concerning the molecular biology of cancer — reflect contemporary understanding.

In the United States, cancer mortality ranks second only to heart failure; treatment of cancer cost over $200 billion in 2020. The most common, potentially fatal cancer is breast cancer, followed in frequency by prostate cancer, lung cancer, skin cancers, and colorectal cancer (Figure 9-12). Although early detection and modern therapies have driven the mortality rates down for most of these cancers, lung cancer is an exception. The lung cancer dilemma stems from intimate proximity to the vasculature promoting metastatic potential very early in the course of the disease whereas symptoms develop later. Similarly for pancreatic cancer, the disease is practically asymptomatic until surgical resection becomes impossible, and the genomics of the disease promote chemotherapeutic escape as the cells rapidly acquire new mutations. Ovarian cancer, although rare, also suffers from a dismal prognosis because the ovaries float in a body cavity that enables cell migration and ovarian cancer cells respond to hormonal influences. So, it becomes apparent that a study of carcinogenesis must expand beyond the cellular scale to include the entire organism.

9.5.1. *Cancer in situ*

Organs are defined by their form and function. If we consider a tumor to be an organ-like system (see Chapter 8, Section 8.5), then we must acknowledge first that the form is variable, and second that the function may be stage dependent. A dozen or so hallmarks — detectable indicators of carcinogenic cells — serve to define cancer, and they can be organized into four subsystems: those enabling carcinogenesis, those contributing to the emergence of cancer, those defining the cancerous state, and those interconnecting the tumor with the organism. Some of these hallmarks drive cancer and others are passenger traits.

9.5.1.1. *The hallmarks of cancer*

We have discussed the hallmarks enabling cancer already. They are the following:

- Replication stress,
- DNA damage,
- Genomic instability.

The hallmarks of emerging cancer include the following:

- Deregulation of cellular energetics,
- Immunosurveillance avoidance.

The rapid mitotic cycling and increased metabolic demands of cancer cells require unusual energy resources. The deregulation of cellular energetics primarily concerns glucose metabolism. Normal, nontransformed mammalian cells use glucose as a fuel source through the multistep biochemical process of glycolysis. The product of glycolysis, pyruvate, is oxidized within the cellular mitochondria via the Krebs cycle to generate adenosine triphosphate (ATP) and NADH (the cell's energy exchange currency). In cancer cells, much of the pyruvate is captured by lactate dehydrogenase to create lactate instead of ATP. Normally, hypoxic cells select this alternative mechanism, but one also can see the advantage for cancer cells that are subject to early onset transient hypoxia. Preferentially selecting the anaerobic pathway under aerobic conditions has been termed the Warburg Effect. This preferential uptake of glucose by cancer cells can be utilized to target tumors during therapy and radiology. Also, cancer cells use glutamine to produce pyruvate and to replenish Krebs cycle intermediates. Glutamine can have the effect of promoting cell proliferation creating a closed loop wherein the cells become addicted.

Neoplastic cells generate distinctive antigens, sometimes (in the case of viral carcinogenesis) arising from expression of the transforming viral genome. These tumor-associated antigens often reside in the cell membrane presenting potential targets for immunological attack. In fact, neoplastic cells are routinely phagocytosed. Every living person has an average of four deficient tumors at any given time. The tumor-generated failure of immunological surveillance mechanisms that eliminate cells expressing tumor-associated antigens is clearly important in the development of successful tumors. Several methods appear to be responsible for immunosurveillance escape. Tumor cells can "mask" by modifying the tumor-associated antigens such that they cannot be recognized by the immune system, or by altering antigen processing to prevent cell surface presentation (internalizing or releasing into the extracellular space). The tumor cells may secrete immunosuppressive products that inhibit the normal biological immune response, or they may fail to induce inflammatory activators. Co-stimulatory factor expression, such as major histocompatibility class (MHC) I, may be downregulated, lost, or altered.

Definitive hallmarks of cancer include:

- Self-sufficient growth signal regulation,
- Evasion of senescence,
- Evasion of apoptosis,
- Limitless replicative potential: immortality,
- Mitotic stress.

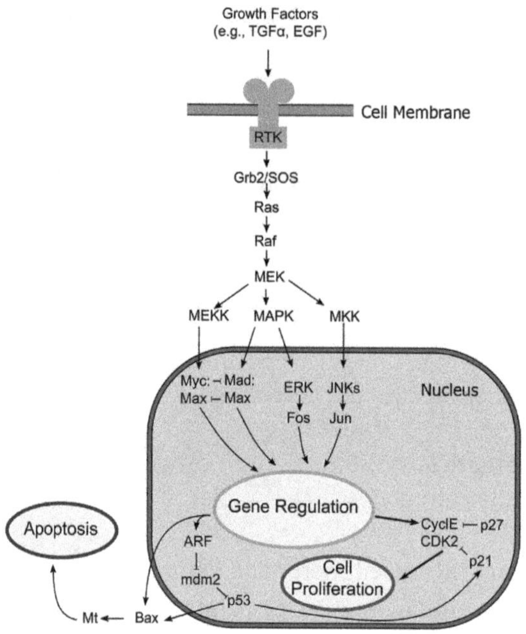

Figure 9-13 An example of a signal transduction pathway. In this example, a growth factor receptor spans the cell membrane exposing the active site to the extracellular space. When the growth factor ligand matching the receptor binds, the configuration of the intracellular tail changes, allowing sites to become phosphorylated by the proto-oncogene GTPase Ras. Through multiple alternative protein activation/inactivation steps, appropriate genes are upregulated or downregulated to drive DNA replication and cell cycling. DNA damage causes p53 phosphorylation leading to p21 regulated cell cycling block or apoptosis (if Bax is upregulated). Arrows represent open paths and "T-bars" represent blocked paths. Reproduced and modified with permission from Signal transduction v1.png, CC BY-SA 3.0, https://commons. wikimedia.org/w/index.php?curid=12081090.

Neoplastic cells acquire three related but distinctly different behaviors to advance from neoplasia to tumor to cancer: cells must develop self-sufficiency in growth signal regulation, they must escape senescence and apoptosis, and/or they must become immortal. Two of these modifications have in common disruptions to growth signal transduction. Figure 9-13 illustrates a typical growth factor signaling pathway. Tumorigenic mutation of any protein along the pathway including the receptor (extracellular or intracellular subunits) and the cytosolic activating enzymes (mostly not shown), the signaling proteins within the cytoplasm, or the cofactors within the nucleus may cause the pathway to be constitutively active absent mitogen binding. Notice that at the point of p53 activation, the pathway bifurcates leading either to apoptosis or cell cycle arrest. Recall that apoptosis (programmed cell death) is an entirely cell-intrinsic mechanism — that is, the genome is degraded in a stepwise process that eventually evokes cell death without inflammatory stimulation. This biochemical pathway triggers off ATM recognition of DNA damage (Figure 4-18) and invokes p53 phosphorylation control as a restriction point between cell cycle arrest and apoptosis, dependent upon degree of damage. Whereas cell cycle arrest is reversible allowing cells to return to mitotic division, senescence is a more persistent loss of proliferative capacity. For a number of years it was believed that senescence was permanent and could not be revoked. In other words, it was believed that the senescent cell, like the apoptotic cell, had "one foot in the grave" and could not recover. However, although that is true for normal wild-type cells, data now suggests that senescent cancer cells, which tend to form after mitotic failure and therefore are frequently multinucleated "giant" cells, can and do undergo a multipolar mitosis that results in a viable clone that can continue to grow and divide. In terms of relative importance, cell immortality is clearly one of the most important transformation drivers because it allows proliferative cells to divide beyond the normal Hayflick limit of 50–60 divisions. The Hayflick limit reflects the systematic shortening of telomeres (RNA chromosome endcaps) during replication. Immortality disables this aging clock in proliferating cells through reactivation or overexpression of telomerase reverse transcriptase (*TERT*), which adds ribonucleosides and maintains telomere length. Recalling that cells have a natural mutation rate, logically an immortal cell has a higher probability of eventually accumulating damage and thus, immortalization expresses as a proliferative advantage that promotes secondary mutations. To transform, a cell either must be able to achieve transformation traits in a short time (through CIN or genomic instability), or it must have sufficient opportunity to slowly develop transformation through immortalization. Since we haven't explored the last two of these carcinogenic drivers in detail, let's consider their evolution.

Cell cycle arrest can advance to senescence; the two mechanisms share a common signaling pathway. For example, say you cut your finger. Epidermal growth factor (EGF) is released by the damaged tissue to arouse quiescent skin cells to reproduce and heal the wound. In Figure 9-13, growth factor binds to the EGF receptor (EGFR) triggering a configurational change that allows receptor stem sites to be phosphorylated, initiating a chain of phosphorylation reactions activating enzymes that eventually pass into the nucleus. There, specific genes are either transcribed or silenced dependent upon the situational requirements and the affiliated enzymes. If DNA damage is detected (via Mre11, Rad50, and Nbs1 [MRN]) during transcription, ATM initiates the cycling controls we have seen previously involving p53 GTPase activity. As Figure 9-13 indicates, p53 activates p21 that binds to and inhibits the CyclinD-CDK4/6 complex leading to cell cycling pause. As a matter of course, cyclin-dependent kinases (CDK) phosphorylate serine and threonine residues to activate proteins that enable progression of specific mitotic phases. The CDKs are themselves activated by cyclins (e.g., CyclD), a scheme that allows high constitutive levels of CDKs to be triggered with precise timing. Alternatively, p53 can transcriptionally activate BAX to induce apoptosis (if BAX protein has been upregulated in response to severe damage, see Figure 9-13). During cell cycle arrest, if growth signaling is persistent or aberrant, the cell can undergo a process known as geroconversion through the expression of $p16^{INK4a}$ (also called CDKN2A) which is a CDK inhibitor, leading to senescence. Thus, the pathway to senescence is an extension of the cell cycle arrest pathway. Notice that p53 controls *proliferative potential*, that is, p53 stalls proliferation during cell cycle arrest (which is reversible maintaining full proliferative potential), it suspends proliferation in quiescent cells (that may be induced to reenter the cycle from G_0, maintaining full proliferation potential), and it eliminates proliferative potential through apoptosis and senescence (cells cannot regain the ability to proliferate). Geroconversion seems to be triggered by time. If p21 is inactivated within approximately 3 days, cells in cycle arrest will return to cell division. If p21 is inactivated after 3 days, cells will express the CDK inhibitor $p16^{INK4a}$ plus β-galactosidase, forcing the cells to become senescent. The pathology leading to senescence sometimes produces "giant cells" with multiple cell nuclei resulting from failed mitosis as described earlier.

Cycling cells and G_0 quiescent cells are nonetheless mortal; the number of times these cells can divide is limited to approximately 50. Cancer cells express limitless replicative potential; they are immortal. How do they do that? Chromosomes have protective RNA endcaps called telomeres. Telomere RNA nucleosides get dropped during each DNA replication cycle due to replication fork enzyme mechanics. When the telomeres shorten enough to fail DNA protection during replication, the ensuing DNA damage signals cell cycle arrest leading to senescence. Under special, normal physiological conditions, unique cells need to replicate an extraordinarily large number of times, for example, a zygote (fertilized egg) must divide many, many times to produce all the pluripotent progeny of the fetus. Therefore, the genome codes for an enzyme capable of extending the telomeres to compensate for unavoidable ribonucleoside losses. This gene, *TERT*, encodes the active subunit of a ribonucleoprotein complex, telomerase, that expresses in pluripotent cells but is silenced in other cell types, almost without exception. The *TERT* promoter contains several *MYC* binding sites permitting a carcinogenic *MYC* oncogene product to stimulate transcription and upregulate telomerase in cancer cells.

The final definitive hallmark of a carcinogenic cell expresses mitotic stress. Recall that mitosis is a complicated and highly choreographed biological phenomenon only part of which involves DNA replication (see Chapter 4, Section 4.2). During normal mitosis, unresolved DNA dsb prompt cell cycle arrest; persistent breaks drive mitotic failure and perhaps senescent arrest. But during this biological ballet, DNA repair and other functions can go wrong. For example, suppose that the spindle fibers fail to form correctly, or fail to connect to all centromeres, or fail to contract, or dissolve anachronistically. Spindle fiber failures produce aneuploidy — an inappropriate number of chromosomes sorted to daughter cells — or multinucleate cells resulting from "mitotic slippage." Both outcomes indicate mitotic stress. In healthy cells, a properly performing spindle formation/function checkpoint (Figure 5-4) arrests the cell cycle if a problem is detected, and this pause, if extended, commonly progresses to senescence. Because cancer cells escape senescence, defective progeny can continue to cycle, or at least attempt to do so. Even when senescence prevails, the aneuploid or multinucleate cells survive metabolically as noncycling cells. Thus, mitotic stress is a ubiquitous hallmark of cancer.

Systemic hallmarks of cancer include:

- Sustained angiogenesis,
- Tissue invasion and metastasis.

The high metabolic demands and continuous mitotic cycling exhibited by cancer cells require energy and oxygen. Blood vessels deliver fuel (glucose and glutamine) and oxygen to the tumor, often through neovasculature elicited from the surrounding normal vessels or bone marrow by the cancer cells themselves. In general, tumors originate near or adjacent to existing blood vessels but as they increase in volume, the cells at about 100 μm distance begin to starve for oxygen, becoming hypoxic (Figure 8-14). Hypoxia upregulates anaerobic metabolism of glucose causing cells to secrete lactose into the microenvironment. Hypoxia and coincident glucose starvation also trigger upregulation of angiogenic factors such as vascular endothelial growth factor (VEGF), fibroblast growth factor (FGF), and interleukin (IL)-8. These factors act on the existing vessel walls to promote vascular budding and stimulate non-tumor cells within the microenvironment to excrete additional vascular stimulating molecules. For example, FGF autoinduces VEGF expression in tumor cells and VEGF receptor expression in endothelial cells. IL-8 is mitogenic; it increases tumorigenicity and neovascularization. Hypoxic tumor cells induce ANG2 expression in the budding vessel endothelial cells as well, causing vessel plasticity and VEGF-mediated growth. Unfortunately, as the buds extend into vessels, the resulting neovasculature is poorly organized and suboptimal. The vessel walls are thin, lack pericytes, develop gaps, and the lumen diameter is irregular. Neovascular vessels are convoluted and tortuous. And ominously, tumor cells may integrate among the endothelial cells in the vessel walls. This hallmark of cancer, this hypervascularization, is integral to the most devastating hallmark of cancer — metastasis.

Metastasis — the shedding of individual cancerous cells from the original mass and subsequent invasion of these cells into distant tissues where they imbed and thrive to create secondary masses — presents a tremendous challenge to successful treatment. The process remains somewhat puzzling, but the aforenoted presence of tumor cells within the endothelial wall possibly reveals part of the mechanism. Models and evidence exist for both single cell migration and multicellular migration wherein cell-cell adhesion persists during breach of the tumor basement membrane, and some evidence suggests that migratory tumor cells may be facilitated by other cell types such as fibroblasts and macrophages. Although unclear whether tumor cells are incorporated into the vascular wall during angiogenesis or migrate through the porous vessel walls into the bloodstream, the vascular system provides transport for tumor cells throughout the human body. In addition to angiogenesis, tumors can also induce lymphangiogenesis — the formation of new lymphatic vessels to support the removal of tumor waste products — and this offers up a supplementary metastatic passageway. Vascular escape presents another barrier to metastasis, with two possible solutions. One hypothesis suggests that tumor cells may breach the vascular wall in a manner resembling the way they entered the vessel; another hypothesis proposes that the tumor cells may burst the capillary when the colony volume exceeds the lumen diameter. We know that IL-8 is chemotactic, and that overexpression increases invasiveness. We also know that heparinase overexpression renders nonmetastatic cell lines metastatic *in vivo*, and that heparinase is upregulated in cancer. And finally, we know this: tumors are composed of heterogenous populations of cells. We suspect that a subpopulation of these cells has the ability to escape the tumor mass and invade the vasculature, and another subpopulation of these cells has the ability to seed a new tumor by inducing a conducive stromal bed. Possibly, some overlapping subset of tumor cells has the ability to invade *and* seed a new tumor. In other words, metastasis may be a combination of cellular and extracellular characteristics that enable secondary tumor growth. Several oncogenes have been identified that correlate with the ability of tumor metastases to seed in specific organs, and just as tumor suppressor genes protect against neoplasia, metastasis suppressor genes protect against secondary tumor evolution. It may be useful to think of the process of metastasis as a feed forward loop rather than a linear cascade. Because hypoxia upregulates factors enabling metastasis, tumor aggression can be estimated by measuring persistent tumor hypoxia. Oxygen-deprived tumors become more metastatic, become radiation resistant (as we have seen), and postsurgical prognosis worsens. In the 1970s, Judah Folkman proposed that antiangiogenesis therapy was the silver bullet for cancer — if you starved a solid tumor by denying it blood supply and therefore glucose and oxygen, it would stop growing or collapse. The concept was greeted with incredible enthusiasm and was shown to succeed in some mouse model systems, but eventually failed when attempted in humans because, in humans, oxygen-deprived tumors become more metastatic, become radiation resistant, and postsurgical prognosis worsens.

Clearly, a lot of things have to go wrong before cancer emerges. Let's set some boundary conditions to illustrate how driver mutations might motivate cancer. We can limit the number of genes by considering only those identified by the Consortium in 65 selected tumors (Figure 9-7) thereby reducing the number of potential driver mutations from >600 genes to 24. Then, we can select 4 of the 12 hallmarks of cancer that represent definitive and systemic traits: self-sufficient growth regulation (6), immortality (9), angiogenesis (11), and metastasis (12). We have reduced these variables somewhat capriciously to facilitate an example scenario restricted to three dimensions. The hallmarks of cancer are numbered below with the hallmarks of immediate interest indicated.

1. Replication stress
2. DNA damage
3. Genomic instability
4. Deregulation of cellular energetics
5. Immunosurveillance avoidance
6. *Self-sufficient growth signal regulation*
7. Evasion of senescence
8. Evasion of apoptosis

9. *Immortality*
10. Mitotic stress
11. *Angiogenesis*
12. *Metastasis*

Table 9-2 recalls the genes identified in Figure 9-7, adds a brief description of the gene product function, and identifies the hallmarks of cancer that are envoked through mutation of each gene. The hallmarks of immediate interest appear in bold font. We now can represent (in two dimensions) a three-dimensional map of the sets of mutated genes, each gene of which drives one or more of our selected hallmarks (Figure 9-14).

Notice that in Figure 9-14, *P53* resides in a bubble that encloses four genes that when mutated may drive angiogenesis (hallmark 11). Three of these genes (*SMAD4*, *ERG*, and *VHL*) also drive hallmark 6, and so they additionally reside in the bubble containing *RAS*, *RAF*, *NF1*, *ATM*, *MAPK*, *RB*, and *CCNE1*. *CDK* and *CTNNB1* mutations may drive metastasis (hallmark 12), as will *PIK3*, *PTEN*, and *MYC* — gene mutations that coincidentally drive self-sufficient growth (hallmark 6) and so reside in that bubble as well. All these sets (bubbles) overlap within the right two-thirds of the base of the *TERT* (hallmark 9)

Table 9-2 Description of mutated genes identified in the 65 tumors of Figure 9-7

Gene	Hallmarks	Notes
TP53	7,8,10,**11**	(tumor) p53; Figure 9-9, Figure 9-13
CDKN2	10,**12**	CDK2; Figure 9-13
ARID1A	3	Chromatin remodeling; may be early and transient
K-RAS	**6**	RAS-MAPK pathway; Figure 9-13
PTEN	3,**6**,7,8,**12**	Phosphatase
TERT	**9**	See Section 9.5.1.1
SMAD4	1,4,**6**,**11**	TGF-β pathway (growth signaling)
PIK3CA	**6**,7,8,**12**	Phosphoinositide 3-kinase; enzyme activation/inactivation
RB1	1,2,3,**6**,7	Rb; retinoblastoma gene
B-RAF	4,**6**	RAS-MAPK pathway; Figure 9-13
CTNNB1	2,7,8,**12**	β-catenin; protein component of gap junctions
ERG	**6**,8,**11**	Loss of cell polarity and cell adhesion
MYC	4,**6**,8,**12**	Upstream gene regulation
NF1	**6**	RAS-MAPK pathway; Figure 9-13
CCNE1	**6**, 10	Controls passage from G_0 to G_1
VHL	4,**6**,**11**	Heritable von Hippel-Lindau disease
KMT2D		Methyltransferase; hallmarks vary with targeted genes
APC	**6**,8,**12**	First mutation in colorectal cancer progression (Section 9.4.2)
PBRM1	7,8,10,**11**	Acts through p53
MCL1	8,4	Amplified in acute myeloid leukemia
CCND1	1,2,3,**6**,7	MAPK; acts through Rb
MAP2K	**6**,8	Figure 9-13
CREBBP		Acetyltransferase; hallmarks vary with targeted genes
ATM	1,2,3,**6**,10	Recognizes DNA dsb; Figure 4-18

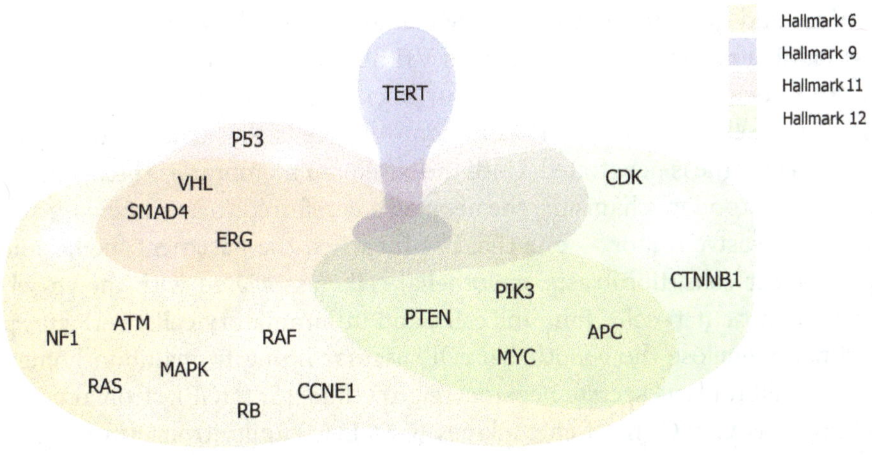

Figure 9-14 Graphical representation of driver mutations resulting in a tumor. Four hallmarks are selected from 12 total, and only genes identified in Figure 9-7 are included. The colored bubbles represent sets of genes capable of producing a particular hallmark phenotype according to the figure legend. Genes contained within overlapping spaces drive both hallmarks. All four hallmarks overlap in the space at the base of the hallmark 9 bubble. See text for detail.

bubble. Thus, within that space a fifth gene mutation motivating deregulation of cellular energetics (hallmark 4), for example, will likely result in carcinogenesis. Notice in Table 9-2 that *SMAD4, VHL, RAF, MYC,* and *MCL1* motivate hallmark 4, but representing a hallmark 4 bubble in Figure 9-7 would obfuscate the image. So, to extend the map into the third dimension, imagine a fifth bubble that resembles a hand, palm facing the page and fingertips engulfing *SMD4, VHL, RAF,* and *MYC.* The heel of the hand, looming above the page, contains the hallmark 4 unique gene, *MCL1.* The "thumb" of the hand-shaped bubble would extend into the space at the base of the hallmark 9 cloud where all four bubbles overlap, representing a carcinogenisis-positive space. Now, look back at Table 9-2 and try to imagine an *n*-dimensional relationship map that includes all the hallmarks. If you can do that, try to imagine one that includes all ~600 known oncogenes and tumor suppressor genes. In hyperspace, a multitude of unions will exist establishing various combinations of driver mutations. Recognizing that some genes reside within multiple bubbles, fewer than x genes are required to establish x hallmarks. An *n*-dimensional hypergraph incorporating all possible driver mutations and all possible hallmarks would represent all possible pathways to cancer.

9.5.1.2. *The Cancer Microenvironment*

Tumors consist of cancer cells and stromal cells. We already have discussed the requirements for embedded vasculature, and supportive cells along with their secretions clearly play a role in metastasis. A supportive relationship must form between cancer cells and tissue stromal cells to enable progression from neoplastic transformation to tumor. Evidence of this comes from experiments showing that tumor cells extracted from a tumor *in situ* and transplanted to a new site in the same fully competent organism will not form a new tumor. However, if tumor cells are transplanted along with some of the primordial tumor stroma, a new tumor can arise. Furthermore, some tissues and organs are completely resistant to neoplasia although primary cultures of these tissues can be transformed *in vitro.* Thus, a tumor may be considered an organ with the stroma contributing to functions analogous to organ stroma.

Let's return to the colon polyp example. Polyps grow outward into the colon lumen because that is the physical path of least resistance. Beneath the polyp lies a basement membrane composed of two layers. The layer to which the epithelial cells attach, the basal lamina, is composed of mucopolysaccharides and glycoproteins both of which are secreted by the epithelial cells themselves. Beneath the basal lamina resides a layer of collagen VII fibrils and microfibrils that provide attachment for the underlying connective tissue that binds our organs together. The connective tissue structure consists of elastin and collagen fibers, fluid, and nontumor cells that contribute to the tumor stroma once the basement membrane is penetrated. Until the basement membrane is breached through complex and only partially understood mechanisms, the neoplasm is referred to as *carcinoma in situ*.

The stroma of an invasive tumor — one that has breached the basement membrane — consists of several cell types: cancer cells, fibroblasts, endothelial cells (associated with the vessel walls and free floating), pericytes, adipose (fat) cells, immune cells, and inflammatory cells. Pericytes perform several functions but primarily enclose the endothelial cell vascular lining to maintain homeostatic pressure within the capillaries. Fibroblasts secrete hepatocyte growth factor (HGF), stromal cell-derived factor 1 (SDF1) — also known as C-X-C motif chemokines (CXCL) — and vitronectin (VTN). Inflammatory response cells include lymphocytes, macrophages, neutrophils, and mast cells. These stromal cells secrete components of the extracellular matrix (ECM) including growth factors, adhesion molecules, integrin signals, cytokines, and chemokines. Under the influence of neoplastic cells, the stromal cells release tumor necrosis factor alpha (TNF-α) and prostaglandins that upregulate the growth of cancer cells and encourage angiogenesis. The cytokines and chemokines released by fibroblasts and neoplastic cells recruit lymphocytes, macrophages, and neutrophils. And neoplastic cells influence the fibroblast phenotype by altering fibroblast miRNA regulation; miR-155 is upregulated, miR-214 and miR-31 are downregulated. The presence of extracellular lactate waste from cancer cell glycolysis influences fibroblast transformation to cancer-associated fibroblasts (CAF). These CAFs are pivotal to tumor progression; they release TGF-β1 that binds to receptors on endothelial cells, immune cells, inflammatory cells, and adipose cells. Figure 9-15 provides a graphical discussion of the impact CAF cells on the microenvironment and the feed forward loops initiated in other stromal cell types. Considering the extent to which CAFs support the tumor and promote aggressive behavior, models that include CAFs among the metastatic cluster seem attractive. At the very least, fibroblast transformation to CAF must be an early step in metastatic implantation and secondary tumor growth. In this case, the metastatic cell(s) must be capable of inducing this transformation whether migrating alone or in a cluster.

9.5.1.3. *Supportive Physiology*

Beyond the tumor stroma microenvironment, tumors and cancers exist within an organism and cannot be fully understood until interconnectivity at the scale of the organism is considered. Physiological factors in the evolution and sustainability of cancer include the following:

- Hormonal activity
- Organism immunocompetence
- Organism age

The endocrine (or hormonal) system consists of the pineal and pituitary glands located in the brain, the thyroid, the adrenal gland at the top of the kidney, parts of the pancreas, and the testis (in men) or ovaries (in women). Hormones serve several physiological purposes, but our interest centers on hormone-activated cell cycle signaling; in that function, they act as growth factors upon cells expressing hormonal receptors. Typically, the beta cells of the pancreas secrete insulin in response to glucose as a metabolic control mechanism. Insulin acts through tyrosine kinase cascades which (as we have seen) can contribute to uncontrolled growth in cancer cells (Figure 9-13). Insulin-induced upregulation has been identified in colon, prostate, pancreatic, and breast cancers. Nonetheless, estrogen and progesterone seem to be the promoting hormones most frequently involved in breast cancer progression. One estrogen,

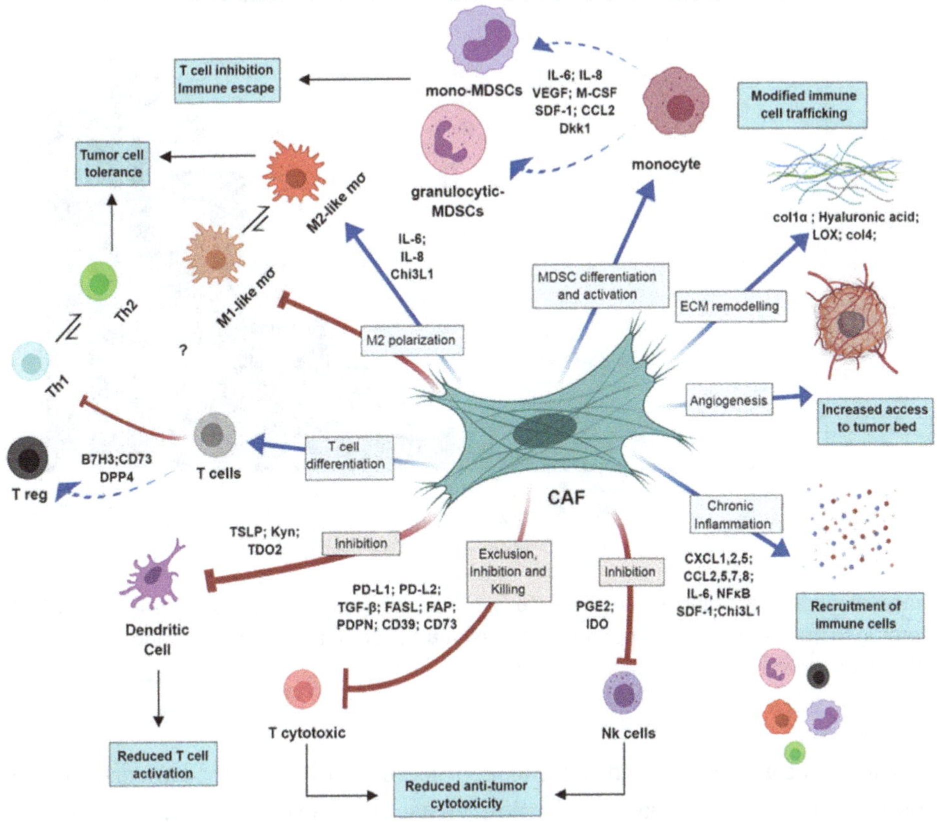

Figure 9-15 The influences exerted by CAF cells on the tumor environment. Tumors consist of cancer cells and multiple other cell types within the tumor stroma including monocytes (myeloid derived suppressor cells [MDSC]), T-cells, dendritic cells, macrophages (including M1 and M2), vascular endothelial cells, pericytes, natural killer (NK) cells, and cancer-associated fibroblasts (CAF). Arrows indicate induction and bars indicate blocks. The biochemistry regarding specific activated cytokines provides more detail than is necessary for our purposes. Reproduced with permission through CC-BY from The Dark Side of Fibroblasts: Cancer-Associated Fibroblasts as Mediators of Immunosuppression in the Tumor Microenvironment. Lea Monteran and Neta Erez, *Frontiers in Immunology*, 2019 | https://doi.org/10.3389/fimmu.2019.01835.

β-estradiol, stimulates the growth of ovarian carcinoma cells that have upregulated the expression of hormone receptors as a mechanism for obtaining growth advantage. In some cancers, escape from growth regulation control depends upon hormonal stimulation. For example, prostate cancer is dependent upon androgens like testosterone. In these cases, hormonal therapy — actively suppressing hormone release — can be quite successful, but because hormones have multiple functions, the side effects can be somewhat uncomfortable. Hormones have been implicated in the initial proliferative advantage of breast, prostate, uterine, ovarian, testicular, thyroid, and bone cancers.

Figure 9-15 includes some prominent immunological influences within the tumor stroma. At the lower right of that graphic, inflammation generated by tumors recruits immunological cells that actively recognize, attack, and destroy cancer cells. However, CAFs have the ability to inhibit NK and T-lymphocytes, they convert monocytes to myeloid derived suppressor cell (MDSC), and they transform m1-like to m2-like mσ cells. Remember, fibroblasts must first be converted to CAFs to enable this cancer cell-protective mechanism. In persons affected by immunosuppression through drug therapy for transplant antirejection (lung and kidney cancers appear more frequently within the immunosuppressed population) or chronic system failure, cancer cells achieve protection without CAF transformation allowing an alternative pathway with expedited and increasingly efficient tumor shielding.

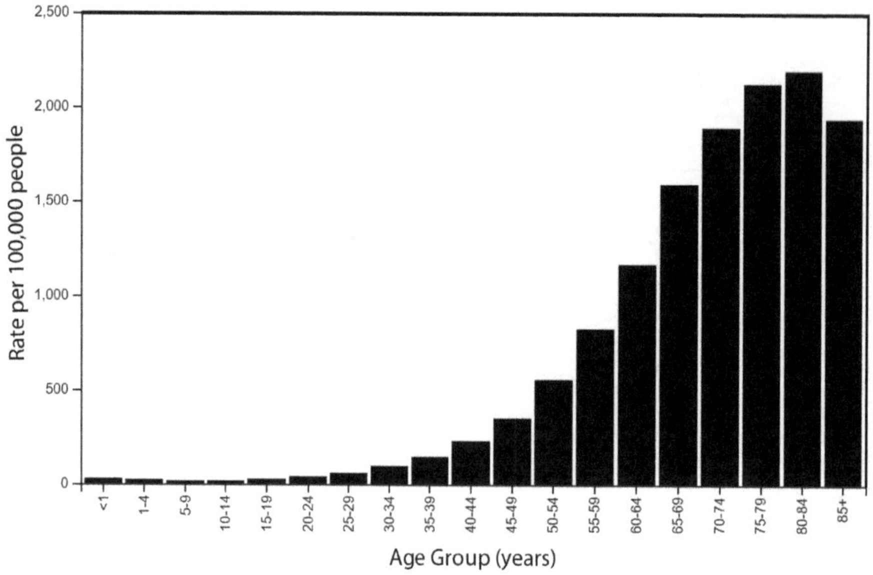

Figure 9-16 Cancer diagnosis according to age. The risk of experiencing cancer begins to increase at 20 years of age, as might be expected according to the average latency period. Real acceleration begins around 40 years and increases almost linearly to 75 years of age. Graph provided by the CDC; data acquired in 2018.

Immunosuppression factors heavily into the oncogenesis of virally induced cancers. To circumvent the elaborate human immune defense against infection, viruses have evolved strategies to evade destruction and these strategies serve equally well to protect virally induced tumors. Several oncoviruses disrupt the presentation of MHC-bound viral peptides on the cell surface by interfering with the transport of protein to the membrane, by preventing binding of the viral epitomes to MHC, or by downregulating MHC production. Interferon (IFN) represents a different immunogenic strategy. The release of IFN by an infected cell stimulates the neighboring cells to produce protein kinase R (PKR) which reduces protein synthesis by phosphorylating elements of signaling pathways. Oncoviruses target IFN signaling pathways. For example, some high-risk HPV subtypes inhibit IFN response pathway component transcription causing a functional loss of the entire pathway. HBV inhibits both IFN production and IFN signal transduction. HCV inhibits PKR and additionally antagonizes IFN by inducing IL-8.

Finally, cancer is an age-related disease. Statistics provided by the Centers for Disease Control (CDC) clearly indicate that age represents a significant risk factor for cancer presentation (Figure 9-16). The motivation for correlation between cancer incidence and aging becomes transparent when we compare the 12 hallmarks of cancer with the 9 hallmarks of aging.

Hallmarks of aging:

- Loss of proteostasis,[7]
- Deregulated nutrient sensing,
- Stem cell exhaustion,
- Mitochondrial dysfunction,
- Altered intercellular communication,
- Increased cellular senescence,

[7] Biological pathways within cells that control the biogenesis, folding, trafficking, and degradation of proteins present within and outside the cell.

- Telomere attrition,
- Epigenetic alterations,
- Genomic instability.

The last three points clearly increase the likelihood of DNA damage and overlap with the enabling and emerging hallmarks of cancer. Two other hallmarks of aging promote tumorigenesis. Deregulated nutrient sensing and mitochondrial dysfunction can stimulate glycolysis. Apparently, aging predisposes cells to transformation.

9.5.1.4. *The Macroenvironment — Nature vs Nurture*

Whatever the genomic or epigenetic drivers, the clinical stages of carcinogenesis present as a totally predictable progression. Within normal tissue, a single cell experiences an initial transforming mutation. Some years later (colon cancer: 5–20 years, head and neck cancer: 4–10 years, cervical cancer: 9–13 years, prostate cancer: 20 years), visible or palpable neoplasia becomes apparent (colon cancer: adenoma, head and neck cancer: dysplastic oral leukoplakia, breast cancer: ductal carcinoma *in situ*, prostate cancer: prostatic intraepithelial neoplasia). If untreated, the disease will progress to invasive cancer over the next several years (colon cancer: 5–15 years, head and neck cancer: 6–8 years, breast cancer: 10–20 years, prostate cancer: 10 years but may remain encapsulated for an additional 3–15 years). As we discussed earlier, the durations of these latencies likely reflect the order in which the mutations driving the phenotype occur. So then, what are *cancer promotors*? Cancer promotors either stimulate clonal proliferation or cause inflammation, thus accelerating mutations within a susceptible population of cells. An inflammatory response recruits lymphocytes that release cytokines, growth factors, and reactive oxygen species (ROS), all of which increase the probability of mutation. Cancer promotors reduce latency periods. For example, the tars in cigarette smoke stimulate cell division if certain mutations (notably ras and myc) have arisen in epithelial cells within the lungs. The smoke particulates irritate the lung lining (causing coughing and mucus production), inducing inflammation. Cancer promotors fall into six categories:

- Hormones,
- Drugs,
- Infectious agents,
- Chemical agents,
- Physical trauma,
- Chronic irritation.

Notice that these promotors are largely voluntarily (a matter of conscious choice) or environmental (external to the organism). Although hormones are essentially internally controlled and thus were included in our previous discussion of cancer supportive physiology, the endocrine system responds to the environment through control by the hypothalamus. The hypothalamus links the nervous system — including the central nervous system — and the endocrine system through the pituitary, thus responding to intangibles like mood or fear. For example, stress may lead to insulin resistance and hyperinsulinemia. On this subject . . . American obesity prompted by our calorie-rich Western diet and sedentary lifestyle has resulted in a metabolic stress syndrome associated with hyperinsulinemia. Changes in insulin level may partly explain the increased prevalence of both cancer incidence and mortality within the obese population. Furthermore, the uptick in cancer diagnoses with age can at least partially be explained by preventable chronic conditions, avoidable cancer inducing or promoting exposures, and modifiable risky behaviors. Thus, we must consider not only the genomic and proteostatic mechanisms of carcinogenesis (nature), but also the environmental factors (nurture).

9.6. Radiation-Induced Carcinogenesis

Now let's return our focus to radiation biology, specifically to the pathway in Figure 9-2 leading from mutation to cancer. This alternative pathway has been firmly established; radiation was identified as the causative factor in postexposure skin cancers as early as 1902, and serious *in vitro* and animal studies of radiogenic transformation and tumorigenesis began following World War II. Radiation therapy–induced secondary cancers now comprise the sixth most common cohort of malignancies as 5%–7% of successfully treated radiation oncology patients will develop a *de novo* secondary cancer. So, radiation-exposed cells may acquire the capacity to express tumorigenic potential. In what ways is radiation-induced carcinogenesis unique and how is the process not unique compared with nonradiogenic cancer evolution?

Application of the fundamental physiochemistry of radiation biology as described in Chapter 3 helps us understand the potential mechanisms of radiation carcinogenesis. When cells are exposed to ionizing radiation, the freed energetic electrons interact with water to create free radicals that chemically react with local nuclides to produce DNA ssb. The charged particle tracks also deposit energy directly within the DNA molecule, disrupting DNA structure (especially in the case of higher LET particles) to produce DNA dsb that can sometimes be sufficiently complex to cause chromosomal damage. How do we get from the initial radiation insult to the complex events required for cell mortality and carcinogenesis? We saw that DNA dsb induction does not correlate strictly with cell death because biochemical response factors such as DNA repair, cell cycle checkpoints, and other cell physiology recovery factors increase the survival. These larger scale factors logically influence transformation efficiency as well. Herein lie the misplaced and incompatible puzzle pieces that we need to understand radiation carcinogenesis. Observation and experimentation have led us hypothesize several — not necessarily exclusive — mechanisms of radiogenic carcinogenesis, several of which we will present in this chapter. Microscopic observation discloses that cells surviving acute x-ray exposure *in vitro* manifest aberrant, delayed phenotypic and genotypic characteristics including delayed reproductive cell death exemplified by a persistent reduction in plating efficiency compared with unirradiated cells. We also observe giant cell formation, cell fusion, and reduced cell adhesion. Examination of the genomes of surviving irradiated cells reveals spontaneous mutation, clonal heterogeneity, chromosomal instability, and a tendency toward increased chromosome numbers (aneuploidy). The timing of the observed phenomena tells us a lot. DNA damage occurs within 10^{-3} seconds of irradiation. Following radiogenic damage, cells respond to breakage signaling and repair commences calling into action cell cycle checkpoints. Within about 24 hours, by the time exposed cultured cells complete their first postirradiation mitosis; repair, misrepair, and chromosomal aberrations reflective of specific repair insufficiencies persist (Figures 4-11 through 4.15). Days later, only after several mitotic cycles, delayed phenotypic traits manifest. Contemporary research seeks to disclose the drivers of these delayed traits; the processes that presumably are triggered by prompt radiogenic damage and DNA repair, and the processes that require clonogenic generations to express.

Investigative work performed by Marc Mendonca and colleagues (Pirkkanen *et al.*, 2017) employs the GCL1 hybrid cell line — a nontumorigenic fusion of HeLa cervical cancer cells and normal human skin fibroblasts — to disclose some of these drivers. HeLa cervical cancer cells have obviously acquired all the drivers necessary to become fully competent cancer cells. Normal skin fibroblasts presumably have none. Fusion of these two cell types inhibits the HeLa oncogenic drivers, presumably through the expression of one or more fibroblast tumor suppressor genes. Mendonca and his colleagues isolated and queried the genomic changes in radiation-induced clones that reverted to the neoplastically transformed phenotype within 21 days postirradiation. The GCL1 cells were exposed to 7 Gy ionizing radiation, allowed to repair for 6 hours, and then plated into transformation (T-75) flasks at about 50 viable cells per cm^2. The chromosome losses and sequence deletions occurring in the neoplastically transformed clones were then mapped and clear evidence of fibroblast chromosome 11 loss and deletions in the 11q13 region in a tumor suppressor locus was revealed. These experiments also allow investigation of the potential mechanisms involved in radiation-induced neoplastic transformation of CGL1 cells. After 9 days, most nontransformed CGL1 cells express a delayed increase in p53 and pro-apoptotic BAX along with decreased anti-apoptotic *BCL-XL* expression resulting in delayed apoptosis that lasts several days (refer to Figure 9-13). Mendonca and others attribute this delayed death/apoptosis to the induction of genomic instability in the majority of nontransformed cells. It is during this timespan that neoplastically transformed clones arise through overexpression of BCL-2 anti-apoptotic protein. Therefore, at least in the GCL1 radiation carcinogenesis assay, we have evidence of a persistent postirradiation instability that lasts for up to 3 weeks resulting in the delayed loss of a tumor suppressor gene through chromosomal loss, gene deletion, and/or epigenetic silencing by promoter methylation.

According to Mendonca's work, a cellular state change enables the dysregulation of genes and gene products leading to the expression of a known genomic driver of carcinogenesis through the inhibition of apoptosis. What might the progression from irradiation to driver mutation look like? For illustrative purposes, let's propose a feasible scenario. Proposed: DNA damage invokes repair activity that infrequently introduces chromosomal aberrations. The aberrations destabilize chromosomal segregation, eventually provoking chromosomal instability after several cell cycles. Because chromosomal instability disrupts sequence fidelity, oncogenes, gene deregulation, and mutated tumor suppressor genes elaborate, driving transformation. Significantly, in this hypothetical scenario, the radiogenic, acutely induced mutations are irrelevant to the delayed phenotypic traits; the transformation driver mutations arise only after chromosomal instability presents several generations later. So, our proposed "blue sky" hypothesis is supported by the temporal evidence. Nonetheless, molecular confirmation has not been presented. Figure 9-17 represents a generalized scheme for radiogenic carcinogenesis. Look first at the furthest right where the attainment of genomic instability creates a global feedback loop disrupting genetic integrity through physiological dysfunction and chromosomal instability. This late radiogenic pathology is entirely consistent with sporadic and inherited carcinogenesis. Early and persistent genetic changes invoked by several radiogenic hypotheses also resemble the attributes of nonradiogenic carcinogenesis. We will examine these later. But just now, let's interrogate the controversial very early initiating events that drive transformation.

9.6.1. *Initial Transformation Events*

Because radiation exposure introduces ssb, dsb, and chromosomal aberrations in the DNA of exposed cells, it would intuitively seem that radiation provides the genetic mutations responsible for

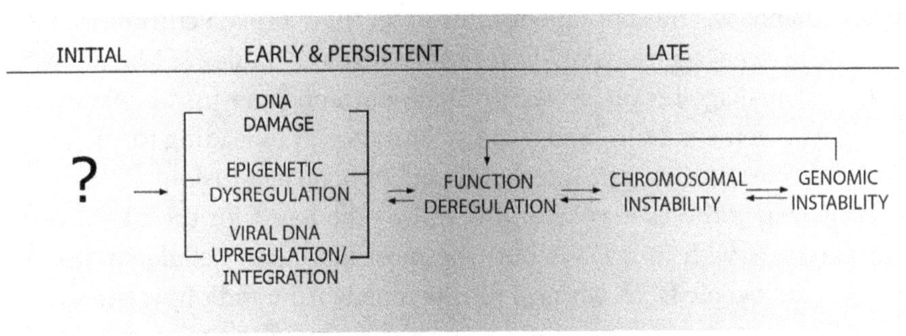

Figure 9-17 Induction of genomic instability. Cancers exhibit a generalized disruption of normal cellular function(s) via one of three genomic mechanisms: DNA damage, epigenetic deregulation, or the insertion and activation of viral DNA. The aberrant behavior activates a series of feed forward and feedback loops that increase cellular pathology culminating in genomic instability, an unstable state responsible for increased expression of delayed radiogenic and carcinogenic traits. Several hypotheses attempt to delineate the initial radiogenic events that drive cell transformation/neoplasia.

carcinogenesis. Alas, intuition fails us; scientific evidence strongly suggests this straightforward initiation mechanism is at best marginally involved in creating driver mutations. First, a number of investigators have shown that the progeny of irradiated cell populations exhibit persistently lower plating efficiencies (clonogenic capacity) induced in 40%–50% of surviving progeny. Because radiation-induced DNA strand breaks are rapidly resolved in the irradiated population, radiogenic damage cannot explain persistent reduction in clonogenic capacity postirradiation beyond the first generation. The gene (or set of genes) responsible for persistent reduced plating efficiency would have to be specifically and frequently targeted during irradiation and data indicate that mutation induction occurs at frequencies of $\sim 1 \times 10^{-6}$, so significant perpetuation of reduced clonogenic capacity by up to 50% cannot be explained by prompt DNA mutation. It would therefore seem that radiation-induced physiochemical damage leads to an alternative mechanism responsible for the delayed *in vitro* traits.

An alternative, nonspecific, radiogenic, DNA damage mechanism capable of generational passage does exist, and one current hypothesis points to it as — at least — partially responsible for the delayed transformational cellular characteristics. Radiation-induced chromosome breakage creates terminal and/or interstitial deletions resulting in acentric chromosome fragments (see Figure 4-15). These free-floating chromosomal fragments lacking centromeres can be enveloped by nuclear membrane during telophase and present as tiny micronuclei present in one daughter cell. Although micronuclei may form at the first postirradiation cell division, the production and the loss of micronuclei in irradiated cell progeny persists over time for many cell generations due to replication during mitosis and disruption of functions related to lost genomic segments.

Another prompt DNA damage–centric hypothesis suggests that persistent traits might be triggered by radiogenic destabilization of "hot spots" within the cellular genome. Sequences of DNA bases complimentary to the folded telomere chromosomal end caps frequently occur within the noncoding segments of DNA (introns). We refer to these $(TTAGGG)_n$ repeats as "telomere-like sequences," and they seem to facilitate chromosome recombination in a manner evoking the couplings between railroad cars. These telomere-like sequences form discrete bands at interstitial chromosome sites — the juncture between exchanged chromosome fragments. Possibly, telomere-like sequences were once the termini of short chromosomes belonging to a distant ancestor and human chromosomes may have elongated by connecting primordial chromosome chunks containing novel genes. At any rate, these telomere-like sequences facilitate breakage and fusion during mitotic chromosomal recombination. The hypothesis proposes that radiation destabilizes these hot spots enabling breakage-fusion cycles, particularly during homologous recombination (HR) when the interchromosomal strands form bridges between damaged and undamaged sites. Initial breakage-fusion generates dicentric chromosomes that promote chromosome breakage at first anaphase, and subsequent refusion of these broken chromosomes in progeny cells initiates the fusion-bridge, breakage-refusion cycle. John Murnane (Bailey & Murnane, 2006) and others have shown that these breakage-fusion cycles produce chromosome instability exemplified by gene amplification, gene rearrangement, LOH, and acentric chromosomes leading to aneuploidy. Gene amplification also produces dicentrics, firing up another round of breakage-fusion.

DNA damage–centric hypotheses may be failing to "see the forest for the trees." Recall that ionizing radiation primarily interacts with (not DNA but) the most abundant cellular molecule: water (Figure 3-2). Prompt (low LET) radiogenic DNA damage results mostly from radiolysis chemistry (see Equations 3.1 through 3.4), not direct interaction with ionizing particles. Radiolysis releases ROS (formerly referred to in this text as "free radicals") throughout the cytosol as well as within the nuclear volume — a small volume pertinent to our previous discussions. The acute radiogenic release of cellular ROS is even shorter-lived than prompt DNA damage but nonetheless, Charlie Limoli and his colleagues (Limoli, *et al.*, 2003) have shown that irradiated cells (exhibiting radiation-induced DNA dsb and chromosome rearrangements) go into a persistent state involving release of ROS, cytokine induction, and DNA damage that feeds back on itself and can last months after the initial radiation exposure. The persistent generation of ROS is critical to the hypothesis that this physiochemical manifestation of cell irradiation, not prompt DNA damage, drives cell transformation. Bill Morgan (Morgan, 2003) has suggested that ROS released by radiolysis could stimulate cytokine production, which in turn could produce more free radicals. His hypothesis contends that TNF-α, IL-1β, and IFN-γ induce intercellular ROS production through these known mechanisms: TNF-α increases mitochondrial ROS production; IL-1β induces ROS production via NADPH oxidase; and IFN-γ induces ROS through both mitochondrial ROS and ROS via NADPH oxidase. The ROS are then released to the extracellular environment. In addition to this ROS/cytokine cycle, remember that non-apoptotic cell death invokes an inflammatory response (see Chapter 4, Section 4.6) and releases cytoplasmic components that influence the biochemical state of local surviving cells. The cytoplasmic components include the abovementioned cytokines that bind to or are absorbed by neighboring cells in addition to activating CAFs (Figure 9-15), thereby contributing to the perpetuation of ROS production. Mendonca and others have also shown that the cytokines TNF-α and TGF-β are involved in this cellular crosstalk, referred to as the "bystander effect" (discussed in more detail in Chapter 6, Section 6.4).

9.6.2. *Early and Persistent Neoplastic Transformation Events*

Transformation genotypic attributes (clonal heterogeneity, chromosomal instability, and aneuploidy) appear only after a several generation delay. The drivers for these *de novo* genotypes presumably arise earlier than their expression and persist into the tumorigenic state, as these attributes continue to increase in tumor and cancer cells. Furthermore, each of these traits results from accumulation of novel mutations. This logic implies that cyclical ROS and acute DNA damage repair lead to mutations and chromosomal aberrations that provoke erroneous metabolic functions such as inappropriate gene regulation, loss of cell cycle and growth control, replication stress, cell death avoidance, and so on. The intertwined cascades invoke chromosomal and genomic instabilities late in the transformation process. But there are two other pathways to cellular dysfunction worthy of discussion.

If DNA repair genes, proto-oncogenes, and tumor suppressors do not specifically express mutation early during radiogenic neoplasia, but the expression of these genes is nonetheless upregulated or downregulated, perhaps epigenetic remodeling plays a role in the radiogenic initiation event. Paquette and Little (1994) showed that while genomic rearrangements were expressed in tumors *in situ*, the cultured cells used to derive those tumors express epigenic modification without genomic rearrangements *in vitro*. In fact, radiation is a strong epigenetic agent. Several cell lines exhibit dose-dependent decreases in global methylation 24–72 hours after radiation exposure. Gamma exposure of mice yields both global and repetitive element hypomethylation. Changes in DNA methylation within murine bone marrow persisted for at least 22 weeks following exposure to 0.4 Gy doses of ^{56}Fe. A general decrease in methylation may be explained by reduced enzyme activity. Methyltransferases enable DNA methylation, and it has been shown that ^{60}Co γ-radiation exposure *in vitro* decreases methyltransferase activity. Alternatively, several miRNAs that specifically target DNA methyltransferase mRNA (miR-29 family, miR-141, and miR-152) leading to methyltransferase mRNA degradation are indeed upregulated following *in vitro* irradiation or bystander triggering. Alternatively, radiogenic DNA dsb trigger protective mechanisms including cell cycle arrest and replication fork stall. Perhaps during this stall, methyltransferases may be reassigned from replication-related DNA methylation patterning to DNA repair processes. Finally, *in vivo* radiogenic cell death–provoked repopulation may outstrip methyltransferase capacity. Each of these mechanisms would manifest decreased methylation, perhaps several of the mechanisms act in parallel.

We must tread carefully here when we generalize to cell types, tissues, or animal models other than the subjects of a specific experiment. For example, radiation-induced alterations in epigenetic modification can differ between experimental conditions and even among strains of the same species. It has been shown that following x-ray exposure, repetitive element methylation patterns vary according to strain, tissue, sex, and time point post *in vitro* irradiation. Furthermore, epigenetic alterations may be transiently induced by radiation insult whereas tumorigenesis requires stable, heritable modification of methylation patterns (as seen in murine bone marrow). Time course studies are required. In general, *in vivo* evidence has shown that exposures to doses higher than 1 Gy generally result in the loss of global DNA methylation in hematopoietic tissues including thymus, spleen, and bone marrow, and mammary tissues; however, radiation-induced alterations in global methylation within muscle or lung has not been reported. Evidence of altered methylation after radiation exposure in humans includes (1) detection of hypermethylation of DNA encoding the cyclin-dependent kinase p16 and the methyltransferase MGMT in uranium miners, and (2) increased methylation (3.5-fold) of p16 and GATA5 in lung carcinomas of plutonium-exposed workers and civilians at the MAYAK facility (long-term radiation leakage event in Russia; for more information see Chapter 11, Section 11.5.3). Interestingly, hypermethylation-induced silencing of tumor suppressor genes, hypomethylation-induced activation of oncogenes, and alterations in DNA methylation of repetitive elements have been described in virtually all human cancers.

Finally, recall that viruses can induce cancer through a variety of mechanisms. Think about this: our genome is full of incorporated viral DNA integrated over the millennia including viral bits acquired during our lifetimes. In general, viral DNA gets inserted into the genome randomly, some fortuitously downstream of promotor regions and some within noncoding regions. In the late 1980s and early 1990s, several groups hypothesized that radiation exposure and subsequent chromosomal rearrangement invoke the upregulation of endogenous viral DNA replication destabilizing the genome and leading to genomic instability. However, more recent work (Toth *et al.*, 2003) examining the effectiveness of oncolytic, episomal, viral vectors used in cancer therapy indicates that radiation does not increase viral DNA load, but rather the upregulated mitotic proteins (e.g., BAX) act in concert with the viral cytotoxicity. Thus the mechanisms of action may be situationally dependent.

9.6.3. *Late Neoplastic Transformation Events*

Radiation exposure introduces prompt DNA ssb, dsb, and chromosome breaks (about six per cell per Gy) triggering DNA repair mechanisms in irradiated cells. DNA repair is a nuclear/mitotic physiological function and evidence suggests that dysregulation — primarily downregulation — of genes involved in DNA dsb repair (tumor suppressing caretaker genes) drives chromosomal instability. If high-fidelity DNA repair is impeded by downregulation of transcripts coding for repair enzymes, then error-prone repair mechanisms become kinetically favored. Error-prone repair promotes a high mutation rate in the genomes of clonal progeny. Mutations and chromosomal rearrangements subsequently multiply each generation presenting as increasing chromosomal instability. We have seen that radiation exposure triggers epigenetic pattern modifications (e.g., methylation) that downregulate DNA repair genes (Table 9-3). Table 9-4 lists repair genes that commonly experience altered regulation resulting from radiogenic mutation of promotor sequences or mutation of activating transcription factor genes in transformed cells. In addition to radiogenic prompt DNA strand and chromosome breaks, radiation exposure triggers an unidentified soluble cofactor that shocks exposed cells as well as nonintercepting cells (the bystander effect) into an altered response state that reflects transmuted cellular function leading to deletion/

silencing of genes controlling stability, induction of genes stimulating instability, and/or activation of endogenous viral DNA.

In Section 9.5.1, we discussed the importance of a clonal growth advantage and involvement of growth signal pathway dysregulation during sporadic carcinogenesis. Activation of signal transduction functionality resulting from or leading to increased phosphotransferase activity of cytoplasmic protein kinases also typifies radiogenic response. Both genomic upregulation of protein kinases and oncogenic *ras* expression in several radiogenic cancers raise the possibility that an early nuclear event triggering

Table 9-3 Genes exhibiting altered methylation patterns following radiation exposure *in vitro*, listed by DNA repair mechanism

BER	MMR	NER	dsb and Interstrand Crosslinks
DNA glycosylases	*Base mismatch, insertion, deletion*	*Global*	*HR*
OGG1	MLH1	RAD23B	ATM
NEIL1	MSH2	XPC	BRACA1,2
MBD4	MSH6	ERCC1	XRCCs
	PMS2	ERCC4/XPF	RAD51
			FANCF
			WRN
AP endonucleases		*Transcription coupled*	*NHEJ*
		ERCC1	KU80
		ERCC4/XPF	DNA-Pkcs
			MMEJ
			FEN1
			LIG3
			PARP1
			Nibrin
			MRE11
			XRCC1

Table 9-4 Genetically deregulated genes identified in cells irradiated *in vitro*, listed according to DNA repair mechanism

BER	MMR	NER	dsb and Interstrand Crosslink
DNA glycosylases	*Base mismatch, insertion, deletion*	*Global*	*HR*
MUTYH	MLH3	TFIIH	BLM
UNG	MSH3,4,5	XPD helicase	MUS81
SMUG1	PMS1	XPB ATPase	MRE11
MPG	PMSL3	RPA	RAD50
TDG			RAD52
AP endonucleases		*Transcription coupled*	*NHEJ*
APEX1		CSAX	KU70
APEX2		CSB	XRCC4
DNA POL b		XPG	
PNKP		AB2	
APLF			

activated signal transduction pathways may be partially responsible for inducing chromosomal instability. The more cycles a cell undergoes, the greater the opportunity for chromosomal destabilization in the progeny. Self-sufficient growth signal regulation in irradiated cells appears to be independent of p53 status (active or inactive) but dependent upon XRCC2/XRCC3 and the catalytic subunit of DNA protein kinase. Coincident with driving mitosis, signaling pathways stimulate alterations in genomic expression (upregulation or downregulation of specific genes; Figure 9-13) creating an altered cytoplasmic environment that in turn drives subsequent genomic responses. Downregulation of genes involved in DNA dsb repair drives chromosomal instability in irradiated cells, as described earlier. In other words, growth signaling increases the opportunity for genomic mutation and remodels the cellular state, predisposing cells to chromosomal instability.

Finally, let's extrapolate our examination of *in vitro* radiogenic neoplastic transformation to *in vivo* carcinogenesis by referring back to Figure 9-17, Altered gene expression (upregulation or silencing) initiated by chromosomal instability feeds back to function deregulation through newly induced proteins and/or growth factors, unstable chromosome sorting, replication stress, or telomere (and telomere-like repeat) instability. Chromosomal instability motivates the pathway leading forward to genomic instability through the inappropriate segregation of chromosomes and chromosome fragments during mitosis, enhancing the defects in DNA replication that promote genome-wide mutation. Genomic instability loops back to provoke the hallmarks of cancer involving stability functions: replication stress, self-sufficiency in growth signal regulation, evasion of senescence, evasion of apoptosis, mitotic stress, and immortality. Genomic instability feeds forward to produce the remaining hallmarks of cancer: deregulation of cellular energetics, immunosurveillance avoidance, sustained angiogenesis, tissue invasion, and metastasis.

9.6.4. *High LET Radiation-Induced Transformation and Carcinogenesis*

As we have seen, high LET radiation achieves greater cell killing than low LET radiation at equivalent doses (RBE, Chapter 7) due to densely ionizing track structures within cell nuclei (Figures 7.17 and 7.18). One might hypothesize that track structure also could cause increased and/or more complex DNA and chromosomal damage leading to increased transformation potential, or one might argue that the increased cell mortality would eliminate cells prior to the induction of instability, reducing the transformation frequency. But then, perhaps before the cells die, they release soluble (bystander) factors that trigger ROS and instability in the unirradiated local population, increasing the number of potentially transformed cells. Let's see what evidence the scientific literature provides.

First, we cannot assume that high LET radiogenic cell transformation initiates via the same mechanisms as low LET radiogenic cell transformation without evidentiary support. In the mid-1980s, Tracy Yang, a biophysicist and colleague at Lawrence Berkeley Laboratory, firmly established a significant increase in the effectiveness of transformation using high LET radiation compared with low LET radiation (Figure 9-18). These early studies used confluent cultures of a C3H embryonic mouse 10T1/2 fibroblast cell line exposed to carbon, neon, silicon, argon iron, and uranium ions with LET values ranging from 10 to 1900 keV/μm. Transformation efficiency was found to be dependent upon both dose and LET, yielding RBEs for transformation (RBE_{tr})[8] ranging from 1.0 (carbon) to 3.0 (Si, Ar, Fe). At the time of her investigation, Tracy acknowledged that experimental results may depend upon cell type and culture conditions (e.g., confluence). More recent investigations have shown that many other factors also influence not only transformation, but several biological markers for transformation and tumorigenesis.

[8] Compared with 225 kVp x-rays

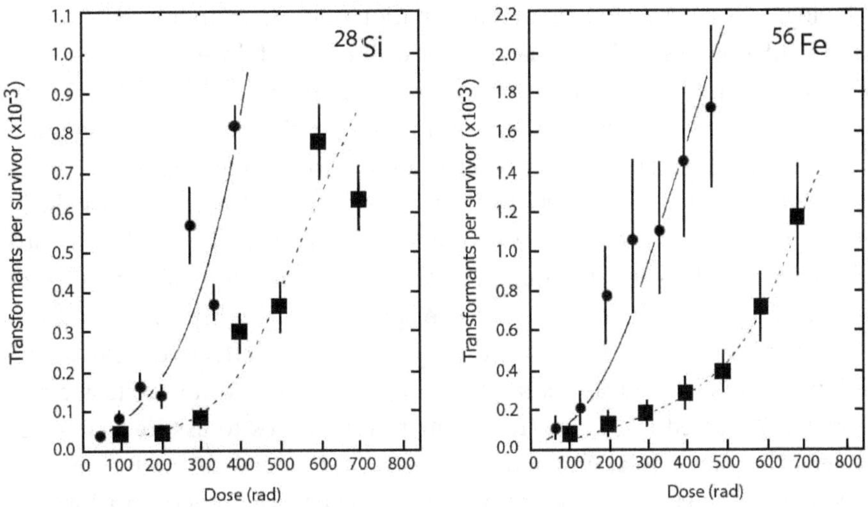

Figure 9-18 Effect of LET on cell transformation. The number of transformants per surviving cell are plotted vs. dose for 50 keV/μm ^{28}Si or 300 keV/μm ^{56}Fe (•) and 225 kVp x-rays (▪). The calculated RBE for silicon is 2.3 and the calculated RBE for iron is 2.6. Reproduced and modified with permission of the Radiation Research Society from "Neoplastic cell transformation by heavy charged particles." TC Yang, *et al.*, *Radiation Research,* 1985; 8:S177–S187.

As always, the methods for calculating LET and RBE_{tr} are critical, and much of the confusion in this field results from trying to "compare apples with oranges"; thus, we note that these early results calculated RBE_{tr} at both 10% and 50% survival doses and LET was calculated as track-averaged (Equation 7.23). Like RBE results for cell killing, RBE_{tr} for transformation peaks around 100–200 keV/μm and then declines. Now, this is interesting. If the physical interaction dependent upon track structure determines both cell killing and transformation, and if both cell killing and transformation efficiencies increase and decline according to LET, it seems likely that similar biological functions govern both end points. Simple point mutations appear not to be common during high LET transformation as demonstrated in a popular point mutation assay examining ouabain-resistance. However, high LET transformed clonal colonies do express chromosomal exchanges and large deletions, as well as evidence of epigenetic alterations causing deregulation of tumor suppressor genes. Data suggest that this high LET complex DNA damage, also known as "clustered damage," may be more difficult to correctly repair and therefore may result in greater biological effects such as mutation, gene loss, and/or genomic instability leading to increased transformation per unit dose.

Data from low LET gamma or x-ray and high LET charged particle irradiation of *in vitro* cell cultures and *in situ* tissues or *in vivo* animal studies have validated the pervasiveness of transformation study results and have shown some overall common themes. While low LET studies show some large-scale structural events, in general low LET irradiation induces point mutations associated with oxidative stress. By contrast, high LET mutations seem to be characterized by intense focal damage due to the passage of the heavy ions through nuclear DNA and include small intragenic alterations, multilocus deletions that vary in size but could be up to 100 Mbp of DNA, mitotic recombination involving homologous chromosomes, large deletions, chromosome loss, and translocations. It is important to remember that these mutation endpoints depend on cell type and mutation locus being studied.[9] In general, although the mutation induction frequency may differ for *in vitro* versus *in vivo* systems, high LET par-

[9] See work by Amy Kronenberg, Mitchell Turker, Charles Waldren, and Allan Balmain for in-depth background.

ticles are more biologically effective at inducing mutation leading to increased expression of genomic instability versus low LET x-rays and gamma rays per unit physical dose.

Here is another important consideration pertinent to our discussion of high LET radiation and cell transformation. The reader should pay close attention to the phrases, "an average of one particle per cell," and "exactly one particle per cell." The "average" implies that some cells experience one, two, three, or more tracks, whereas many cells experience no tracks. Richard Miller and colleagues working at Columbia University found that transformation frequency arising from *exactly* one α-particle track per cell did not differ significantly from spontaneous transformation frequency in unirradiated cultures.[10] However, an *average* of one particle per cell has been shown to be highly effective at introducing transformation. In other words, exactly one particle track per cell is less effective biologically than a mean of one particle track per cell. What does this imply? Well, there are two nonexclusive possibilities: either more than one track is required to induce transformation (perhaps to assure that a track traverses the nucleus), or the cells that have escaped particle transit are induced through a bystander effect. This second possibility is intriguing. Recall that only *surviving* cells experience transformation. A relatively small fraction of cells survives a single α-particle track and additional tracks simply increase the percent mitotic death. Experimental evidence supports this. When 20% of human-hamster hybrid A_L cells were exposed to 20 α-particles per cell, less than 1% of the exposed cells survived. Furthermore, the mutation frequency in surviving, unirradiated cells was four times background. In primary human fibroblasts, 100 keV/μm helium ions also revealed significantly more damage than expected based on direct effects of radiation. If we assume that mutation and DNA damage are indicators of transformation potential (Figure 9-17), then these results support the hypothesis that transformation frequency is influenced by the bystander effect. Following high LET exposure, transfer of medium provides similar results as those obtained using low LET x-ray or gamma ray irradiation but with increased effectiveness, although the stimulation threshold for the bystander effect remains at about 1 mGy using α-, proton-, or iron-particle radiation. Two mechanisms of intercellular bystander factor transport are possible. The bystander factor may freely traverse the cell membrane into the extracellular environment, or it may pass through intercellular communication gap junctions that form between adjoining cells. When exposed to limited α-particle flux, sparsely plated cells — cells not in contact and without gap junctions — can express a radiation response in cells not traversed by particle tracks. Apparently, the first mechanism seems plausible. To investigate the second mechanism, lindane — a gap junction inhibitor — was added to the exposed culture medium prior to transfer to the unirradiated cell culture. In this case, the bystander effect was significantly diminished but not eliminated. Evidently, the bystander factor passes into unirradiated cells via both mechanisms.

So, could a bystander effect be responsible for high-LET-induced cell transformation? Sure. Suppose the bystander factor is indeed a collection of ROS — peroxides, superoxide, hydroxyl radical, and singlet oxygen (Chapter 3, Section 3.3.4) — as proposed by Charlie Limoli, Bill Morgan, and others. We have seen how low LET radiation — that is, high-energy electrons set in motion through ionization — creates energy deposition tracks in water (Figure 2-19) releasing ROS along the track (Figure 3-3). Not surprisingly, high-LET particle tracks through water similarly create local ROS (Figure 2-21), our previous obsession with high LET direct damage to chromosomes notwithstanding. The increased density of high LET energy deposition creates very high local concentrations of ROS, the theoretical lifetimes, distribution, and yield of which have been calculated by Yusa Muroya and colleagues (Muroya *et al.*, 2006). The ROS diffuse beyond 100 μm within microseconds after which the concentration of each

[10] For these experiments, target cells must be grown as monolayers on Mylar and thus the energy deposited per traversal is less biologically effective.

species is primarily determined by mitochondrial network dynamics. Coincident with DNA radical formation (Chapter 3, Section 3.3.4) to form ssb, ROS induce a rapid depolarization of the mitochondrial inner membrane potential with subsequent reduction of oxidative phosphorylation of ADP. More importantly, an ROS-induced ROS release (RIRR) positive feedback loop has been identified that greatly amplifies the concentration of ROS through the mitochondrial production of additional ROS. RIRR transpiring in one mitochondrion can affect neighboring mitochondria eventually propagating an ROS surge throughout the entire cell. RIRR-produced ROS surges are now recognized as important signaling mechanisms. Bill Morgan's proposed ROS/cytokine feedback loop may be responsible for the delayed perpetuation of genomic instability.

The exposure of cells *in vitro* to high LET radiation results in both prompt chromosomal aberrations from direct DNA damage and numerous delayed chromosomal aberrations appearing several generations after clonogen exposure. It is significant that these aberrations are delayed. Recall that the bystander factor fails to evoke prompt chromosomal aberrations although Bill Morgan's proposed feedback loop invokes genomic instability that that induces CIN (Figure 9-17), rationalizing the delayed appearance in that system. Aberration induction efficiency varies with cell type and there does not seem to be any obvious correlation between cell type and chromosomal aberration characteristics. Furthermore, both dose and LET, as well as genetic background, influence the delay between exposure and initiation of chromosomal aberration manifestation. When comparing experimental results, it also is important to keep in mind the cell type–specific temporal pattern of aberration expression. Let's look at some specific investigational examples. When thin layers of hematopoetic cell suspensions were exposed to a parallel beam of ^{238}Pu α-particles (LET = 121 keV/μm) clones presented both delayed chromosome and chromatid aberrations. The lesion configurations and locations were random within the progeny of a single clonal population, such that it is mathematically probable that each colony arose from a single cell that survived the passage of one or more α-particle tracks. In this experiment, the outstanding feature was the frequency of karyotypic abnormalities; whereas x-rays produced perhaps two aberrations in 100 colonies, the α-particles produced around 40–60 aberrations per 100 colonies. Furthermore, α-particles cause sister chromatid exchanges whereas x-rays and γ-rays rarely do. Chromosome aberrations presented at the first postirradiation metaphase 10 times more often following high LET exposures than low LET exposures (presumably reflecting direct DNA damage) and high LET irradiation generated more than twice as many prematurely condensed chromosome breaks. The chromosomal aberrations are likely due to misrepair (incorrect rejoining) rather than increased *de novo* lesions calling to mind a repair malfunction similar to that suspected early during low LET transformation (Figure 9-18), however, the malfunction likely arises because the damage is more complex, not because of repair malfunction.[11] Sentaro Takahashi and his colleagues used cell-fusion-based premature chromosome condensation (PCC) to show that chromosome breaks in high-LET iron ion irradiated cells had significantly slower repair kinetics and a higher number of residual chromosome breaks compared with x-ray irradiated cells. This indicates that complex high LET damage is more difficult to repair and unrepaired or resolved DNA dsbs are lethal to cells. In another experiment, high-LET iron ions introduced CIN at a rate of approximately 4% per Gy compared with approximately 3% per Gy induction with x-ray radiation.[12] The yield of chromatid-type aberrations in mouse embryo cells, human epithelial cells, and fibroblasts was found to be greater than the yield of chromosome-type aberrations at various times after irradiation, and it was cell type dependent. In these experiments, both high- and low-LET radiation induce chromosomal instability. In general, high-LET irradiation produces more chromatid aberrations, whereas high-dose, low-LET

[11] See Munira Kadhim and colleagues for details.

[12] See Charles Limoli and colleagues for details.

irradiation results in more chromosome aberrations. The variability presented by examination of CIN and the nonstatistical correlations between CIN and cell transformation admonish us against relying on structural influences over neoplastic transformation and encourage us to consider the resulting genetic and epigenetic mutations that drive the hallmarks of carcinogenesis.

So, several *in vitro* markers of transformation exhibit increased relative biological effectiveness (RBE_{tr}) following high LET radiation exposure. Furthermore, transformation induction following the application of exposed culture medium to unexposed cells (bystander effect) consistently yields RBE_{tr} greater than one, regardless of the conditioning high LET particle identity, including protons and helium. Based on this evidence, it is reasonable to propose that the higher RBEs observed for high LET carcinogenesis may be due to a combination of more complex resultant focal DNA and chromosome damage, induction of bystander effects, and genomic instability that drives the genomic and epigenetic changes resulting in the accumulation of the driver mutations required for cancer induction.

Do we have evidence that *in vitro* high LET irradiation-induced transformation and expression of genomic damage has bearing on *in vivo* neoplasia? The best documented *in vivo*, high LET tumorigenesis experimental system uses the Harderian gland found in the ocular orbit of most land vertebrates, including female CB6F1 mice (but not humans). The gland is composed almost entirely of epithelial cells and secretes an oily fluid to lubricate the eye. Ed Alpen initiated this work in the 1980s and 1990s, accumulating an impressive data trove. More recently, the work was extended by Polly Chang, Ellie Blakely, and their colleagues (Chang *et al.*, 2016) to include lower dose data, the Alpen and Chang data sets were combined, and mathematical models were derived to search for best fit to the number of tumors arising 16 weeks after irradiation. The first of Chang's models includes only targeted effects (TE) — it assumes that direct interaction between radiation particles and the cell nuclear material accounts for tumorigenesis. Here, tumorigenesis at 16 weeks postexposure is modeled as linear-quadratic, but the quadratic term (β) is applied only to low LET radiations (x-ray, γ-ray, protons, or helium ions), which we have seen exhibit a small RBE for mitotic cell death and some measures of repair (Chapter 7, Section 7.5.1). The formalism for targeted effects looks like this:

$$P_{TE} = P_0 + [\alpha(L)D + \beta(L)D^2]\, e^{-\lambda(L)D}, \tag{9.1}$$

where P_0 is the sporadic occurrence of gland tumors and L = LET. The negative exponential term accounts for decreasing tumor occurrence with increasing cell death as dose increases; γ is an LET-dependent term. However, we saw earlier that radiation-induced tumorigenesis responds to a soluble signal, aka the bystander effect, inducing further cell death as well as tumorigenesis. Equation 9.1 can be modified to incorporate these nontargeted effects (NTE) by including an additional term,

$$P_{NTE} = P_0 + \{[\alpha(L)D + \beta(L)\, D^2] + \kappa(L)\Theta(D_{th})\}e^{-\lambda(L)D}. \tag{9.2}$$

Here, the added bystander function includes a new LET-dependent term, κ, and an LET-dependent but dose-independent step function constant, Θ, that is applied above a certain dose threshold. The Θ term is included to account for our inadequately understood radiological response at very low doses or dose rates; it allows for a threshold effect — that is, a dose below which the soluble trigger is either not released or the concentration is too low to stimulate unirradiated cells. The natural tumor frequency is accounted for, so the number of tumors above that baseline occurrence are assumed to be induced by radiation. Although hundreds of mice were included in these experiments, the biological variability is always large when working with animal models and the number of induced tumors small, so uncertainty, expressed as error bars in Figure 9-19, is substantial. Nonetheless, a second NTE model (NTE2) wherein

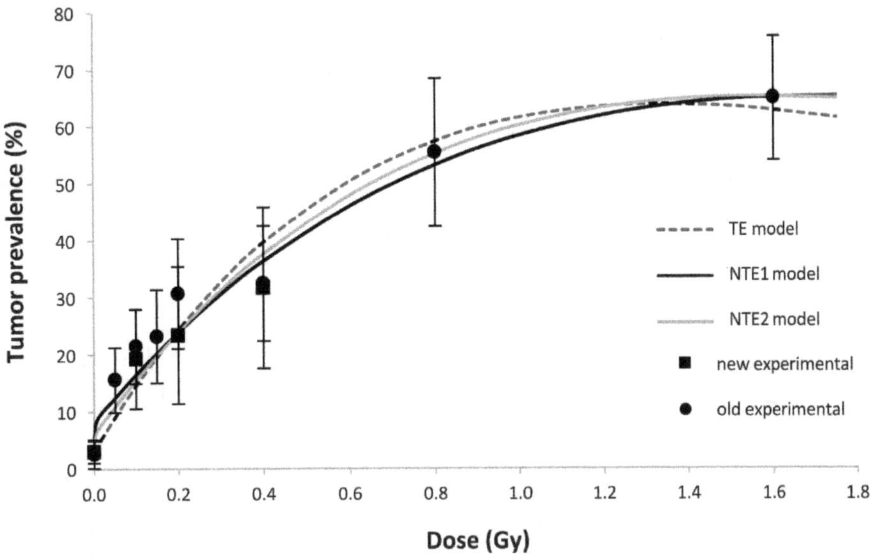

Figure 9-19 Influence of high LET radiation dose on tumor incidence. Data from Alpen (old) and Chang (new) acquired from irradiation of mouse Harderian gland (either head or whole body exposure) using iron ions at ~193 keV/μm. The data are fit using three models: TE, NTE1, or NTE2. NTE1 (Equation 9.2) provides the best fit. Reproduced and modified with permission of the Radiation Research Society from Chang *et al.*, (2016).

κ is allowed to be independent of LET and Θ is set at a threshold dose below 1 mGy, fit significantly less well than NTE (NTE1 in Figure 9-19). Both NTE models fit significantly better than TE, implying that the bystander effect is both real and important. However, note that TE fits better below 0.17 Gy, implying that the bystander effect may be diminishingly small at low doses.

The question of repair competency following high LET irradiation remains unclear. In Chapter 7, we showed that the cell survival establishes a small or absent initial curve ($n \approx 1, D_q \approx 0$) indicating that cells lack DNA dsb repair following high LET exposure *in vitro* (Figures 7-21 and 7-22). Nonetheless, several *in vivo* studies assessing tumorigenesis suggest that tumorigenic cells remain repair competent following high LET exposure. In Ed Alpen's studies, he found that six doses separated by 2 weeks are about 1.5 times as effective at producing tumors as the same dose delivered in a single fraction, likely due to increased cell survival. Another experiment exposed the mouse Harderian gland to fission neutrons at 0.25 Gy per fraction; it contradicted Alpen's results showing a fractionated regimen over a dose range of 0.5–40 Gy did not produce more tumors than single doses. And low dose data show no dose rate effect. However, in BALB/c/An NBd mice, 0.5 Gy given in two fractions separated by 30 days induced significantly more lung adenocarcinomas than a single dose. So, what is going on? The effects seem to be dependent upon higher scale impact factors and are therefore organ- and organism-dependent. Repair competency is important because our hypotheses for both sporadic tumors and low LET radiogenic tumors invoke DNA repair/misrepair as fundamental for early and persistent generation of carcinogenesis (Figure 9-18). Here's the relevant point: upregulation of repair enzymes and activation of repair pathways persist regardless of whether the complex damage generated by high LET radiation can be repaired. In fact, the misrepair of complex damaged sites possibly increases tumorigenic prevalence by introducing chromosomal instability.

Perhaps it might be more useful to explain high LET tumorigenesis in terms of microdosimetry. Polly Chang, Ellie Blakely, and colleagues help us out here. They plotted Ed Alpen's data from proton (H), helium, neon, iron, niobium, lanthanum, and gamma irradiation as well as their own data from silicon,

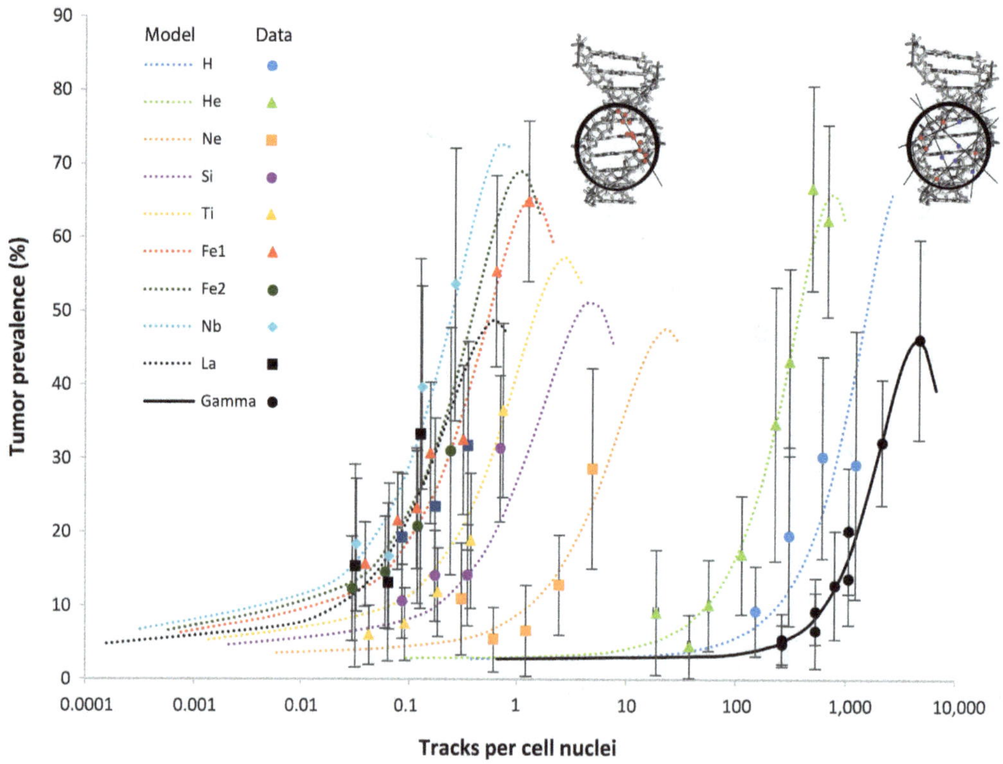

Figure 9-20 Influence of track structure on tumor incidence. Data herein is derived from Alpen and Chang at a dose of approximately 1 Gy. The high-energy particle data sets are modeled using Equation 9.2 for NTE. The high LET particles are assumed to deposit a single track per cell nucleus; for tracks less than 1 per cell nucleus the reader may envision 1 track per 10 nuclei, 1 track per 100 nuclei, and so on. The inset images from Figure 7-18 remind the reader of the difference in dose deposition patterns for high LET vs lower LET radiations at equal doses. Reproduced and modified with permission of the Radiation Research Society from Chang *et al.*, (2016).

titanium, iron, and gamma irradiation (all exposing Harderian gland tumors in CB6F1 female mice). The derived average number of tracks (see Chapter 7, Section 7.5.1; specifically, Equations 7.23, 7.24, and 7.25) are plotted against percent tumor prevalence in Figure 9-20. NTE analysis of each data set presents curves that reveal some interesting characteristics. High LET radiations exhibit a strong correlation between track structure (i.e., LET or energy) and tumor prevalence up to a saturation at 50%–70%. Lower LET radiations — the set of curves representing more than 10 tracks per nucleus: protons, helium, and γ-rays — demonstrate a similar LET dependence with a similar saturation.[13] Within each particle type (i.e., track structure), the greater the number of tracks, the more frequently tumors arise. A small range of tumor prevalence saturation maxima emerges for all radiations (50%–70%), despite an LET range of 175 keV/μm to 953 keV/μm. Nonetheless, at an average of one track per nucleus, tumor prevalence ranges from 5% (Ne) to 70% (Fe), indicating the importance of track structure on tumorigenesis.

Recall that the RBE for cell killing ranges from 1.2 to 3.8 for protons, dependent upon LET, but the RBE for cell killing using neutron radiation is 6.8. We can calculate relative biological effectiveness for tumorigenesis (RBE_{tg}) at the NTE crossover dose (D_{cr}),

$$D_{cr} = \frac{\kappa(L)}{\alpha(L)}, \tag{9.3}$$

[13] Gamma LET is derived from electron track stopping power as shown in Chapter 7.

as,

$$RBE_{tg} = \frac{\alpha(L)}{\alpha_\gamma} + \frac{\kappa(L)}{\alpha_\gamma D_L}, \quad (9.4)$$

where $\alpha(L)$ is taken from the NTE solution for the test radiation and α_γ is taken from the LQ solution for the standard radiation (^{60}Co in Chang's work); $\kappa(L)$ is taken from the NTE solution, and D_L is the test radiation dose. Chang *et al.* derived these values from the linear portions of the model slopes analogous to the way RBE is calculated (Figure 7-23) and found the values presented in Table 9-5. RBE_{tg} does not

Table 9-5 Derived RBE_{tg} for particle irradiations. Data from Chang *et al.* (2016).

Ion	LET (keV/μm)	RBE_{tg}
Proton (H)	0.4	1.78 + 0.92
Helium (He)	1.6	2.10 ± 0.98
Neon (Ne)	25	7.86 ± 2.07
Silicon (Si)	70	16.3 ± 3.81
Titanium (Ti)	107	21.1 ± 4.86
Iron (Fe)	175	26.2 ± 6.13
Iron (Fe)	193	26.9 ± 6.36
Iron (Fe)	253	28.0 ± 6.87
Niobium (Nb)	464	23.3 ± 6.89
Lanthanum (La)	953	8.61 ± 3.92

resemble RBE for cell death; the values are generally larger and exhibit greater track structure dependence.

From these results, we conclude again that it is reasonable to propose increased RBEs observed for high LET carcinogenesis are multifaceted and likely reflect a combination of the introduction of more complex focal DNA and chromosome damage in cells, induction of a transcellular bystander state change, and genomic instability that drives the genomic and epigenetic changes resulting in the accumulation of the driver mutations required for cancer induction.

9.7. Summary

- Transformed cells exhibit specific phenotypic abnormalities including:
  - Loss of cell–cell contact inhibition — piling up into dense aggregates
  - Loss of anchorage-dependent growth characteristics
  - Loss of normal dependence on exogenous growth factors
  - Chromosomal abnormalities
  - Expression of unique surface antigens
  - Escape from normal regulation of cell growth, division, and differentiation
  - Immortalization
- Tumorigenic transformation is identified by:

  - Tumorigenic growth in recipient animals under appropriate conditions
- Transformation, like cancer, is stochastic.
- The probability of transformation decreases with increasing dose because the number of surviving cells decreases with increasing dose.
- Because there is a baseline natural mutation rate and a baseline cancer incidence, induced mutations and cancers must rise above these baselines in a statistically significant way.
- There are three tumor origin hypotheses:
  - Monoclonal origin
  - Clonal evolution
  - Cancer stem cell hypothesis
- DNA or chromosomal reconfigurations can drive carcinogenesis
  - DNA sequence mutation
  - Gene inversion
  - Gene deletion
  - Chromosomal translocation
- Mutated genes that induce neoplasia belong to one of three metabolic classes:
  - Caretaker
  - Gatekeeper
  - Landscaper
- DNA sequences that code for protein products (30,000 genes) or enzymatically active RNA molecules account for about 1.5% of the genome.
- There are currently about 600 genes known to participate in carcinogenesis through mutation.
- Genes and regulatory RNA are also deregulated through mutation of noncoding regions of DNA such as promotor regions.
- Epigenetic pattern modification is associated with tumors at least as frequently as genetic mutation.
- Carcinogenesis may be enhanced by posttranslational modifications resulting from physiological state changes such as miRNA dysregulation, ROS upregulation, or postprocessing disruption, none of which can be detected in mutation assays.
- Some viruses are cancer promotors, albeit inefficient. These viruses:
  - Mutate coding and noncoding DNA during insertion into the host genome
  - Deregulate cellular functions (especially proliferation) and may induce genomic instability
  - Upregulate ROS
  - Promote immortality
  - Evade immunological suppression
  - Degrade extracellular tissue
- Viruses can acquire human oncogenes.
- Often, particularly in colorectal cancer, a growth advantage that may be expressed as a clonal field expansion represents the first phenotypic expression of carcinogenesis.
- The mutator phenotype hypothesis is exemplified by colorectal cancer wherein there is an orderly progression of mutation events leading from clonal expansion to polyp (adenoma) to carcinoma in situ to cancer.
- CIN describes changes in chromosome structure or number associated with metaphase checkpoint failure.
- Genomic instability is characterized by an extremely high mutation frequency throughout the entire genome. This instability promotes CIN and destabilizes the cell physiological state resulting in driver mutations leading to carcinogenesis.

- Replication stress including replication fork collapse, a functional disruption that produces excessive point mutations, results from initial mutation of replication enzymes or cofactors.
- Mitotic stress including failed chromosome segregation and spindle fiber disruption, evokes CIN.
- Kataegis (60% of cancers), presents as a cluster of nucleotide substitutions biased toward one of the two strands.
- Chromoplexy (18% of cancers) translocations produce shuffled chains of rearranged genes.
- Chromothripsis (22% of cancers) presents as hundreds of clustered chromosomal rearrangements presumably arising from a single localized event resulting in amplifications (58%), homozygous deletions (34%), and structural variants (8%) within genes and promoters.
- The lag between the carcinogenesis initiation event and the discernable presentation of a tumor is called the latency. Latency for solid tumors in humans is approximately 20 years and latency for blood-borne tumors is approximately 2–12 years. The period is presumably the amount of time required for a progenitor cell to accumulate 2–5 driver mutations and acquire immune avoidance with possible metastatic capabilities.
- The hallmarks of cancer:
    - Replication stress
    - DNA damage
    - Genomic instability
    - Deregulation of cellular energetics
    - Immunosurveillance avoidance
    - Self-sufficient growth signal regulation
    - Evasion of senescence
    - Evasion of apoptosis
    - Limitless replicative potential: immortality
    - Mitotic stress
    - Sustained angiogenesis
    - Tissue invasion and metastasis
- Acquired ability to divide without restraint is key to tumorigenesis; each division provides an opportunity for driver mutation.
    - Growth signal independence, cell death and senescence avoidance, and immortality are potent initiators of carcinogenesis.
- The Hayflick limit, enforced by loss of telomere RNA, is 50–60 cell divisions.
- Tumors require a complex environment of supportive cells and tissues to thrive.
- Organism physiology influences carcinogenic potential.
    - Hormonal activity
    - Organism immunocompetence
    - Organism age
- The evolution of cancer driver mutations, cell state alterations, and physiological tumor support can be accelerated by environmental conditions that largely result from free will choices.
- The most common cellular responses to radiation exposure are death or senescence.
- The fraction of cells that survive radiation exposure display the same genetic responses expressed by terminal cells, although the severity of damage and the cellular state may be different.
- The progeny of irradiated cell populations exhibits persistently lower clonogenic capacity that is induced in 40%–50% of surviving progeny.
- Several initiation mechanisms have been proposed to explain radiogenic cell transformation.

- Formation of micronuclei
- Activation of telomere-like hot spots
- Viral DNA activation
- A radiolytic hydrolysis–induced cytokine-ROS cycle that is implicated in the bystander effect

- Although a few cancers appear to be induced by two to three driver mutations, the majority of solid cancers require at least five driver mutations to advance from transformation to cancer.
- Radiation exposure induces persistent epigenic remodeling; alterations in methylation patterning predominates.
- The feed forward and feedback loops involving CIN and genomic instability also appear to promote radiogenic carcinogenesis.
- High LET radiations are not only more effective at cell killing (RBE > 1.0), heavy ions are also more transformation efficient (RBR_{tr} > 1.0).
- Two hypotheses prevail
  - High LET radiation introduces large, complex DNA damage sites that are repair-resistant
  - High LET tracks through cytosolic water release high concentrations of ROS
- Application of conditioned medium from irradiated cells does not induce chromosomal aberrations in unirradiated cells.
- Polly Chang's mathematical model for high LET-induced cell transformation indicates a role for bystander effects.
- RBE_{tg} (for tumorigenesis) is greater than RBE for cell killing.

9.8. Problems

1. CGL1 hybrid cells are irradiated with 0 Gy, 3 Gy, 5 Gy, or 7 Gy of 250 kVp x-rays and plated for surviving fraction and neoplastic transformation frequency assay. The CGL1 plating efficiency or PE = 0.75 (75%). The surviving fractions (SFs) are 0 Gy =1.0, 3 Gy = 0.75 ± 0.07, 5 Gy = 0.50 ± 0.06, and 7 Gy = 0.15 ± 0.03. Plot the SF versus x-ray dose on a semilog plot. The data for the transformation frequency assay are as follows:

Radiation Dose	# of Cancer Foci	# of Viable Cells at Risk	Transformation Frequency
0 Gy	2	150,000	?
3 Gy	9	150,000	?
5 Gy	30	150,000	?
7 Gy	80	150,000	?

   Calculate the Transformation Frequencies at each dose and plot TF versus Radiation Dose on a semi log plot.

2. The metaphase chromosomal spreads from human-hamster hybrid GM10115 cells reproduced here (right) appeared in a publication: Nontargeted and Delayed Effects of Exposure to Ionizing Radiation: I. Radiation-Induced Genomic Instability and Bystander Effects In Vitro, authored by William F. Morgan, in *Radiation Research*, volume 159, pp. 417–431, 2003. During this study, cultured cells were exposed to 5 Gy x-ray radiation and metaphase chromosomal spreads were prepared from single cells. The human chromosome 4 was painted (light gray in the micrograph to

the right) and the hamster chromosomes counterstained (dark gray). The spread labeled "A" was prepared from an unirradiated cell.

(a) Describe the aberrations exhibited by chromosome 4 in panels B, C, D, E, and F.

(b) Cells yielding B, C, D, E, and F were selected from a single colony. What conclusions would you draw regarding the generation of these mutations?

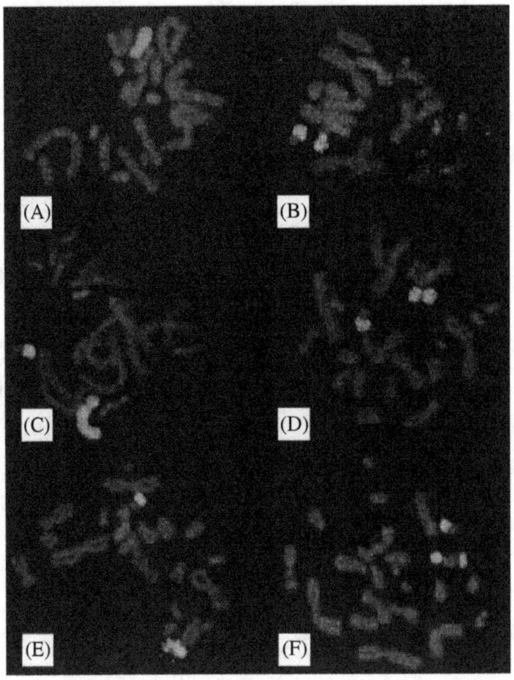

3. For each of the following mutation types listed, provide an example of one cancer that expresses this mutation, the mechanism of neoplastic transformation (e.g., amplification), and the preferred treatment protocol.

   Genetic mutations
   - o Disruption within the promotor region
   - o Nucleotide substitution or loss
   - o Chromosomal rearrangement
   - o Chromosomal deletion
   - o Chromosomal inversion
   - o Chromosomal translocation
   - o Gene deletion
     - Heritable epigenetic mutation
     - miRNA mutation
     - Viral infection

4. In Figure 9-17, the initiating (immediate) events leading to sporadic cancers are not specified. How might the oncogenic mutations and epigenetic mutations leading to collapse originate? Provide at least three possible scenarios.

5. The bar chart depicting the number of transformed foci per petri dish reproduced here (right) appeared in a publication: Influence of Noncarcinogenic Secondary Factors on Radiation Carcinogenesis, authored by John B. Little, in *Radiation Research*, volume 87, p. 240–250, 1981. In this experiment, sparsely plated cells were exposed to either 4 Gy or 6 Gy of x-ray radiation with or without antipain (AP) or 12-O-tetradecanoyl-phorbol-13-acetate (TPA) and allowed to subsequently form clonal colonies. The number of colonies exhibiting transformed cell behavior were tallied. The "control" and "AP" treatments were not irradiated but otherwise subjected to conditions identical to the exposed cells. TPA is known to strongly promote carcinogenesis; AP is a protease inhibitor. (a) What happens when TPA is added following radiation exposure? (b) What happens when AP is added following radiation exposure? (c) What effect does AP have on TPA-induced foci formation?

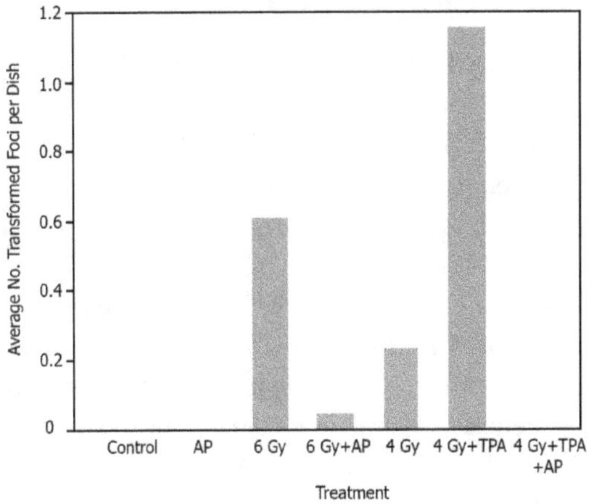

(d) As the researcher performing this experiment, what conclusions would you draw regarding transformation promotion by TPA?

6. ROS stimulate a metabolic adaptation to hypoxia activating anaerobic metabolism. Anaerobic metabolism increases low LET radiation resistance (OER), but this accommodation does not affect high LET radiation response. Explain why.
7. The bar chart depicting S1 mutation rates in human-hamster hybrid A_L cells reproduced here (right) appeared in a publication: Nontargeted and Delayed Effects of Exposure to Ionizing Radiation: I. Radiation-Induced Genomic Instability and Bystander Effects In Vitro, authored by William F. Morgan, in *Radiation Research,* volume 159, pp. 417–431, 2003. The treatment conditions were as follows: (1) 4 α-particle tracks through the cell nucleus, (2) 4 α-particles through the cell cytoplasm, (3) 8% dimethyl sulfoxide (DMSO) treatment for 10 minutes pre- and postcytoplasmic exposure, (4) 8% DMSO control (unirradiated cells), (5) treatment of cytoplasmic exposed cells with 10 mM of BSO for 18 hours, (6) BSO control (unirradiated cells). DMSO increases cell membrane permeability; buthionine-S-R-sulfoximine (BSO) reduces intracellular glutathione.

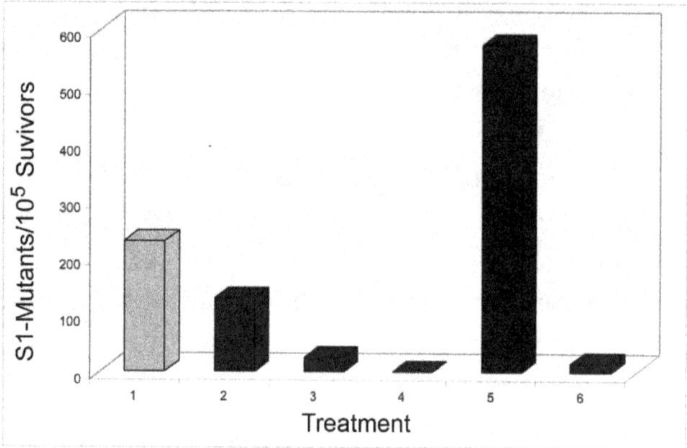

Explain these results — what does the mutation frequency tell you about the treatment? What would you conclude about the impact of ROS (free radicals) on S1-mutation?

9.9. Bibliography

Bailey, S. M. & Murnane, J. P., 2006. Telomeres, chromosome instability and cancer. *Nucleic Acids Research,* 34, pp. 2408–2417.

Bedford, J. S. & Dewey, W. C., 2002. Historical and current highlights in radiation biology: has anything important been learned by irradiating cells?. *Radiation Research,* 158, pp. 251–291.

Chang, P. Y. *et al.,* 2016. Harderian gland tumorigenesis: Low-dose and LET response. *Radiation Research,* 185, pp. 449–460.

Das, A. & Denekamp, J., 2000. Inducible repair and intrinsic radiosensitivity: a complex but predictable relationship?. *Radiation Research,* 153, p. 279–288.

Erfle, V. *et al.,* 1986. Activation and biological properties of endogenous retroviruses in radiation osteosarcomagenesis. *Leukemia Research,* 10(7), pp. 905–913.

Huang, L., Snyder, A. R. & Morgan, W. F., 2003. Radiation-induced genomic instability and its implications for radiation. *Oncogene,* 22, pp. 5848–5854.

ICGC/TCGA Pan-Cancer Analysis of Whole Genomes Consortium, 2020. Pan-cancer analysis of whole genomes. *Nature,* 578, pp. 82–93.

Kim, R., Emi, M. & Tanabe, K., 2007. Cancer immunoediting from immune surveillance to immune escape. *Immunology,* 121, pp. 1–14.

Limoli, C. L. *et al.,* 2003. Persistent oxidative stress in chromosomally unstable cells. *Cancer Research,* 63, pp. 3107–3111.

Martínez-Jiménez, F. *et al.,* 2020. A compendium of mutational cancer driver genes. *Nature Review Cancer,* 20, pp. 555–572.

Miousse, I. R., Kutanzi, K. R., & Koturbash, I., 2017. Effects of ionizing radiation on DNA methylation: from experimental biology to clinical applications. *International Journal of Radiation Biology,* 93(5), pp. 457–469.

Morgan, W. F., 2003. Is there a common mechanism underlying genomic instability, bystander effects and other nontargeted effects of exposure to ionizing radiation? *Oncogene,* 22, pp. 7094–7099.

Morgan, W. F. *et al.,* 1996. Genomic instability induced by ionizing radiation. *Radiation Research,* 146, pp. 247–258.

Mukherjee, S., 2010. *Emperor of All Maladies: A Biography of Cancer.* New York: Scribner.

Murnane, J. P., 1996. Role of induced genetic instability in the mutagenic effects of chemicals and radiation. *Mutation Research,* 367, pp. 11–23.

Muroya, Y. *et al.,* 2006. High-LET ion radiolysis of water: visualization of the formation and evolution of ion tracks and relevance to the radiation-induced bystander effect. *Radiation Research,* 165, pp. 485–491.

Negrini, S., Gorgoulis, V. G. & Halazonetis, T. D., 2010. Genomic instability — an evolving hallmark of cancer. *Nature Reviews Molecular and Cellular Biology,* 11, pp. 220–228.

Papetti, M. & Herman, I. M., 2002. Mechanisms of normal and tumor-derived angiogenesis. *American Journal of Physiology-Cell Physiology,* 282, pp. C947–C970.

Paquette, B. & Little, J. B., 1994. *In vivo* enhancement of genomic insta- bility in minisatellite sequences of mouse C3H/10T1/2 cells transformed *in vitro* by X-rays. *Cancer Research,* 54, pp. 3173–3178.

Pirkkanen, J. S., Boreham, D. R. & Mendonca, M. S., 2017. The CGL1 (HeLa x Normal Skin Fibroblast) Human hybrid cell line: a history of ionizing radiation-induced effects on neoplastic transformation and novel future directions in SNOLAB. *Radiation Research,* 188, pp. 512–524.

Toth, K. *et al.,* 2003. Radiation increases the activity of oncolytic adenovirus cancer gene therapy vectors that overexpress the ADP (E3-11.6K) protein. *Cancer Gene Therapy,* 10, pp. 193–200.

White, M. K., Pagano, J. S. & Khalilia, K., 2014. Viruses and human cancers: a long road of discovery of molecular paradigms. *Clinical Microbiology Reviews,* 27(3), pp. 463–481.

Chapter 10

Radiation Therapy

10.1. Introduction

Radiation biology and radiation biophysics acquire practical application in two fields: the delivery of radiation for medical purposes, and radiation protection. In this chapter, we shall examine the former.

Physicians began treating neoplastic tissues with radiation at the turn of the century,[1] starting in 1896. Isotope sources and early x-ray tubes emitted relatively low-energy x-rays — ^{60}Co providing the most penetrating radiation at approximately 1 MeV — resulting in the preferred use of radiation therapy for superficial benign and cancerous lesions. Skin was a favorite target. Treatment of deeper tumors, such as breast cancer, often lead to serious skin wounding, scarring, and deleterious late effects in non-targeted tissues, but nonetheless irradiation frequently provided the best contemporary treatment option. In the second decade of twentieth century, William Coolidge developed a cathode ray tube capable of emitting relatively high-energy x-rays, and so by the roaring twenties, more effective radiation delivery techniques were rapidly evolving. Nonetheless, another decade passed before modern clinical techniques could be practiced — first it was necessary to accurately measure the delivered dose at depth. The application of a dosimetry device used in radiation physics since the turn of the nineteenth century, the ionization chamber, to medical physics was pivotal. An ionization chamber measures uncharged radiation by capturing x-ray energy within a metal anode centrally housed in an evacuated tube. The anode releases ionized electrons that are accelerated to the cathode coating of the vacuum chamber and collected for counting. The charge collected by the cathode is proportional to the x-ray dose. Ionization chambers can measure radiation exposure in air or be placed at depth in water to measure dose. Through fastidious use of ionization chambers, it became possible to designate a treatment dose and compare one treatment to the next. So, the first half of the twentieth century saw enormous advances in external beam radiation delivery and brachytherapy — the use of isotopic radiation sources surgically placed in tumors.

The second half of the twentieth century saw the advancement of three fields that immediately impacted the clinical application of radiation: imaging, computer science, and subatomic physics. The accessibility of rapid computational algorithms enabled depth dose modeling on a timescale appropriate for patient treatment planning. Within a few hours, medical physicists and dosimetrists could now develop, model, and test several delivery schemas and optimize the treatment for conformality, avoiding organs at risk (OAR) and bathing the target with a uniform prescribed dose. High-resolution computational imaging techniques such as computed tomography (CT) and magnetic resonance imaging (MRI) support planning, visualization of dose deposition, and follow-up assessment of tumor shrinkage or eradication. In 1947, Robert Wilson hypothesized that subatomic charged particles could be used in lieu of x-rays to kill cells, and that the Bragg peak characteristics of these particles would provide superior dose deposition conformality. The physical and biological advantages of particles were soon embraced, enabling more effective control of large, anoxic tumors, more precise targeting of small tumors, and

[1] The phrase "turn of the century" or "fin de siecle," refers to 1890–1900, although the more recent rollover occurred in 2000. That is referred to as the new millennium.

increased relative biological effectiveness (RBE) for controlling radiation-resistant tumors. Currently, in the twenty-first century, radiation oncologists have a toolbox full of delivery techniques including brachytherapy, external x-ray beam therapy, and particle therapy.

We certainly will not attempt to cover all aspects of radiation oncology and medical physics in this text. Several excellent references are provided at the end of this chapter. We hope to be able to illuminate some of the more impactful radiation biology central to the hypotheses guiding effective radiation therapy.

10.2. The Therapeutic Ratio

In Chapter 8, we introduced the concept of NTCP — the normal tissue complication probability — resulting from functional subunit (FSU) inactivation in serial and parallel organs. As always, reality is a little more complicated. Sometimes, FSU is a mathematical construct that is difficult to define physically (as in the central nervous system) or it may not be confined to a structure and may be free to migrate (as are vascular endothelial and epidermal skin cells). Furthermore, some organs are neither strictly serial nor parallel, but rather are a combination of the two (Figure 10-1). The tolerance dose for a tissue can be defined by the most sensitive serial component; the percent composition of serial and parallel elements therefore results in a volume effect. For example, in the brain, tissues have specialized regions with discrete functions. Exposing additional glial cells results in more extensive demyelination and necrosis and exposing multiple volumes of high functioning brain structures results in an increasing number of cognitive and functional deficits. Irradiating volumes above tissue tolerance can result in severe failure. Lung failure occurs in stages, the greater the number of damaged FSU, the more seriously impaired a lung becomes. Thus, each organ expresses a dose sensitivity and a volume sensitivity. Obviously, parallel dominant organs demonstrate a greater volume effect than serial dominant organs. Both the exposure dose sensitivity and the exposure volume sensitivity determine the NTCP of irradiated tissues and organs.

Logically, tumors — which we described as organs composed of a single FSU — must also present a curve analogous to NTCP. The tumor control probability, TCP, can be determined from the experiments described in Chapter 8, Section 8.5.2. In principle, all tumors can be annihilated — locally

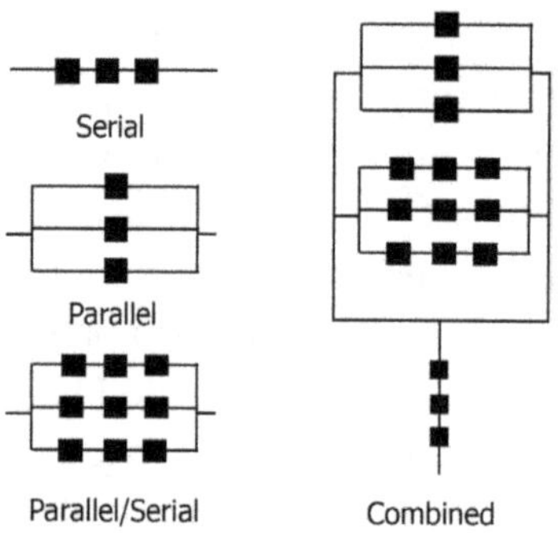

Figure 10-1 Organ tolerance to radiation exposure; serial and parallel behavior. Organs that respond in a serial manner, such as spinal cord, suffer morbidity if a single functional unit is damaged. Organs that respond in a parallel manner, such as lung, suffer progressive functional impairment as units are eliminated, but only experience failure if all FSUs are damaged. Parallel-serial organs, such as heart, lose function in multiple, local tissues if one unit is damaged, but organ failure requires a majority of tissue subsets be damaged through single or multiple FSU hits. Combination organs, such as the kidney, express functional loss if a serial FSU is damaged or if multiple parallel FSU are damaged, or the majority of serial-parallel FSU are damaged.

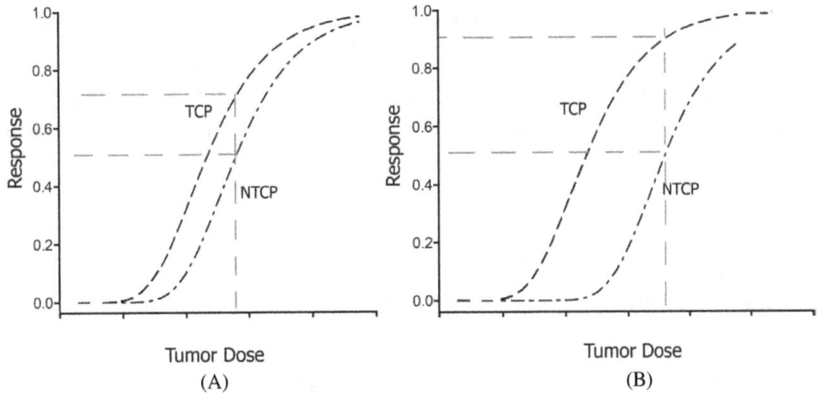

Figure 10-2 The therapeutic ratio: relationship between NTCP and TCP. The outcome probabilities for tumor control and normal tissue complications are plotted against dose for a single fictitious tumor using two different clinical delivery protocols. In (**A**) the protocol fails to protect the normal tissue, whereas in (**B**) a different protocol provides normal tissue protection.

controlled — if sufficient radiation dose can be delivered. However, recall that survival curves (e.g., Figure 7-10) encompass several logs of linear survival and thus the elimination of the last tumor cell is a mathematical probability. Tumors *in situ* present an even more challenging limitation dominating the determination of maximum treatment dose. The therapeutic ratio — the difference between the NTCP curve and the TCP curve for a given dose — determines the maximum dose that can be delivered to a given tumor *in situ*. The driving mandate of radiation therapy is to find ways to separate the two curves, and thereby increase the therapeutic ratio. Figure 10-2 illustrates the enhancement of therapeutic ratio, at the level of 50% negative impact to normal tissue, by reducing the response of normal tissue. In panel A, 70% of tumors can be controlled while in panel B, 90% of tumors are controlled. In practice, a 50% adverse complication rate is generally unacceptable, and oncologists strive for tissue damage probabilities well below 30%, optimally 5%.

Unlike the derivation of the expression for NTCP, which relied on the probability of inactivating some number of FSU, the derivation of an expression for TCP relies on the probability of cell death, because the entire tumor represents a single FSU and all cells of this FSU must be eliminated to prevent regeneration from a single cell. Also, because fractionated regimens dominate clinical delivery protocols, we will assume small fraction sizes of about 2 Gy (as we promised previously, we will explain the choice of 2 Gy later). Starting with Equation 7.18:

$$S = e^{-(\alpha nd + \beta nd^2)}, \tag{10.1}$$

and rearranging,

$$S = e^{-\left[\alpha nd + (\alpha nd)\left(\frac{\beta}{\alpha}d\right)\right]} = e^{\left[-\alpha D\left(1+\frac{\beta}{\alpha}d\right)\right]}. \tag{10.2}$$

As before, S is survival fraction, d is dose per fraction, n is the number of fractions, and D is total dose. Because the tumor is composed of a single FSU, every cell must be killed (or mitotically inactivated) to achieve tumor control. We impose the Poisson statistic for that case:

$$P(N,y) = \frac{e^{-N}N^y}{y!}, \tag{10.3}$$

that is, the probability of the occurrence of exactly y events given N, the mean number of events. Notice that if there is an *average* of one cell remaining ($N = 1$), the probability of *no* cells remaining is $1/e$ or 37%. This exemplifies the difficulty of killing the last cell of a tumor, as demonstrated by the survival curve. Our goal of killing all tumor cells (attaining tumor control) requires that we set y equal 0 events,

$$TCP = P(N, 0) = e^{-N}. \tag{10.4}$$

Because the survival fraction, $S = N/N_0$, where N is the number of surviving cells and N_0 is the original number of cells, we can substitute for N using Equation 10.2:

$$TCP = \exp\left\{-N_0 e^{-\alpha D\left(1+\frac{\beta}{\alpha}d\right)}\right\}. \tag{10.5}$$

Equation 10.5 is idealized. Corrections must be added to account for cell repopulation, uneven dose distribution, hypoxia, and tumor heterogeneity. Nonetheless, Equations 10.5 and 8.7 provide a good handle for the generation of TCP and NTCP curves. The steepness of the TCP curve depends on the underlying Poisson relationship, the tumor type, within a single tumor type all forms of variability (cell radiation sensitivity, tumor radiation sensitivity, repopulation kinetics, percent hypoxia, etc.), and inter-patient variation in dose delivery. The task of the radiation biologist is to understand the underlying mechanisms and mathematics. We shall leave it to the medical physicist to apply these principles to the precise, complicated clinical situation.

A comparison of the expressions for TCP (Equation 10.5) and NTCP (Equation 8.6; also see question 5 in Chapter 8) reveals that whereas TCP relies on cell killing in a straightforward manner, NTCP is determined by the probability of inactivating FSU (via cell killing) and the configuration of FSU (serial, parallel, serial-parallel, or combined) which determines the number of FSU that must be inactivated to impact function. To simplify things, we can generalize organ response by realizing that cell killing remains at the heart of NTCP (Equation 8.9). Recall that tissue sensitivity was described in Chapter 4 as dependent upon tissue cell cycle kinetics. Tissues with rapid cell turnover are more radiation sensitive, that is, responsive. Tissues with slow cell turnover, or a minimal number of cycling cells, suffer from latent effects (e.g., as modeled by Casarett) and damages that manifest only when cells are recruited from G_0. This damage happens much more slowly than acute cell killing — on the scale of weeks to months or even years. Therefore, we can describe tissue and tumor damage as "early effects" or "late effects." Tumors, which depend on constant cell cycling, suffer early effects. Brain, a tissue with few cycling cells, expresses late effects. Some normal tissues express early effects, for example, vascular, mucosal, skin, intestinal, and so on. Several tumors present late damage, such as prostate cancer and breast cancer. Nonetheless, in general, tumors are early responders and normal tissues are late responders.

The classification of tissues into early and late responders provides a useful simplification. Early responders have a wide range of larger α/β ratios. Late responders have a narrow range of smaller α/β ratios. Early responders have larger D_0, and are better fitted by the linear quadratic (LQ) model, whereas late responders have smaller D_0 and are often modeled well by MTSH. Figure 10-3 illustrates archetypal curves for the two tissue types. Early responders (A) produce dose response curves with a pronounced initial slope — evident in the gray extrapolation — and large α/β ratio — represented as a dashed gray line. The dose response curve for late responding tissue (B) exhibits a nearly flat initial slope approaching zero that results in a very small α/β ratio, again represented by gray lines. The differences between the linear (α) and quadratic (β) components of the two tissue types contribute to the separation between the TCP and NTCP curves, although these factors represent only the cell killing components and do not

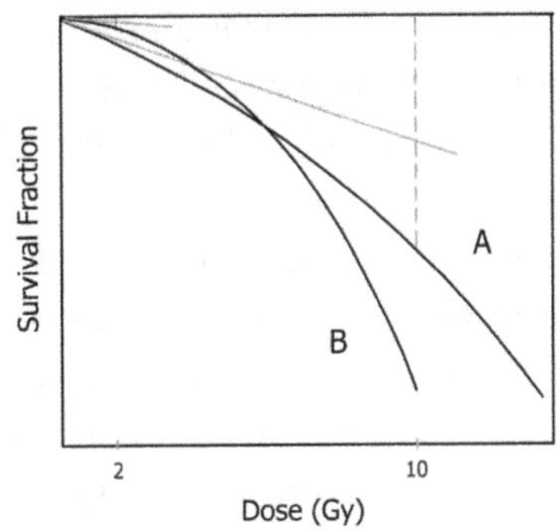

Figure 10-3 Early and late responding tissues. Early responding tissues (**A**) yield large α/β of approximately 10 Gy. They are generally fit by LQ modeling well. Late responding tissues (**B**) exhibit small α/β of approximately 2 Gy. They are generally fit well using MTSH modeling.

consider FSU inactivation or serial vs parallel behavior. Nonetheless, identification of early responding tissues and late responding tissues is relatively easy and, in most cases, so clear cut that this classification system has dominated clinical radiation biology.

The simplification is not without validity. Fertil and Malaise (1981) found good agreement between published, *in vitro* tumor cell line results and reported clinical results for the tumors from which they originated. The findings were verified and expanded by Deacon *et al.* (1984). In their study, tumor types were grouped into five curability categories ranging from most curable to least curable; they are shown in Table 10-1. In truth, the clinical tumor curability findings even within specific tumor types exhibit a broad range of survival after 2 Gy not apparent in the table, and individual tumors often violate the classification. The range of survival fractions seen in the second column reflect cell lines derived from different donor tumors of the designated type. The trend toward more resistant cells can be garnered

Table 10-1 Tumor curability categories; a comparison of *in vivo* to *in vitro* results.

Tumor	Cell Line S/S_0 (at 2 Gy)
Lymphoma	0.08–0.4
Melanoma	
Neuroblastoma	
Medulloblastoma	0.1–0.38
Small-cell lung cancer	
Breast	0.2–0.56
Bladder	
Cervical carcinoma	
Pancreatic cancer	0.12–0.87
Colorectal carcinoma	
Squamous cell lung cancer	
Melanoma	0.18–0.9
Osteosarcoma	
Glioblastoma	
Renal carcinoma	

from the higher end of each range, except for the second group, wherein the most sensitive cell line is significantly more resistant than the most sensitive cell line of the first group. As tumors are early responding tissues composed largely of cycling cells, the agreement between tumor curability and survival fraction seems logical.

Is there also good agreement between normal tissue effects and component cell kinetics, despite tissue organization influences? The correlation is less clear. Table 10-2 organizes several normal tissues in order of radiation sensitivity (column 2), and provides the total dose delivered in 5 fractions (TD) at which effects are readily detectable by both patient and physician. Most normal tissues exhibit both early and late effects. The table notes the earliest expressed effects (column 4). The organ sensitivity ranks roughly according to cell turnover (compare column 2 to column 3) advancing from days to years, with several exceptions. The most sensitive set experiences cell turnover on the scale of days, except for the ovaries and the lens of the eye, each of which are composed of nonmitotic cells. Radiation damages the ovaries by inducing apoptosis, that is, interphase death. Cells need not be dividing. This biological adaptation prevents genetically mutated fetuses. Although the lens is composed of nonmitotic fiber cells, these cells derive from a thin layer of slowly dividing epithelial cells that differentiate into lens cells. Unlike skin, lens cells are never removed; a cell damaged during differentiation fails to form a transparent fiber cell and becomes the nucleus of a cataract. This effect is not reflective of cell killing. It is sufficient to damage a small number of cells, making the lens of the eye radiation sensitive. The moderately sensitive set of organs exhibit cell turnover rates on the scale of months, with the exceptions of intestine and stomach. Exposure of these organs induces nausea and vomiting at relatively low doses, but physical damage to the organ requires the loss of a large number of cells resulting in peptic ulcers and villus denudation (villus loss requires that every crypt cell is killed to prevent recovery). The final set of normal tissues exhibits the range of cell turnover rates experienced by organs that are relatively resistant. Epithelial stem cells in skin and rectum migrate to offset areas of damaged cells. This physiology compensates for rapid cell turnover. In this set of tissues, radiation sensitivity depends on additional factors and does not track with cell cycle rates exclusively.

The far right column of Table 10-2 reports the approximate onset of untoward effects. This information can be used to determine whether a tissue is an early responder or a late responder. Notice that this is independent of cell radiation sensitivity (column 2 vs column 4). However, onset often correlates with cell turnover kinetics (column 3 compared with column 4), verifying that the acute effect results from

Table 10-2 Normal tissue radiation sensitivity, time to response, and cell kinetics.

Tissue	TD (Gy)	Average Cell Turnover	Approximate Time to Effect
Testes	1	60 days	1 week
Ovaries	2.5	Lifetime	days
Bone marrow	2.5	37 days	16 days
Eye lens	10	Lifetime	Months to years
Lung	17.5	1 month	2–6 months
Kidney	23	9 months	years
Liver	30	4.5 months	Few months
Intestine	40	9 days	4 days
Stomach	50	5.5 days	6 months
Skin	55	20 days	3 weeks
Rectum	60	3.5 days	days
Muscle	60	7.7 years	months
Bladder	65	2 months	months

cell death in those cases. Sterility results within a week or two of testicular radiation exposure — as soon as the radiation-resistant mature sperm have been removed from the pool. The apoptotic death of differentiating ovum evokes hormonal imbalance and sterility in a matter of days. Loss of blood cells becomes evident as soon as the pools of precursor cells empty. The eye lens, however, forms cataracts slowly because lens cells are nonmitotic; it is a late responding tissue. The other late responding tissues expressing in weeks to months (or even years) invoke Cassarett's mechanisms. Notice that skin is relatively late responding, but rectum, that expresses as bleeding, responds acutely. Similarly, intestine responds to crypt cell killing and is an early responder, but the stomach responds gradually to loss of acid neutralization by columnar cells (killed) and so is a late responding tissue. Because most tumors are early responding and most normal tissues are late responding, some exhibiting an early response as well, generalizations about these are useful. Nonetheless, NTCP — which considers both cell killing and FSU structure — is the gold standard for establishing specific organ response curves.

The task then is to find ways to discriminate between early responding tumors and late responding normal tissues, thereby separating the TCP curve from the NTCP curve (Figure 10-2), thus providing increased dose to the tumor without introducing increased normal tissue damage (morbidity or sequelae). Notice that biological techniques attempting to differentiate between tumors and adjacent early responding tissues — techniques like caffeine-induced radiation resistance — will not be so straightforward because of delivery constraints, and physical techniques will not be so straightforward since the response curves for tumor and early responding tissue would essentially sit right on top of one another. Also, late responding tumors present special challenges. We will begin, therefore, by describing some physical techniques available to medical physicists. These techniques emphasize dose conformality — the delivery of radiation exclusively to the tumor avoiding dose to normal tissue.

10.3. Dose-Limiting Techniques for Normal Tissue

10.3.1. *External Beam X-Ray Radiation Volume-Limiting Techniques*

At the turn of the century, collimation devices restricted external beam radiation to the target tissue. The technique protected the normal tissue surrounding the neoplasm reasonably well, and because the x-ray energy was low, minimal dose was delivered beyond the target. The dose distribution was far from ideal; dose falls off with depth, so the target was not uniformly irradiated. Additionally, the perpendicular field was limited to a set number of roughly circular aperture diameters inaccurately held in place (Figure 10-4).

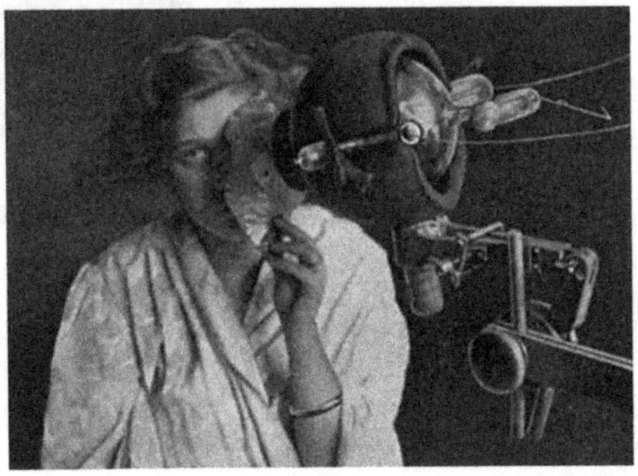

Figure 10-4 Emile Grubb treated superficial growths using orthovoltage, external beam radiation. The Crooke's tube is shielded to reduce stray emissions, and the patient holds a beam shaping, lead shield that allows x-rays to reach only the targeted tissue. Image reproduced through PD-US-expired.

As megavoltage linear accelerators came into use (Figure 2-22), deep seated tumors became accessible using higher energy external beam radiation. Collimators continued to be useful for cross-plane conformality, now manufactured one-off in a machine shop to match the lesion profile, but high-energy x-rays deposit energy in healthy tissue upstream of the tumor and also beyond the tumor — the so-called exit dose. Recall from Chapter 2 that x-rays deposit dose at or near the skin surface through Coulombic interactions, losing energy with each subsequent collision as photons scatter forward until the remaining energy is sufficiently low to result in a final photoelectric interaction. Most of the x-ray energy is deposited near the skin with dose deposition falling off with depth. Simple lateral collimation fails to protect healthy tissue in the plane of the beam. Figure 10-5A illustrates depth dose deposition in a homogeneous absorber, such as water. The horizontal strip represents the graph below. The dose builds up at the surface (skin) where the majority of photons travel a short distance before interacting with the atoms of the absorber.

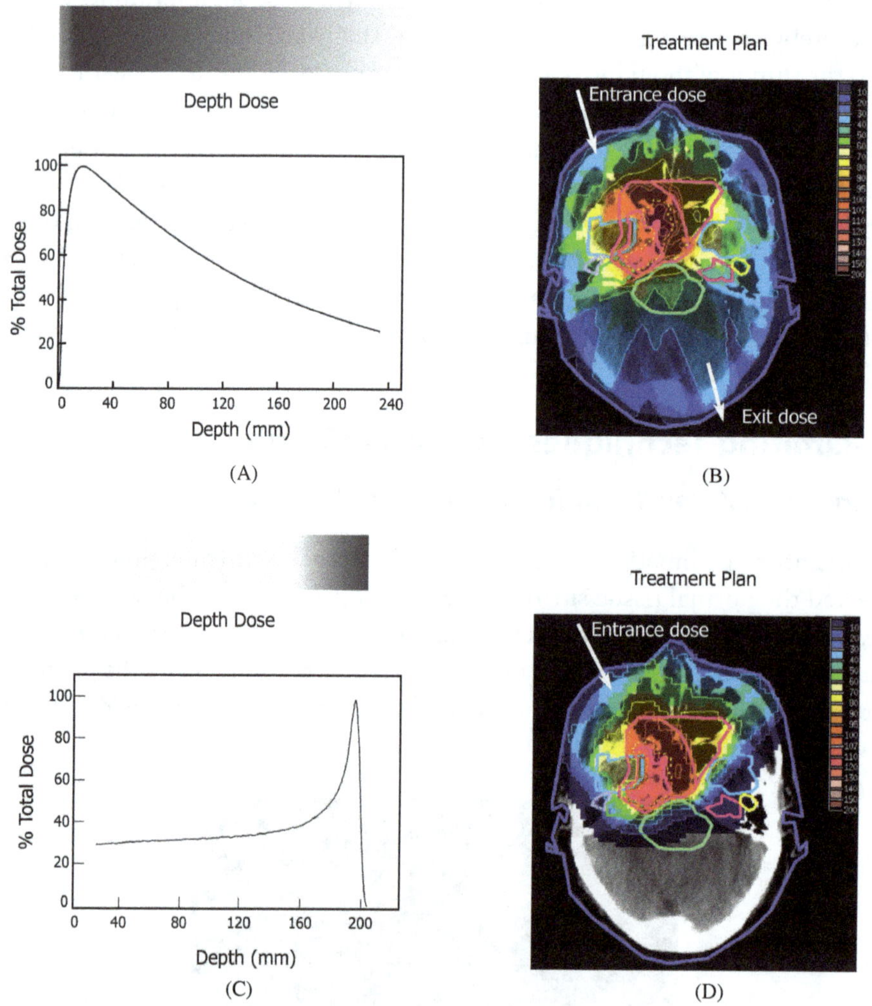

Figure 10-5 Protecting healthy tissue by physical means. (**A**) X-ray depth dose in a homogeneous absorber. The strip illustrates the deposition quantified in the plot below. (**B**) Conformal 3D plan showing the dose deposition within one slice of the CT stack. The dose scale is shown to the right. (**C**) Proton radiation depth dose in a homogeneous absorber. The strip illustrates the deposition quantified in the plot below. (**D**) Conformal 3D proton plan showing the dose deposition within one slice of the CT stack. The dose scale is shown to the right. Treatment plans are reprinted with permission as CC-BY.

The photons then experience Coulombic interactions and scatter forward, the energy being reduced following each subsequent interaction. Lower energy photons scatter at larger angles, reducing the local energy deposition. If a strip of film was placed in the beam, edge on in the plane of the beam, it would look similar to the gradient in Figure 10-5A.

One solution to this dilemma is a technique known as three-dimensional (3D) conformal therapy. This technique delivers beams from several angles, each beam contributing a fraction of the total prescribed radiation. The beams cross at the tumor, ideally bathing the tumor with an even dose. The total dose delivered to healthy tissue remains equal to single beam delivery, however, the dose is spread over a larger volume so that the dose to any given cubic centimeter of tissue is reduced. The holistic effect of total integrated dose vs partial local dose is a topic of contemporary radiation biology. Resulting morbidity appears to be tissue dependent, generally expresses as late effects, and may be undetectable in many cases. Figure 10-5B shows an example of a 3D conformal treatment plan. Briefly, the process involves acquiring a CT of the patient's anatomy including the tumor and using the information encrypted in the CT gray scale to predict therapy dose deposition. Let's look at the physics of the process. An x-ray radiograph (colloquially, an "x-ray") is produced by exposing a film (or equivalently an electronic digital panel) to low-energy x-rays — ideally a large field of uniform energy photons. The photons excite silver bromide crystals in the film emulsion to create foci that turn black when certain chemicals (developer) are introduced. If an object is placed between the x-ray source and the film, photons will not reach the crystals and the film will not react with the chemicals. Because the x-rays are low-energy, photoelectric interactions are favored in high Z materials. Therefore, if a person's hand is placed between the x-ray source and the film, bones will shield the film (light exposure) and tissue will pass scattered photons (gray exposure) and air will pass photons unobstructed to activate the crystals (black exposure). The image is a two-dimensional (2D) projection of everything in the person's hand. In the image, bones near the x-ray source lay on top of the bones near the film, in 2D. A CT machine rotates the x-ray source around the patient to produce a projection at every angle. The data collected, however, is not the projection. Instead, the volume of the scanned object is parsed into voxels and the energy absorbed by each voxel is computationally determined by back projecting along individual rays. We leave the details of the technique to some excellent cited references. The result is a 3D, virtual space mapping to the patient in which each voxel represents a gray value corresponding to the dose absorbed in that voxel. Using Beer's Law, the density of the tissue can be derived from the energy of the x-ray at that depth. Once the density is known, the dose absorbed by any energy x-ray (e.g., megavoltage therapy) can be determined. Treatment planning software uses the CT tissue density map to display absorbed dose for any external beam that the dosimetrist designs. Figure 10-5B shows a typical 3D conformal plan for a brain tumor. The volume of the head has been divided into many slices in a stack (at the console, the dosimetrist can scroll up and down to view each of the slices) viewed from the top of the head; the nose points upward in the image. Figure 10-5B shows one slice. The OAR (brain stem) has been encircled in green. The tumor is enclosed in a red outline, some of the tumor does not reside in this slice. In this plan, the beams enter anteriorly and laterally — from the face and temples. The proximal bones receive about 40%–60% of the total prescribed dose, and the tumor receives 100%–115%. Several exit dose tails can be seen at the back of the head. Remember that the beams may be oblique to the plane of the slice, so the short, centered tail likely penetrates downward. Modern conformal therapy can be enhanced by using both dose and dose rate to paint the tumor during intensity modulated radiation therapy (IMRT). Adjusting the dose rate to specific voxels, for example, hypoxic tissue, can increase the relative biological effectiveness (RBE).

10.3.2. *External Beam Charged Particle Radiation*

10.3.2.1. *Volume Limiting Techniques*

Electrons, protons, and carbon ions are commonly used for therapy, the frequency of application occurring in that order. Electrons find practical use for superficial treatments. They are accelerated in clinical megavoltage linacs, so electron therapy maximum energy is about 30 MeV. Also, electrons scatter excessively at high energies creating unacceptable penumbra; the practical limit is more like 16 MeV for treatments requiring conformality. Nonetheless, these three charged particles share two common advantages: they exhibit a finite range, and stopping power dictates the Bragg peak phenomenon. Figure 10-5C represents the depth dose characteristics of protons (~200 MeV). The majority of dose is delivered not at the entrance to the absorber, but at the end of range. Because the entrance dose is reduced about 60% compared with megavoltage x-rays at 15 mm depth (Figure 10-5A and C), fewer beams are required to achieve the same dose to the target at the same local dose to healthy tissue. Most significantly, the beams deposit no exit dose. These advantages can be seen clearly by comparing Figure 10-5D to Figure 10-5B. Most obvious is the absence of exit dose to the posterior brain. A closer examination discloses that much of the cyan-colored entrance dose (40%–60%) depicted in the x-ray plan converts to deep blue (20%) in the proton plan. Much of the yellow (70%) dose to healthy tissue is reduced to cyan (40%–60%).

10.3.2.2. *RBE and NTCP*

Charged particle therapy invokes additional radiobiological considerations impacting tumor-adjacent normal tissue. In Chapter 7 (Section 7.5.1), we examined linear energy transfer (LET) and we discussed the increased RBE (in Section 7.6) derived from more efficient cell killing achieved by high-LET particles (Figure 7-21). Recall that LET increases as energy decreases, reaching a maximum just short of the end of range (where $E = 0$), at the Bragg peak falloff. For protons, that maximum is approximately 100 keV/μm. For carbon ions, the maximum LET is about 200 keV/μm (Figure 7-24). However, the pristine Bragg peak shown in Figure 10-5C would not suffice to treat a tumor with a diameter greater than 1 cm. If applied to a 4.5-cm diameter tumor, 18 cm deep, the distal centimeter of the tumor would receive 60% more dose than the proximal portion. To compensate, the Bragg peak is "spread out" to cover the entire depth of the target (Figure 10-6). A spread out Bragg peak (SOBP) can be created by either of two methods. Originally, rotating "propellers" composed of tissue density, plastic wedges of various thicknesses were placed in the proton

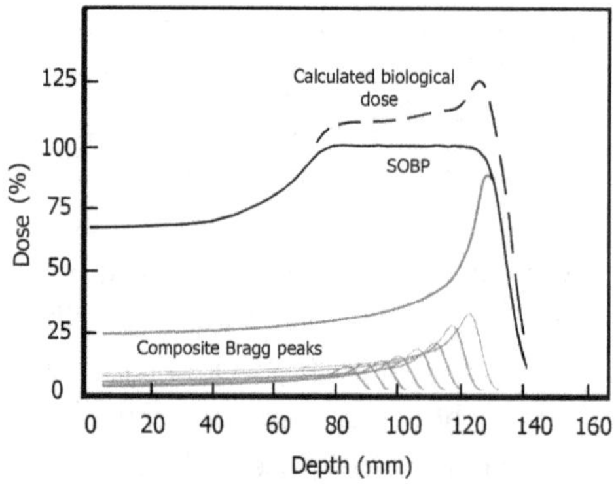

Figure 10-6 Proton radiation biological depth dose calculation. An SOBP (black, solid line) is created by reducing the entrance energy of the proton beam in steps. Several Bragg peaks, shown in gray, are delivered. The dose of each is determined by the fluence of protons at that energy. The total dose, the sum of the curves at each point, results in the SOBP. If the RBE is determined at each Bragg peak depth, the biological dose (dashed line) can be calculated.

beam. Spinning the propeller fanned the Bragg peak back and forth as more or less tissue depth was added in front of the patient. Modern machines use scanned beams that paint a raster pattern. These scanned beams paint each layer, like a card in a deck, by adjusting the incident energy of the particles. The shape of the scan matches the profile of the target at the depth of the slice (e.g., painting only the spade at the center of the ace of spades). Scanning improves conformality and extends the Bragg peak to cover the tumor volume. The initial dose plateaus are superimposed on one another as the layers are painted, and the Bragg peaks are superimposed on initial plateaus from subsequent layers. At each depth, an energy mix of protons is created from the plateaus and peaks (see Figure 7-16). Therefore, only the distal falloff of the SOBP is pristine, achieving the maximum RBE. When the RBE is calculated for the energy mix at the depth of each of the composite Bragg peaks, and the biological dose is determined as physical dose × RBE, the resulting dashed curve represents the biological dose to tumor — the x-ray irradiation dose that would result in the same tumor response. Biological dose is reported in centigray equivalent (CGE). For example, a proton radiation dose of 60 Gy would be equivalent to an x-ray dose of 66 Gy (60 Gy × 1.1), or 0.66 CGE. In this way, exposures of various radiations can be compared, and equivalent treatments can be prescribed. Say, for instance, a physician wants to treat a pediatric tumor with protons rather than x-rays to avoid exposing sensitive, developing tissues. She knows that 87 Gy x-rays delivered in 50 fractions will control that tumor in an adult. She can assume that 0.87 CGE, or 79 Gy (87 Gy/1.1) of proton radiation, will be equally effective. A universal RBE of 1.1 has been recommended for clinical use because a compilation of *in vivo* proton irradiation experimental data loads heavily toward an RBE of 1.1–1.2 (Paganetti & Giantsoud, 2018) inclusive of a variety of parameters. However, the full range of RBE reported values for protons is 1.1–1.7.

More than two and a half decades of proton therapy experience has indicated that employing an RBE of 1.1 has resulted in satisfactory outcomes for tumor control and, to a lesser degree, late effects. Nonetheless, evidence reveals an indisputable "'hot spot" at the distal portion of the proton SOBP indicating that caution must be taken when overlapping beams or approaching OAR. Paganetti warned that, "There is a clear need for prospective assessments of normal tissue reactions in proton irradiated patients and determinations of RBE values for several late responding tissues in laboratory animal systems, especially as a function of dose/fraction in the range of 1-4 Gy" (Paganetti *et al.*, 2002). Uncertainty is greater for carbon ions because the *in vivo* range RBE for carbon ions is 1.04–5.04 (Karger & Peschke, 2018). Delivering a dose five times as damaging as intended represents a much greater concern. For this reason, carbon therapy is always planned using biological dose (Figure 10-7). A robust plan is one that does not

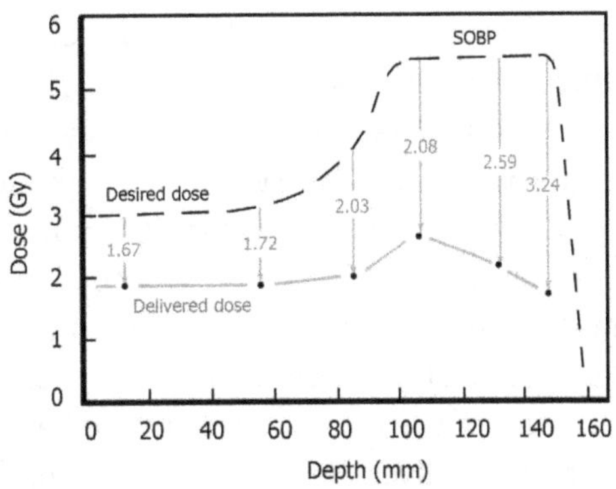

Figure 10-7 Carbon therapy biological planning. The desired depth dose for a carbon ion beam in homogeneous tissue is represented by the dashed line (- - -). Dividing the desired dose by the depth-dependent RBE (shown in gray) at each arrow provides the calculated physical dose to be delivered (•). The calculated points have been connected to indicate the shape of the physical depth dose curve.

Table 10-3 Uncertainties threatening high-LET treatment regimens prescribed using biological treatment plans.

Uncertainty	Effect	Mitigation	Disadvantage
Target motion	Marginal miss	Plan larger margins	Increased dose to healthy tissue
Large margins	Increased NTCP, transformation and cancer induction	Plan tighter margins	Marginal miss (decreased TCP)
Tumor type	Miscalculation of RBE	Biopsy, *in vitro* analysis	Delays therapy initiation, costly, low plating efficiency
Hypoxia	Miscalculation of RBE	Pretreatment oxygenation assessment	Transient hypoxia, costly

cause overdosing of OARs if the RBE is higher than expected, or underdosing of the target if RBE is lower than expected (Paganetti & Giantsoud, 2018).

Radiobiological treatment planning remains in its infancy for several reasons (see Table 10-3). In addition to the difficulty of delivering the planned dose precisely to the designated anatomy due to patient motion, organ motion, and setup inaccuracies, RBE value is dependent upon tissue. High LET radiation tends to result in higher RBE for late effect damage (low α/β) than early effect damage (higher α/β). In other words, the RBE is higher for normal tissues than for tumor tissue. RBE is also higher at low doses, as in fractionated dose delivery (see Figure 7-23A). RBE is increased for mutation and transformation compared with cell killing. Evidence suggests that a single high-LET particle traversing a cell nucleus is sufficient to induce transformation. Fractionation tends to increase the incidence of high-LET-radiation-induced cancers. Because tumors express unique survival curves reflecting various α/β, RBE is tumor type dependent. As we have seen previously, RBE increases with hypoxia, so it is tumor size dependent. If radiobiological treatment planning is employed, these uncertainties must be accounted for, making the process of radiobiological treatment planning labor-intensive.

10.3.3. *Brachytherapy: Volume-Limiting Techniques*

Thus far, we have insinuated the use of external beam radiation produced by machines from Crooke's tubes and low-energy radiography devices to megavoltage linear accelerators (Roentgen's legacy). However, radioisotopes (the Curies' legacy) can be used internally to great advantage, as first suggested by Alexander Graham Bell. Although these techniques are invasive and only about 5% of tumors are accessible for brachytherapy, the use of internal radioisotopes eliminates entrance and exit dose and minimizes the marginal miss that arises from tumor motion. Two types of brachytherapy dominate clinical practice: low dose rate brachytherapy (LDR; <2 Gy/h) and high dose rate brachytherapy (HDR; >12 Gy/h). Currently, these two techniques comprise approximately 70% of all brachytherapy treatments. The remaining brachytherapy treatments involve ^{90}Y microsphere implantation (Selective Internal Radiation Therapy; Y90-SIRT) and electronic brachytherapy. In SIRT, the radioisotope is bound to glass (20 μm–30 μm diameter) or resin (20 μm–60 μm diameter) microspheres and then injected into an artery leading to a vascularly enriched tumor — usually in liver but recently applied in brain. The microspheres aggregate and form an embolism in the capillaries of the tumor. ^{90}Y is a pure beta emitter with a 2.6-day half-life; the average tissue penetration is 2.5 mm in liver. Because the radiation biology of SIRT and permanent implants are similar, we shall examine the interstitial brachytherapy in detail and allow the reader to extrapolate to SIRT. Electronic brachytherapy is in fact a hybrid of external beam and isotope implantation. In this case, one or more applicators are placed in a cavity, generally created by surgical removal of a tumor, and a mobile linear accelerator is used to deliver electrons into the applicator(s). Escaping radiation sterilizes the tumor bed, reducing the probability of tumor recurrence in the margins. The radiation biology of electron therapy has been described previously in this text. So, our examination of brachytherapy will begin by describing the physical deployment of radioisotopes in the practice of brachytherapy (see Table 10-4) and then investigate the radiation biology principles guiding the practice.

During LDR, small radioactive sources, called seeds, are surgically implanted inside a tumor (interstitial brachytherapy) or they are contained within an applicator that is placed in a cavity next to the tumor (intracavity brachytherapy; common for gynecological tumors). The seeds are generally a few millimeters long and a few tenths of a millimeter in diameter. Each seed is a radioisotope source encapsulated in a submillimeter coating of titanium or platinum. The seeds can be regularly spaced within a wire or placed into pockets of a shaped, shielded mold (surface brachytherapy), or simply implanted individually to create a unique pattern. Seeds can be positioned for several days (usually 1–4) and then removed once the desired dose is achieved, or they can be left in place permanently in the case of isotopes with shorter half-lives. Fractionated LDR can be delivered using an afterloader following the same procedures used for HDR; that process is described below in the discussion of HDR techniques. The patient's tumor is located using several 3D imaging techniques (e.g., ultrasound, CT, or MRI), and a treatment plan is developed indicating optimal seed placement and implantation technique. Because the seeds are radioactive, special precautions must be taken to protect the patient and medical personnel in the room during implantation. Radiation protection is a serious topic that we shall discuss later in this book. Following implantation, a verification image is taken, usually either orthogonal radiographs (Figure 10-8 shows one image of a pair) or ultrasound, and, if necessary, the seeds can be repositioned. Seed placement is determined by the type of radiation emitted by the isotope, the radiation energy, and the density of the tissue. Assume that each seed is a point source. The photon dose deposition will be spherical with the highest photon dose at the center (seed location). Electrons deposit dose nearly uniformly throughout their range due to scatter (Figure 2-24). Ideally, the dose spheres overlap, establishing a relatively even dose without cold (underdosed) or hot (overdosed) spots.

HDR brachytherapy protocols insert radioactive seeds into a tumor temporarily for minutes to hours, depending on the activity of the sources. HDR is a popular therapy for cancers of the prostate, breast, esophagus, cervix, skin, bile duct, bronchus, brain, and head and neck. Two methods are common: intracavity applicators and remote afterloading. Applicators are designed to provide shielding, absorbing

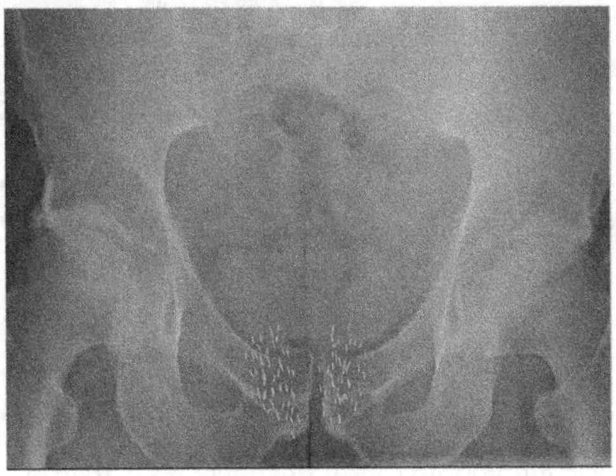

Figure 10-8 Radiograph of implanted seeds for LDR treatment of prostate cancer. The metal seeds are radiopaque. The volume of the prostate and location of the urethra can be visualized from the seed placement, however, the distance between seeds cannot be determined from this projection image. At least two images must be provided to determine seed placement in 3D. Image reproduced with the permission of James Heilman, MD, through CC BY-SA.

material, and cavity conformality such that when isotopes are placed in the designated locations, uniform dose is delivered to the cavity wall adjacent to the tumor. Applicators are common for esophageal, oral, cervical, vaginal, and other cancers residing in or neighboring anatomical cavities. An afterloader is a portable device that houses a lead "pig" (a hollow, cylindrical storage container for radioactive isotopes), motors, up to 40 flexible tubes that connect to large bore needles (also called applicators), a cable driving mechanism for each tube, and a timing mechanism. The procedure involves installation of the desired sources into the delivery housing at the base of the flexible tubing, in the pig. The sources are aligned in a "train"; the number of sources determined by the tumor size and the spacing between sources determined by the radiation type. Treatment needles are positioned in the patient's tumor as required by the treatment plan, and the delivery tubes are securely connected to the needles. Radioisotopes are then delivered from the pig into the needles remotely. The train sits on a flexible cable that pushes the sources through the tube and into the needle. The length of the cable extension from the afterloader to the needle determines the location of the train within the tumor, and it is predetermined by the treatment plan. The technician operates the afterloader from a shielded room adjacent to the treatment room. A preset timer withdraws the radioisotopes when the treatment is completed and returns them to the storage pig. The tubes are disconnected, the needles are removed, and the patient can return home. In general, the sources are stationary during treatment, but at least one manufacturer provides for a single source to oscillate within the needle delivering a cylindrical dose distribution rather than a series of spherical distributions. An HDR source placement plan for prostate cancer is illustrated in Figure 10-9. The rectum and bladder are considered OAR. Placement of sources near or within the urethra also should be avoided. This placement plan does not indicate a dose distribution as represented in Figure 10-5 (B and D), but the advantage of afterloaders is that the manufacturers provide treatment planning software that is linked to the software controlling timing and source train placement. Several provide "inverse planning" software. Inverse planning allows the physician to outline the region to receive ≥ 100% dose and designate the OAR; the software then calculates the source placement required to deliver that prescription. The energy of the emitted radiation predicts the depth dose. The type of radiation determines the depth dose profile (Figure 2-23).

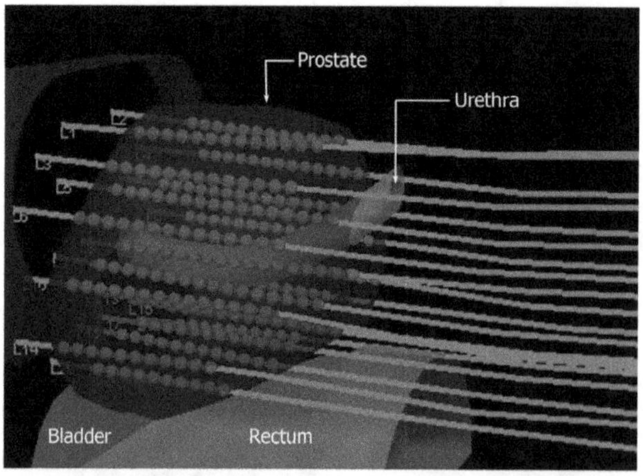

Figure 10-9 HDR source placement plan for prostate cancer treatment. Each train of sources (spherical beads) rest within a single needle. The drive cables can be seen extending from the needle tips to beyond the frame, back to the afterloader. Reproduced with permission of Springer from High Dose Rate Brachytherapy as Monotherapy for Localised Prostate Cancer: A Hypofractionated Two-Implant Approach in 351 Consecutive Patients. N. Tselis *et al.*, *Radiation Oncology*, 2013; 8:115.

10.4. Dose Protraction in Brachytherapy

Brachytherapy pursues the same goals as external beam therapy: minimize NTCP and maximize TCP. Physically, the dose to healthy tissue must be minimized, and the dose to tumor must be optimized, but optimizing the dose to tumor — creating a flat dose at 100% prescription — requires that seeds be placed close to the margins. That placement exposes adjacent healthy tissue to something very close to 100% prescribed dose. The adjacent healthy tissue can be protected (minimizing NTCP) by selecting radiobiologically favorable dose rates. The α/β for tumors is larger (~10) than for normal tissue (~3); therefore, normal tissue is more sensitive to dose rate. If given time, normal cells will stop cycling, repair SLD, and return to cell division. For example, in sarcomas of the head and neck, 80% of tumors are controlled at a rate of less than 0.5 Gy/h. At that rate, only 27% of patients experienced adjacent healthy tissue necrosis. Therefore, the first radiobiological consideration in brachytherapy is source selection. The physical attributes of each source are the size of the source, the type of radiation that the source emits, the energy of the emitted radiation, and the half-life of the isotope. The physical size of the source is determined by specific activity — the number of disintegrations per gram. The half-life determines the activity of the source at the time of implantation, calculated from the age of the source. Half-life also establishes how often sources need to be replaced to maintain prescribed dose rates. And half-life determines whether a source is useful as a permanent implant.

A quick review of radiation physics might be helpful, here (see Chapter 1, Table 1-1). As can be seen in Table 10-5, three radiations are popular: electron capture, beta, and gamma. During electron capture, an inner shell electron is absorbed by the positively charged nucleus. The negative charge converts a proton to a neutron, and the new element remains neutral. The vacancy in the K-shell is filled by a higher orbital electron and a gamma is emitted, the energy of which is the difference in binding energies between the K-shell and whichever shell the new K-shell electron previously inhabited. For this reason, gamma rays of several different energies are emitted from these isotopes. The table presents an average of all possible energies. During electron capture, the gamma energy may be passed to a higher orbital electron, ejecting it from the atom. Because electronic stopping power is high in metals, most emitted electrons, either Auger or beta, do not penetrate far beyond the encapsulating metal. An electron with an energy between 1 MeV and 1.5 MeV will penetrate tissue with a steep fall off at about 4 mm from the source. The first three isotopes in Table 10-5 may be thought of as gamma-emitting sources. ^{137}Cs decays to ^{137m}Ba 94.6% of the time, emitting a beta particle with a maximum energy of 0.512 MeV. The metastable barium immediately decays to a stable form by emitting a 0.662 MeV gamma ray. The remaining 5.4% of emissions are 1.174 MeV betas directly delivering the final stable barium product. ^{60}Co decay follows a similar pattern. A 0.31 MeV beta is released 99.88% of the time, followed by the release of a 1.1732 MeV gamma ray. The unstable nickel product releases another gamma of 1.3325 MeV to reach the

Table 10-4 Classifications of brachytherapy placement motifs.

Class	Description
Intracavity	Applicators containing radioactive sources are placed a cavity adjacent to a tumor
Interstitial	Individual seeds planted within a tumor surgically
Molded (surface)	Seeds are placed in a shield shaped to the surface of the target
Intralumenal	Sources are inserted into a tubular organ (e.g., blood vessel)
Intraoperative	Sources or applicators placed during surgical tumor removal
Intravascular	Vascular brachytherapy, restenosis prevention

Table 10-5 Common brachytherapy radioisotopes.

Isotope	Radiation Emission	Half-Life	Energy	Common Use
^{131}Cs	Electron capture	9.7 days	<30.4 keV>	LDR permanent implant
^{125}I	Electron capture	59.6 days	<31.3 keV>	LDR permanent implant
^{103}Pd	Electron capture	17 days	<21.0 keV>	LDR permanent implant
^{137}Cs	β-, γ	30.17 years	0.512 MeV; 0.662 MeV	Intracavity applicators
^{60}Co	β-, γ	5.26 years	0.31 MeV; 1.17, 1.33 MeV	Intracavity applicators
^{192}Ir	β-, γ	73.8 days	668.4 keV; <360 keV>	Afterloading (HDR and LDR)
^{106}Ru	β-, γ	1.02 years	105.9 keV, <0.38 MeV>	Uveal melanoma, LDR
^{198}Au	β-, γ	2.697 days	1.37 MeV; <138.5 keV>	LDR interstitial
^{226}Ra	α	1599 years	<4.7 MeV>	Historical use only

eventual stable product. The remaining 0.12% of disintegrations produce a 1.48 MeV beta leading to the release of the remaining 1.3325 MeV gammas. ^{192}Ir is a favorite isotope for HDR but also can be used for LDR when the activity drops to appropriate levels. The secondary use in fractionated LDR extends the isotope's useful lifetime, which is helpful because the half-life is short. ^{192}Ir decays 95.24% of the time through beta emission and 4.76% of the time by electron capture, resulting in 29 different gamma ray energies plus various characteristic gammas. The five emitted betas have a maximum energy of 668.4 keV, and the gammas range from 206 keV to 1.378 MeV. ^{192}Ru decays to ^{192}Pt, releasing beta particles 95.13% of the time. The remaining 4.87% of the time, ^{192}Ru decays through electron capture to ^{192}Os, releasing gamma rays with an average energy of 0.38 MeV. ^{198}Au undergoes 1.37 MeV beta decay. Three gammas are emitted: 1087 keV, 676 keV, and 411 keV. The second set of isotopes in Table 10-5 can be considered gamma-emitting sources, except for gold, for which the beta emission should also be considered. Radium sources were used from the beginning of internal radiation until around three decades ago when they fell into disfavor because of fears that radiation could leak from the source. ^{226}Ra decays to ^{219}Rn (5.78 MeV), which decays to ^{215}Po (6.88 MeV), and then to ^{211}Pb (7.53 MeV), and then to ^{207}Ti (6.62 MeV) and ^{201}Pb (7.45 MeV), all emitting alpha particles (see Figure 1-1). These alphas are massive and charged so they lose energy over extremely short distances.

Figure 10-10 shows the inverse square dose rate fall off produced by gamma photons emitted from a seed in a homogeneous medium. The inset of Figure 10-10 illustrates the physics behind this behavior. The fluence of photons at the inner ring is greater than the fluence at the outer ring because surface area expands with increasing radius. Fluence implies dose rate, and because the time of exposure is fixed, fluence indicates dose. Dose, dose rate, and depth dose in tissue are all described by Figure 10-10.

Now, we know that tumor control benefits from HDR. Figure 8-13 demonstrates that acute doses control mouse mammary tumors better than protracted doses (in 10 fractions). Dose protraction favors normal tissues that benefit from active cell cycle checkpoints, as demonstrated by Jack Fowler's investigation of pig skin irradiation (Table 8-1). Normal tissue sparing is also a major benefit of brachytherapy isotope treatment since the fluence or dose rate from the radioactive source falls off by $1/r^2$. Therefore, for seeds residing at the interface of tumor and healthy tissue, the dose rate drops to minimal in a matter of millimeters. The mandate here, then, is to select an isotope that provides the appropriate dose rate to the tumor but does not exceed NTCP guidelines a few millimeters away. Dose rate at the source is a consequence of source activity. To determine the activity at the time of the procedure, recall Equation 1.16:

$$A(t) = A(0)e^{-\lambda t}, \tag{10.6}$$

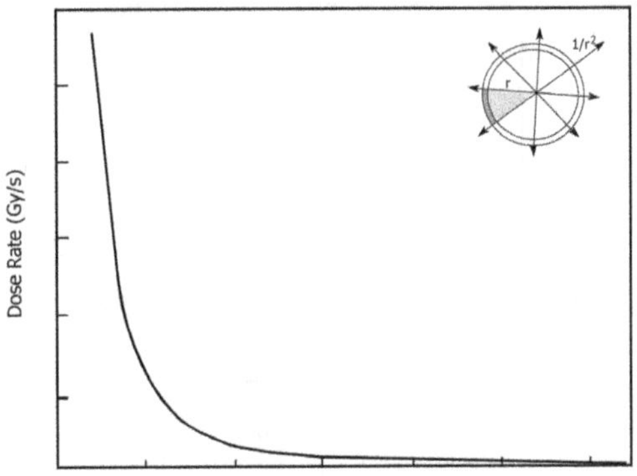

Figure 10-10 The inverse square dose rate falloff from a generic isotope emitting photons. The inset demonstrates the $1/r^2$ reduction of photons per surface area as measured at a given distance from the seed. Isotopes are isotropic sources, emitting uniform numbers of particles in all directions. The graph shows the $1/r^2$ falloff in dose rate per distance from the source.

where $A(0)$ is the activity measured prior to shipping, and t is the time passed between that measurement and implantation. If the half-life is short, such as in the case of ^{192}Ir, the source will arrive at the clinic with diminished activity. Typical ^{192}Ir seed activities range from 20 Ci to 4 Ci. The dose rate for gammas can be approximated from this activity as,

$$\dot{D} = (1.5 \times 10^{-13}) AE\left(\frac{n}{r^2}\right) \tag{10.7}$$

where $\dot{D}$ is in Gy/h, A is the activity of the isotope in disintegrations per second at time t, E is the energy of the emission in MeV, n is the number of gammas emitted at that energy, and r is the distance from the source in meters. Best clinical practice requires that source strength be measured on a regular basis and dose be empirically determined. Dose rates are additive for the radiations emitted by each isotope implanted in geometrical proximity. In other words, total dose rate at a given point depends not only on the source activity but also on the number of sources and their arrangement.

It is important to be aware of the dose rate sensitivity of both tumor and adjacent normal tissue because brachytherapy dose rates lie within the range of steep biological response effects (see Figure. 7-11). The prescribed dose should consider both dose rate effects and radiation type effects. For example, to achieve tumor control using interstitial implants, a large total dose of 160 Gy is often prescribed to compensate for dose protraction effects. However, if the medical physicist recommends ^{125}I for the treatment, the RBE for ^{125}I is 1.5, and thus the actual effective dose is 1.5 × 107 Gy = 160.5 Gy. As noted previously, even large doses can be rendered ineffective if fast cycling cells compensate through repopulation. Consequently, LDR may not be effective in fast growing tumors. Remember also that tissues may be modeled as parallel or serial or some combination of the two. Therefore, irradiated volume may be as important as dose rate for the protection of normal tissue. Because there are many interrelated factors, biological equivalents must be empirically derived. So called "isoeffect" curves for various isotopes, tumors, and late effects have been derived from clinical data and they agree reasonably well with *in vivo* experimental data (e.g., skin burns or hind limb paralysis). Frank Ellis (1968) used clinical data to derive equivalent doses and dose rates for interstitial therapy. The plot in Figure 10-11 shows that similar late effects develop as dose protraction increases if the total dose increases. In other words, suppose you have a source of ^{192}Ir that will have a dose rate of 0.357 Gy/h at the time of implantation. To achieve 60 Gy

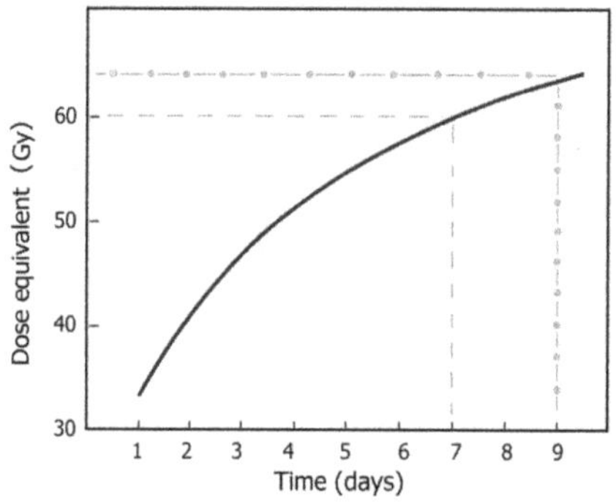

Figure 10-11 Biologically effective dose and dose rate for late effects during brachytherapy. The curve follows isoeffect for normal tissue. For example, if the standard of care is 60 Gy delivered over 7 days, 68 Gy over 9 days would produce indistinguishable outcomes in the exposed healthy tissue. Data from Ellis (1968).

total dose, the implant will remain in place for 1 week. Then, as things will happen, the patient informs the clinic that she or he will be unavailable on Tuesday but can come in to have the implant removed on Thursday. Can the physician expect to see increased NTCP? The answer is no, provided the additional dose is equal to or less than 8 Gy.

In addition to radiation biology, anatomical and physiological parameters need be taken into consideration. For example, cervical carcinoma has traditionally been treated using LDR intracavity brachytherapy. Recently, many clinics have begun using several fractions of HDR intracavity delivery instead. It turns out that the benefit of protracted dose to the healthy tissue is not critical in this case. The dose-limiting organs — bladder and rectum — receive reduced dose and dose rate because of anatomical distance; manual organ displacement can further reduce exposure. Although brachytherapy has been used to control neoplastic growths for over a century, it remains more art than a science. There is considerable need for protocol standardization and radiobiological investigation, modeling, and clinical guidance. Brachytherapy is a terrific exercise in radiation biology.

10.5. Fractionation and Biological Equivalence

We saw that dose protraction during brachytherapy facilitates separation of NTCP from TCP. Obviously, HDR radiation can be fractionated to produce dose protraction. Because the dose rate delivered in a single fraction has biological effects dependent upon cellular α/β, but the overall fractionation scheme has biological effects dependent upon α/β, cell turnover kinetics, and organ (or tumor) physiology, the radiation biology of fractionation demands a bit of scrutiny. Fractionation provides therapeutic advantage in many cases, but not all.

What do we know about fractionation? Fractionation allows tumors to reoxygenate (Chapter 8, Section 8.5.3). Fractionation allows healthy cells to repair (Figure 7-4, Tables 8-2 and 8-3), whereas cancer cells have defective cell cycle checkpoints and reduced repair fidelity. And . . . fractionation suppresses β (Figure 7-12), such that Equation 10.2 can be approximated as

$$S = e^{-\alpha D}. \tag{10.8}$$

The effective survival curve is almost entirely dependent upon α. This expression holds true for small fraction radiotherapy when α/β is very large, and for LDR brachytherapy where β approximates 0. This is all well and good, but ultimately clinicians require a method for determining equivalent fractionation protocols, ones that will provide the same (or greater) tumor control and the same (or reduced) late tissue damage.

In Chapter 7, we introduced the concept of RBE and pointed out that different types of radiation produce different ratios of direct DNA damage to indirect DNA damage. The percentage of indirect damage determines the requirement for oxygen fixation in the conversion of transient damage to fixed damage, making low LET radiations less effective against hypoxic cells. In RBE, we examine biological effects and establish equivalencies; a certain dose of radiation A produces the same effect as a different dose of radiation B. In our discussion of brachytherapy, we introduced a new measure of biological equivalence: the equivalent dose (EQD). Clearly, when dose is protracted, the total dose must be increased to achieve the same degree of cell killing as the acute dose (Figures 8-13 and 10-11). In these equivalencies, radiation biologists focus on the target; "I did *something*, and the result is *this*." Then, "When I do this *other thing* . . . the result is *the same*." In these cases, the action is variable; the result remains constant. The biologically effective dose (BED) is the generalization of cell survival (the biological effect) along the minimal dose per fraction limit; in other words, *this* protocol produces *that* biological effect.

10.5.1. *Isodose, BED, and EQD*

Physicists are always happy to discover that some phenomenon follows a power law relationship. Ernst Witte made this happy discovery in 1939 when he realized that the appearance of skin erythema compared with dose rate could be described using a power law relationship. This led Magnus Strandqvist to state the power law for skin reddening in terms of overall time:

$$D = kT^{1-p}, \tag{10.9}$$

where D is dose, k is a constant, T is total time, and p is the power (which is < 1). Strandqvist found from examining 280 carcinomas of the skin and assuming that a single fraction was delivered in 0.35 days, that the exponent $(1-p)$ was approximately 0.22. Figure 10-12 shows a plot of the Strandqvist relationship for skin erythema. According to this relationship, if an acute dose of 10 Gy produces faint erythema, 20 Gy delivered over 12 days will also produce erythema, as will 30 Gy over 60 days.

The popular but simplistic Strandqvist relationship ran into some difficulties. When he examined cancer recurrence using Strandqvist's assumptions, Bernie Cohen found squamous cell carcinoma also yielded an exponent of 0.22, but he derived a different exponent for normal skin (0.33; when he assumed

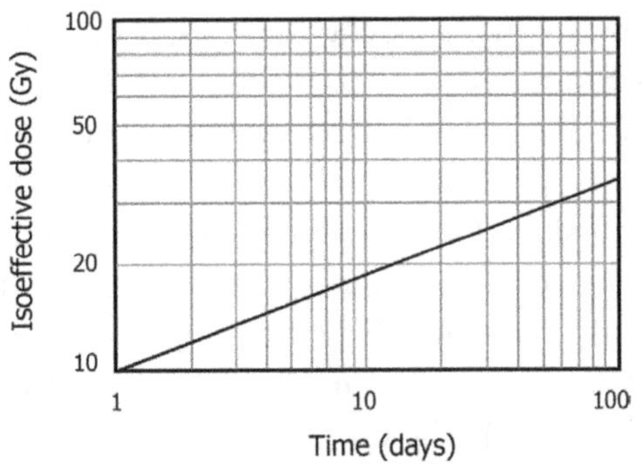

Figure 10-12 Strandqvist power law relationship for equivalent dose. Using the exponent of 0.22 for the initial appearance of skin erythema, the relationship between isodose and total dose can be derived for any treatment delivered over a period of 100 days or less.

the first fraction lasted 1 day). Also, in Equation 10.9, Strandqvist assumed that the primary parameter is overall time, that the number of fractions and the distribution of fractions has insignificant effect. But we know from Jack Fowler's work that overall time is relatively unimportant, whereas the number and size of individual fractions determine isoeffect (Table 8-2). So, Frank Ellis proposed a new power law formulation based on both the number of fractions and the overall time,

$$D = (\mathrm{NSD})\, T^{0.11}\, N^{0.24}, \tag{10.10}$$

where D is dose, and T is total time, and N is the number of fractions. Ellis named the constant: the nominal standard dose (NSD), requiring empirical determination. Ellis determined the two exponents of Equation 10.10 by adjusting Strandqvist's exponent of 0.22 to 0.24 to compensate for the delivery of five fractions per week (clinical practice) and by taking the difference between Strandqvist's exponent and the one derived by Cohen (0.33 – 0.22 = 0.11). Although the details were disputed, the Ellis formula found widespread clinical application. NSD lookup tables for specific organs and tumor types provide an advantage and ease of use, and each clinic can derive NSD for their specific equipment and methods. The NSD formula was based on skin response, which is an early response, and therefore, the formula fails in general for late responding tissues. For example, Ellis' formula underestimates both acute effects and late effects in lung following breast irradiation. Furthermore, the NSD technique could account only for tissue tolerance, not partial tolerances. So, Ellis produced a series of time, dose, and fractionation factors (TDF) independent of NSD that can be added together to develop a treatment that reaches normal tissue tolerance. These factors also are available in lookup tables. Nonetheless, power laws cannot account for biological processes. Power laws imply a more rapid recovery resulting from repair early in the fractionation scheme whereas, in fact, initial recovery is slow. The laws also fail to account for genomic repair variability. And, amazingly, the power laws assume that tumors do not proliferate. Finally, repopulation tends to lag initially (for human skin, the lag is about 4 weeks), then escalates rapidly, and then subsides (see Figure 8-11). Consequently, radiation biologists began to seek models that explained both the biological underpinnings and physiological behavior at higher scales.

Let's review the basic tenets of the survival curve as it pertains to fractionation. As we have interpreted it, the initial slope indicates nonreparable damage induced by high LET radiation (including the terminal segment of an electron track): complex, single event, multiple damages. According to LQ formalism, this part of the curve is characterized by α. The quadratic slope, driven by β, indicates dual event, sublethal damage (SLD) that is reparable. Potentially lethal damage (PLD) presents complicated and somewhat unpredictable impacts on fractionated repair. In general, PLD repair decreases the slope (increases D_0), requiring increased dose to maintain isoeffect, but this type of repair exhibits a threshold beyond which changes in the slope are independent of dose. For this reason, we shall dispense with PLD for now. A close examination of Figure 7-12 discloses some interesting characteristics. There are limits to biologically effective fractionation. Figure 10-13 reexamines Figure 7-12. Clearly, an acute dose defines the lower limit of fractionated doses: one dose equal to the total prescribed dose. The quadratic slope of the acute dose defines the limiting slope for large fractions (a small number of fractions each at a high dose). If additional fractions are added to the regimen, the surviving population at the commencement of each fraction can be determined from Figure 7-12 at intervals of the selected dose per fraction. A line connecting 100% survival at 0 Gy dose to the surviving fraction at the selected dose per fraction in Figure 10-13 defines the effective survival relationship for the fractionated protocol. A linear extrapolation can be thus determined for any dose per fraction between the total prescribed dose and some diminishingly small dose. In the limit, the extrapolated slope for fractionated delivery is determined by the acute dose survival curve initial slope. So, these define the limits of fractionation: the single dose quadratic slope and the initial slope — that is, the limits reflect β and α.

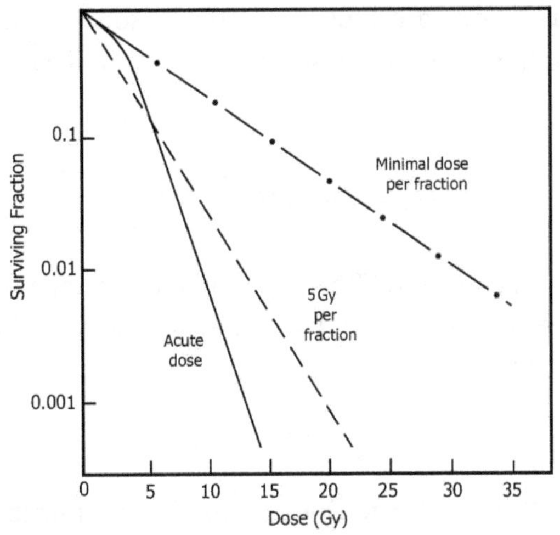

Figure 10-13 Limits on fractionation. The single fraction (acute) dose response curve is represented as a solid line. The quadratic slope of this curve defines the lower limit for the number of fractions that can be delivered (one). The dashed line (– –) indicates the effective survival for a fractionation protocol of 5 Gy per fraction. Curves for smaller doses per fraction can be found by "pinning" the dashed line at 0 Gy, 100% survival, and swinging the terminal end upward like the hand of a clock (the point of intersection with the acute dose curve will be found at decreasing doses). At the extreme, the dashed line will superimpose on the extrapolation from the initial slope of the acute dose curve (– • –) when the dose per fraction becomes diminishingly small.

The implication of Figure 10-13 indicates a fractionation model based on LQ formalism. Starting with our survival equation for a fractionated dose (Equation 7.18):

$$S = e^{-(\alpha nd + \beta nd^2)}, \tag{10.11}$$

we can derive the total biological effect for n fractions:

$$S^n = \left[e^{-(\alpha d + \beta d^2)}\right]^n, \tag{10.12}$$

take the natural log of both sides and simplify to get:

$$n \ln S = -\alpha D - \beta dD. \tag{10.13}$$

As a reminder, S = survival fraction, n = number of fractions, d = dose per fraction, and D = total dose. The effective survival during fractionation, the dashed line in Figure 10-13, can be represented as,

$$\ln S = -\alpha D - \beta dD = -(\alpha + \beta d)D = -\alpha_{eff} D, \tag{10.14}$$

where α_{eff} determines the effective slope, the dashed line of Figure 10-13.

Jack Fowler and B. G. Douglas observed these relationships using Fowler's mouse skin system and determined that if equal amounts of repair occur after each identical dose (in a specific tissue), and if the biological outcome depends on the surviving fraction of clonogenic cells (tricky for late effect outcomes), then every fraction will have the same biological effect. Mathematically,

$$(\alpha nd + \beta nd^2) = \text{constant}. \tag{10.15}$$

In other words, ln S represents a constant biological effect (isoeffect) under the specified conditions (e.g., cell death). Therefore, a simple function exists that delivers ln S when the function is multiplied by dose:

$$\ln S = (\mathit{function})D = \left(\frac{F_e}{a}\right)nd. \tag{10.16}$$

F_e is the Fowler function and α is a constant. When repopulation is not a factor, F_e is the reciprocal of the total dose ($1/nd$) that produces an isoeffect. F_e plotted as a function of dose per fraction results in a straight line, again mathematically,

$$F_e = b + cd. \tag{10.17}$$

Substitution for F_e in Equation 10.16 yields the Douglas and Fowler formulation for isoeffect:

$$\ln S = n\left(\frac{b}{a}d + \frac{c}{a}d^2\right). \tag{10.18}$$

This should look mighty familiar. Recall the linear form of the LQ equation (Equation 6.26),

$$\ln S = -(\alpha D + \beta D^2) = -n(\alpha d + \beta d^2). \tag{10.19}$$

If the isoeffect is related to S (if the effect results from clonal stem cell depletion), then (−ln S) can be designated the biological effect (E). Rewriting Equation 10.19 and dividing by nd results in a linear relationship equivalent to Equation 10.17:

$$E = n(\alpha d + \beta d^2) = nd(\alpha + \beta d). \tag{10.20}$$

Dividing by E yields

$$1 = \frac{nd(\alpha + \beta d)}{E}, \tag{10.21}$$

and dividing by nd yields

$$\frac{1}{nd} = \frac{\alpha}{E} + \frac{\beta}{E}d. \tag{10.22}$$

As before, we can plot the reciprocal of total dose for isoeffect ($1/nd$) against the dose per fraction (d) to achieve a linear plot with the y-intercept yielding α/E, and the slope providing β/E. Although it is not possible to derive α and β, the ratio of $(\alpha/E)/(\beta/E)$ provides α/β — the determinant of radiation damage repair capacity, according to the LQ model (see Figure 8-6). Equation 10.22 provides a good fit for data from many normal tissues, indicating that the fundamental basis for late effects in these tissues is clonogenic cell death (e.g., crypt cells). Knowing the repair capacity of normal tissues provides information regarding the potential benefits of fractionation. The range of normal tissue α/β values extends from 1.5 to 7, with a mode of about 3 to 5. Tumors, on the other hand, exhibit α/β values from 5 to 30, with a mode of around 10–20. Figure 10-13 indicates a limit on effective dose per fraction determined by the initial slope of the acute survival curve. This extrapolated line also defines α/β, as illustrated in Figure 6-8A. Rodney Withers named the limiting dose per fraction the "flexure" dose, the dose beyond which smaller doses per fraction yield no additional protection. Withers considered Equation 10.12 and derived a simple formula to determine the new total dose that would be required to achieve isoeffect for change in dose per fraction:

$$\frac{D}{D_{ref}} = \frac{(\alpha/\beta) + d_{ref}}{(\alpha/\beta) + d}, \tag{10.23}$$

where D is the new total dose, D_{ref} is the original total dose, d is the new dose per fraction, and d_{ref} is the original dose per fraction. Thus, if a physician prescribed a total dose of 60 Gy in 2 Gy fractions but discovered that the patient had to leave town prior to the last three treatments, the physician could calculate a new total dose required for isoeffect given 27 fractions as opposed to the original 30 fractions.

Suppose that E/α is defined as BED — the total dose required for a given isoeffect under the condition of infinite fractions and complete recovery between fractions (i.e., the upper limit in Figure 10-13) — then,

$$\text{BED} = \text{dose} \times (\text{relative effectiveness}). \tag{10.24}$$

Thus, BED is not unlike RBE, or risk (Table 7-2). Rearranging Equation 10.20 to reflect Equation 10.24,

$$\text{BED} = \frac{E}{\alpha} = (nd) \times \left(1 + \frac{d}{\alpha/\beta}\right). \tag{10.25}$$

Although BED is useful, EQD, or normalized total dose (NTD), provides a direct comparison between protocols by comparing the relative effectiveness of each. For example, to compare a standard 2 Gy/fraction treatment of tumor (assuming $\alpha/\beta = 10$),

$$\frac{\left(1 + \frac{d}{\alpha/\beta}\right)}{\left(1 + \frac{2}{\alpha/\beta}\right)} = \frac{\left(1 + \frac{d}{\alpha/\beta}\right)}{\left(1 + \frac{2}{10}\right)} = \frac{\left(1 + \frac{d}{\alpha/\beta}\right)}{(1.20)} \tag{10.26}$$

and for late responding tissue (assuming $\alpha/\beta = 3$),

$$\frac{\left(1 + \frac{d}{\alpha/\beta}\right)}{\left(1 + \frac{2}{\alpha/\beta}\right)} = \frac{\left(1 + \frac{d}{\alpha/\beta}\right)}{\left(1 + \frac{2}{3}\right)} = \frac{\left(1 + \frac{d}{\alpha/\beta}\right)}{(1.67)}. \tag{10.27}$$

Unfortunately, Equation 10.25 fails to resolve effective dose for early responding tissues and fast growing tumors because it does not include a consideration of overall time. Repopulation is time dependent. This shortcoming becomes apparent when the equation is applied to hyperfractionation schemes, protocols that call for two or more fractions to be delivered in a single day. Assuming that cell proliferation remains constant (which it does not), the number of cellular clones at any time is an exponential relationship identical to that of radioactive decay (Equation 1.5), where λ is a constant representing doubling time rather than the probability of decay. Thus,

$$\lambda = \frac{0.693}{T_{pot}}, \tag{10.28}$$

where T_{pot} is the potential doubling time of the cells composing the tissue of interest. The biological effectiveness (e.g., cell killing) is reduced by repopulation. If cell division is constant (it is not), repopulation can be represented by the time available for division, t, multiplied by the cell doubling time, λ. Remember that cell repopulation experiences a lag following the initiation of fractionated radiation delivery, so t is not necessarily equal to the overall time since the initiation. Equation 10.25 can be modified accordingly:

$$\text{BED} = \frac{E}{\alpha} = (nd) \times \left(1 + \frac{d}{\alpha/\beta}\right) - \left(\frac{0.693}{\alpha}\frac{t}{T_{pot}}\right). \tag{10.29}$$

Obviously, to solve this approximation, one must know α for the irradiated tissue and the duration of the lag prior to repopulation. The final term represents a reduction in dose effectiveness due to repopulation.

LQ modeling adequately handles genomic repair and considers repopulation through Equation 10.29. These are the two prominent radiobiological factor determinants with respect to fractionation isoeffect. However, in addition to repair and repopulation, reoxygenation and cell cycle reassortment can impact BED calculations. For example, prostate cancer is commonly regarded as a late responding tumor indicating that fractionation and dose protraction may not be beneficial for discriminating between tumor and normal tissue. However, the case has been made that prostate tumors are also often hypoxic, responding favorably to fractionation as it enables reoxygenation and facilitates reinitiation of cell cycling. Reoxygenation is one facet of reassortment — cycling cells, cells not in G_0, are more radiation sensitive. Additionally, for rapidly dividing cells, fractionated delivery preferably catches cells in late S phase, when they are most resistant. Slowly cycling cells also present resistant phases, G_0 and an extended G_1 phase. Because of repopulation lag, the rate of cell division varies throughout the fractionation scheme. Also, radiation insult may induce increased cell division. Fortunately, these concerns are subordinate to repair and repopulation. Because of vagaries in the requisite variables (α, repopulation lag, α/β, etc.), attempts to correct for reoxygenation and reassortment more often result in poorer outcomes than application of simpler methods. Thus, despite shortcomings, the community has come to general agreement that LQ formalism reliably calculates isoeffect during fractionation. David Brenner (2008) makes a persuasive argument.

> "LQ has the following useful properties for predicting isoeffect doses: First, it is a mechanistic, biologically-based model; second, it has sufficiently few parameters to be practical; third, most other mechanistic models of cell killing predict the same fractionation dependencies as does LQ; fourth, it has well documented predictive properties for fractionation/dose rate effects in the laboratory; fifth, it is reasonably well validated, experimentally and theoretically, up to about 10 Gy/fraction, and would be reasonable for use up to about 18 Gy per fraction. To date, there is no evidence of problems when LQ has been applied in the clinic."

It was in 1982, when Howard Thames and his colleagues recognized a consistent difference between early responding tissues and late responding tissues with respect to fractionated radiation delivery isoeffect curves. Late responding tissues produce steeper curves — they are more sensitive to changes in fractionation protocols. In Equations 10.26 and 10.27, this separation of TCP from NTCP can be readily seen. The BED for early responding tissues (tumor) depends on α/β of ~10, whereas the BED for normal tissue (late effects) depends on α/β of ~3. Fractionation (as opposed to acute dosing) protects healthy tissue but has less impact on tumor. For this reason, optimizing the fractionation protocol to protect healthy tissue often maximizes the therapeutic advantage. Early responders, however, cannot be neglected. Although repopulation represents a minor consideration for most normal tissues, early responders exhibit rapid cell division, and the additional factor of Equation 10.29 becomes important. In other words, fraction size dominates late responding tissue reaction, whereas fraction size and protraction time both determine early responding tissue reaction.

Customarily, contemporary fractionated radiation therapy is delivered once daily, 5 days a week at ~2 Gy per fraction. A while back, when low-energy, kV machines dominated clinical practice, fractionation schedules were designed based on skin erythema — if the schedule caused skin reddening, reduced dose per fraction was indicated. It was also discovered that delivering two doses per day (bis in die) reduced erythema, but only at low energy. When high-energy machines became available, fractionation schemes were still empirically determined, still evaluated on the basis of erythema. During this period, the standard of ~2 Gy daily doses, 5 days a week, became established. High-energy photons scatter secondary electrons forward. Forward scatter results in a "buildup" where energy transfer equilibrium is not established at the skin surface but rather at some depth beyond the skin, dependent upon the photon energy (review Figure 2-18). This "skin sparing" allowed higher fraction sizes delivered once per day. Nonetheless, once Thames established that late tissue response limited the fractionation, as skin once did, the standard protocol persisted. Equation 10.29 left radiotherapists feeling empowered. If only the constants could be established, ideal fractionation schemes could be devised.

Hyperfractionation protocols emerged first. The theory here is that the fraction size can be reduced by delivering two fractions per day, while increasing the total dose to compensate, and holding the overall treatment time constant (or nearly so). A clinical trial was held in Europe at the end of the previous millennium; it investigated the benefits of hyperfractionated radiotherapy for oropharyngeal cancer. During this study, patients received 80-Gy doses of radiation in 1.15-Gy fractions delivered over 7 weeks. Local tumor control at 5 years was compared with that of patients receiving the standard protocol: 70 Gy delivered in 35 fractions over 7 weeks. Local tumor control improved from 40% of tumors controlled to 59% of tumors controlled, with no increase in late sequelae. These results were promising and launched several and continuing studies in hyperfractionation involving increased numbers of fractions per day, other cancers, and charged particle radiations. A similar idea, accelerated radiotherapy, aimed to provide equivalent therapy over a shorter schedule by increasing the number of fractions per day. NTCP becomes the limiting factor, so fraction size must be reduced, or a break must be added to the schedule; both extend the overall treatment time, but it remains shorter than the standard treatment. The theory here is to reduce the impact of repopulation in the early responding tissue. Unfortunately, clinical trials were not encouraging. Although local control of head and neck cancers was improved, survival was not, and late effects were exacerbated; some were lethal. The oversight likely involved incomplete repair within the normal tissue. Large fractions delivered multiple times per day possibly also synchronized the cell population, making it more sensitive. The two protocols were then combined, introducing continuous hyperfractionated accelerated radiation therapy (CHART). The theory of CHART recalls the benefit of short overall treatment time to control repopulation, and low dose per fraction to control late effects. In a clinical trial employing ~1.4 Gy fractions delivered three times a day to a total dose of ~52 Gy, early effects were greatly enhanced (good for the tumor, less so for healthy tissue) and late effects often reduced. Patients reported that they appreciated the abbreviated 12-day treatment (consecutive, no weekends off) and that the early effects were tolerable. No loss in tumor control was reported at this reduced total dose, likely because repopulation was diminished. More

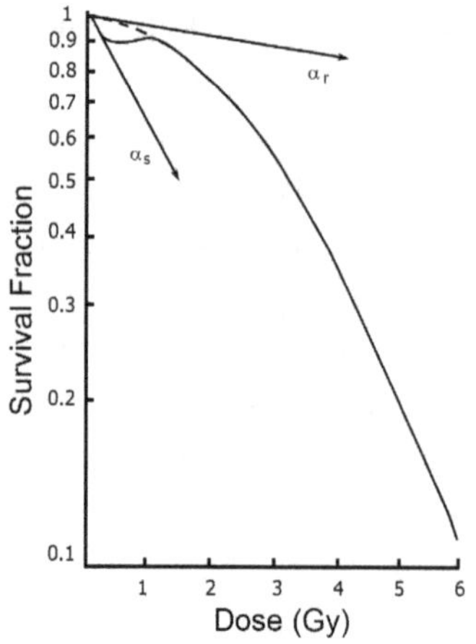

Figure 10-14 Illustration of the low-dose hyper-radiosensitivity phenomenon. The LQ fit to dose response at doses <1Gy is modeled as a dashed line. The IndRep fit is modeled by the solid line. The initial slopes: α_r (LQ) and α_s (IndRep), are indicated.

recently, investigations into hypofractionation have increased, despite the cautionary results reported from accelerated therapy. Remember, low α/β tissues — late responders — are more sensitive to large fractions. Although both accelerated protocols and hypofractionation increase the dose per fraction, when the volume of exposed normal tissue is controlled, it becomes feasible to deliver a smaller number of large fractions to the tumor. For example, charged particle therapy and IMRT provide tighter margins. When large fractions are restricted to the tumor, the results are promising.

A final caveat regarding fraction size is not apparent in Figure 10-13 because the dose response for acute delivery in that figure follows LQ formalism. It turns out that the clonal survival response to doses less than 1 Gy cannot be fit using either LQ or MTSH. At extremely low doses, cell killing is enhanced, introducing a sharp negative slope into the survival curve (Figure 10-14). Killing efficiency falls off and then falters allowing the curve to rejoin the standard LQ dose response curve at just under 1Gy. This shallow "U" shape at the beginning of the survival curve, this low-dose hyper-radiosensitivity, must be modeled uniquely. Michael Joiner (2002) modified the LQ equation to arrive at the induced repair (IndRep) model:

$$S = \exp\{-\alpha_r D[1 + (\alpha_s/(\alpha_r - 1) \times \exp(-D/D_c)] - \beta D^2\}, \tag{10.30}$$

where α_r determines the initial slope of the LQ fit and α_s determines the slope of the IndRep fit. D_c is the transition dose from hyper-radiosensitivity to increased radioresistance at about 0.2 Gy. The behavior is difficult to assess under clonal cell culture conditions, and it has yet to be identified at higher scale, but, nonetheless, predicting EQD at less than 1 Gy fractions should be approached with caution.

The foregoing formulas consistently assumed complete repair between fractions. Incomplete repair between fractions may result in severe late effects as experienced by patients treated using accelerated fractionation. In this situation, following the first fraction, normal cycling cells arrest to repair DNA damage. If the second fraction is delivered too soon, the genome gets hit by a second radiation assault while halted in sensitive phases of the cycle. Our friend, Jack Fowler noted three important factors for avoiding late tissue damage. First, do not treat incomplete repair. Second, the half-life of repair is likely to be biexponential at 4 hours and 0.3 hours. Third, a high repairable fraction of damaged DNA is desirable. If $d/(\alpha/\beta)$ represents repairable damage, the repairable fraction can be represented as,

$$P_{\text{rep}} = \frac{\left[\dfrac{d}{\alpha/\beta}\right]}{\left[1+\dfrac{d}{\alpha/\beta}\right]}. \tag{10.31}$$

Table 10-6 The percent incomplete recovery damage to late responding tissues predicted by Equation 10.31.

3.85 Gy/f	2 Gy/f [2 f/day]	1.6 Gy/f [2 f/d]	1.2 Gy/f [2 f/d]
13%	9%	7.9%	6.4%

For normal tissue ($\alpha/\beta = 3$) with an assumed repair of 50% at 0.3 hours and the remaining 50% at 4 hours, Fowler's predicted late damage determined from Equation 10.31 appears in Table 10-6. Figure 10-15 presents a pictorial representation of the fractionation protocol development timeline.

Here is a sticky problem: how does one account for interpatient heterogeneity? The TCP is a negative exponential function of the number of cells exposed. In a fractionated treatment, the number of cells exposed at each fraction is a function of the original number of cells, the proliferation function, and the (LQ) survival function. Mathematically,

$$\text{TCP} = e^{-N_f}, \tag{10.32}$$

where

$$N_f = N_0 \times S_f \times F_{\text{prolif}}. \tag{10.33}$$

The proliferation function (F_{prolif}) equals 1 at any time prior to the onset of proliferation; at proliferation onset and beyond, the function is,

$$\exp\left[\frac{\ln 2}{T_{\text{pot}}} t\right]. \tag{10.34}$$

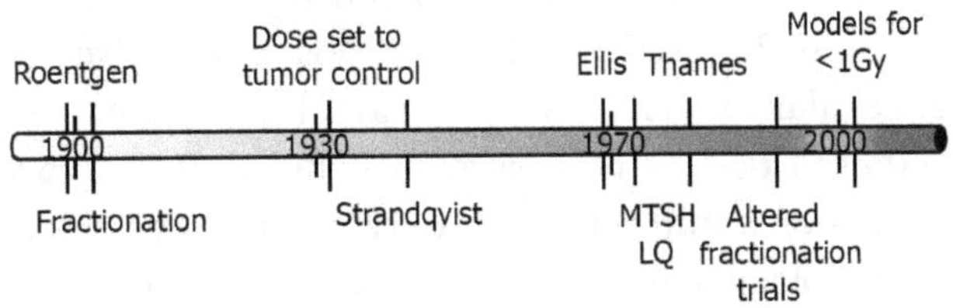

Figure 10-15 Radiotherapy timeline. From this representation, the prompt recognition of the benefits derived from fractionation is clear. Erythema avoidance drove fractionation protocols until the 1980s, when late tissue damage became the marker for tissue tolerance.

Thus, for an *individual*, tumor control can be expressed as,

$$\text{TCP} = \exp\{-N_0 \exp[-n(\alpha d + \beta d^2)]\ F_{\text{prolif}}\}. \tag{10.35}$$

For a *population*, each individual is assigned a weighting factor (w), to yield the expression,

$$\text{TCP} = \sum_i w_i TCP_i, \tag{10.36}$$

describing the behavior of a generalized tumor response for a given population. Defining the weighting factors is the tricky part.

Another low dose phenomenon expresses as increased radiation resistance to subsequent exposures, as if small radiation doses prime DNA repair by upregulating requisite enzymes. This inducible repair response (IRR) has historically been modeled as a linear relationship between the *in vitro* survival characteristics at doses greater than 1 Gy compared with those at doses lower than 2 Gy. However, more recently it has been demonstrated that the relationship is better modeled as the ratio of the alpha coefficients from the initial slope of the hypersensitive low-dose survival curve (Figure 10-14) and the initial slope of the classical LQ curve (Figure 6-8) — α_S/α_R — although the relationship is more complex than indicated by this simple model (Das & Denekamp, 2000).

10.6. FLASH Radiotherapy

A novel radiotherapy technique involving the delivery of large fractions over less than a second has recently garnered the attention of several translational radiation biologists. Ultra-HDR radiotherapy (FLASH-RT) delivers photons or charged particles at rates exceeding 100 Gy/s (up to > 1000 Gy/s) compared with typical clinical external beam dose rates of 0.2 Gy/s. Although interest in ultra-HDR dates back almost 50 years, over the last two decades theoretical and technical advancements have reinvigorated the discussion. To date, this protocol remains experimental; it is being extensively tested on animals and model tumor systems including treatment of companion animals (cats and dogs) with cancer. Nonetheless, the results look promising and the hypotheses for improved therapeutic ratio present an interesting radiobiological discussion. The initial predominant hypothesis (Spitz *et al.*, 2019) is based on radiation physics (Chapter 2, Section 2.5), radiation chemistry (Chapter 3, Section 3.2), and cell redox metabolism (Chapter 4, Section 4.3).

A FLASH pulse (~1.8 μs) compared with a 6-MV machine standard pulse (~5 μs, repeated every few milliseconds), results in 6.2×10^{19} eV vs 1.75×10^{15} eV deposited per kilogram of tissue, respectively. The deposited energy releases aqueous electrons and hydrolysis free radicals that react with oxygen within nanoseconds (G ≈ −0.2 μM Gy^{-1}; for a review of this chemistry, see Chapter 3, Section 3.2). A FLASH pulse might consume as much as 25 μM O_2; a standard dose rate pulse might consume as much as 15 μM O_2. Based on normal cellular oxygen tension, a FLASH pulse effectively depletes cellular oxygen completely, depriving the exposed tissue (the treatment field) of O_2. Replacement of oxygen through diffusion from capillaries occurs on a millisecond timescale, too slow for recovery from FLASH depletion, but too quickly for cell physiology to respond (i.e., cell cycle arrest). To see how this O_2 consumption might be leveraged — late vs early responding tissues — let's examine some radiation chemistry.

Protection of late responding tissues as described previously derives from modification of intensity, volume, or fractionation. Fractionation theory follows the logic that differences in cell cycle checkpoint activity and genomic repair rates — late vs early responding tissues — provide therapeutic advantage. The hypothesis for normal tissue protection observed during FLASH-RT calls upon completely different mechanisms. In this case, the difference between free radical oxidative metabolism in normal tissue vs tumor justifies the separation of TCP from NTCP. Reactive oxygen species (e.g., superoxide, hydrogen peroxide, organic hydroperoxides) are formed after free radicals, produced by radiation hydrolysis, bind with oxygen. Cells respond to the formation of reactive oxygen species through redox reactions wherein the superoxide product releases Fe_2^+. Because tumor cells have 2- to 4-fold higher levels of soluble iron, reactions of Fe_2^+ with ROOH and H_2O_2 increase free radical chain reactions in these cells. The theoretical process can be summarized as follows. FLASH dose rates consume local oxygen by forming organic hyperoxides. Normal cells have lower pro-oxidant concentrations and more efficient enzymatic reduction of resultant hydroperoxides. Because normal tissues also contain lower levels of soluble Fe_2^+ and transferrin receptor, they can more effectively regulate and sequester Fe_2^+, limiting subsequent damaging reactions. The hypothesis claims: FLASH-RT leverages the difference between the decay rates of redox products, and the difference in Fe_2^+ concentrations, in normal vs tumor tissues.

More recent data has now called the oxygen depletion hypothesis for FLASH normal tissue protection into question. Investigators are now testing the roles of lipid peroxidation and mitochondrial damage, for example, to help explain FLASH normal tissue protection.

10.7. Cancer Recurrence

Approximately, 1 in 20 patients treated for cancer with radiation, systemic therapy, or a combination of therapies, will experience cancer recurrence. There are several reasons for this outcome. We know that many cancers are influenced by individual behavior and personal choices such as cigarette smoking, obesity, excessive exposure to ultraviolet (UV) light (sunbathing), diet, and so on. It is not uncommon for patients to choose to continue risky behaviors following cancer therapy. Other patients have a genetic predisposition, a family history of specific cancers. Colon cancer and breast cancer are infamous for plaguing certain families, but the list of heritable cancers while rare is still long. Breast cancer will often recur in a second breast and retinoblastoma tends to affect the second eye. Additionally, physiological depression of a patient's immune system, or medically induced immunosuppression, increases the risk for several cancers. Cancer occurrence may result from marginal miss during radiation therapy — if undifferentiated tumor is not included in the 100% dose field, or if organ motion takes the tumor in and out of the radiation field reducing the total dose to parts of the target, tumor cells may recover and the tumor reappear. Microscopic metastatic disease may eventually take hold and flourish. Finally, radiation exposure can itself induce cancer (see Chapter 9). This is particularly problematic for children, as the rule of thumb for an induced solid cancer requires about 20 years from initiation to presentation.

Can recurrent cancers be retreated with radiation? Yes, under certain circumstances. Several organs exhibit lifetime limits on radiation tolerance, and recovery factors into others. For example, both spinal cord myelin and lung recover about 50% within 6–12 months after completion of radiation therapy. The bowel, bladder, and rectum suffer extensive, permanent late damage. Kidney damage also does not recover for the following reason. In parallel organs or partially parallel organs, function is diminished as FSU are destroyed. Previously irradiated organs with FSU loss must be considered at extreme risk from retreatment radiation exposure. The initial treatment volume limits subsequent treatment. Surgery is recommended, if possible, but if reirradiation appears to be the best option, hyperfractionation or a combination of external beam and brachytherapy are preferred. Charged particle therapy provides a unique opportunity for radiation retreatment because conformality can be extremely tight.

10.8. Summary

- Organs respond to radiation as if FSU were connected in serial, parallel, or some combination.
- Tumors represent a single FSU wherein every cell must be eliminated to avoid recovery.
- A comparison of NTCP with TCP at a given dose provides the therapeutic ratio.
- TCP is derived based on cell killing (survival fraction) and repopulation.
- For a fractionated dose regimen, $TCP = \exp\left\{-N_0 e^{-\alpha D\left(1+\frac{\beta}{\alpha}d\right)}\right\}$, where $\left(1+\frac{\beta}{\alpha}d\right) = \left(1+\frac{d}{\alpha/\beta}\right)$ = *relative effectiveness.*
- NTCP must consider cell survival fraction, repopulation, and kinetics: cell cycle duration, tissue kinetics, and repair rates. Organ to organ variation also includes FSU structure.
- In general, normal tissues may be considered late responders — tissues that exhibit damage weeks or months to years after exposure — although many also express early damage.
- Tumors are generally early responders.
- Early responders have large α/β values (5–30) with a commonly used approximation of 10.
- Late responders have small α/β values (1.5–7) with a commonly used approximation of 3.
- Therapeutic advantage can be maximized by limiting the volume of normal tissue that is exposed.
- Conformal 3D radiotherapy and IMRT reduce the dose to normal tissue by distributing the same total dose to a larger volume of normal tissue.
- Charged particle therapy (because of the Bragg peak and finite range) and brachytherapy (because of $1/r^2$ dose falloff) provide conformality advantages, thus limiting the volume of normal tissue exposed.
- To uniformly deliver dose the tumor, proton and carbon ion Bragg peaks are spread out (SOBP). This reduces RBE.
- Because RBE increases as dose deposition approaches the distal end of the SOBP, biological treatment planning may be necessary.
- Because interstitial brachytherapy sources reside within the tumor, and because the dose falloff is steep with depth of tissue (as $1/r^2$), brachytherapy is highly conformal and relatively target motion independent.
- Brachytherapy is delivered at HDR (>12 Gy hr^{-1}) or LDR (<2 Gy hr^{-1}).
- Radioisotope selection depends on the activity, the size of the source, the type of radiation emitted, the energy of the emitted radiation, and the half-life of the isotope.
- Dose protraction favors normal tissue recovery kinetics (primarily repair). HDR increases TCP. However, early responding tissues are less sensitive to changes in dose rate than late responding tissues.
- Increased dose protraction requires increased total dose to achieve the same effect — the isoeffect.
- Fractionation can be used to achieve dose protraction.
- Ellis's NSD power law formulation for isoeffect is $D = (NSD)T^{0.11} N^{0.24}$. It does not account for number and size of individual fractions.
- Equivalent total doses for isoeffect during fractionation may often be derived based on LQ formalism because cell survival drives both early effects and late effects (albeit to a lesser degree).
- According to Douglas and Fowler, *if* the biological effect (E) is proportional to stem cell survival, $\frac{1}{nd} = \frac{\alpha}{E} + \frac{\beta}{E}d$. This is true for late responding tissues that regenerate FSU from a stem cell population.
- In cases where the biological effect is proportional to stem cell survival, $\frac{D}{D_{ref}} = \frac{(\alpha/\beta)+d_{ref}}{(\alpha/\beta)+d}$ provides a simple method for deriving the new total dose required to achieve isoeffect, given a change in dose per fraction.

- The BED is the total dose required for a given isoeffect under the condition of infinite fractions and complete recovery between fractions (the flexure dose), or $\text{BED} = (nd) \times \left(1 + \frac{d}{\alpha/\beta}\right)$, where $\left(1 + \frac{d}{\alpha/\beta}\right)$ is the *relative effectiveness.*
- The EQD is the original dose multiplied by the ratio of the relative effectiveness of the original treatment to the relative effectiveness of the new treatment, or $\text{EQD} = \frac{(\text{BED})}{\left(1 + d'/_{\alpha/\beta}\right)}$.
- BED must be modified for instances involving repopulation: $\text{BED} = (nd) \times \left(1 + \frac{d}{\alpha/\beta}\right) - \left(\frac{0.693}{\alpha}\frac{t}{T_{\text{pot}}}\right)$.
- Hyper- and hypofractionation techniques are being actively investigated as methods to improve the therapeutic ratio. Advances in theory and technology have enabled and encouraged this line of research.
- Fractions sizes below 1 Gy do not appear to follow MTSH or LQ formalism.
- FLASH-RT delivers between 100 Gy s^{-1} to 1000 Gy s^{-1} doses. In this case, therapeutic advantage may arise from the difference between the decay rates of redox products, lipid and mitochondrial damage, and the difference in Fe_2^+ concentrations, normal tissue vs cancerous tissue.
- Radiation retreatment of recurrent cancer is organ and volume dependent.

10.9. Problems

1. Suppose you are about to treat a 1-cm diameter pediatric tumor (100×10^6 cells) and the plan indicates that you will expose the lumbar region of the spinal cord to 10% dose. The treatment calls for 80 Gy delivered to the tumor in 40 equal fractions. You would like to know the therapeutic ratio for this treatment. The lumbar region of the spinal cord is 18 cm long and each FSU is estimated at 8 mm. The spine is a serial organ with approximately 50×10^6 nuclei, and α/β of roughly 2 (assume $\alpha = 0.3$). If 70% of the cells are glial cells (mitotic), what are the predicted NTCP and TCP for this treatment? Do you feel confident in the plan? Why or why not? (You will need the formula for NTCP from Chapter 8).
2. What is the approximate dose rate of a 4 Ci ^{192}Ir seed at 10 mm from the center of the seed in tissue?
3. The text proposed that you have a source of ^{192}Ir that will have a dose rate of 0.357 Gy/h at the time of implantation. To achieve 60 Gy total dose, the implant must remain in place for 1 week. However, the implant was removed after 9 days. There will be no increase in NTCP if the additional dose is equal to or less than 8 Gy. Is it? Find the additional dose.
4. Derive NSD for skin erythema according to Fowler, at the scale of 2.0 (refer to Table 8-2). Does it hold true for all Fowler's schema?
5. Show that Ellis' formula implies that repair is more rapid early in the fractionation cycle by plotting the isodose vs time, using Ellis' formula, at NSD = 18 for 30 fractions. Compare to the typical empirical behavior in the figure to the right.

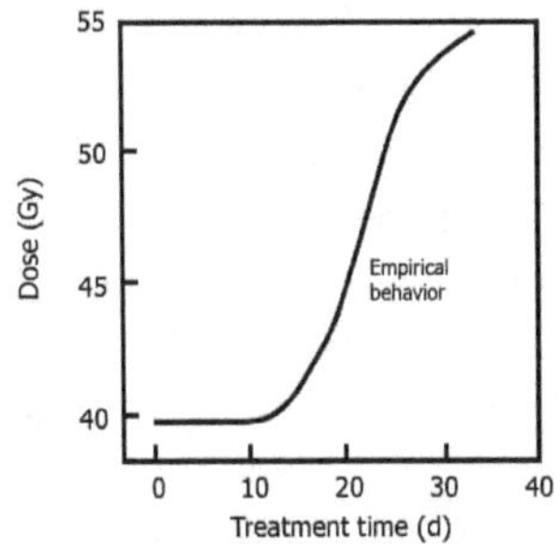

6. Using LQ dogma, predict the required dose at 1.5 Gy/f to achieve isoeffect for tumor control if the original total dose was 62 Gy delivered in 2 Gy/f.
7. For the same total dose and original dose per fraction, find the BEDs for a tumor and the surrounding normal tissue. Do not account for repopulation. Estimate α/β. Why is the BED larger than the acute delivery dose?
8. What happens to the BED in question 7 if tumor cell repopulation is considered? Assume α for tumor is 0.3. For this tumor, a lag of 25 days occurs before repopulation impacts cell loss. Tumor

cells cycle rapidly every 2–25 days, but cell loss compensates for some of this. Most commonly, a value of 5 days is assumed for tumor cell kinetics.

9. Suppose you are treating a carcinoma of the head and neck. For this tumor, the cell repopulation lag is 21 days and cell cycle kinetics indicate cell doubling every 3 days. For a total dose of 72 Gy, 2 Gy/f, what is the BED? What is the EQD for the exposed mucosa ($\alpha/\beta \sim 6$) if the mucosal cells turn over every 2.5 days with a repopulation lag of 7 days, and $\alpha = 0.35$? Even though the mucosa is an early responder, the α's are similar and the α/β's are large, the mucosa appears to be protected. How?
10. Stereotactic radiosurgery, for example, gamma knife treatments, are delivered in a few large fractions (1–5). Because the treatment time is short, repopulation can be ignored. Suppose the treatment course is delivered in three equal fractions, one per day, a total of 50 Gy. Estimate the EQD compared to an acute dose.

10.10. Bibliography

Brenner, D. J., 2008. Point: The linear-quadratic model is an appropriate methodology for determining iso-effective doses at large doses per fraction. *Seminars in Radiation Oncology,* 18(4), pp. 234–239.

Bushberg, J. T., Seibert, J. A., Leidholdt, Jr, E. M. & Boone, J. M., 2021. *The Essential Physics of Medical Imaging.* 3rd ed. Philadelphia: Lippincott, Williams, Wilkins.

Das, A. & Denekamp, J., 2000. Inducible repair and intrinsic radiosensitivity: A complex but predictable relationship? *Radiation Research,* 153, pp. 279–288.

Deacon, J., Peckham, M. J. & Steel, G. G., 1984. The radioresponsiveness of human tumours and the initial slope of the cell survival curve. *Radiotherapy and Oncology,* 2, pp. 317–323.

DeLaney, T. F. & Kooy, H. M., 2008. *Proton and Charged Particle Radiotherapy.* Philadelphia: Lippencott Williams & Wilkins.

Douglas, B. G. & Fowler, J. F., 1976. The effect of multiple small doses of X rays on skin reactions in the mouse and a basic interpretation. *Radiation Research,* 66(2), pp. 401–426.

Ellis, F., 1968. Dose, time and fractionation in radiotherapy. In: M. Elbert & A. Howard, eds. *Current Topics in Radiation Research.* Amsterdam: North-Holland Publishing Company, pp. 359–397.

Fertil, B. & Malaise, E. P., 1981. Inherent sensitivity as a basic concept for human tumour radiotherapy. *International Journal of Radiation Oncology Biology and Physics,* 7, pp. 621–629.

Fowler, J., Dasu, A. & Toma-Dasu, I., 2015. *Optimum Overall Treatment Time in Radiation Oncology.* Madison: Medical Physics.

Hall, E. J., 1996. Clinical and radiobiological research. In: R. A. Gagliardi & J. F. Wilson, eds. *A History of the Radiological Sciences: Radiation Oncology.* Reston: Radiology Centennial, pp. 129–164.

Hall, E. J. & Giaccia, A. J., 2019. *Radiobiology for the rRdiologist.* 8 ed. Philadelphia: Wolters Kluwer.

Joiner, M. C., 2002. Models of radiation killing. In: G. G. Steel, ed. *Basic Clinical Radiobiology.* London: Arnold, pp. 64–70.

Karger, C. P. & Peschke, P., 2018. RBE and related modeling in carbon-ion therapy. *Physics in Medicine and Biology,* 63, pp. 1–35.

Mayles, P., Nahum, A. & Rosenwald, J.-C. eds., 2007. *Handbook of Radiotherapy Physics — Theory and Practice.* 1 ed. Boca Raton(Florida): Taylor & Francis Group.

Paganetti, H. & Giantsoud, D., 2018. Relative biological effectiveness uncertainties and implications for beam arrangements and dose constraints in proton therapy. *Seminars in Radiation Oncology,* 28, pp. 256–263.

Paganetti, H. *et al.*, 2002. Relative biological effectiveness (RBE) values for proton beam therapy. *International Journal of Radiation Biology, Oncology and Physics,* 53(2), pp. 407–421.

Spitz, D. R. *et al.*, 2019. An integrated physico-chemical approach for explaining the differential impact of FLASH versus conventional dose rate irradiation on cancer and normal tissue responses. *Radiotherapy Oncology,* 139, p. 23–27.

Withers, H. R. & Peters, L. J., 1980. Biological aspects of radiotherapy. In: G. H. Fletcher, ed. *Textbook of Radiotherapy.* Philadelphia: Lea and Febinger.

Chapter 11

Radiation Safety

11.1. Introduction

When studying radiation biology, in addition to purposeful radiation exposure, students also must consider incidental radiation exposure. Researchers using radioisotopes and irradiating samples, technicians validating machine operation, medical physicists preparing brachytherapy seeds, cardiologists viewing fluoroscopy, food sterilization machine operators . . . all these folks must be mindful to minimize their inadvertent exposure to radiation. And above all else, the unaware public must be kept safe. Fortunately, living systems evolved on the Earth where cosmic radiation and activated elements abound. Our sun emits ultraviolet (UV) radiation and solar protons. Cosmic rays ionize in the atmosphere; you can see it happening if you are fortunate enough to observe the Northern Lights. Radioisotopes partly compose our planet (we use ^{14}C activity to date artifacts) and our food (the potassium content in a banana is 120 ppm ^{40}K). The higher one travels into our thin layer of atmosphere, the greater the radiation exposure (e.g., Denver), and the deeper one ventures into our planet, the greater the radiation exposure (e.g., the Mponeng gold mine in South Africa). In between, we call the nominal level of radiation "background radiation." The biological mechanisms that we have examined in this book, from DNA repair to cell cycle control and immunological recognition of transformed cells, these mechanisms evolved to protect organisms from background radiation detriment. Because there is no place on Earth or in space that is free from ionizing radiation, levels are always compared to background, which is assumed to be safe. Our mandate as radiation biologists requires that we use our insight to safeguard and care for those who might be unintentionally exposed to higher than background levels of ionizing radiation.

In this chapter, we will examine three topics concerning unintended radiation exposure. First, we will discuss the protection of professionals — those who must be safeguarded against potential exposure during routine tasks required by their vocation — and associated nonprofessionals (patients and family members) at risk in any space where radioisotopes or machine-produced radiation may exist. Our second topic considers the evaluation of risk — how is risk calculated, what precautions are necessary to control risk, and once exposed, what is the risk to one's health? This field is generally referred to as "health physics." The final topic presents an uncomfortable conversation about mass radiation exposure such as might occur following a nuclear power plant malfunction or a terrorist attack involving radioactive materials. Although our government has mechanisms in place to respond to such events, their activation will be slow compared to the immediate needs of survivors and innocents, the requirement to control and contain contamination, and the assistance of first responders. This may become your responsibility — you, the knowledgeable radiation biologist at the scene.

11.2. Radiation Protection

Every facility licensed to use radioactive materials or (ionizing) radiation-producing machines employs at least one *radiation safety officer* (RSO). These facilities include hospitals, universities, industry, military instillations, and so on. The RSO provides the first and last word regarding the use of ionizing radiation. He or she must be familiar with federal and state regulations and is ultimately responsible for any infractions or accidents occurring at his or her facility. Regulatory oversight falls under the jurisdiction of an alphabet soup of agencies, including international agencies — the International Atomic Energy Agency (IAEA), the International Commission on Radiation Units and Measurements (ICRU), and the International Council on Radiation Protection (ICRP) — federal agencies — the American National Standards Institute (ANSI), the American Society for Testing and Materials (ASTM), the Department of Energy (DOE), the National Council on Radiation Protection (NCRP), the U. S. Nuclear Regulatory Commission (NRC), the Department of Transportation (DOT), and the Environmental Protection Agency (EPA) — and individual state regulatory laws requiring licensing and oversight. For example, in our state, the Indiana Administrative Code (410 IAC 5) and the Indiana State Department of Health Code of Federal Regulations (10 CFR 20) provide guidance for all facilities handling radioactive materials or generating ionizing radiation. Although regulatory details are terribly important, we shall limit our discussion here to the fundamental radiation biology of radiation protection and refer the reader to the excellent websites provided by the agencies cited, as well as the reader's state regulatory codes. For our immediate objectives, radiation safety precautions follow a simple principle: keep exposure of workers and the public as low as reasonably achievable (ALARA). ALARA can be accomplished through four mechanisms: minimizing the time of exposure, maximizing the distance between the source and subject, providing engineering controls, and enforcing administrative (regulatory) controls.

11.2.1. *Radiation Protection in Radiotherapy*

Although there are many vocational uses of radioisotopes and machine-produced radiation, let's exemplify the clinical environment associated with Chapter 10 as a typical case requiring radiation protection. Radiation-producing machines — linacs, computed tomography (CT) machines, x-ray machines, catheterization labs — are placed in rooms that are also vaults (see Figure 11-1). The walls of the vault are constructed of low Z, that is, low-density material (e.g., concrete). The principle of ALARA (as specified by administrative regulations) determines the wall thickness according to the half value layer (HVL;

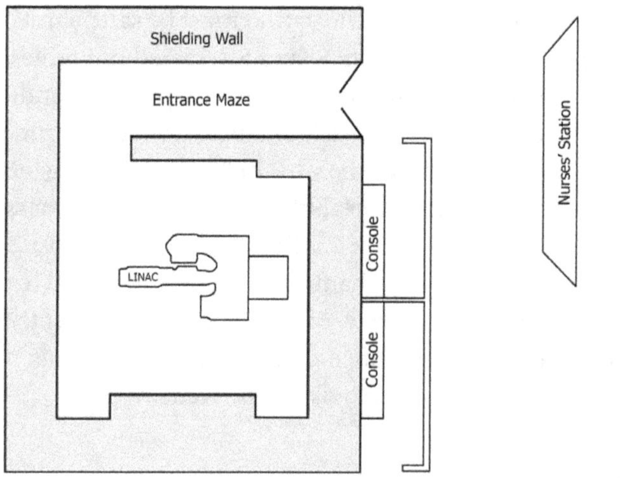

Figure 11-1 Typical linear accelerator vault floor plan. The walls of the vault are thicker (add the inner maze wall to the exterior wall) directly opposite the rotating linac head to control for the use factor (*U*). Although not shown, design engineers must consider floor and ceiling shielding if the clinic includes multiple floors. To minimize cost, vault walls are never thicker than ALARA requires. In general, interlocked doors are located at the maze entrance so that if anyone attempts to enter the maze while the machine is operating, the delivery is interrupted — another engineering control.

Equation 2.9) of the wall material at the highest energy x-rays emitted by the machine. For example, suppose that adjacent to the linac vault the therapist operations console is occupied by personnel 100% of the time when the machine is operating. And suppose a hallway borders where patients and family wander freely to and from treatment areas and lounges. The passersby are in danger only when the machine is operating, which is probably something like 20%–25% of operating business hours. Additionally, the public is unlikely to stand outside the vault for any significant time; they pass by rather quickly. Nonetheless, the allowable exposure for the unwitting citizen is considerably less than the allowable exposure for a trained, occupational worker. So, it is a good idea to check for both cases. Regulatory limits set on exposure to x-ray radiation are determined by risk — a topic to be covered later. To calculate the required wall thickness, the engineer must simply multiply the relevant factors, thus:

$$P = (WUTB)/d^2, \tag{11.1}$$

where B is the barrier transmission limit — the minimum required thickness of the shielding wall. The fraction of time the machine produces radiation is represented by W (the workload), and because the head of the linac rotates to deliver beams from any direction in a plane, the fraction of time that the beam intercepts the wall in question is represented by U (the use factor). The occupancy factor is the fraction of time (T) a person resides at the location of interest, and because radiation dose rate falls off as $1/r^2$, the permissible transmission (P) is multiplied by the inverse of the distance squared ($1/d^2$) between the source and the person. One week is a convenient unit of time (P is 0.2 mGy/wk–1.0 mGy/wk) due to workflow considerations. Therefore, in this case — protection from radiation produced during machine operation — ALARA is enforced using engineering controls (wall shielding) informed by administrative (regulatory) standards.

You may have wondered, "What is *reasonably achievable*?" Practically, ALARA reflects cost. Because each HVL of concrete absorbs half of the impinging photons, no wall can be constructed sufficiently thick to stop every photon, at least in theory. So, when is enough, enough? The facility will stop pouring concrete when state and federal regulations tell them that P is sufficiently low. To reduce wall thickness, could architects increase the distance between the machine and the therapist? Yes, but then the therapist would spend much of the day walking back and forth between the console and the patient. Patient throughput would suffer, and revenue would decline. Nonetheless, conscientious pursuit of ALARA through every action and every design optimally protects workers and public.

Now, let's think about brachytherapy. Here is a treatment that requires intimate contact with radioactive sources. High dose rate (HDR) brachytherapy design engineering allows a technician to operate an afterloader remotely from a shielded room, observing through leaded glass or via a camera and monitor. Although HDR brachytherapy is a marvel of engineering controls, someone must load the sources into the afterloader, and someone must verify and validate (V&V) the radioactivity of the sources, the location of the sources along the cables, and the accuracy of the cable travel to a predetermined position. LDR treatment requires loading seeds into molds or appliances and placing them precisely in position (e.g., suturing molds to the back of an eyeball) or dropping seeds into needles to be accurately situated interstitially, one at a time. Procedures such as these rely on minimizing the time of exposure and maximizing the distance between professional and source. Minimizing the time of exposure obviously reduces the dose to the worker as the activity (proportional to dose rate) of the radioactive material determines the dose delivered. And because the dose rate falls off as $1/r^2$, providing as much space between source and worker provides protection. In LDR brachytherapy, distance is limited by the length of the worker's arms. Furthermore, because the attending professional's hands receive maximum dose, forceps can provide significant protection. Engineering controls are also useful. Seeds can be temporarily stored in shielded containers facilitating individual seed removal and loading, and the technicians, physicians and medical physicists can wear lead aprons to protect their more radiation-sensitive organs. Here is a simple rule of thumb: in any radiation-involved situation in which you feel uncomfortable or uncertain, move away as quickly as possible — without jeopardizing the safety of other individuals.

All radiation workers must abide by administrative controls (regulatory restrictions). Compliance is monitored through engineering controls: radiation workers wear at least one radiation detection device. The majority of these accumulate dose over a designated time and are surrendered to a licensed reporting agency for dosimetry readout. For example, radiation technicians may wear a film badge clipped to the breast pocket of their scrubs. The sealed, plastic badge case encloses a square of film that becomes exposed proportional to radiation dose, and includes a small, high Z disk that limits the exposure of the underlying film to penetrating radiation. Thus, the percentage deep dose can be calculated. If the film is exposed to doses greater than allowed for the use interval by state and federal regulations, the radiation worker cannot enter any controlled radiation area until the dose rate limitation has expired. For example, the dose limit for any radiation worker is 5 cGy per year, so if the worker's total dose from several badges collected over the past months nears or exceeds 5 cGy, the worker is reassigned for the remainder of the year. Alternatively, real-time devices can be worn that report instantaneous radiation exposure. These are convenient for visitors and temporary workers when single or limited numbers of exposures are anticipated. Real-time devices often look like ink pens and accumulate dose until a small limit is reached, at which time the device emits an alarm, and the worker must leave the area immediately. These devices generally collect and retain charge — secondary electrons generated by radiation interaction with metal. The collected charge can be read out when the worker leaves the premises, or simply cleared. Dosimetry engineering is a fascinating topic, unfortunately well beyond the scope of this book. Other engineering controls are ubiquitous in clinics, and you have seen them. Radiology technicians stand behind clear shielding, often lead impregnated Plexiglas, or stand in control rooms protected by leaded windows and shielded walls. Whereas a patient gets exposed once per imaging protocol, a technician attends dozens of procedures a week and ALARA must inform all his or her actions. Dental assistants tend to simply leave the exam room during x-rays because these procedures are extremely low dose and distance suffices to reduce the exposure to acceptable limits. During fluoroscopy, workers wear leaded protective aprons and make every possible effort to increase distance and decrease exposure time (whole-body doses less than 0.001 cGy per procedure). Nonetheless, attending physicians' and technicians' arms and hands may receive distressingly high doses (up to 0.4 cGy per procedure) and limits are sometimes enforced through regulation by rotating personnel through the interventional procedures room on a frequent basis.

11.2.2. *Patient Radiation Protection in Radiology*

Patients are protected during external beam radiation therapy using high dose rates of high-energy radiation (typically MV) through conformal treatment planning and a thorough understanding of normal tissue complication probability (NTCP)/tumor control probability (TCP). Patient protection from low dose, lower energy (typically in the hundreds of keV) ionizing radiation during imaging concerns NTCP exclusively. At molecular and cellular scales, the radiation biology of low energy particles does not differ from high-energy radiation. At larger scales, the energy deposition distribution is shallower, and tissues and organs respond differentially to lower energy radiation, particularly regarding late effects. Although radiology delivers significantly lower doses than therapy, the volumes are larger (often whole body). Nonetheless, in radiology, we have the same two concerns: obligatory exposure of the patient and incidental exposure of nonpatients. To establish ALARA requirements for radiation safety, we need to know the anticipated level of exposure and the associated risk.

Table 11-1 Typical patient doses delivered during x-ray procedures.

Imaging Procedure	Dose (cGy)
X-ray (radiograph)	0.01–0.6
Computed tomography	0.2–1.6
Fluoroscopy (interventional radiography)	0.5–7.0

Table 11-1 presents typical patient doses delivered during various radiology procedures. Actual doses vary widely depending on the available technology and a clinic's established protocols. A skilled medical physicist can significantly reduce patient exposure through administrative controls. For example, imaging protocols should limit field size to the smallest, adequate diagnostic image, and translational couch speed should compensate for CT exposure rates. Fluoroscopy clearly presents the greatest exposure to the patient; this exposure can be minimized through judicious reduction of real-time visualization. Patient exposure from repeated procedures adds risk for the same reasons discussed previously for repeated radiation therapy. The radiologist must consider the total dose from all previous procedures, the recovery time(s) between exposures, and tissue tolerances. This has important implications for children who are being imaged by CT, for example, who could potentially live for 60 or 80 years after exposure and therefore have a much longer time to express late effects such as cancer induction. This concern has led to the "Image Gently Campaign" that instructs radiologists and CT technicians to lower the energy and exposure time of the CT x-rays when imaging children; because of their small size, they do not require the larger adult imaging settings. Undoubtedly, radiological imaging has been of enormous benefit to medical science. Prior to Roentgen's discovery, exploratory surgery provided the only means available to physicians to diagnose internal disorders. Nevertheless, unwarranted use of ionizing radiation to examine all sorts of complaints has led to a concerning excessive exposure of adults and children to potentially harmful chromosomal and tissue damage. More than 1% of patients develop erythema following interventional radiology, and about a quarter of those exhibit late skin damage. Assessment of increased cancer incidence due to increased number of radiology procedures is complicated, but as routine imaging has increased, so has the incidence of cancer. In the United States, the average dose per person resulting from ionizing radiology procedures now equals the dose received from background. In other words, on average, the general population receives twice the radiation dose as just a few decades ago.

In addition to radiology performed using x-ray sources external to the patient — such as x-ray machines, CT devices, and fluoroscopy irradiators — patient imaging methodology includes procedures historically grouped under the heading "nuclear medicine." The name reflects the manufacture of radiochemicals administered to patients to enable passive anatomical and physiological imaging. A nuclear medicine procedure necessitates internalization of soluble, radiolabeled chemicals. The internalized radiochemicals sequester within a tissue of interest according to biological affinities for the tagged biomolecule. The radioactive tag possesses physical characteristics that optimize detection and safety. For example, patients presenting with cancerous tumors often undergo positron emission tomography (PET). The PET process begins by selecting a molecule that will be taken up preferentially by cancerous tissue. Because cancer cells require nutrient to support growth and cell division, they absorb glucose at a significantly higher rate than normally metabolizing cells. ^{18}F-labeled fluorodeoxyglucose (^{18}FDG)[1] is a popular positron-emitting radiochemical for cancer imaging. The differential uptake of ^{18}FDG by cancer cells

[1] ^{18}[F]-2-deoxy-2-fluoro-D-glucose.

selects biologically for tumor visualization. The positron-emitting isotope tag (^{18}F) facilitates detection two ways: coincidence detection and time of flight discrimination. Recall that in positron emission (Figure 1-11; Figure 2-5; Chapter 2, Section 2.5.1), a positively charged electron escapes the nucleus and travels a maximum of 2 mm, transferring energy along the way, before it annihilates through collision with a negatively charged electron. The annihilation of mass results in the creation of two photons, each with the energy equivalent of the mass of an electron: 0.511 MeV. These two photons escape the annihilation event at 180° separation, opposite directions along a straight line. If the patient resides on the central axis of a cylinder composed of small detectors, an emission event will register as two (nearly) simultaneous photon collections along a straight line. Single detection events can be discarded as noise, greatly reducing background. The detection events are *nearly* simultaneous because the source of the annihilation event is (in general) not centrally located within the detector ring; it is off-center. Thus, it will take very slightly longer for the more distant photon to reach the detector ring than the closer photon. The time it takes the annihilation photon to reach the detector determines the *time of flight*, and by taking the difference between the two times for the opposing photons, the location of the annihilation event can be determined. In other words, the emissions can be mapped to anatomy (Figure 11-2). The half-life of ^{18}F is 110 minutes, so prolonged internal exposure, even if localized in a tissue long term, is avoided. Unfortunately, the short half-life couples manufacture and use (2-day shipping would result in the loss of 13 half-lives of activity prior to ^{18}FDG arrival at the clinic) so, fluorine must be activated — through proton bombardment of oxygen-18 ($^{18}O(p,n)^{18}F$) — at a local cyclotron.

Radiation from isotopes that decay through gamma emission, such as ^{99m}Tc ($^{99m}Tc \rightarrow {}^{99}Tc + \gamma$), also proves useful for imaging via single photon emission computed tomography (SPECT). The "gamma camera" of this device often resides at one or more static locations, collecting photons emitted from a tissue-specific radiochemical. The resultant images are frequently superimposed on film or digital radiographs, but camera rotation and three-dimensional reconstruction is also common. Gamma cameras

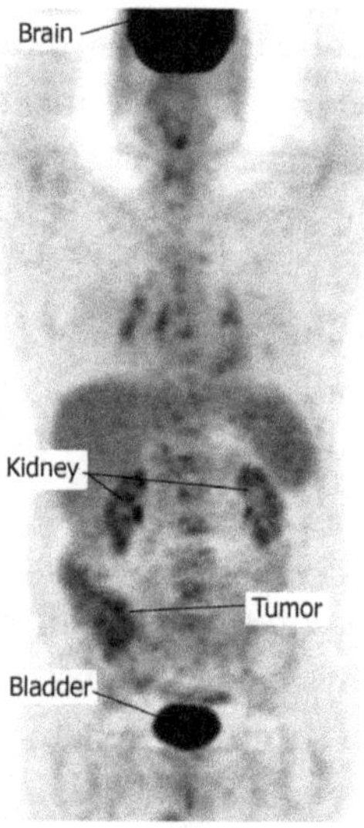

Figure 11-2 ^{18}FDG-PET scan revealing brain (glucose metabolism), kidney (filtering and excretion), bladder (collection and excretion), and tumor (glucose metabolism) sequestration of positron-emitting labeled glucose. Reproduced and modified with permission from Automated Measurement of Uptake in Cerebellum, Liver, and Aortic Arch in Full-Body FDG PET/CT Scans. C. Bauer, S. Sun, W. Sun, J. Otis, A. Wallace, B. J. Smith, J. J. Sunderland, M. M. Graham, M. Sonka, J. M. Buatti, R. R. Beichel, *Medical Physics*, 2012; 39:3112–3123.

Table 11-2 Popular radiopharmaceuticals and their uses.

Radioisotope	Active Biomolecule(s)	Medical Interrogation Of
^{99m}Tc	Medronate (MDP), Etidronate, Oxidronate (HDP)	Bone density
	Microaggregated Albumin (MAA)	Lung perfusion (inhaled)
	Pentetate (DTPA), Succimer (DMSA), Mertiatide (MAG3)	Kidney function
	Sulfur colloid	Liver function, lymph nodes
	Sestamibi, Tetrofosmin	Cardiac perfusion (stress test)
	Exametazime, Bicisate	Brain perfusion
	Mebrofenin, Disofenin, Lidofenin	Gall bladder function
	Sodium Pertechnetate, sodium iodine	Thyroid, salivary gland
^{18}F	Fluorodeoxyglucose	Brain function, cardiac, kidney function, ovaries, spleen, thyroid, tumor
	Fluorothymidine (FLT)	Rapidly cycling cells within tumors
	Fluoromisonidazole (FMISO)	Hypoxic tumor
	Fluoroazomycin (FAZA)	Hypoxic tumor
^{201}Tl	Chloride	Cardiac (stress test)
^{133}Xe	[inhaled isotope]	Lung function
^{11}C	Choline	Tumors of the brain, colon, esophagus, and lung[a]
^{15}O	H_2O	Brain function, cardiac, kidney function, ovaries, spleen, thyroid, tumor[b]

[a]Difficult to use due to short half-life of approximately 20 minutes.
[b]Difficult to use due to short half-life of approximately 2 minutes.

(radiation detector plus software) are considerably less expensive than PET/CT and ^{99m}Tc can be generated conveniently in the hospital using a process known as "milking the cow." First, ^{235}U fission in a nuclear reactor produces ^{99}Mo. The chemically separated ^{99}Mo ($T_{1/2} \approx 66$ hours) is adsorbed onto cylindrical alumina columns (about the size of a pencil) and shipped to clinics. As the ^{99}Mo decays, saline solution passed over the column will elute the daughter ^{99m}Tc ($T_{1/2} \approx 6$ hours, $\gamma = 140$ keV), which then can be covalently bound to any of numerous molecules that are preferentially taken up by tissues and organs of interest. Because ^{99}Mo continuously decays to ^{99m}Tc, a column (generator or "cow") can be eluted several times a day. This incredibly convenient process motivates two-thirds of all nuclear medicine procedures (see Table 11-2). As before, the radioisotope exhibits a short half-life, thus reducing patient exposure, and the imaging technique is functional: the target of interest metabolically prefers the tagged radiochemical. This assumption is important. If a target fails to take up radiochemical, it will not be visualized. Although this characteristic is key to assessing organ function, it is fraught with peril when identifying neoplastic tissue. Slow growing tumors and hypoxic tumors may escape detection.

Internalized radioisotopes clearly present novel factors to be considered when assessing patient risk. Dose from external sources (isotopic or machine-produced) initiates when the machine is turned on and ceases when the machine is turned off. Brachytherapy sources are sequestered within a specific patient target; the localized dose commences upon seed insertion and ends upon removal or isotopic decay to background. During nuclear medicine procedures, the internalized, soluble, radioactive molecules introduce radioactivity that dilutes throughout the organism and then accumulates in specific tissues as determined by either the chemical host of the radioactive tag (e.g., deoxyglucose) or the attributes of the unbound isotope. Unbound radioisotopes sequester in tissues through administration (e.g., inhaled ^{133}Xe

Figure 11-3 Elements presented in the first three columns of the periodic table. Columns 1 and 2 contain elements that, when available systemically along with their radioactive isotopes, tend to be incorporated into bone: potassium (K), calcium (Ca), cesium (Cs), and strontium (Sr).

localizes to the lungs) or in a manner resembling their bioactive nearest neighbor on the periodic table (e.g., isotopes of potassium, calcium, strontium, and cesium all incorporate into bone; see Figure 11-3). Alternative elements will not be incorporated with the same efficiency as biologically favored elements (e.g., K or Ca). The metabolic difference is called the "discrimination ratio" and it reflects the number of incorporated foreign molecules and the number of incorporated biologically favored molecules when the total available number of each is equal. Furthermore, distribution kinetics throughout the body are complicated. Pharmacokinetics must be considered — at what rates are the biomolecules and isotopes taken up, metabolized, retained, or excreted? And, of course, each isotope has a physical decay constant (half-life) that impacts total dose. For example, consider this: bone absorbs ^{99m}Tc-MDP, where MDP can be retained for decades — the radiochemical ^{99m}Tc-MDP has a very long *biological half-life*. Nonetheless, ^{99m}Tc has a *physical half-life* of only 6 hours; dose rate diminishes quickly. Internalization of radioactivity happens intentionally in the case of nuclear medicine, or internalization may happen unintentionally. Because there exist both controlled and uncontrolled exposures, and because kinetic complexity complicates the computation of risk from internalized isotopes, let's spend a little while considering internalized radioactivities and risks.

11.3. Internalized Radioactivity

Radioactive isotopes enter the human body by three mechanisms: inhalation, ingestion, or injection. Delivery method is an important factor. Inhaled radiation, for example, passes through a sophisticated, physiological filtration system. Particles adhere to the mucous lining our nose, sinuses, and trachea. Tiny hairs continuously waft the mucous up and out of the nostrils to remove contaminants. Unfortunately, this mucous escalator also passes particles into the esophagus where they are swallowed and ingested. Smaller particles better suspend in air and become deposited along the airways of the lungs, and the smallest particles — the aerosols — diffuse to the alveoli where they can inflict cellular damage and be taken into the bloodstream. To determine the fate of inhaled radiation, which may present as an isotope gas or may contaminate particulate matter, the distribution of particle size within the airway must be ascertained. The attribute used to quantify the anatomical distribution of particles in the airway is *the*

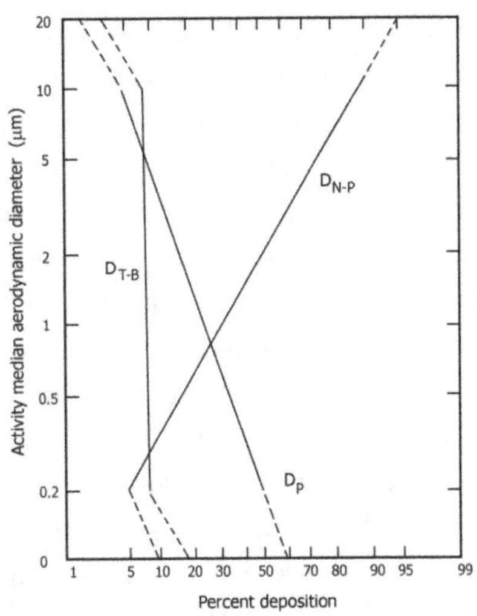

Figure 11-4 AMAD vs percent deposition. Reproduced with permission from International Commission on Radiological Protection (1978).

activity mean aerodynamic diameter (AMAD).[2] If all particles are spherically shaped and each has unit density, AMAD characterizes the diameter of the particle under consideration. So, as we often do in physics, let's assume this description to be sufficiently accurate. Figure 11-4 describes the deposition of inhaled particles based on AMAD. This figure indicates that large particles get trapped in the nasopharynx and the smallest particles diffuse into the parenchyma/alveoli. The percentage of particles captured in the tracheobronchial tree is virtually independent of diameter because the other two compartments compete for large and small particles. The tracheobronchial tree traps about 10% of all particles — fewer of the very large (they get trapped in nose hairs) and more of the very small (uniform distribution increases contact with the airway wall) as indicated by the dashed extrapolations. During ^{133}Xe imaging, over 60% of the isotope gas reaches the lung parenchyma. The introduction of radioactive material through inhalation during an unplanned nuclear event depends entirely upon the size of the particles on which the isotopes reside, from gas to smoke to granular debris. Therefore, respiratory system exposure following inhalation depends on the quantity of radioactivity taken up, the rate of removal by mucous drainage (including coughing and sneezing), the duration of retention in the lung parenchyma, and the physical half-life of the isotope(s). And one more thing . . . So far, we considered photon radiation: γ-rays and x-rays. These radiations are penetrating and prevalent in imaging technologies. Nonetheless, positrons and beta emission have also been encountered in the isotopes discussed earlier. Fortunately, for the sake of our discussion, electrons (and therefore positrons and betas) evoke the same radiobiological effects as photons (γ-rays and x-rays). Why? Because the radiation damage induced by photons results from electron interactions with absorbing materials (see Chapter 2). However, other particles — neutrons, protons, alpha particles — exhibit high LET and introduce considerably more cellular damage over shorter ranges. If a person inhales an isotope that emits one of these particles, we anticipate considerably greater impact. In the authors' state of residence, Indiana, ^{86}Rn gas is ubiquitous in the earth and limestone building materials. Radon decays via alpha emission (see Figure 1-1) and the gaseous daughter product concentrates particularly efficiently in basements. Radon is an example of potential inhalation exposure where the emitted particle type also must be considered.

[2]The precise definition of AMAD is *the diameter of a unit density sphere with the same terminal settling velocity in air as that of the aerosol particle the activity of which is the median for the entire aerosol.*

Radioactive materials become ingested primarily through ignorance; intentional poisoning aside (e.g., Karen Silkwood and Alexander Litvinenko), ingestion of isotopes is accidental. Obvious missteps involve drinking coffee or other beverages in a laboratory where radioactive materials are handled. Not wearing gloves in the lab, or not removing gloves before leaving the lab can pass isotopes into a worker's mouth or onto eating utensils and food. One vaguely nefarious incident occurred when luminous watch dial painters at three production facilities ingested toxic levels of radioactivity between 1917 and 1928 by repeatedly shaping radium paint brushes into fine points using their lips and tongue (resulting in lip and tongue cancers). Possibly the best documented incident of mass isotope ingestion occurred following the Chernobyl nuclear reactor accident in 1986. Vaporized radioactivity was released into the atmosphere followed by smoke for 9 days after the initial explosion. Radioisotopes were carried back to the planet surface by precipitation (70% fell in Belarus due to the prevailing winds). Unwashed greens accounted for a small portion of the ingested radiation. A larger percentage resulted from plant uptake of contaminated water and nutrients. Subsequently, the plants were consumed by people, and primarily by animals. Cows and goats produced milk contaminated with radioisotopes (principally ^{131}I). The chain of contamination was broken once the extent of the accident was understood, and human ingestion of radioactive substances ceased. Ingested isotopes become rapidly distributed into the circulatory system or are excreted from the body.

Nuclear medicine accounts for virtually all injected radiochemicals. For reasons of mathematical expediency, injected radiopharmaceuticals are customarily considered an instantaneous introduction of activity delivered simultaneously throughout the circulatory system. Circulation of radiopharmaceuticals throughout the body involves both blood and lymph, followed by rapid, approximately uniform dilution throughout the extracellular fluid.

Radioactive iodine provides a useful example. Iodine isotopes can be inhaled, ingested, or injected for therapeutic purposes (hyperthyroidism). In the blood, the radioisotopes of iodine rapidly come to equilibrium with naturally occurring physiological iodine by similarly attaching to albumin. The iodine then concentrates in the thyroid where it is incorporated into hormones. As new hormone molecules are synthesized, they incorporate the internalized radioisotope. Because hormone synthesis is slower than establishment of albumin-bound circulatory equilibrium, radioactive iodine flushes relatively easily from the system by swamping circulatory equilibrium with large quantities of stable iodine. Decreasing the relative concentration of radioisotope-bound albumin suppresses the incorporation of radioisotopes into hormones. Therefore, distribution of iodine pills and the use of iodized salt is recommended following radioactive iodine exposure; it reduces the biological retention (i.e., biological half-life) of radioisotope. Now, let's examine how we might quantitatively model the distribution, sequestration, and elimination of internalized radiochemicals.

11.3.1. *Distribution of Internalized Radioactivity*

In the grand scheme, radioactivity is taken into the body, distributed, and excreted (Figure 11-5). The total activity from a given internalized isotope is a function of the radiochemical biological half-life (T_b) and the isotope's physical half-life (T_R). If we assume that the physical decay of an isotope is a constant fraction of the remaining activity per unit time (true) and that the fraction of isotope biologically removed is also a constant over time (almost never true), then the effective decay constant of the isotope is,

$$\lambda_{\text{eff}} = \lambda_R + \lambda_b. \tag{11.2}$$

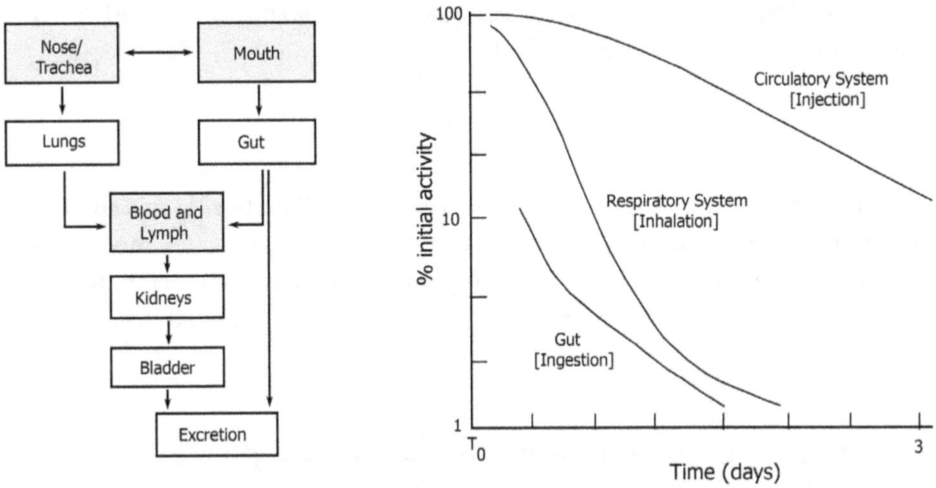

Figure 11-5 Introduction and clearance of radioactive substances. The flowchart on the left illustrates the route of soluble radiochemicals through principle exposed organs. Gray boxes indicate points of entry. A percentage of the radioactivity is taken up from circulation (and extracellular milieu) by cells, where it is metabolized and incorporated into biomolecules that are retained in the body. The remaining radioactivity is eliminated through excretion. The graph on the right provides a representation of a typical case for the clearance of radiation resulting from each mechanism of introduction. Because activity is reduced through nasal and oral expulsion, the initial respiratory activity falls short of 100%. Ingestion is difficult to model short-term because of losses to the trachea, early absorption through the mucosa, and elimination through the mouth. Activity is removed relatively quickly from the gut and lungs but is retained in bodily fluids at the 1%–10% level long term.

The physical and biological half-lives can be derived as usual,

$$\lambda = \frac{\ln 2}{T_{1/2}} = \frac{0.693}{T_{1/2}}. \tag{11.3}$$

Substitution of Equation 11.3 into 11.2 provides an estimation of the effective half-life for the total internalized radioactivity burden:

$$\frac{0.693}{T_{\text{eff}}} = \frac{0.693}{T_R} + \frac{0.693}{T_b}, \tag{11.4}$$

or,

$$\frac{1}{T_{\text{eff}}} = \frac{1}{T_R} + \frac{1}{T_b} = \frac{T_R T_b}{T_R + T_b}. \tag{11.5}$$

Of what use is Equation 11.5 if the assumption regarding biological removal is faulty? Well, it holds true in certain specific cases. Otherwise, Equation 11.5 is sufficient for rapid assessment, which is often required in emergency situations. A more accurate estimate of the time-dependent radioactivity can be derived by using a q-exponential distribution[3] to represent the decay of local activity due to biological

[3] The q-exponential distribution is a probability distribution arising from maximization of Tsallis entropy constraining the domain to be positive; it is a generalization of the exponential distribution.

redistribution. When using the q-exponential, the physical and biological exponentials can be multiplied to arrive at local radioactivity at time t ($A(t)$), thus:

$$A(t) = e^{-\lambda t} \sum_j q_j e^{-\lambda_j t}, \tag{11.6}$$

where all sources and sinks (j) are summed independently. When Equation 11.6 is applied to the left side of Figure 11-5, the three decay curves displayed in the graph to the right (for mouth/trachea/lungs, or blood/lymph, or gut) can be produced. When applying this formula, it is important to account for radioactive daughter products that may be generated *in situ*, as well as internalized isotopes.

Determining Internalized Isotope Distribution

To determine time-dependent organ radiation burden from an ingested radionuclide, we must account for the rate of isotope passage into the organ, and the losses of radioactivity due to physiological removal as well as radioactive decay. If we consider the stomach to be the first organ into which radiation is introduced (ignoring, for now, the mouth and esophagus), then the differential equation describing the rate of change in stomach radioactivity looks like this:

$$\frac{dq_{St}}{dt} = \dot{A}(t) - q_{St}(t)\lambda_{St} - q_{St}(t)\lambda_R, \tag{11.7}$$

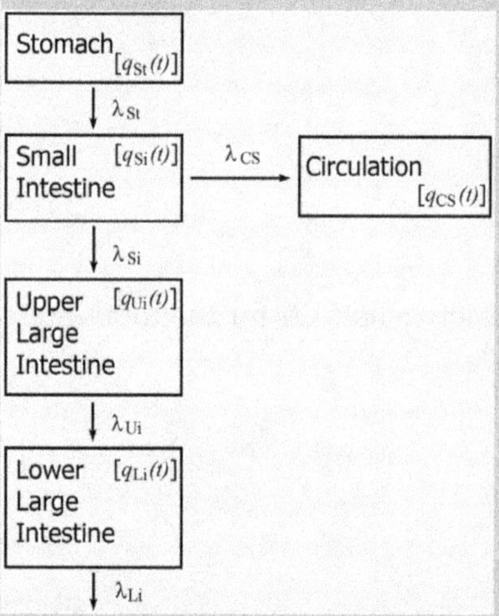

where A_0 is the activity ingested (leading to a rate of ingested activity, $\dot{A}(t)$) and q_{St}(t) is the radioactive quantity in the stomach at time t. The biological decay (removal) constant for the stomach is λ_{St}. If we substitute λ_T ($\lambda_{(total)}$) for ($\lambda_R + \lambda_{St}$) and rearrange,

$$\frac{dq_{St}}{dt} + \lambda_T q_{St}(t) = \dot{A}(t), \tag{11.8}$$

takes the form of an ordinary differential equation (ODE),

$$\dot{q}_{St} + P(t)q_{St} = Q(t). \tag{11.9}$$

ODE can be solved thus:

$$qe^{\int P(t)dt} = \int Q(t)e^{\int P(t)dt} + C, \tag{11.10}$$

where C is a constant. Substituting λ_T for P, and $\dot{A}$ for Q in Equation 11.9, provides the solution:

$$q_{St}(t)e^{\lambda_T t} = \int \dot{A}(t)e^{\lambda_T dt} + C. \tag{11.11}$$

Enforcing the boundary condition that at $t = 0$, $q_{St}(t)e^{\lambda_T t} = A_0$, we arrive at,

$$q_{St} = A_0 e^{-\lambda_T t}. \tag{11.12}$$

Next, let's determine the equation for the small intestine radioactive burden. The biological decay (*removal*) constant, λ_{St}, for the stomach becomes the biological *influx* constant for the small intestine. Furthermore, the small intestine has two physiological routes of removal: the upper large intestine and the circulatory system. Let the biological decay (removal) constants for those be λ_{Si} and λ_{CS}, respectively. For the small intestine, there is one source and there are three sinks:

$$\frac{dq_{Si}}{dt} = q_{St}\lambda_{St} - [\lambda_R q_{Si} + \lambda_{Si} q_{Si} + \lambda_{CS} q_{Si}]. \tag{11.13}$$

Let

$$\lambda_\Sigma = \lambda_R + \lambda_{Si} + \lambda_{CS}. \tag{11.14}$$

Then, Equation 11.13 can be simplified as,

$$\dot{q}_{Si} = q_{St}\lambda_{St} - \lambda_\Sigma q_{Si}. \tag{11.15}$$

Substituting from Equation 11.12, we get the ODE form of Equation 11.13:

$$\dot{q}_{Si} + \lambda_\Sigma q_{Si} = \lambda_{St} A_0 e^{-\lambda_T t}. \tag{11.16}$$

Again, this can be solved as per Equation 11.10 yielding:

$$q_{Si}(t) e^{\int \lambda_\Sigma dt} = \int \lambda_{St} A_0 e^{-\lambda_T t} e^{\int \lambda_\Sigma dt} + C. \tag{11.17}$$

Integrating the exponents leads to

$$q_{Si}(t) e^{\lambda_\Sigma t} = \lambda_{St} A_0 \int e^{-\lambda_T t} e^{\lambda_\Sigma t} dt + C, \tag{11.18}$$

and integrating the expression for decay provides the solution,

$$q_{Si} e^{\lambda_\Sigma t} = \lambda_{St} A_0 \frac{1}{\lambda_\Sigma - \lambda_T} e^{-(\lambda_\Sigma - \lambda_T)t} + C. \tag{11.19}$$

The constant, C, can be determined because the small intestine contains no radioactivity ($q_{Si} = 0$) at $t = 0$. Therefore,

$$(0)(1) = \frac{\lambda_{St} A_0}{\lambda_\Sigma - \lambda_T}(1) + C, \tag{11.20}$$

$$C = -\frac{\lambda_{St} A_0}{\lambda_\Sigma - \lambda_T}. \tag{11.21}$$

The distribution of radioactivity can be similarly determined for the remaining digestive organs. By compartmentalizing the respiratory system including all clearance methods (exhalation, sneezing, coughing, swallowing, and vascular diffusion), and the circulatory system (including elimination through the kidneys and bladder, distribution to vascularized organs, lymph plus glands), this method can be extended to inhaled or ingested radioactivity.

11.3.2. *Determining Dose*

You have probably begun to appreciate the complexity of determining the dose to organs resulting from internalized radiation. Calculations must consider biological kinetics of precursor pool equilibrium, delivery and distribution, biochemical processing, metabolism, organ (or organ structural compartment) sequestration, pharmacokinetics, and physiological elimination. Equation 11.6 provides a method for determining the radioactivity at a specific time (*t*) within an organ (the "source organ"), but we know from Chapter 2 that while charged particles transfer energy over short ranges, neutral radiations exhibit long path lengths; photons and neutrons travel millimeters to centimeters between interactions. Therefore, to determine the dose to the source organ, the fraction of emitted energy that escapes the organ must be established (see Figure 2-18). That fraction of radiation escaping the source organ exposes other organs and tissues — these are the "target organs (Figure 11-6)." For example, when ^{131}I

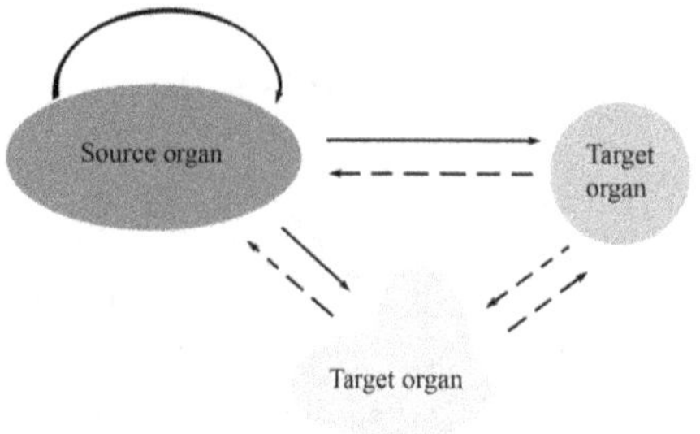

Figure 11-6 Depiction of internalized radiation emission. Any organ of interest that contains radioactivity may be designated the *source*. The source organ exposes itself and other *target* organs within the range of the emitted radiation. Two target organs are indicated by solid arrows. If other organs also contain internalized radioactivity, they become source organs as well, exposing organs within range of the emitted radiation. This exposure is indicated by the dashed arrows. Exposures to sources and targets are computed individually and then added together.

concentrates in the thyroid, emitted γ-radiation exposes the central nervous system (CNS), heart, and lungs. Target organ exposure can be determined by the $1/r^2$ dose rate falloff rule but also must consider differential absorption by biomaterials along the path, e.g., tissue, air (lungs and sinuses), and bone. Again, let's simplify our situation to gain understanding and arrive at some useful approximations. Employing a naïve physical model that assumes a uniform distribution of isotope within an organ, a known number of disintegrations, and a known fractional absorption (International Commission on Radiological Protection, 1978), we can estimate dose to target organ as:

$$\overline{D}(T \leftarrow S) = \widetilde{A}(nE)(\phi m^{-1}). \tag{11.22}$$

According to this formula, the average dose to a target organ emitted from a source organ ($T \leftarrow S$) is proportional to the activity in the source organ integrated over a predetermined time ($\tilde{A}$). The time integrated activity yields the number of disintegrations occurring during a predetermined time, as the unit of activity is disintegrations per second (Bq) and the unit of time is seconds. Because the activity decreases over time due to physical decay (λ_R) and biological removal (λ_b), the activity over time must be integrated. Because activity reflects the number of nuclear disintegrations without consideration of the number of particles and γ-rays emitted per disintegration, n provides the number of emissions per disintegration (the branching ratio). E represents the mean energy per particle or γ-ray. The absorption fraction, ϕ, is the fraction of energy emitted within the source that is absorbed in the target of mass, m. The target organ may be either the source organ itself (e.g., thyroid in our ^{131}I example) or another organ (e.g., lungs in the same example). Because the time-dependent activity is a function of physical and biological decay, it remains the most difficult parameter to ascertain. Biological half-lives have been approximated empirically and estimates of $\tilde{A}$ and $\overline{D}(T \leftarrow S)$ have been tabulated and presented in several publications. The NCRP has published a set of workbooks (National Council on Radiation Protection and Measurements, 1996), for example.

The Integrated Activity Over Time (Ã)

The total number of decays within a source organ is the activity of an isotope residing within an organ over a predetermined time. Thus, from Equation 11.12, the total number of decays occurring within the stomach from an ingested quantity of radioactivity, over the specified time (τ), can be determined as,

$$\widetilde{A}_{St} = \int_0^{\tau} q_{St}(t)\,dt = \int_0^{\tau} A_0 e^{-\lambda_T t}\,dt = \frac{A_0}{\lambda_T}\left[1 - e^{-\lambda_T \tau}\right]. \tag{11.23}$$

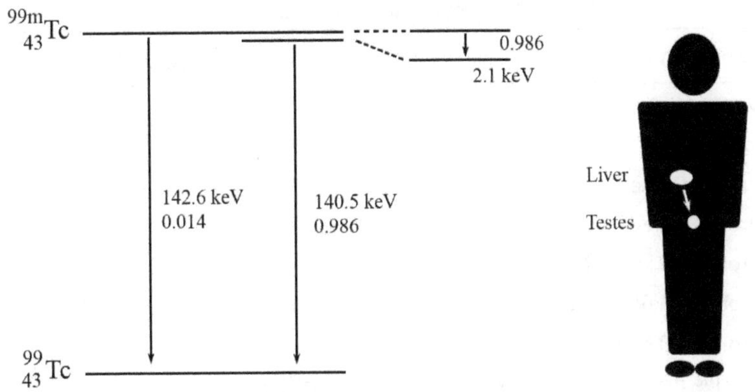

Figure 11-7 Calculation of dose to target organ. The decay scheme for ^{99m}Tc, shown at left, identifies three γ-rays emitted per decay heavily weighted toward the two-step decay noted in the right branch. The total energy released per disintegration (*nE*) is the sum of the individual emissions multiplied by their probability of emission (branching ratio): (142.6 keV) (0.014) + (140.5 keV) (0.986) + (2.1 keV) (0.986) = 142.6 keV. The tissue masses, densities, and organ locations established by the standard man phantom provide the parameters for Monte Carlo calculations to determine the penetration and scattering of each of the three γ-rays. The fraction of radiation (ϕ) reaching the testes (*T*) from isotope distributed uniformly in the liver (*S*) is thus determined.

Equation 11.22 assumes that fractional absorption (ϕ) is known. If we assume a *standard man* — a phantom that reflects the nominal physique of an adult male — we can calculate a typical fractional absorption for every standard target organ for any internalized isotope. Phantoms have been provided by the Oak Ridge Institute of Nuclear Studies representing standard women and standard children as well. Calculation is straightforward. First, the decay products — short-range charged particles or penetrating photons — are determined from the isotope decay scheme. For example, ^{99m}Tc is the most common isotope used in radiology. This isotope emits three primary gammas and no charged particles (Figure 11-7); only a fraction of the total energy released per disintegration (*nE*) is deposited within the source organ and the remainder penetrates beyond. Then, the specific absorbed fraction (ϕ/m) is derived from simulations of transmission — Monte Carlo calculations independently track each emitted γ-ray modeling randomized photoelectric absorption and Compton scatter events. The description of standard phantoms provides interceding tissue densities. Again, just as for $\tilde{A}$ and $\overline{D}(T \leftarrow S)$, values of ϕ/m derived from phantoms have been tabulated and are readily available for estimations of dose to organ.

A committee within the Society of Nuclear Medicine, the Medical Internal Radiation Dose (MIRD) Committee, has provided a method for establishing dose resulting from nuclear medicine imaging procedures. The method was originally published in 1968 and therefore required revision to agree with contemporary nomenclature (Bolch *et al.*, 2009). Recall from Chapter 7, Section 7.5.1, that the biological impact of linear energy transfer (LET) was represented by radiation quality factors (Q) prior to 1991, at which time weighting factors (w_R) for each radiation type replaced Q. The weighting factors are committee-derived values based on experimental and clinical experience. They are not mathematically derived values, but rather politically agreed upon relative quantities. As discussed in Chapter 7, the ICRP (1991) established the concept of "equivalent dose": the average absorbed dose of a given radiation type (subscript *R*) in a specified tissue (subscript *T*), taking the relative radiation weighting factor into account. The formula for *equivalent dose* is thus:

$$H_T = w_R\, D_{T,R}. \tag{11.24}$$

With respect to Equation 11.24, the SI unit for dose is the gray (Gy), the weighting factor is dimensionless, and the SI unit for equivalent dose (H_T) is the sievert (Sv). For those with rigid mathematical

Table 11-3 Radiation weighting factors.

Radiation Type	Weighting Factor (w_R)
x-rays, γ-rays, β^+, β^-, e^-	1
Protons	5
Neutrons (<10 keV)	5
Neutrons (10 keV–100 keV)	10
Neutrons (100 keV–2 MeV)	20
Neutrons (2 MeV–20 MeV)	10
Neutrons (>20 MeV)	5
α-particles, fission fragments, relativistic heavy particles	20

Table 11-4 Values of w_T for calculation of effective dose[a].

$w_T = 0.01$	$w_T = 0.05$	$w_T = 0.12$	$w_T = 2.0$
Hard bone	Bladder	Bone marrow	Gonads
Skin	Breast	Colon	
	Liver	Stomach	
	Esophagus	Lung	
	Thyroid		

[a]NCRP 1993; $w_T = 0.05$ for all unlisted organs.

inclinations, the dimensional analysis of Equation 11.24 is disturbing, but nonetheless accepted within the radiation protection community. Table 11-3 discloses that the various radiation effects are relative to photon radiation, given that w_R for photons (and the charged particles released by photons: electrons and positrons) is assigned a value of 1.0. Because the designated tissue may be exposed to more than one radiation type during an exposure incident, the generalized form of Equation 11.24 evaluates all impinging radiations summed independently as stated by the radiation superposition principle,

$$H_T = \sum_R w_R D_{T,R}. \tag{11.25}$$

Notice that the equivalent dose is defined for a specific tissue. We saw in Chapter 8 that different organs exhibit different radiation sensitivities based on functional subunit (FSU), serial and parallel response behavior, vascularization, and so on. Thus, the attributes of the target organ also contribute to the biological response. Suppose the biological effect under consideration, for example, cancer risk, is at the scale of the organism. Then, the relative radiosensitivity of each organ factors into the assessment. Whole-body *effective dose* is established by incorporating relative tissue weighting factors, w_T, (Table 11-4) for each of the organism's impacted tissues:

$$E = \sum_T w_T H_T = \sum_T w_T \sum_R w_R D_{T,R} \tag{11.26}$$

The tissue weighting factors are chosen to encompass all relative tissue stochastic outcomes — that is, any outcome that has a *probability* of occurrence — thus E can be used to predict the risk from both uniform radiation burden and nonuniform (organ sequestered) radiation burden. These appropriately

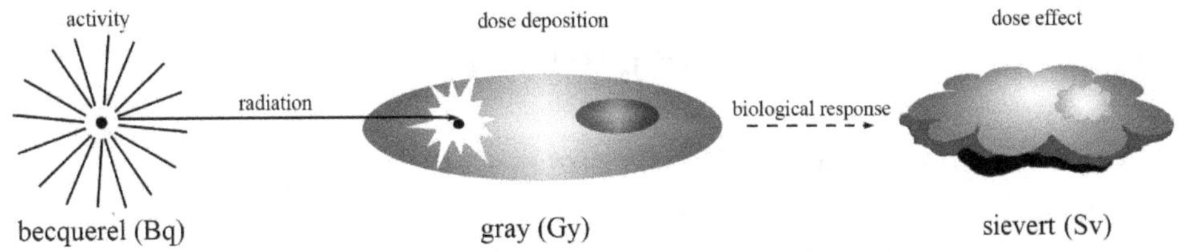

Figure 11-8 Comparison of SI units from physics, radiation biology, and radiation safety. Activity from radioisotopes is measured in Bq. Energy transferred to biomass (e.g., a cell) is measured in Gy. The biological impact of energy deposition in biomass is measured in Sv.

chosen w_T allow the extension of MIRD Equation 11.26 to external radiation exposure, such as that resulting from environmental nuclear accidents. Notice that implicit in Equations 11.25 and 11.26 is that radiations are independent actors (the superposition principle), that tissue response is independent of radiation type, and that radiation type is independent of tissue type. The latter two statements are not rigidly accurate. For example, because of parenchymal structure, an organ may be more sensitive or less sensitive to certain radiation penetration patterns. Also, high-energy emissions may interact with high-density tissues to release secondary particles, for example, irradiated bone may release short-range particles that negatively impact bone marrow. Nonetheless, the tissue weighting factors are sufficiently conservative to provide reasonably accurate assessment of effective dose. The equivalent and effective doses are the doses that result in the same *biological effect* as expressed following x-ray irradiation. This is easy to remember: when you read *sieverts*, you know that biological response has been considered (Figure 11-8).

Equations 11.25 and 11.26 provide dose for steady-state radiation delivered over a finite time (Figure 11-9), such as dose delivered by external beam or a brachytherapy isotope possessing a long half-life compared with the implant dwell time. For internalized radioisotopes, clearly the physical and biological half-lives must be considered (as per Equation 11.22). For internalized radioactivity, the equivalent dose becomes the *committed* equivalent dose:

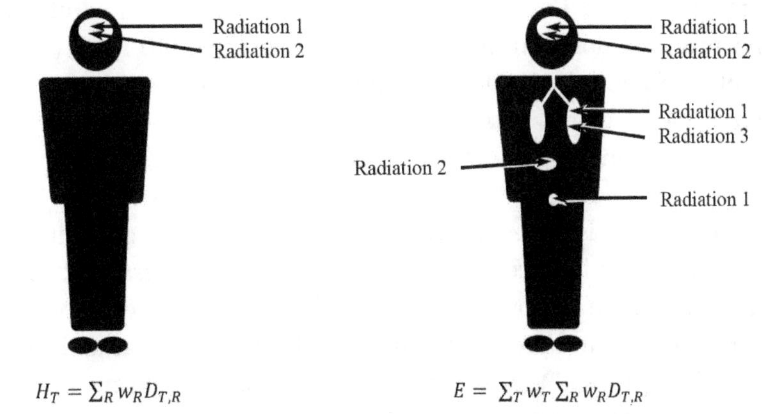

Figure 11-9 Representation of *equivalent* dose and *effective* dose. The equivalent dose, H_T, indicates the sum of the doses in a specified target organ contributed by each radiation type (α, β, photon, etc.) depositing energy in that organ. The biological effect is modified by the weighting factors attributed each radiation. The effective dose, *E*, indicates the sum of doses deposited in all sensitive organs by each radiation type. The biological effect is modified by the weighting factors for each tissue sensitivity and each radiation type. Here, three radiations deposit energy in four different organs. Not all radiation types reach all organs.

$$H_T(\tau) = \int_0^\tau \dot{H}_\tau(t)\,dt, \tag{11.27}$$

and the effective dose becomes the *committed* effective dose:

$$E(\tau) = \sum_T w_T H_T(\tau) = \sum_T w_T \int_0^\tau \dot{H}_\tau(t)\,dt. \tag{11.28}$$

$\dot{H}$ is the equivalent dose *rate* and the values of τ are generally taken to be 50 years for adults and 70 years for children. If radioactive intake continues for a long time, a time approaching τ, Equations 11.27 and 11.28 considerably overestimate the lifetime dose. Also, the integration period must be adjusted — the lower limit must be raised — for an older population.

11.4. Health Physics — The Evaluation of Risk

Humans are incredibly inept at assigning risk because psychological factors such as cognitive bias influence our perceptions. For example, frequently repeated threatening news stories provoke greater perceived risk. Similarly, if a threat persists outside our personal experience, car crashes or lethal viral infections, we diminish the risk. Therefore, we require a mathematical construct to guide our evaluation. Rigidity seems problematic as this chapter has pointed out several mathematical inaccuracies and faulty assumptions. More exist. For example, the LET-derived relative biological effectiveness (RBE) for protons proposes a value of 1.1–1.2, but the ICRP set w_R at 5 (the NCRP values w_R at 2). Nonetheless, approximations are appropriate here. Risk can be safely overestimated, but must never be underestimated. Contrary to radiation therapy and experimental radiation biology, which employ exceedingly precise estimates of dose (better than ± 2.5%), risk estimates generally overestimate by ≤25% of the actual value. Risk characterizes an assessment of worst case scenario. One might inquire, for example, whether we are endangered by our enthusiasm for anatomical and functional imaging. Considering that about three quarters of the total radiological dose delivered to the general population results from radiation-producing machines (Table 11-1) and the remaining quarter of the total is attributable to nuclear medicine, we can approximate average dose per person per U.S. population. Of the nuclear imaging dose, 85% derives from cardiac and 10% from bone scans in which typical administered activities run about 1,000 MBq and resulting doses to organs are on the order of 10 mGy (^{18}FDG for brain scans delivers 10^{-3} cGy per mCi). When we consider the total number of procedures performed in the United States each year, modern radiology has doubled our natural radiation dose — twice the dose of background. Is that dangerous? You might ask, "What is the risk to me if I get this brain scan? Might I get cancer?"

The answer to this particular risk question can be couched only in probabilities. The experimental and clinical radiation responses we have considered thus far are *deterministic* (aka nonstochastic). Given a specific dose, one can determine the outcome. For example, you can refer to a survival curve for a specific cell line (e.g., Figure 7-1), and predict the fractional cell survival at a given dose of x-rays. For subconfluent CHO cells in culture, 2% of the cells will survive the delivery of 8 Gy x-ray radiation. Cancer risk is *stochastic.* Stochastic outcomes either present or do not present following exposure to radiation; cancer either does or does not occur, it is randomly determined. Random probability distributions may be analyzed statistically but may not be determined precisely. Nonetheless, unlike a coin toss, cancer risk increases as dose increases and at some point, cancer is no longer likely (because death will precede the progression of transformation to carcinogenesis). So, we need to rely on statistics of large numbers to determine the risk to an individual versus dose. This leads to the *collective* effective dose,

$$S = \sum_i E \cdot N_i \tag{11.29}$$

where N_i is the number of individuals in the exposed population and the unit for the collective effective dose is the person-sievert. The calculated value of the collective effective dose, S, for the exposure of individuals in the United States resulting from imaging procedures is approximately 900,000 person-Sv. Does this number answer the question, "might I get cancer?" Conversion of person-Sv (a measure of dose) to a probability of developing cancer (risk) is appropriate only in cases where dose versus risk is linear. Let's examine some cases, then, to see in which the relationship is linear.

With respect to radiation risk, three outcomes concern us: lethality, health detriment (such as cancer occurrence and to a lesser extent, heart failure, stroke, asthma, bronchitis, and gastrointestinal conditions), and heritable disorders. Additionally, exposure of the unborn presents a unique concern: the extremely rapidly dividing and differentiating cells of a gestating organism are uniquely susceptible to radiation-induced cell death and genetic mutation. Thus, we will examine radiation detriment during *in utero* development later as a distinct topic. To date, the most complete set of data examining human radiation risks of death, cancer, and heritable disorders continues to be generated from survivors of the atomic bomb detonations over Nagasaki and Hiroshima at the end of World War II (WWII).[4] American and Japanese investigators began collecting data within weeks of the bombing, but those efforts were not intended to continue long term. In November 1946, President Harry Truman approved the creation of a study group, the Atomic Bomb Casualty Commission (ABCC). The ABCC employed nearly 150 American and a thousand Japanese experts continuously over the next several decades with a mandate to detail the health histories of 75,000 atomic bomb survivors and 25,000 unexposed citizens. The two cohorts included male and female participants of all ages and later expanded to include an additional 20,000 subjects. Most (>70%) of the study participants have died, as the youngest survivors now would be more than 75 years old. The ABCC eventually passed its directive to the Radiation Effects Research Foundation (RERF, 2021), an international collaboration that continues the mandate of the ABCC. The RERF particularly focuses on possible effects in generations 2 and 3, among the children of survivors and their children. A suspicion persists that the ovaries and testes of surviving children irradiated *in utero* may have been impacted, thereby passing health detriment on to the third generation of progeny. The RERF maintains biological samples from 30,000 study participants collected over the past 7 decades; these samples await genomic analysis and novel, yet to be determined assessments. Because the actual exposure event was never intended to be a controlled biological experiment, dose and radiation composition are difficult to determine with certainty. Models, estimates, and some incident recreation experiments have endeavored to provide guidance, but real-time dosimetry data are unavailable. Remember also that the radiation source was a bomb, meaning that individuals at the hypocenter mostly died from explosive trauma and not from the approximately 260 Sv absorbed radiation dose. One mile from the hypocenter, the dose to individuals is estimated at 9.46 Sv. Supplemental information regarding human response to doses greater than 10 Sv has been acquired from reports of smaller and more recent incidents. We will rely heavily on ABCC and RERF reports to streamline our discussion of radiation risk.

[4]Exhaustive studies are currently underway to document the health histories of persons exposed to environmental radiation following the Chernobyl and Fukushima Daiichi nuclear power reactor accidents. The dosimetry of these incidents is better known.

11.4.1. *Lethal Radiation Syndromes*

The mortality endpoint resolves relatively quickly and does not require long-term studies, so the first large pool of data was provided by the original WWII investigators predating the ABCC. Regrettably, numerous experiences have elucidated biological progression from radiation exposure to mortality; every reported incident since WWII, and some retrospective information, contribute to our current understanding. Potentially fatal illness resulting from acute radiation exposure can be attributed to cell killing (cytocidal effects) and cellular dysfunction (bystander and multicellular abnormality) leading to tissue FSU loss and organ failure, but morbidity and mortality are better conceptualized as health outcomes at the scale of the organism. *Acute radiation syndrome* (ARS) presents as physiological symptoms leading to radiation-invoked illness and death. ARS partitions into three system-defined subsyndromes: the hematopoietic syndrome, the gastrointestinal syndrome, and the neurovascular syndrome. Recently, a radiation-induced "multiorgan dysfunction syndrome" not restricted to a single functional system also has been proposed. Given your comprehensive understanding of *in vivo* radiation response acquired from Chapter 8, you might predict plausible cellular and tissue scale causes for each of these organism scale syndromes. Figure 8-7 and the accompanying text describe the loss and recovery of red blood cells (RBC), white blood cells, and platelets following radiation exposure. Accordingly, the hematopoietic syndrome expresses in a deterministic, dose- and dose rate-dependent manner but saturates when RBC replacement fails completely at doses that permanently damage bone marrow. Figure 8-3 illustrates the loss of crypts causing depletion of nutrient-absorbing villi in the upper small intestine exemplifying one deterministic, dose-dependent manifestation of the gastrointestinal system. Both neurovascular damage and radiation-induced multiple organ failure may result from endothelial cell loss, in the latter case leading to a systemic inflammatory response mediated by cytokines (Figure 8-16). Although the cytocidal effect of radiation on endothelial cells is deterministic, the inflammatory response tends to run-away; it is unpredictably manifested, and some individuals are hypersensitive. Thus, in general, the severity of ARS is a function of dose and dose rate, but additionally because ARS expresses on an organismal scale, it is also a function of exposed volume, homogeneity of dose distribution, and radiation composition. And . . . as you might anticipate from Figures 8-7 and 8-16, ARS develops over time. We can divide the syndrome time course into phases: the prodromal phase, the latent phase, manifest illness, and the final phase (Table 11-5). The subsyndrome(s) exhibited during manifest illness depend on the exposure circumstances, each subsyndrome presenting only following sufficient acute dose absorption and inductive delivery conditions.

Whole body, acute radiation exposure may induce vomiting at doses as low as about 0.7 Sv. At absorbed doses up to 2 Sv, only 10%–50% of exposed individuals experience vomiting within the first couple of hours following the exposure (Figure 11-10). At doses of 2–4 Sv, an increasing number of exposed persons (70%–90%) experience vomiting over that same period. Following these relatively low

Table 11-5 Acute radiation syndrome time course.

Prodromal Phase	Latent Phase	Manifest Illness	Final Phase
Nausea	Symptom relief	Hematopoietic syndrome [1–8 Sv]	Recovery or death
Vomiting	Apparent recovery	Gastrointestinal syndrome [5–20 Sv]	
Anorexia	Granulocytopenia	Neurovascular syndrome [≥10 Sv]	
Fever	Lymphopenia		
Headache			
Erythema			

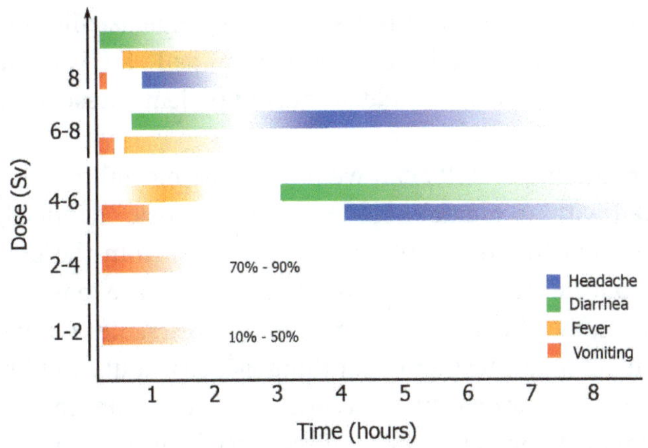

Figure 11-10 Prodromal syndrome time and dose dependencies. Below 4 Sv absorbed dose (dependent upon dose rate, exposed volume, dose distribution, and particle type) the syndrome manifests as vomiting in a population that increases with dose. Doses above 4 Sv induce increasingly severe vomiting and diarrhea, the onset of diarrhea occurs earlier with increasing dose. Fever appears about 1 hour after exposure (correlates with inflammatory response) and headache develops sooner as dose increases.

exposures, collected blood samples reveal a reduced hematocrit and leukocyte chromosomal aberrations. Higher doses result in intense vomiting experienced by all exposed victims, expelling the gut contents increasingly rapidly as dose increases. Despite the reaction correlating with dose, neither incidence nor severity appear linear with dose. Fever expresses only in persons receiving more than 4 Sv, presenting about an hour after exposure. Although temperature increases with dose (and inflammatory response), the initiation is discontinuous, the maximum is bounded, and the duration is somewhat constant. Diarrhea, like vomiting, becomes more severe with increasing absorbed dose and so, the duration shortens corresponding to the rate at which the bowel is emptied. However, unlike vomiting that expresses early regardless of dose (apart from extremely large acute doses wherein vomiting is instantaneous), the onset of diarrhea occurs sooner as dose escalates. This pattern holds true for the appearance of headache, as well. The manifestation of the prodromal syndrome can be used to estimate a patient's radiation exposure and predict prognosis in the same way that the symptoms of any illness may be used to predict disease progression, based on the collective experience and individual resilience, but our formulas for evaluating risk are not useful here.

The latent phase unfortunately hides a cruel eventual outcome. During this phase, the patient will feel as if he or she has recovered, and it is often accompanied by euphoria. Dose and exposure determine the onset and duration of the latent phase that commences a week to a month after irradiation. Following higher doses (>8 Sv), patients may not experience latency at all. Regardless, even after low-dose exposure, a complete blood count (CBC) reveals significant loss in the number of granulocytes (i.e., low hematocrit) and leukocytes and platelets due to bleeding and reduction in the precursor pool (see Figure 8-7 and the associated text).

Following this period of apparent good health, manifest illness emerges. Subjects can experience fatigue, hair loss, fever, infections, bleeding, cognitive dysfunction, hypotension, and disorientation. Not all patients experience all symptoms; the percentage of persons experiencing each symptom increases with dose, dose rate, and exposure volume. Lymphocyte and platelet counts decrease as shown in Figure 8-7 due to precursor pool depletion and stem cell compensatory repopulation failure. Higher exposure more severely diminishes cell count and recovery becomes less likely. Hematopoietic syndrome accounts for nearly all radiation-induced mortality. Neurovascular syndrome usually appears only after very large exposures of more than 20 Sv but may manifest at doses as low as 10 Sv. The causes and pathology of this syndrome are less well studied and less well understood, but patients present with seizures, tremors, and loss of consciousness.

Armed with this understanding of ARS, can we predict the mortality risk — the probability of survival or death — associated with accidental acute radiation exposure? Indeed, we can. Suppose if you will, that human beings are not that different from single cells, from the perspective of an observing alien Tralfamadorian. If triplicate large cohorts of humans were to be exposed to uniform fields of radiation in a dose escalation study, the survival curve might resemble a typical survival curve for a well-described cell culture, although test subject inhomogeneity could muddy the results a bit. Thus, mortality due to radiation exposure is *deterministic*, not stochastic. Thinking about it differently, because most mortality results from hematopoietic stem cell recovery failure, one might expect the human survival curve to resemble the radiation survival curve for cultured granulocytes. Figure 11-11 presents the probability of human mortality resulting from hematopoietic syndrome (the primary cause of death) following whole-body radiation exposure as a function of dose. Because dose rate partially determines the severity of the syndrome, Figure 11-11 presents a range bounded by fatalities resulting from low dose rate (0.2 Sv/h) and high dose rate (>10 Sv/h). The data indicate the LD_{50} for human mortality (due to hematopoietic syndrome) lies between 3.2 Sv and 4.5 Sv. Figure 11-11 tells us that about 20% of people receiving 5.5 Sv acute radiation dose will survive, or that a person receiving 5.5 Sv radiation would have an 80% risk of succumbing to hematopoietic syndrome.

Death is unavoidable and accidents happen. Nonetheless, the modern benefits of radiation to individuals (e.g., medical use) or society (e.g., nuclear energy) outweigh the risks at some balance point; production of ionizing radiation should not be prohibited but the risk must be managed. Health physicists determine acceptable radiation limits for workers and public citizens by evaluating risks (in this case, the risk of death) compared with various other occupations. For example, the Center for Disease Control and Prevention (CDC) reported that 20.4 farm workers died in 2017 for every 10,000 persons working in agriculture. On the other hand, only 0.4 service industry workers died per 10,000 service employees that same year. An acceptable mortality risk for radiation workers should lie somewhere between those two limits. Again, ALARA comes into play. If the limit for workers is set too low, prevention, for example, increased shielding, becomes cost-prohibitive. Regulatory agencies have declared the

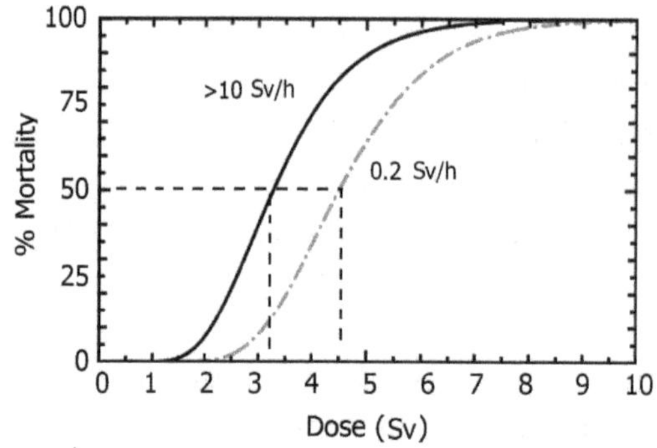

Figure 11-11 Human mortality risk from hematopoietic syndrome following radiation exposure. The deterministic curve for percent mortality vs dose resembles curves for cell mortality as well as NTCP, insinuating causative effects at smaller scales. Because mortality is dose rate dependent, boundary curves at 0.2 Sv/h and >10 Sv/h have been constructed. According to this analysis, the LD_{50} for humans falls between 3.2 Sv and 4.5 Sv.

"negligible individual dose" to be one death per million nonradiation workers (the public at large), an order of magnitude below that reported for service workers — certainly, a conservative limit.

11.4.2. *Increased Cancer Risk*

When assessing cancer promotion by ionizing radiation exposure, two particulars must be held firmly in mind. First and foremost, does the statistic under examination reflect cancer *incidence* or does it report cancer *deaths*? Many more people experience cancer diagnoses than succumb to the disease (for an excellent review of cancer statistics, refer to the National Cancer Institute [NCI] website [National Institutes of Health, 2021]), and this disparity continues to increase as early detection and treatment modalities improve. Furthermore, cancer incidence is optimally sorted into origin categories (e.g., lung cancer, breast cancer, lymphoma) to provide precise analysis. Second, cancer is ubiquitous throughout the world population of humans. One cannot ascertain radiation-induced cancer occurrence; one can only estimate *increased* cancer occurrence above the persistent baseline. The cancer incidence baseline can be terribly difficult to define due to the various cancer promotion determinants outlined in Chapter 9.

Soon after the ABCC began collecting data, it appeared that radiation exposure cancer risk resulted from an increase in the incidence of leukemia. But that was a false indicator. In truth, leukemia developed earlier than other cancers because genomic instability coupled with rapid cell turnover rates reduced the latency period from mutation to carcinogenesis. After a few decades, the ABCC data indicated that breast cancer represented the dominant radiation-induced solid tumor. Other cancers of the stomach, lung, uterus, bone, thyroid, and liver also emerged above baseline among the exposed population. Figure 11-12 illustrates typical data collected and analyzed up to 40 years after bomb detonation. Increased risk (abscissa) is defined as greater incidence among the exposed cohort compared with the unexposed cohort. Notice that the axis of Figure 11-12 is a log scale. Analysis of this data poses complexities. Cancer diagnosis depends on the victim's sex, age at exposure, genetic susceptibility, and lifestyle. For example, women exposed at the onset of puberty appear more susceptible to breast cancer and uterine cancer (diagnosis of each occurs later in life) than those exposed prior to, during, or following puberty. Thus, population risk for which the data is pooled often differs from individual risk assessed at finer granularity.

Interestingly, when we ask the broader question, we find that the risk for solid tumor development following atomic bomb radiation exposure increases linearly with dose (Figure 11-13). The risk is greater

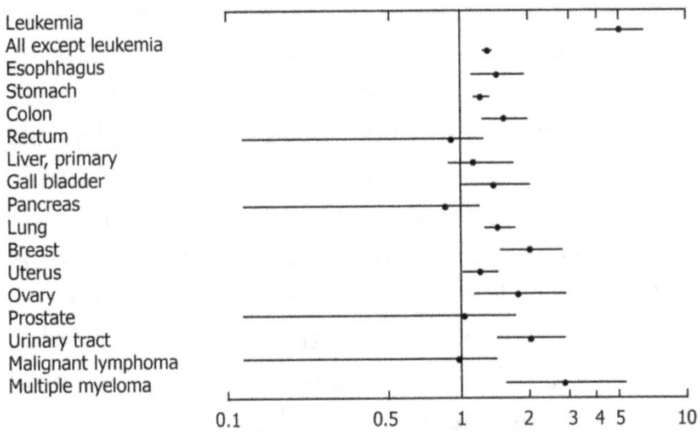

Figure 11-12 Increased risk of cancers following exposure to 1 Gy atomic bomb–produced radiation. Error bars indicate 90% confidence level. Data collected 1950 to 1985. Reproduced and modified with permission of the Radiation Research Society from "Risk of Cancer Among Atomic omb Survivors." Y. Shimizu *et al.*, *Radiation Research*, 1991; supplement 2:54–63.

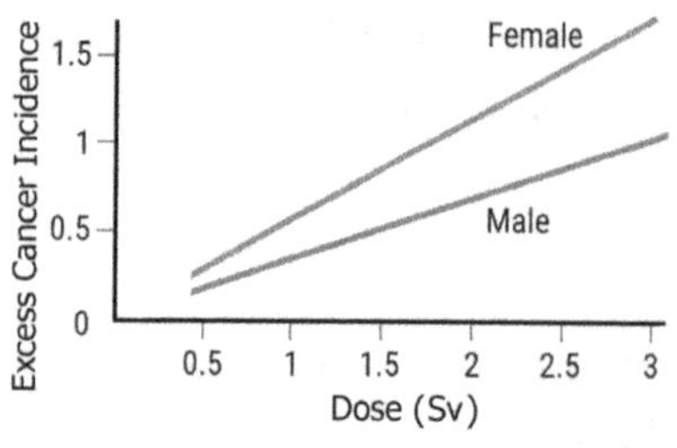

Figure 11-13 Risk of increased cancer presentation following exposure to atomic bomb radiation. Incidence for females is greater than males due to the high incidence of breast cancers. Indicated excess cancer incidence is per 10^4 persons. Data below 0.7 Sv is unreliable but at exposures above that, both sexes exhibit linear increased risk with increasing dose. Reproduced and modified with permission of the Radiation Research Society from Studies of the Mortality of Atomic Bomb Survivors, Report 14, 1950–2003: An Overview of Cancer and Noncancer Diseases. Kotaro Ozasa *et al.*, *Radiation Research,* 2012; 177:229–243.

for exposed women than exposed men because breast cancer is primarily sex-specific. Additionally, whereas uterine and ovarian cancers increase among women following radiation exposure, prostate cancer does not increase in men following radiation exposure. Nonetheless, for both men and women, cancer risk from radiation is both stochastic and linear. Think about this: deaths due to ARS are deterministic (Figure 11-12), but deaths due to cancer induced by radiation exposure (Figure 11-13) are stochastic (if deaths are extrapolated as proportional to incidence).

Finally, something that appears to be straightforward! Alas, even the linear, increasing probability of cancer with increasing dose must be approached with caution. Figure 11-13 utilizes data collected by the ABCC and the RERF following detonation of atomic bombs in Hiroshima and Nagasaki. This data is the gold standard. It is deep and rich and has been analyzed with incredible scrutiny. But bombs represent special circumstances. Detonation releases gamma and neutron radiation in addition to several subatomic particles resulting in exposure, followed by contamination and radioactive isotope internalization through all three processes. Circumstances vary with time and distance and sheltering from detonation. Additional trauma may impact health outcomes. So, the post-WWII data has been compared with more recent data — much of which are better characterized — such as the Chernobyl and Fukushima nuclear accidents, anecdotal incidents involving fewer victims, low dose rate occupational exposures, low dose data from radiographic imaging, and small volume, high-dose data from radiation therapy. The expanded pool of data includes persons exposed at a continuum of dose rates from high dose rate acute doses to continuous low dose rate or multiple doses (fractionated or multiple exposures). A *dose and dose rate effectiveness factor* (DDREF) between 2 and 10 is required to justify combining data from disparate rate cohorts. Radiation risk following high dose rate exposure is reduced by the DDREF to achieve equivalence to low dose rate risk. The ICRP recommends the most conservative value of 2 for the purpose of radiation protection. When the more inclusive data are considered, the linearity of dose versus cancer risk for all experimental cohorts persists for doses above 0.7 Sv and below 10 Sv without substantial divergence from the archetypal ABCC report. From these results, the *linear no threshold* (LNT) model of radiation risk evolved. Following the logic that the "zero dose" boundary limit presents no increased cancer risk above background; one should be able to extrapolate the ABCC slope from 0.7 to 0 Sv. Figure 11-14 (panel A) demonstrates that when this extrapolation is performed — the dashed line — the linearity of the plot remains smooth to 0 Sv. One might assume then, that the extrapolation would also hold at doses greater than 10 Sv (dashed line). Note that the plot in Figure 11-14 describes a logarithmic relationship, emphasizing the lower dose region. An enormous volume of radiation protection and health physics dogma reflects the LNT model. Nonetheless, persons absorbing 10 Sv of whole-body radiation are far

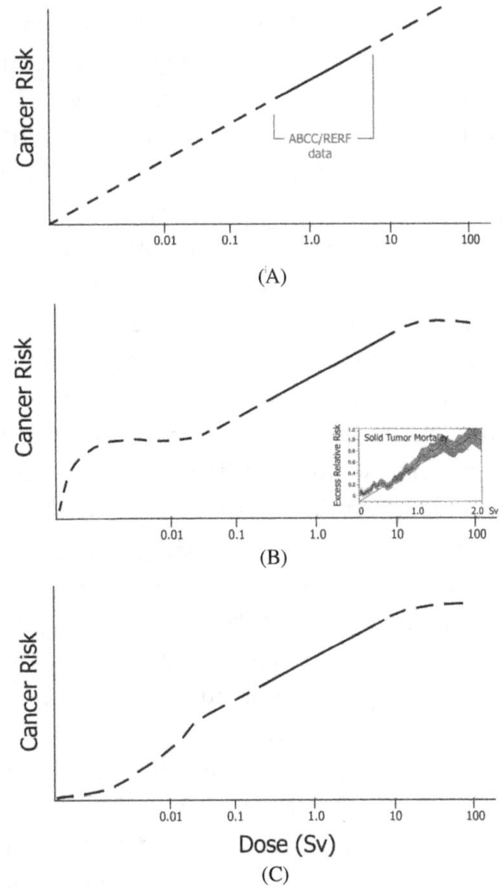

Figure 11-14 Excess solid tumor risk determined from atomic bomb survivors. Panel A shows a linear extrapolation to higher and lower doses extending beyond the "gold standard" ABCC/RERF data. This plot represents the LNT model for stochastic radiation risk. Panel B presents a possible nonlinear, hyper-exaggerated low-dose response resulting from bystander or sensitive subpopulation effects. The inset shows published ABCC solid tumor mortality data suggesting a deviation from linearity at low doses. Panel C proposes an adaptive response reducing cancer incidence at low doses. Inset reproduced and modified with permission of the Radiation Research Society from Radiation Dose Dependences in the Atomic Bomb Survivor Cancer Mortality Data: A Model-Free Visualization. Marianne Chomentowski *et al.*, *Radiation Research,* 2000; 153:289–294.

more likely to succumb to ARS than to develop cancer 10 or 20 years later (Figure 11-11). So, the high-dose extension might be expected to flatten and perhaps curve downward above 10 Sv as projected in Figure 11-14B. The low-dose region of the risk vs dose relationship presents a special dilemma. Questions regarding low exposure circumstances consume vast resources. What are the safe limits for radiation exposure? How many imaging procedures are too many? What is ALARA for a particular circumstance? How unsafe is the 1–2 mSv per day dose absorbed during space travel? Examination of additional studies that have extended statistics to lower doses have garnered intensive scrutiny. Two scenarios emerge. As far back as the early ABCC publications, a slight, nonlinear increase in cancer mortality incidence can be seen at doses below 0.5 Sv. The inset in Figure 11-14B shows the results for mortality vs dose; linearity is indicated by a solid, gray line. Might the low-dose region of the risk vs dose graph actually look more like the dashed line in Figure 11-14B? Radiobiological rationale exists for such a proposal. First, a cellular "bystander effect" has been documented in two ways. Under specific conditions, when a minority cells of a subconfluent culture, are irradiated, the unirradiated cells in the same plate exhibit a radiation response (e.g., upregulation of repair enzymes, cell cycle arrest, etc.). Alternatively, if the medium from a plate of irradiated cells is removed and added to a plate of unirradiated cells, the unirradiated cells express a radiation response. *In situ*, a bystander effect could increase carcinogenesis at low doses. This low-dose deviation from linearity also could be explained by a sensitive subpopulation. For example, rapidly repopulating stem cells might be more readily transformed at low doses. On the other hand, Figure 11-14C illustrates a third possibility. Exposed individuals may experience cancer resistance following exposure to low doses of radiation. Statistics collected from radiation workers indicate that this cohort, for whom doses are tightly controlled, experiences a reduction in cancer diagnoses compared with the general population. Although it is possible that this group is hyperaware and thus more conservative regarding nonvocational exposures (such as radiography), the data is robust. The adaptive response could be explained by an activation of DNA repair mechanisms at low doses that compensate for damage over an extended period, or conversion from mutation to transformation might be inhibited. On a systemic scale, a heightened immune response following low-dose irradiation might inhibit carcinogenesis or even induce prediagnosed cancer remission. Pending definitive resolution of the three low-dose scenarios, the LNT model may provide the best guidance for radiation safety regulation.

Leukemia induction following radiation exposure represents a special case. Onset is early. Preliminary data collection disclosed that leukemia incidence peaked at 5–7 years postexposure, causing the ABCC to report that leukemia presented the most common cancer risk following acute, external radiation exposure. The origins of transformation appear to be different in adults and children, with adults presenting with acute and chronic myeloid leukemia, and children presenting with acute lymphatic leukemia. Risk of developing leukemia following very low dose exposure (<0.2 Sv) appears to be minimal (Figure 11-15), increasing relatively linearly between 0.7 Sv and 3 Sv, and then declining again above about 3 Sv. We can use what we learned in Chapter 9 regarding the evolution of carcinogenesis to develop a couple of models that might fit the data collected by the ABCC for leukemia incidence.

First, let's consider the linear model, as we have cautiously justified reliance on the LNT model for solid tumors induced by atomic bomb irradiation. Suppose that a perfectly homogeneous human population is exposed to increasing doses of acute, external radiation and the resulting incidence of a given cancer occurs directly proportional to dose — that is, the transformation of type-specific irradiated cells

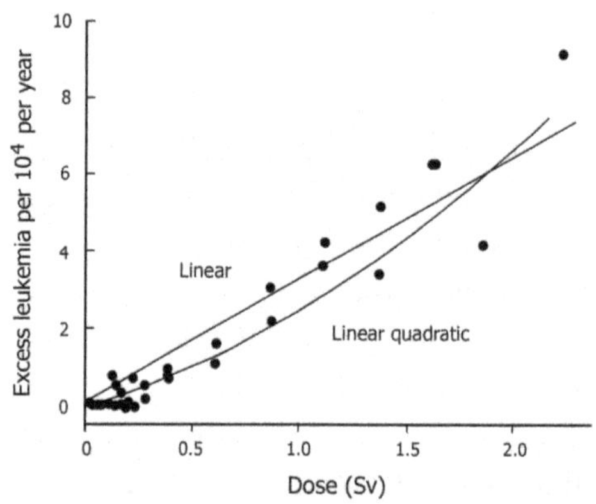

Figure 11-15 Increasing leukemia incidence per 10,000 persons per year with increasing radiation dose. Data collected by ABCC/RERF from 1950 to 2000 for persons aged 20–39 in 1970 has been fitted using the linear and linear-quadratic models for radiation risk. Because leukemia incidence below 0.2 Sv is heavily weighted toward no increase in incidence, the linear-quadratic model appears to provide the better fit at low doses.

depends only upon the number of *hits* that specific cells experience — and a constant percentage of transformed cells convert to carcinogenic. In that case,

$$I_D - I_n = \zeta_1 D, \tag{11.30}$$

where I_D is the recorded incidence at each dose, D, and I_n is the nominal cancer incidence within a matched control group. The nominal rate of cancer incidence must be subtracted to derive the number of cancers induced by a given radiation dose. ζ_1 is a constant of proportionality — the *linear risk coefficient.* Nothing surprising or complicated here. Equation 11.30 represents the linear model for cancer risk due to radiation. The conceptual difficulty here is that mammalian cell transformation does not follow a linear relationship with dose, it follows a linear-quadratic relationship (Figure 9-3). A more reasonable representation of cancer incidence following irradiation, one based on transformation studies, would be linear quadratic:

$$I_D - I_n = \zeta_1 D + \zeta_2 D^2, \tag{11.31}$$

where ζ_2 is the *quadratic risk coefficient.* But we know that this is oversimplified. Cell survival studies (Chapter 6) show that increasing radiation dose results in nonlinear increasing mammalian cell death. Dead cells are removed from the population of cells, so the progression to transformation and thereafter to carcinogenesis must be influenced by cell death. We can include a factor for linear-quadratic cell death thus:

$$I_D - I_n = (\zeta_1 D + \zeta_2 D^2)(e^{-(\alpha D + \beta D^2)}). \tag{11.32}$$

Equation 11.31 only holds true at low doses where cell killing is negligible. Equation 11.32 becomes increasingly influential at high doses (Figure 11-14B,C). With respect to excess leukemia incidence following radiation exposure, Figure 11-15 indicates that the low-dose portion of the data is better fit using the linear-quadratic model because the data are heavily influenced by no increased risk below 0.2 Sv. However, the model fails between 1.5 and 3.0 Sv where the relationship becomes linear. This example is presented as an indication of the complexity encountered when attempting to describe low dose (and low-dose rate) radiation exposure risk. We have almost exclusively presented radiation risk in the context of data collected from survivors of Hiroshima and Nagasaki following WWII. This decision was made deliberately to prejudicially simplify what is an extremely nuanced field of study. For example, one of us (Mendonca) has uncovered evidence that at extremely low exposures extending to doses lower than normal background, the cell genome appears to become unstable. The field of low dose radiation biology offers fascinating opportunities and, again, the interested student is encouraged to pursue these investigations. Initial offerings are presented in the references at the end of the chapter.

This section has played a bit fast and loose with terms such as excess incidence, cancer occurrence, excess radiation risk (ERR), and risk. For the purposes of radiation protection, the reader has been referred to publications providing tabulated risk values, but the definition of risk has remained rather vague throughout. Let's clear that up. Risk is the likelihood of epidemiological evidence for increased occurrence over normal incidence. *Absolute* risk assumes that the risk of illness (or death) following radiation exposure is wholly independent of the natural occurrence of that manifestation under otherwise identical conditions. The risk is additive. If a population is divided into i appropriate subgroups according to sex, age, geographic location, and so on, then the excess risk to subpopulation, i, of disease, b_i, can be found as,

$$b_i = \lambda_{1i} - \lambda_{0i}, \tag{11.33}$$

where λ_1 is the incidence of disease in the exposed population and λ_0 is the incidence of disease in the unexposed population. Absolute risk fails to describe several epidemiological studies. *Relative* risk assumes that the risk factor (radiation exposure) multiplies the nominal rate of incidence. In this case, the relative risk (or the risk ratio), r_i, is defined:

$$r_i = \frac{\lambda_{1i}}{\lambda_{0i}}. \tag{11.34}$$

Notice that expressed in a log form, Equation 11.34 is identical to Equation 11.33, except that each term is the log of the absolute value:

$$\log r_i = \log \lambda_{1i} - \log \lambda_{0i}. \tag{11.35}$$

It is not possible to discriminate between these definitions using the RERF results that we have relied on thus far. Nonetheless, contemporary animal studies and recent epidemiological data have persuaded the health physics community that relative risk provides a more reasonable basis for the purposes of radiation protection. Using relative risk assessment, we can finally answer our question, "What is the risk to me if I get this brain scan? Might I get cancer?" The risk of cancer resulting from one or more brain scans is 5% per sievert.

11.4.3. *Heritable Radiation Risk*

Irreparable somatic cell genomic radiation damage results in one of three outcomes: the cell continues to metabolize and divide without remarkable change in phenotype; it dies (or becomes mitotically inactive), or it transforms and becomes neoplastic. The risks associated with these outcomes have been discussed. When a reproductive cell, a haploid gamete, experiences irreparable genomic damage, it either dies or it replicates the mutation. Germ cells do not divide until they are fertilized by a haploid gamete of the opposite sex to create a zygote. Following fertilization, the zygote begins rapid cell division (if the genomic damage is not fatal; Figure 4-4), replicating and distributing the mutated gene to every cell in the developing embryo. A radiation-induced mutation in a gamete may increase the risk of cancer incidence in the recipient child or adult, depending on the latency and the probability of transformation. Alternatively, transcription of the mutation may produce an aberrant translated protein affecting fetal development or resulting in any number of maladies later in the life of the affected child. These risks are *inherited* from the radiation-exposed parent through their radiation-damaged germ cell. Contrary to the lore of Godzilla, radiation exposure of meiotic cells cannot induce novel phenotypes; the human genome, composed of 46 chromosomes with $>10^4$ encoded genes, can be corrupted through mutation, but not reimagined. Heritable risks include only those that result from genetic modifications. These fall into three categories: mendelian, chromosomal, and multifactorial. Mendelian disorders include predictable outcomes that can be traced directly to DNA mutations (point or frameshift). These predispositions can be traced through "pedigrees" or family trees; a familial predisposition to colon cancer is an example of a mendelian disease. Chromosomal disorders arise from chromosomal structural changes including rearrangement, loss, and crossover. These configurations can be visualized in a chromosomal spread (Figure 4-16). Multifactorial detriments have a genetic component but may involve other factors, such as environmental stimuli. Birth defects (e.g., neural tube malformation) and adult-onset disorders (e.g., heart disease) are representative of multifactorial disorders attributable to germ cell irradiation. Although familial "clustering" may be identified, multifactorial maladies do not follow mendelian patterns and are likely due to shared lifestyle. In addition to the life span study (LSS) that generated data for mortality and cancer risks, the ABCC commissioned a study examining heritable risks, the F1 generation study.

The F1 generation study followed 31,150 children born to survivors exposed within 2 km (~250 Sv to 8 Sv) of the bomb hypocenter. All children born to identified exposed parents were included in the study, so children born 9 months to more than 30 years after detonation were followed throughout their lifetimes. Children *in utero* at the time of detonation were not included as these effects would not be heritable, but rather developmental (more on that later). More than 41,000 Japanese children of unexposed parents comprised the unirradiated control group, born around the same time, and living under similar environmental conditions. Assessment included congenital defects, growth abnormalities, male/female gender redistribution (bias away from 50/50), emergence of genetic abnormalities, cancer occurrence, and of course, life expectancy. In addition, some 30 proteins and enzymes were tested for electrophoretic and functional defects through standard, periodic blood analyses. Here is the amazingly fortunate conclusion: no correlations were identified between parental dose absorption and heritable risk. Not a single anticipated defect could be shown to have statistically significant change in frequency compared with the control group. A possible explanation invokes *in vivo* cell survival studies conducted using animal testis (Figure 8-4) and ovary. Germ cells are extremely susceptible to radiation-induced cell death — a mechanism destined to protect against birth defects and heritable disorders. It is likely that parental exposure results in short- or long-term infertility, with only (or primarily) unaffected germ cells surviving. The likelihood of an effected germ cell becoming fertilized is diminishingly small. Well done, evolution! Nonetheless, a trend suggesting a dose response was identified — the descendants of at least one parent receiving significant dose exhibited increasing detriment with increasing absorbed dose, compared with unexposed parental control children. Although not statistically significant, this trend was used to establish the dose required to double heritable risk (the "doubling dose") at about 2 Sv. The RERF continues to examine the F1 generation with an eye toward epigenetics, specifically modification through methylation and chromosome compaction.

Again, although we have presented conclusions derived from acute exposures following atomic bomb detonations, we remind the reader that considerable additional data derived from both acute and chronic exposures in animal studies and other unintended human radiation exposures have contributed to our understanding of heritable radiation risk. Here are some of the notable findings. Inherited detriment looks stochastic and dose/dose rate dependent. The appearance of a particular genetic disorder correlates with mutation frequency, which correlates with the size of the gene (number of base pairs) — that is, the volume of the target and the probability of a *hit*. In the male mouse model, heritable risk decreases with the delay between exposure and conception — presumably relating to DNA repair and elimination of damaged spermatogonia.

Before closing out this discussion, we caution students to be particularly attentive to the statistical description of the heritable indicator. Some statistics are reported per total population, although not everyone in the population is capable of reproduction. The subset composed of persons of childbearing age is identified as the "reproductive population" that spans 30 years of an average 75 year life span (30/75 = 40% of the total population). Radiation safety regulations often refer to the working reproductive population, a subgroup that omits fertile persons younger than 18 years of age. The specifics are important. The number of incidents quoted per total population will be much smaller than the number of incidents quoted per reproductive population.

11.4.4. *Fetal Detriment*

Radiation sensitivity of embryonic tissue represents a unique population differing from other continuously dividing cell populations for several reasons. First, no resting, G_0 faction can be recruited into the mitotically active population to compensate for radiation-induced cell loss. Second, cell differentiation progresses from totipotent to pluripotent, to multipotent, to oligopotent, to unipotent stem cells. One lost totipotent stem cell may be replaced by an additional mitotic division of any surviving totipotent stem cell. Thus, one or two cells can be physically removed from a 4 to 8 cell morula without apparent consequence to normal fetal development. On the other hand, the loss of a unipotent neural progenitor cell might result in serious malformation of the CNS. Because the extent of and degree of stem cell differentiation critically affects radiogenic *in utero* detriment, prenatal development is generally divided into three radiosensitivity stages: preimplantation, embryogenesis, and fetal development. Whereas heritable genetic risks appear stochastic with no threshold dose, *in utero* detriment appears deterministic and follows cell survival characteristics. Rare cases of birth defect arising from stem cell radiogenic mutation may occur, however, these events are anatomically indistinguishable from detriment arising from cell death and may be included as a radiation *hit* for purposes of radiation safety risk assessment. Also, just as for cancer risk assessment, a background incidence of nonradiogenic birth defects exists in the human population at about 5%–8% of births.

11.4.4.1. *Preimplantation*

From fertilization to uterine implantation, a human zygote undergoes about four cell divisions as a morula (solid sphere of cells) to eventually become an implanted blastocyst (cell mass within an inner cavity encircled by a single spherical layer of cells) composed of 200–300 cells. Radiation-induced cell death during this stage generates one of two outcomes. The loss of perhaps half of the morula cells results in no apparent effect on normal development, although the success rate of implantation may be impaired (the probability of this outcome is difficult to establish). More severe cell loss generates failure of the morula to develop a blastocoel, successfully implant in the uterine wall, or survive. Therefore, with extremely rare exceptions possibly reflecting single somatic cell genetic mutation, radiation exposure during the first several weeks of pregnancy produces one of two outcomes: the infant suffers no detectable untoward effects, or the pregnancy terminates early on. The ABCC statistics indicate a reduction in birthrate for women exposed during the first 4 weeks of gestation. In other words, significantly fewer babies were delivered within the exposed population 36–40 weeks following detonation. Studies irradiating rodents in utero reinforce these conclusions and suggest that doses above 10 cGy impose conceptus loss risk to humans.

11.4.4.2. *Embryogenesis*

Following implantation, embryonic cells begin to differentiate and create organ systems. Stem cells become pluripotent or multipotent with coincident radiation sensitivity impact. As progenitors of each physiological system transition toward more restrictive mitotic products, irreplaceable cell loss following irradiation becomes more impactful. Thus, the appearance of congenital abnormalities follows the natural pattern of organogenesis (Figure 11-16). Exposure of an 18-day-old human embryo coincident with CNS organogenesis promotes neurological abnormalities, whereas exposure at 48 days postfertilization, coincident with musculoskeletal genesis, may manifest as limb malformation. Thus, the relevant radiation risk factors during embryogenesis logically depend on the absorbed dose and the phase of organogenesis. Returning to our understanding of deterministic cell survival models, we can predict the risk of fetal abnormality resulting from radiation exposure during organogenesis. Consider, for example,

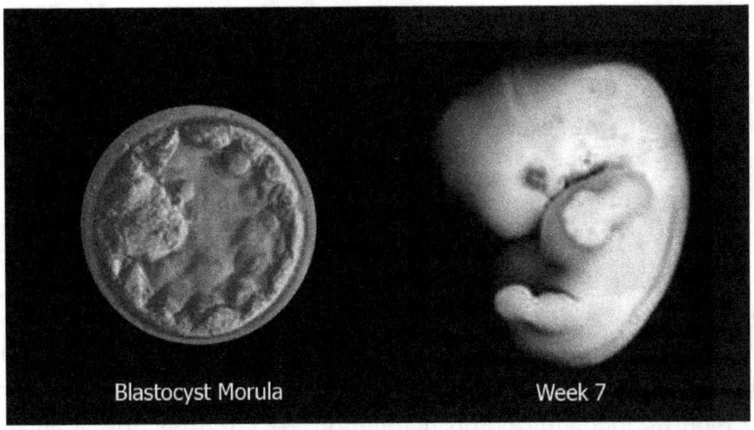

Figure 11-16 Early *in utero* human development. The preimplantation sphere of totipotent cells differentiates into a single cell layered shell enclosing a pleuripotent mass of cells within the cavity, or blastula. The miocrograph on the left shows a cross section of an implanted blastocyst at about 2 weeks postfertilization (original image source: nobelprize.org). The 7-week-old embryo on the right reveals the prominent primordial head containing the forebrain, midbrain, and hindbrain, as well as an eye, the spine and the heart. Limb and digit progenitors are also apparent at this early stage. Image provided by Dr. Mark Hill, https://embryology.med.unsw.edu.au.

that an embryo has reached a stage of development where several anatomical systems are composed of fewer than a dozen multipotent stem cells. If the embryo can be considered an anlage of equally susceptible cells, then the probability of killing one of the critical progenitor cells may reflect MTSH cell survival as per Equation 6.10:

$$\frac{S}{S_0} = 1 - \left(1 - e^{-D/D_0}\right)^n. \tag{11.36}$$

A "back-of-the-envelope" approximation can be achieved by assuming mammalian tissue culture experimental constants of $D_0 = 100$ cGy with an extrapolation number, $n = 4$. For example, at an absorbed dose, D, of 50 cGy, the survival fraction would be 0.976, or the probability of killing a critical cell would be 0.024 (2.4%). If every cell loss manifested as a birth defect (worst case scenario), this dose would lead to an unacceptable risk on the same order of magnitude as typical birth defect occurrence. Severe malformation of critical organs can, of course, result in fetal death and cell loss spread over several systems late in organogenesis may simply result in low birth weight.

The collection of birth defects data following Hiroshima and Nagasaki proved challenging for several sociological and radiation dosimetric reasons. The numbers of identified children irradiated *in utero* were low (less than 100), with the presentation of individual malformations extremely infrequent. Evidence for three birth defects emerged: low birthweight (growth retardation), microencephaly (small head circumference), and mental retardation. Microencephaly dominates for embryos exposed before 7 weeks of gestation. The frequency of microcephalic infant births appears to be linear with dose (stochastic), although this trend relies on more recent data collected from animal studies and (historical or unintended) medical exposures. During early embryonic growth, brain organogenesis entails precerebral neuronal expansion without glial proliferation, causing the primitive brain to be relatively radiation resistant. This may explain the minimal occurrence of mental retardation in children exposed prior to week 7 postconception. However, during early embryogenesis, the head and neurons experience extremely rapid growth, presenting a large population of mitotic differentiated stem cells and dendritic branching activity — plentiful radiation-sensitive targets. Indiscriminate cell loss suppresses head growth. In fact, because embryogenesis represents a growth phase as well as organogenesis, growth retardation constitutes the second most common birth defect. For embryos exposed to radiation, organ-specific risks are diluted among the several anatomical systems. The larger the percentage of cells killed (i.e., the higher the dose), the more severe growth retardation becomes and the greater the risk of low birthweight becomes. Interestingly, there appears to be a threshold for the linear growth suppression dose response at around 0.2 Sv, implying perhaps that a minimum number of cells must be lost before the effect manifests.

Late embryogenesis, 8–15 weeks postfertilization, accompanies increased organ refinement with continuing growth. The frequency of and dose response for growth retardation, including microencephaly, remain unchanged from early embryogenesis. However, the risk of mental retardation increases significantly. During the 9 months following bomb detonation, 62 children presented with microencephaly (3.9% of births) of which 30 children presented with both microencephaly and severe mental retardation (1.9%). In other words, half as many children suffered from mental retardation because the duration of vulnerability lasts half as long. During late embryogenesis, the primitive brain develops a cortex at the forebrain through migration of neurons and embellishment of the glial population. Glial cells are radiation sensitive, and their dose-dependent loss disrupts the requisite neuronal shepherding responsible for cortex structure. Although radiation-induced mental retardation risk persists from week 15 to week 25 of gestation, the risk is reduced approximately fourfold beyond week 16. The risk for microencephaly reduces to background if exposure transpires later than 16 weeks postfertilization, presumably because head growth slows as the rest of the body catches up. Data gleaned from historical and unintended medical exposures of 11- to 16-week-old embryos have disclosed subsequent deformities of the eye, skeleton, and genitalia. Each of these organs undergo structural refinement during late embryogenesis.

11.4.4.3. *Fetal Growth*

Fewer organs continue morphogenesis during the fetal stage of organism maturation, so the variations of abnormality resulting from radiation exposure are greatly reduced. This stage is principally one of growth. Radiation detriment therefore expresses as functional disorders and low birth weight, although large doses may result in similar effects as those expressed in ARS (Table 11-5) without prodromal manifestation. Lethal doses tend to be expressed postpartum rather than *in utero.* Growth retardation develops following exposure at any time postimplantation with the severity of reduction correlating with the percentage of cells killed. In other words, the younger (smaller) the conceptus is at the time of exposure, the greater the impact of growth retardation at birth. It is not possible to predict all possible risks from *in utero* human exposure to radiation from the existing compilation of exposures. The matrix of individuals at various stages of development exposed to escalating doses from 5 cGy to 5 Gy is too large, as is the variety of possible responses. To achieve statistical significance, we must rely on animal studies, most of which have been conducted using male mice — an imperfect system, at best. These results should be applied to the human population with a healthy soupcon of skepticism.

Having briefly discussed cell death and consequential developmental risks, we now must address radiation-induced genomic mutation in utero. Might mutation progress to cell transformation and eventual cancer? Again, the F1 cohort was small and therefore statistics present large uncertainties, but it appears that above about 10 cGy there exists a significant risk of neoplastic transformation accompanying *in utero* exposure (including embryonic and fetal phases). The absolute risk of childhood cancer subsequent to *in utero* radiation exposure is about 6% per Sv, with leukemia incidence dominating. Furthermore, the effect can be multifactorial; in these cases, the neoplasms behave more like natural cancers (requiring induction by environmental promotors) than radiation-induced cancers. Cancers expressing *in utero* and apparent at

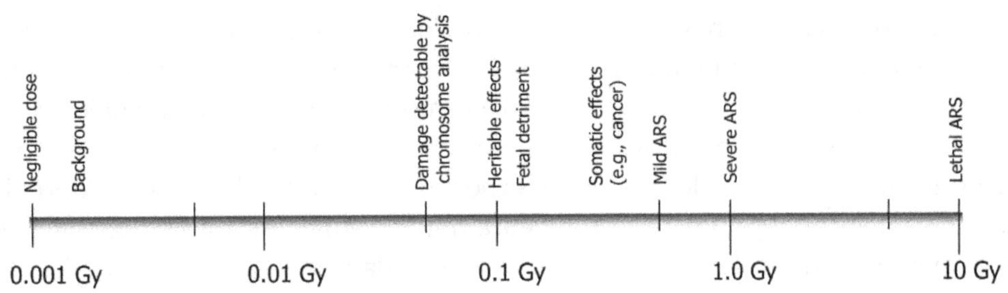

Figure 11-17 Radiation dose and health risk. Human exposure to doses less than 1 mGy are considered negligible and are not included in annual or lifetime dose limit calculations. Natural background radiation levels on Earth fall between 1 mGy and 3 mGy. Above 5 cGy, the frequency of detectable leukocyte chromosomal rearrangements corresponds with dose. Human heritable abnormalities begin appearing at more than 20 cGy and somatic effects arise between 0.2 Gy and 1.0 Gy. ARS expresses from 0.5 Gy to above 10 Gy. Note that the dose is presented logarithmically.

delivery indicate radiation is a teratogen. Growth retardation and teratogenesis correlate. From experiments performed using pregnant mice, we learned that neoplasms of almost any type can be induced by *in utero* radiation exposure. Neoplastic origins are technique and species-dependent, and as anticipated, the frequency of cancer occurrence generally increases with increasing dose. Nonetheless, there is a saturation effect caused by cell killing at higher doses. The latency of tumor appearance depends primarily on the tumor type. Figure 11-17 summarizes organism health risks initially expressing at various radiation doses.

11.4.5. *Health Physics Regulations*

An exhaustive review of radiation safety regulations is well beyond the scope of this text. Nonetheless, a brief summary may be useful — something to provide frameworks for personnel protection and urgent response to incidents putting persons at risk. The NCRP succinctly states the objectives of radiation protection:

1. Prevent deterministic radiation effects
2. Limit stochastic radiation effects.

Table 11-6 lists a sampling of limits that the NCRP (National Council on Radiation Protection and Measurements, 1993) has put in place to achieve these two goals. Because this text primarily has presented data acquired from survivors of the atomic bomb detonations over Nagasaki and Hiroshima, it is important to note that both external and internalized exposures contributed to the effects described. In addition to blast-emitted photons, neutrons, subatomic particles, and contaminants, survivors breathed

Table 11-6 NCRP recommended dose limits.

Exposure	Limit
Prenatal embryo or fetus	0.5 mSv per month gestation
Adult public	1 mSv per year chronic exposure or 5 mSv per year acute exposure
Eye (to prevent deterministic cataracts)	15 mSv per year
Lifetime occupational stochastic effects	10 mSv x age in years
Annual occupational stochastic effects	50 mSv per year

in, swallowed, and through puncture wounds injected radioisotopes. Doses were approximated as closely as possible, as a total of external and internal exposure. In reference to radiation protection, the annual limit on *intake* is the activity (expressed in Bq) taken into the body during a single year that would provide a committed equivalent dose to a standard person equal to the absolute limit set by the NCRP. The absolute limits in Table 11-6 include the total exposure from both internal and external sources.

Exposures exceeding the recommended limits are not alarming, but rather are cause for concern. Human beings engage in a myriad of risky behaviors. The fundamental operative in radiation protection is a cost-benefit analysis. If a physician suspects that a patient has a life-threatening condition that can be diagnosed only through radiology, the physician and patient might well determine that the risk of additional dose to the patient, even if the patient has exceeded the recommended limit, is well worth the additional risk of radiation detriment. Consider this. A coworker slips and falls in the lab while carrying radioactive liquid in a flask that breaks upon impact. The worker is injured and bleeding and unconscious. Radiation protection recommends that first aid should be administered without regard to radiation risk to the responder. The radiation risk to the responder (and injured party) should be attended to at the earliest opportunity once the injured party is stabilized. This is another example of cost-benefit analysis. But the methods used to attend to the injured coworker should minimize radiation exposure to the unconscious victim, the responder, and other persons that may be impacted later (EMT, hospital employees, public, etc.). Let's review some historical radiation exposure events and consider the appropriate, safe, and ethical responses.

11.5. Mass Radiation Exposure Events

11.5.1. *Chernobyl Nuclear Reactor*

In late April 1986, the number 4 reactor at the Chernobyl Nuclear Power Plant outside Pripyat, Ukraine, experienced an uncontrolled nuclear chain reaction. The postmortem analysis of the events leading up to the meltdown is incredible, involving a planned safety test and an unprepared operations crew conflated with several engineering failure points. Numerous summaries of the incident are available with varying degrees of detail. They are, without exception, fascinating and read like best-selling suspense novels. Here, we will concentrate on the radiation safety ramifications of the disaster but strongly recommend the reader browses the failures leading up to the Chernobyl environmental radiation disaster. Pertinent to the lessons learned from the incident, the Soviet Union failed to alert neighboring nations to the ongoing emergency, thereby releasing an expanding radiation cloud for 2 full days, impacting the citizens of both the Soviet Union and surrounding nations. The morning of the second day, radiation alarms began to sound at the Forsmark Nuclear Power Plant in Sweden, 1,000 km distant from Pripyat. Once determined that radiation was not emanating from within Forsmark, the source was tracked to Chernobyl. Although it may seem patently obvious that radiation accidents must be reported immediately and action taken to care for exposed individuals while preventing additional exposures, incidents on both global scale and laboratory bench repeatedly reinforce that human nature motivates perpetrators to obfuscate accidents. This principle cannot be overemphasized: radiation accidents must be reported, and actions necessary to reduce impact must be initiated immediately. The establishment of a community that promotes prompt incident reporting is essential.

During the most dangerous period of the episode following the steam explosion, physicists estimate the exposure in the reactor building at 5.6 roentgens per second, reaching LD_{50} in about a minute for any human in the building. In the early hours, the beta to gamma radiation ratio measured 30:1, indicating that radioisotope emission rather than radiation exposure (Table 11-7) represented the primary health

Table 11-7 Primary radioisotopes released from Chernobyl.

Isotope	Half-Life	Form(s)	Activity Released ($\times 10^{12}$ Bq)
^{132}Te	78 h	Vapor	1.15
^{133}Xe	5 d	Gas	5.20
^{131}I	8 d	Vapor, particles, organic compounds	1.76
^{90}Sr	28.8 y	Particles	0.01
^{134}Cs	2 y	Aerosol, particles	0.08
^{137}Cs	30 y	Aerosol, particles	0.085

detriment threat. The reactor crew chief and his team remained at their stations throughout that day and night trying to get the situation under control, as they mistakenly believed the reactor remained intact and radiation levels were much lower than actual (personal dosimeters saturated at 0.001 roentgen per second). The explosion killed two of the operations staff. Most of the remaining crew died from ARS within 3 weeks. Of the 134 emergency response personnel, 28 died within days to months and 14 presented with excess cancers within 10 years. Including subsequent chronically exposed firefighters, cleanup crews, and other radiation workers, a total of 237 persons suffered from ARS, 31 of whom died within 3 months. Evacuation of Pripyat commenced 36 hours after the explosion; within 10 days, the evacuation zone was expanded to 30 km. Today, the evacuated area is referred to as the "exclusion zone" — 350,000 persons were permanently relocated. Estimates based on the LNT model conclude that as many as 16,000 individuals eventually may succumb to radiation exposure health detriment.

Distribution of radioactive material released in the initial steam blast and subsequent fire depended on weather conditions including microclimates. Prevailing winds concentrated the radioactive fallout over Belarus, Ukraine, and Russia (in order of decreasing radioactive levels), although Europe and the Nordic countries were impacted to a lesser extent. The Soviet air force seeded rain clouds over Belarus to deter fallout over more populated areas. Pockets of contamination also resulted from atmospheric condensation over mountains (e.g., the Alps and Scottish Highlands). Gas, aerosols, and vapor traveled the farthest and inflicted harm through parenchymal lung uptake as well as deferred ingestion. Dust (10 μm particles) settled out locally near Chernobyl; in turbulent air, dust will settle at a rate of 5 vertical feet in less than 10 minutes. Small particles remained in the atmosphere longer than dust having a settling rate of 5 vertical feet in 40 hours for 0.3 μm particles, to 5 vertical feet in 10 hours for 1.5 μm particles. These particles contaminated plants and soils throughout the afflicted countries with the percentage of initial release falling off with distance from the origin.

Longer-surviving ARS victims died most often from bacterial infections of extensive radiation skin burns arising from a lack of radiation protective attire and prolonged exposure to radioisotope-saturated clothing. This finding has prompted a recommendation that survivors presenting with burns should be isolated to prevent infection (standard protocol for burn victims). The deposition of particulate, long-lived isotopes of strontium and cesium require continued abandonment of the exclusion zone, and the reactor itself is enclosed in a sarcophagus to contain the remnant radiation. The local, evacuated population suffered from inhalation of gaseous and vaporized isotopes presenting with 15 thyroid tumors to date (iodine composed 40% of local radioisotope load). The Chernobyl Forum estimates 3,935 additional incident-related solid tumor and leukemia deaths will occur among the 200,000 emergency workers, 116,000 evacuees, and 270,000 residents within 30 km of Chernobyl at the time of the reactor explosion. Within the general population of the broader impacted area, ^{131}I represents the most prominent health risk. Because iodinated albumin concentrates in milk glands, the isotope made its way into the dairy food chain. The dairy industry necessarily shut down until ^{131}I ground contamination reached levels safe for grazing. On the other hand, long-lived cesium continues to contaminate wild game at unacceptable levels for consumption. An independent study predicts a total of 41,000 excess cancers eventually may present among exposed Europeans (0.01% of all cancers in the region).

11.5.2. *Fukushima Daiichi Nuclear Power Plant*

In March 2011, an earthquake and subsequent tsunami off the coast of Okuma, Japan, devastated the Fukushima Daiichi Nuclear power facility. Briefly, the disaster unfolded as follows. Engineered radiation safety protocols shut down the reactors in units 1, 2, and 3 automatically when the earthquake was detected; earthquakes are not unusual events in Japan. Units 4, 5, and 6 had been taken offline for inspection prior to the earthquake but retained their spent fuel pools. To provide the requisite continued reactor core and spent fuel cooling, emergency diesel generators cut in when the electrical power grid suffered disruption — also a result of the earthquake. So far, so good. Then the 46-foot high tsunami, generated by interface friction between the quaking suboceanic floor and the body of sea water off the coast of Tohoku, breached the protective sea walls constructed off the coast adjacent to the power plant to prevent just such an event. Unfortunately, a tsunami of this extraordinary size was unanticipated. The sea water swamped the generators and they failed. Three nuclear meltdowns, three hydrogen explosions, and the release of radioactive materials from three reactors (units 1, 2, and 3) transpired over the next 4 days. As rapidly as possible, 154,000 inhabitants within 20 km of the plant were evacuated to a safer environment. In addition to 10^{17} Bq of released airborne contaminants, an estimated 18×10^{15} Bq of ^{137}Cs washed into the Pacific Ocean during the immediate crisis. An additional 1,320 tons of radioactive waste was intentionally released into the ocean from the subdrain pits of units 5 and 6 to prevent further damage. The ^{137}Cs radioactive contamination of the ocean continued for more than 2 years at an estimated rate of 30×10^9 Bq per day. Japanese engineers constructed a wall onshore to contain the flow of radioactive pollution, but because the depth of the molten core cannot be determined, an unknown amount of contamination continues to seep into the ocean. Plant cleanup is estimated to take 30–40 years.

Unfortunately, TEPCO[5] administrators minimized the significance of the emergency in the short term. The drift and composition of the radioactive plume was inaccurately described (despite receiving mapping data from the U.S. Airforce) resulting in the evacuation of some people from relatively safe zones to highly contaminated zones. In a later postmortem analysis, the United Nations Scientific Committee on the Effects of Atomic Radiation estimated that 100–500 PBq[6] of ^{131}I and 6–20 PBq of ^{137}Cs were released into the air, 80% fortunately settled into the Pacific Ocean between Asia and North America. The committee confirmed that 1 week after the earthquake, soil radiation levels 25 miles north of the power plant had reduced to 125 mSv/h. Increased levels of ^{131}I were detected at 18 water purification plants between Tokyo and Okuma. Radiation contaminated tea leaves, spinach, milk, fish, and beef as far as 320 miles from Fukushima. In defense of the Japanese government's contamination control failures, it should be remembered that this radiation crisis unfolded following the worst earthquake in the history of Japan. The massive earthquake (9.0 on the Richter scale) and subsequent record tsunami killed 15,899 people through blunt force trauma and (mainly) drowning. More than 6,157 people were seriously injured and 2,529 remain missing. The World Bank estimated the total recovery cost at $235

[5] Tokyo Electric Power Company, operators of the Fukushima Daiichi power plant.
[6] Petabecquerel, 10^{15} Bq.

billion. In addition to the damage inflicted at the Fukushima Daiichi plant, four other nuclear facilities sustained damage: Fukushima Diani, the Onagawa Nuclear Power Plant, the Tokai Nuclear Power Plant, and the Rokkasho Reprocessing Plant. Nonetheless, not a single death resulted from ARS due to radiation exposure. The hydrogen explosion in unit 1 injured 5 people, the hydrogen explosion in unit 3 injured 11 people, and 51 elderly or hospitalized patients died during the evacuation. To put the incident into perspective, the amount of radiation released into the atmosphere was 10%–40% of that released during the Chernobyl accident (Chernobyl emission data being speculative at best), with about 10% as much land surface impacted. No tested seafood has disclosed radiation levels exceeding Japan safety standards since April 2015. A government subsidized thyroid study group reported that 36% of children living within 20 km of Fukushima at the time of the earthquake have "abnormal growths" on their thyroid gland. According to the LNT model, 130 excess cancer deaths are anticipated, although accurate statistics require a minimum 10 years before postevent follow-up, which is only now possible.

11.5.3. *Additional Mass Radiation Exposure Events*

The second most significant environmental radiation release transpired in 1957 at the Mayak Production Association reprocessing plant in Siberia, Russia, but because of the plant's remote location, the event ranks third in severity, behind Chernobyl and Fukushima. The plant has a troubled history. Originally constructed to produce plutonium in support of the Soviet atomic bomb program, the plant dumped 76 million cubic meters of toxic chemicals and 3.2×10^6 Ci of high-level radioactive waste into the Techa River between 1949 and 1951. The radiation dose in the river measured 35–50 mGy/h in 1951 (equivalent to lifetime dose in 1 week) when the government began relocating people from towns along the river, eventually abandoning 23 towns. The largest town, Muslumovo, was not evacuated and has become an interesting study in chronic, low dose rate human exposure. The Muslumovo townspeople were chronically exposed to emissions from ^{137}Cs, ^{106}Ru, ^{90}Sr, and ^{131}I. The same isotopes were internalized and deposited in organs including bone marrow according to natural biological uptake. The frequency of birth defects and neonatal death reached 300% nominal rate. In 1953, 200 cases of radiation poisoning were identified among 587 patients (out of 28,000 Muslumovo residents). Then in 1957, a poorly maintained storage tank at Mayak exploded, releasing 50–100 tons of high-level radioactive waste (740 PBq) into the atmosphere. During the immediate aftermath, more than 750 km^2 (9 mile radius) were contaminated by airborne radiation, effectively ending the chronic, low dose rate study. In all, the Mayak facility has been responsible for 35 environmental radiation release incidents to date, including the 1957 third-ranking nuclear disaster.

Prior to 1945, most radiation exposures were single victim laboratory or medical mishaps. Ignorance represented the dominant cause of radiation poisoning until the burgeoning nuclear industry (including military) presented the opportunity for mass exposure events. Wilhelm Roentgen, his wife Anna, and Pierre and Marie Currie all succumbed to radiation-inflicted ailments. Marie Curie's lab papers are so contaminated that they are stored in a lead box. Eben Byers, a wealthy socialite, innocently ingested 1,400 prescribed bottles of "Radithor"; he is buried in a lead lined coffin. Although we can learn from anecdotal instances, mass radiation exposure events provide unique opportunities to derive statistical inferences establishing quantitative radiation safety limits and safe medical procedures. Admittedly, viewing these humanitarian crises as poorly designed scientific experiments is unprincipled at best. Nonetheless, a failure to derive useful information from these events only diminishes the victims' suffering. Table 11-8 lists some of the more significant reported mass radiation occurrences.

Table 11-8 Significant accidental radiation exposure events.

Date	# Persons Affected	Description	Location
1958	8	In the Y-12 criticality incident, highly enriched uranium diverted into a steel drum resulted in a fission reaction lasting about 15 minutes.	Oak Ridge, TN
1977	*	An explosion released a toxic and radioactive waste mixture through a disposal shaft.	Dounreay, United Kingdom
1979	*	An earthen dam failed, releasing 100×10^6 gallons of liquid radioactive mining waste into the Puerco River.	Church Rock, NM
1982	~7000	^{60}Co contaminated recycled steel was used in construction of buildings throughout Taipei. The overall cancer risk was significantly *reduced*. Specific cancer incidence increased (leukemia, thyroid).	Taiwan
1983	*	A ^{60}Co therapy machine was scrapped and recycled along with (eventually) the contaminated transport truck. The metal (at ~11 TBq) was used to construct table legs for homes, bars, and restaurants. All border crossings now employ radiation detection equipment.	Ciudad Juarez, Mexico
1985–1987	>6	Software with inadequate safety protocols allowed excessive therapy doses (up to hundreds of Gy) to be unintentionally delivered by the Therac-25 linear accelerator. Six patients received visible injury, three died. Medical software is now highly regulated.	(Various — manufactured in Canada)
1987	250	A 40 TBq ^{137}Cs source scavenged from a therapy machine in an abandoned clinic was sold as a "glowing curiosity." Four people died of ARS.	Goiania, Brazil
1989	23	A capsule of ^{137}Cs was discovered mixed into concrete and embedded in an apartment wall. Six residents died of leukemia.	Kramatorsk, Ukraine
1990	27	Following maintenance, a linear accelerator was returned to service without appropriate quality assurance (QA). The incorrectly calibrated electron accelerator overdosed 27 patients; 11 died.	Zaragoza, Spain
1992	>100	A construction worker took home a 370 GBq ^{60}Co source found in a well at an abandoned environmental monitoring station. Three people died of ARS.	Xinzhou, China
1996	116	An improperly calibrated and aligned therapy machine overdosed patients.	San Jose, Costa Rica
1998	18	An unidentified ^{60}Co source was sold as scrap. Seven children were hospitalized. Ten patients were treated for ARS.	Istanbul, Turkey
2000	28	A revised a protocol for prostate and cervical cancer therapy data entry caused the treatment planning software to miscalculate delivered dose, resulting in lethal levels. Mandatory validation and verification are now required for all protocol design changes.	Panama
2005–2006	7,500	Treatment errors including misapplication of dynamic wedges and daily portal imaging lead to overdosing prostate cancer patients. In all, 700 significant overdoses, 24 severe injuries, and 5 deaths occurred. The hospital was closed and reopened under new management and staff.	Epinal, France
2008	*	45 GBq ^{131}I gas escaped over 2 days from an isotope production facility. Authorities restricted farm products within a 5 km radius. The ECURIE alert system notified EU countries 5 days after the leak was detected.	Fleurus, Belgium
2008–2009	206	CT software miscalculation resulted in patients receiving approximately eight times higher dose than anticipated. Although recommended, medical physicist oversite of radiology has not been universally implemented.	Los Angeles, CA
2013	33	Accidental leakage of radioisotopes at the J-PARC nuclear research facility was mishandled following machine malfunction. The contamination was mostly contained within the site.	Tokai-mura, Japan

*Unknown or unrecorded.

11.6. Environmental Risk Remediation

Prompt remediation of risk becomes urgent following any accidental release of radiation. Medical triage and care protocols are specified for exposed persons and we will address these, but prevention of subsequent chronic exposure is equally pressing. In the simplest cases, for example, medical misadministration, there is no risk of subsequent chronic exposure. For other medical cases — such as an accident at the Indiana Regional Cancer Center (in 1992) where an ^{192}Ir pellet remained in a patient following HDR brachytherapy and the patient returned to his or her communal care facility — both the victim and others may be chronically exposed. During the Indiana incident, the pellet eventually dislodged. Had it not been discovered, identified as a radioactive source, and properly controlled, the pellet could have continued exposing patients, workers, and the public in unpredictable ways. Certainly, chronic exposure avoidance becomes a priority following radiation escape into air or water (and subsequently soil). Historical radiation events have provided a framework for development of remediation goals and models. The EPA recommends a remediation goal — the limit of risk for ongoing radiation control and cleanup — of 10^{-4} to 10^{-6} cancer risk in an affected population. Again, because the topic of environmental protection is too expansive to be comprehensively covered here, straightforward introductory references have been provided at the end of this chapter.

Two questions dominate the health physicist's concern following a radiation accident: "Does the current state pose a danger to individuals?" and "Will the scenario evolve to pose increasing or decreasing risk (or perhaps through dispersion, a diminished risk to an increasing number of persons)?" These questions can be answered, *comme d'habitude*, through modeling. Environmental remediation models use situational information to perform calculations representing incident scenarios. Model outputs are risk or dose — which can be used to derive risk. The guiding assumptions essential for modeling can be generic, allowing preliminary assessments of risk predating unavailable more precise data, or they can be comprehensive to achieve site-specific estimates. All remediation models consider the radioisotope composition, the material contaminated, the exposure scenario, the biological absorber, and the designated output. Often, models include chemical contaminant input options because the combination of radioisotope and chemical contamination is (unfortunately) not uncommon. Accordingly, cancer risk can be derived for a multicomponent scenario. Remediation assessment methodologies include online calculators, software, spreadsheets, and tabular reports. These approaches consider either contaminant distribution kinetics or steady-state assessments.

The EPA's Preliminary Remediation Goals (PRG)[7] electronic calculator[8] takes user input parameters and calculates, via programmed equations, remediation goals for residential situations, industrial settings, groundwater, and agricultural products that have been exposed to contaminants by soil, tap water, and living organisms (Figure 11-18). Note that this program cannot address ongoing airborne plume diffusion or water dispersion in rivers, streams, or oceans. PRG is limited to steady-state scenarios. The EPA recommends that the remediation manager initially identifies preliminary goals using default input options and then modifies them later in the calculation process to align more closely with site-specific parameters. The calculator user enters basic information in online data entry fields — the radionuclides present, the contaminated media (water, soil, air), land use, and human exposure assumptions. Effective dose assessments for each radionuclide convert to carcinogenic potential using cancer risk slope factors (as in Figure 11-14) provided by the Oak Ridge National Laboratory (ORNL) Center for Radiation Protection Knowledge. A second EPA electronic calculator, the Dose Compliance Concentration (DCC) calculator, differs from PRG in only three ways: DCC uses target dose rate rather than cancer risk so, the

[7] The full title of the calculator is "Preliminary Remediation Goals for Radionuclide Contaminants at Superfund Sites."

[8] https://epa-prgs.ornl.gov/radionuclides/.

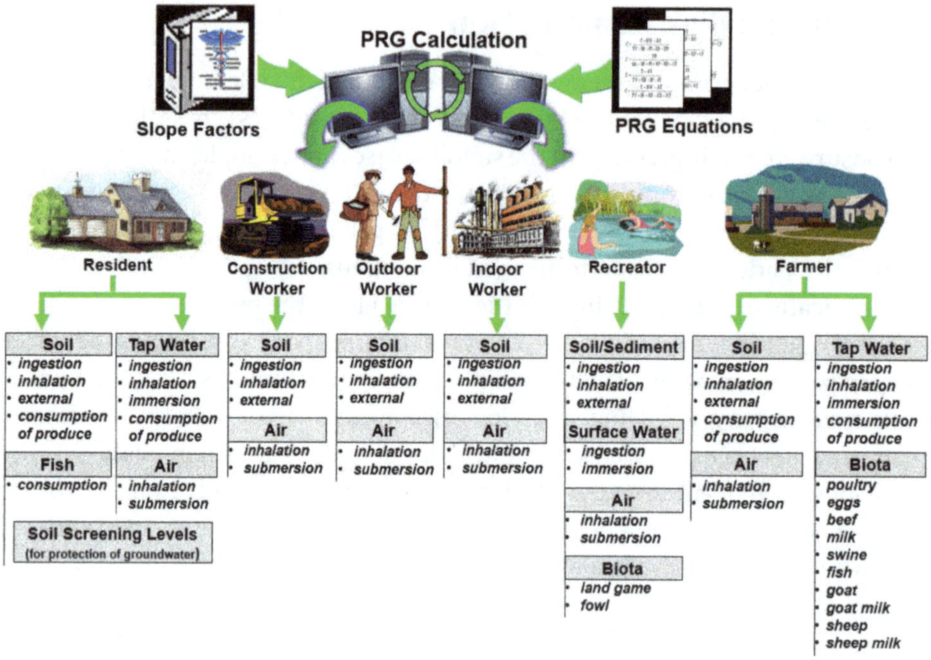

Figure 11-18 Flow diagram for the EPA's PRG calculator. Calculations are performed determining the cancer risk derived from each bullet point and then totaled. Reproduced with permission of the illustrator, Fredrick Dolislager, (Shubayr, 2017).

slope factor of the PRG is replaced by a dose conversion factor, and the exposure is calculated for 1 (peak dose) year.

Software remediation programs resemble online calculators except that the user downloads the program prior to use. The DOE published one such program, the Residual Radioactive Material Guideline (RESRAD-ONSITE),[9] in 1989 (Figure 11-19). DOE maintains and regularly updates this software through the Argonne National Laboratory. RESRAD-ONSITE considers *external exposure* from contaminated soil and *internal exposure* from inhalation of airborne radionuclides (including radon), and from ingestion of foods (grown in contaminated soil including soil irrigated with contaminated water), meat or milk (livestock fed with contaminated fodder and/or water), and fish (residing in contaminated water). The code supports both deterministic and probabilistic analyses. Site-specific soil input parameters include chronological data, contaminated zone characteristics, active hydrological and cover reduction, percent hydrological saturation, occupancy, inhalation, radiation type, ingestion parameters (such as diet), radon concentration, storage factors, ^{14}C data, and biota factors. RESRAD-ONSITE provides a standalone utility program, the Dose Conversion Factor (DCF) Editor, that permits selection of ICRP dose coefficient (e.g., equivalent dose), transfer factor (efficiency of radioisotope absorption from ingested material), and slope factor (e.g., mortality risk from external exposure). The output report (in text and graphics) provides a summary of steady-state ground, food, and water pathways, time-dependent contamination concentrations, cancer risks, and an uncertainty analysis. In addition to the United States, other nations offer online remediation software as well. NORMALYSA,[10] maintains a set of selectable modules (sources, cover layers, transports, receptors, and doses) that exchange information. HGC-NORMALYSA can provide flow distribution dynamics and radiological impacts from contaminated land and surface waste

[9] http://www.evs.anl.gov/resrad/.
[10] From NORM and LegacY Site Assessment, http://project.facilia.se/normalysa-hgc/software.html.

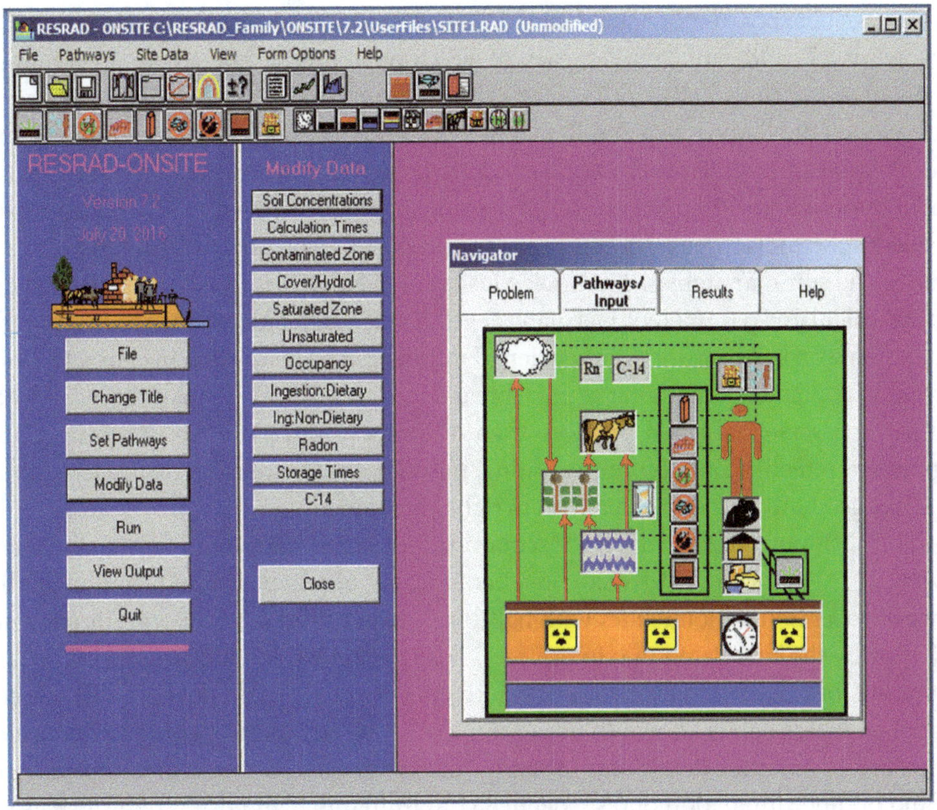

Figure 11-19 Homepage of the RESRAD-ONSITE software package. Reproduced with permission of Argonne National Laboratory, managed and operated by UChicago Argonne, LLC, for the U.S. Department of Energy under Contract No. DE-AC02-06CH11357.

(such as uranium tailings) by inputting groundwater radionuclide concentrations and flux simulations output by the HGC TPC software (National Central University, Taiwan). The user accordingly obtains a time series of radionuclide concentrations and fluxes at the interface between the contaminated groundwater and target environments. Lastly, the Polytechnic University of Madrid developed CROM,[11] a code performing assessments in rivers, lakes, and marine environments as well as in contaminated atmosphere and soils. The CROM databases are derived from IAEA and ICRP publications.

Excel-based models deliver exactly what you would expect — interactive Excel spreadsheets. In the United Kingdom, the Department for Environment, Food and Rural Affairs (Defra) recommends using the Radioactively Contaminated Land Exposure Assessment (RCLEA)[12] Excel-based calculator to verify compliance with Defra Part 2 regulatory requirements. Interactive pages provide buttons (e.g., reset or calculate) and informative text along with modifiable spreadsheets. The mathematical models and input data calculate radiation doses from radionuclides in soil to establish the level of contamination and provide remediation guidance. As in the previously presented software and calculators, a preliminary assessment can be formulated using preset values and then later refined when site-specific data become available. Four land use options are presented: residential (with or without home-grown vegetables), allotments (less frequently occupied acreage), and commercial/industrial. Situational factors include building type (timber

[11] ftp://ftp.ciemat.es/pub/CROM.

[12] https://www.gov.uk/government/publications/rclea-software-application

framed or brick), age of exposed individual (adult, infant, or child), and sex of exposed individual. The user may select "worst case" or "default" modes, each of which populate the spreadsheets with predetermined entries. Individual spreadsheet cells can be overwritten to provide site-specificity. Likewise, the state of New Jersey developed the Radioactive Soil Remediation Standards (RaSoRS)[13] workbook to support rapid compliance evaluation and standards development. New Jersey intended the worksheet to be used inhouse, but it is available online and illustrates another example of spreadsheet-based remediation appraisal. RaSoRS considers six radioisotopes (^{238}U and selected progeny), two construction modes (with basement or slab on grade), and summary tables. The book consists of five sheets (tabs): results, assumed parameters, subchain factors, calculated parameters. Germany developed a "Calculation Guide to Mining" that assesses public and radiation worker exposure from accidental mining radiation release. In characteristic twentieth-century Germanic contrition, the name of the spreadsheet — WISMUT—pays homage to Saxony and Thuringia, regions adversely affected by more than 40 years of unrestrained uranium processing. The worksheet applies to remediation, decommissioning, installation, and reuse. The model is not available in English and is not generally accessible due to copyright agreement.

In 1999, the NCRP published a report, "Recommended Screening Limits for Contaminated Surface Soil and Review of Factors Relevant to Site Specific Studies," aka Report No. 129. The publication provides guidance for over 200 radionuclides with half-lives greater than 30 days, with limits determined by dividing 0.25 mSv by the calculated maximum total dose per unit soil concentration. The derived doses are not appropriate for calculating estimates of health effects, rather the report provides screening limits that can be applied where soil is known to be contaminated. Nonetheless, Report 129 considers external radiation exposure (β-ray skin dose), and ingestion of contaminated foodstuffs, ingestion of soil, and both indoor and outdoor inhalation of resuspended contaminants. Land use is described as agricultural, heavily vegetated pasture, sparsely vegetated pasture, heavily vegetated rural, sparsely vegetated rural, suburban, no food suburban, and construction/commercial/industrial. Figure 11-20 summarizes several available remediation models and classifies them according to type.

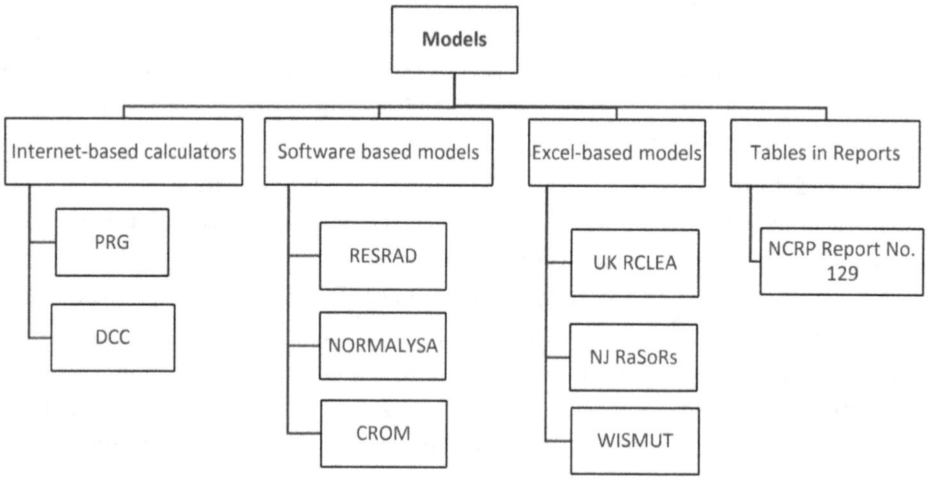

Figure 11-20 Summary of radiation remediation computational methodologies. Reproduced with permission of Shubayr (2017).

[13] http://www.nj.gov/dep/rpp/rms/agreedown/NJrasorsver60.xls

11.7. Radiation Protection and Exposure Remediation

While health physicists concerns themselves with environmental remediation, RSOs and medical professionals scramble to protect first responders and those suffering from ensuing health detriment. First responders are individuals responsible for preserving life as well as protection of property, evidence, and the environment during the earliest stages of a mass casualty event or other emergency. We designate healthcare workers in hospitals or other facilities where victims arrive as first receivers. Clearly, the actions of first responders and first receivers must be elegantly choreographed to protect not only the public, but also themselves. Let's now look into the processes and rationale for personal protection and health remediation. Recall that the increased risk of cancer among Japanese atomic bomb survivors is estimated at 8% per sievert. Extrapolating to a generic radiation emergency, for LD_{50} around 5 Sv, the anticipated elevation in cancer incidence among surviving responders roughly doubles. The NCI estimates the natural occurrence rate for all cancers in the United States at 442.4 new cases annually per 100,000, or 0.44%. When doubled, this risk remains below 1%. Thus, during an ongoing crisis, the eventual prognosis of cancer is neither urgent nor high priority. Radiation workers and survivors are put at risk through three mechanisms: *exposure* to photon or subatomic particle irradiation, *contamination* by radioactive debris, or *internalization* of radioactive substances. Exposure often occurs without warning and thus radiation safety protocols — administrative, engineering, and procedural — must be relied upon to first order. Should these preventive measures fail, first receivers are called upon to administer appropriate treatment to exposed victims. Radiation workers (including first responders) mediate contamination and internalization hazard reduction through personal protective equipment (PPE), including respiratory and dermal protective ensembles. Radiation workers are trained in the proper use of this equipment and are required to comply with all regulations regarding the selection of equipment and its use. The level of personal protection required depends on the duties of the responder.

Unfortunately, the United States endorses two independent classification systems for PPE levels of protection. The Occupational Safety and Health Administration (OSHA)[14] together with the EPA classify PPE as Level A to Level D (most protective to least protective). The National Fire Protection Association (NFPA)[15] approach uses Class 1 to Class 4 (most protective to least protective). The U.S. military uses their own internal third system, a mission-oriented protective posture (MOPP) scale: MOPP 4 to MOPP 0. For the sake of brevity, we shall describe the OSHA/EPA system in some detail and leave it to the interested reader to investigate the parallel NFPA or MOPP schemes. Level A PPE includes a positive pressure, full-face, self-contained breathing apparatus (SCBA) or a supplied air respirator (SAR) with SCBA-type auxiliary escape respirator. Body covering requires a totally encapsulating chemical- and vapor-protective suit with two sets of gloves, inner and outer, made of chemical-resistant material (e.g., latex). Steel toe and shank, chemical-resistant boots are also required. Openings, such as sleeves and pant cuffs, are sealed to gloves and boots using duct tape (it's a good idea to fold the end back on itself to create a tab for eventual PPE removal!). The requirements for Level B inhalation protection, boots, and gloves remain the same. Clothing requirements are less severe allowing chemical-resistant overalls with a jacket, or coveralls. Level C PPE may be used only when airborne radionuclide concentrations do not pose an imminent threat to life or health. A half mask or negative pressure air purifying respirator (APR), overalls or a two-piece chemical splash suit (may be disposable), and disposable outer boot covers may be worn. Double gloving is still required. Level D PPE requires a face shield or goggles (if potential contamination persists), disposable surgical gowns or scrubs, shoe covers, and gloves. Openings should be secured closed with tape at all levels. A dosimeter must be worn above the waist on

[14] https://www.remm.nlm.gov/osha_epa_ppe.htm.

[15] https://www.remm.nlm.gov/nfpaclasses_ppe.htm.

the outside of the PPE ensemble for obvious reasons, and it is a terribly good idea to wear an adhesive identification label, such as your name and vocational position written in Sharpie on tape. As the level of protection passes from A to D, the ensemble becomes less restrictive, heat insulating, and psychologically impactful, encouraging responders to comply with the lowest safe level of protection. First responders may rely on Levels A and B for protection; first receivers use Level C or Level D ensembles.

We have learned much about the treatment of ARS since Marie Curie and Wilhelm Roentgen succumbed to its effects. The Radiation Emergency Medical Management (REMM) website (U.S. Department of Health and Human Services, 2020) offers a comprehensive guide for the most recent recommendations from the CDC. Most treatments are approved by the Food and Drug Administration (FDA) but in cases of emergency such as mass casualty events, emergency use instructions can be provided by the CDC permitting treatment without FDA emergency use approval. Recommended materials and doses always carry the caveat, "if sufficient supplies are available." For exposures of less than 2 Gy (full body or significant partial body), the CDC recommends patient observation with treatment reserved for specific symptoms, such as skin burns that should be treated with antibiotics to avoid infection. Blood transfusions are contraindicated as they delay regeneration of blood-forming organs. Exposure dose can be estimated through any of three ways. The onset of vomiting provides a rough estimation (the higher the dose, the more common, rapid, and severe the response). A CBC reveals a dose-dependent reduction in white blood cells (Figure 8-7) but may not be accurate for more than 24 hours. Lymphocyte counts of 0.8–1.5 g per liter indicate a dose of less than 2 Gy; the count drops to 0.5–0.8 g per liter at 2 Gy and decreases to 0.3–0.5 g per liter at 4–6 Gy. The most precise biodosimetry can be achieved using lymphocyte chromosomal analysis, but this technique is expensive, time consuming, and impractical during an ongoing crisis (International Atomic Energy Agency (IAEA), 2001).[16] When practical, a technician draws a blood sample, separates, and prepares the lymphocytes for chromosomal (PCC) analysis, scores the average number of dicentric chromosomes per lymphocyte, and compares that number to a standard, empirically derived dose response curve depicting dicentric chromosome occurrence. Although the clinical sensitivity of the assay achieves about 0.2 Gy (200 mSv) resolution (experimental conditions may achieve sensitivity as low as 0.05 Gy), that dose is two orders of magnitude above recommended safety limits. Furthermore, natural background aberration occurrence depends on patient age. A healthy young adult can be expected to experience 8.2 dicentric abnormalities per 1,000 cells following a 0.25-Gy exposure. High LET radiation introduces increased chromosomal aberration efficiency. Partial body exposure results in localized pools of affected lymphocytes; homogeneous redistribution of affected cells requires about a day.

For exposures of greater than 2 Gy, the treatment protocol rule of thumb recommends: (1) treat only specific symptoms, (2) in case of skin wounding, isolate victims to avoid infection and provide antibiotics, and (3) avoid physical trauma to prevent bleeding and provide plasma infusion if hemorrhage occurs. Again, transfusions are to be avoided as that treatment delays blood-forming organ regeneration. The CDC recommends prophylactic myeloid cytokine administration. Myeloid colony stimulating factors (Neupogen, Neulasta, and Leukine) have been shown to improve survival, shorten the duration of severe neutropenia, and minimize neutropenia-associated complications including infection. The CDC also recommends administration of a platelet cytokine, Romiplostim, which was approved by the FDA for hematopoietic ARS in 2021. For irradiation doses above the LD_{50}, stem cell transplantation is advised only if the bone marrow is unlikely to recover and the procedure is medically and situationally feasible.

[16] Available online at https://www-pub.iaea.org/MTCD/Publications/PDF/TRS405_scr.pdf.

11.8. Nuclear Terrorism

Cautionary provisions have established agencies and guidance responsive to any mass radiation casualty event, including civilian accidents such as nuclear power plant mishaps or the remote likelihood of a nuclear terrorist attack. Thus, the reactions and remedies described in this section provide a continuation of the previous discussion regarding radiological accidents applied to a specific cohort of events. Nuclear terrorist events include the detonation of stolen weapons as well as improvised nuclear devices (IND) and radiological dispersal devices (RDD or "dirty bombs"). The NRC's mandate expanded to respond to these events as well as radiological exposure events (not including reportable radiation therapy mishaps), accidents involving transportation of nuclear materials, and nuclear power/nuclear reactor incidents. In this context, an exposure event includes instances of solid sources placed surreptitiously in public settings — the purposeful exposure of unsuspecting individuals to noncontaminating sources of radiation. Now, because within its regulatory capacity the NRC supervises the use of radionuclides in medicine, accidents such as the one at the Indiana Regional Cancer Center are reportable to the NRC as well, urgently if, for example, the source has gone unnoticed and has ended up in the laundry, and subsequently in the public sewer system. Most often, NRC reporting is performed by responsible authorities — managers of facilities, RSOs, military officers, or law enforcement officers. Nonetheless, anyone made aware of a radiological or nuclear incident should contact the NRC immediately. An occurrence may be obvious. An explosion at a nuclear power plant would indicate the possible presence of radiation. The discovery of a brachytherapy pellet on a bus would cause alarm. However, without the assistance of radiation detection equipment, most scenarios initially appear to be non-nuclear events. Often, radiation exposure is recognized only after a cluster of ARS victims has presented at local hospitals. Although the U.S. Department of Homeland Security (DHS) provides backpack radiation detectors and personal radiation detectors (PRD) to interstate transportation companies (e.g., AMTRAK) as well as to local police securing high visibility events (e.g., visiting dignitaries), routine deployment of real-time radiation detection is scattershot, at best. For most police departments, accurate high dose rate radiation detection equipment is prohibitively expensive. In its report to congress in 2020,[17] the Countering Weapons of Mass Destruction (CWMD) Office reported they were "exploring approaches to make low-operational burden [radiation] sensors available for a wide variety of law enforcement vehicles," to "fuse radiation data with contextual sensors and computer vision algorithms that continuously identify, track, and classify objects in the scene into categories useful for radiation propagation modeling." The CWMD Exploratory Research Program has initiated pilot feasibility evaluations of automated sensor reporting. The program's objectives aim to decrease the time to detect and identify radioactivity and the cost of components and operation. Because detection equipment currently is neither continuously monitoring nor ubiquitously present, and because state and federal resources are not instantaneously available, the first 100 minutes of any terrorist attack is a critical period for protection of life and property, crime scene preservation, and environmental containment (Figure 11-21).

Once notified of a radiological emergency, the NRC alerts the Federal Radiological Monitoring and Assessment Center (FRMAC) within the National Nuclear Security Administration (NNSA) under the Department of Energy (DOE). A broad collaboration of advisory agencies contributes to the FRMAC brain trust including the EPA, DHS, Department of Health and Human Services (HHS), CDC, FDA, Department of Agriculture (USDA), NRC, and perhaps a dozen other agencies. FRMAC maintains a data analysis and consequence management team that assesses, using internally developed software programs, the appropriate response level to the emergency. Does the incident involve exposure,

[17] https://www.dhs.gov/sites/default/files/publications/cwmd_-_research_and_development.pdf.

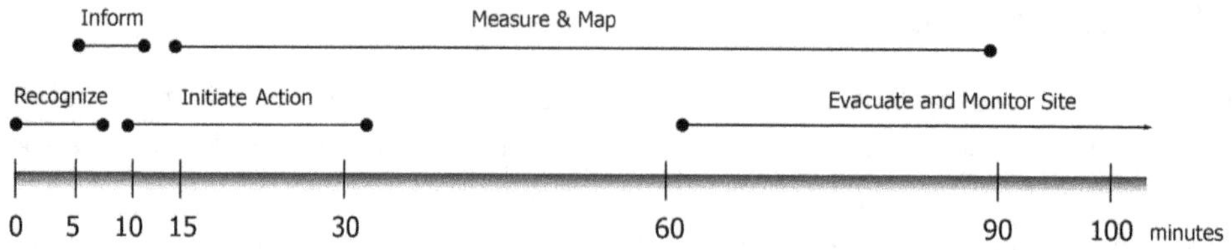

Figure 11-21 RDD mission response timing recommendation from the DHS. Recognition of a radiological event takes place within 5 minutes. Responders, public, state, and local agencies are informed of the event and ensuing precautions within 5–10 minutes. Lifesaving rescue and crime scene control actions begin no more than 10 minutes into the crisis, prior to complete understanding of radiation status. Radiation levels are measured and mapped over the next hour, starting no later than 15 minutes after detonation. Decontaminated and stabilized victims are ready for transport to hospitals approximately an hour postdetonation. Radiation and situation monitoring continue until no longer required logistically or for safety.

contamination, internalized radioactivity, or a combination of circumstances? According to the situational requirements, FRMAC rapidly deploys manpower and equipment to the incident site, where it taps state, local, and tribal agencies as appropriate for consultation with the advisory team at FRMAC. Deployed equipment includes heavy machinery, radiological and geological (including meteorological) monitoring equipment, fixed wing and rotary aircraft, communications command centers, field hospitals, and mobile testing laboratories. Deployed manpower expertise spans heavy equipment operators, field scientists, information technology (IT) professionals (including document control), medical technicians and physicians, administrative personnel, and radiation scientists. The field team is in constant contact with the crisis management command center at FRMAC. Although the logistics resemble military action, FRMAC only advises and assists governmental and law enforcement agencies charged with protection of public interests. The Center is not charged with any policing action mandate.

Nonetheless, state and federal crisis support is unlikely to arrive at the scene of a radiological incident for at least several hours with full deployment requiring more than a day. DHS released a document (National Urban Security Technology Laboratory, 2017) intended to help local agencies identify gaps in their preparedness and to guide their actions during the initial hours following RDD detonation, but the report guidance can also inform response to other explosive radiological events. According to this document, both goals and timing are key to achieve containment and minimize harm. Potential radiation hazard should be identified within the first 5 or 6 minutes of an explosion. If radiation detection equipment is unavailable in the short term, the event should be considered a radiological event. The presence of radioactivity should be confirmed or rejected within the next 5 minutes and upon confirmation, the scene must be secured and managed to prevent unnecessary contamination dispersion. Immediately following declaration of a radiation event, command authorities should be notified. The first two of these recommendations require first responder periodic training in anticipation of an extremely rare event and routine deployment of broad dose rate range, high dose, calibrated detectors. The cost-benefit analysis is tricky for cash-strapped local authorities. Life- and limb-saving rescue activities should begin no more than 5 minutes after the initial explosion. Additional assistance should arrive at the scene within 20 minutes and these responders can begin measuring radiation and mapping dispersion. The arrival of equipment and medical personnel will enable onsite victim triage and phased evacuation of the injured, followed by release of decontaminated, ambulatory persons. Transfer of the injured to local hospitals may be delayed by more than an hour while contamination containment takes precedence. The command authority, usually the Chief of Police, should release public statements and perform other protective actions as soon as he or she receives the radiological incident report from the field (no more than 15 minutes into the crisis). The NRC, Federal Bureau of Investigation (FBI), and other partner agencies should be notified immediately thereafter. Finally, the crisis management command center should initiate documentation and data management activities. Within 100 minutes, the local authorities should be completely in control of the situation, ready to partner with state and federal agencies upon their arrival.

Fortunately, we cannot present a historical terrorist radiological emergency to illustrate how this scenario might unfold. Purely as a mechanism for enlightenment purposes, let's employ the 2013 Boston Marathon bombing tragedy and *imagine* that radioactive material had been included with the hardware components of the two pressure cooker bombs detonated at the finish line. The possibility is not far-fetched; much of the radioactive material utilized in industry throughout the world is poorly managed. In addition to the solid source misappropriations noted in Table 11-8, a 23-kg radioactive ^{192}Ir source used for industrial radiography went missing from the back of a pickup truck in Malaysia in 2018. This is only one example of several reported missing sources that have not resulted in known radiation injury. So, let's fabricate a tale wherein some ^{232}Th obtained from lantern mantels, ^{241}Am found in smoke detectors, ^{3}H extracted from gun sites, several purloined brachytherapy pellets, and perhaps an abandoned ^{60}Co source were tossed into the pressure cookers along with gunpowder, ball bearings, and nails. Without real-time radiation detection equipment, victims, witnesses, and first responders could not have determined the radiation status of the crisis immediately following the explosion, however, most large city police departments like Boston routinely deploy radiation safety equipment at soft target events and periodically conduct terrorist training. The proposed imaginary radioactive materials would emit primarily gamma radiation along with alpha and beta particles, producing a mixture of effective dose w_R and w_T factors dependent upon exposure pathway. The two bombs were detonated at 2:49 in the afternoon. Because the marathon was a large crowd, soft target event in a large city, at least one police vehicle would have carried radiation detection equipment. Patrolmen at the scene would have immediately

Figure 11-22 Boston Marathon 2013 bombing aftermath. Classifications of potentially contaminated persons prior to confirmation of an RDD event: (1) injured victims, (2) race spectators, (3) police, (4) good Samaritans, and (5) emergency responders. Photo courtesy Aaron Tang from Cambridge, MA — DSC03156, CC BY 2.0, https://commons. wikimedia.org/w/index.php?curid=25613055.

recognized that the explosion was potentially an RDD and would have retrieved two detectors to take duplicate readings at different locations within 20 m of the center of the explosions. Once the presence of radiation was confirmed, the designated command officer would have informed all enforcement personnel at the scene including those deployed for marathon safety (firemen and EMT) and would have reported the radiation hazard to the task force at headquarters. The task force personnel would have then informed all dispatched responders in transit to the scene that a radiation hazard existed. Command at headquarters would also have notified the NRC and would have issued (preapproved) public messaging to notify persons within 500 m to shelter in place and others to stay away from the scene. Meanwhile, emergency responders at the scene — police, EMT, firemen, and volunteers — would have begun emergency life- and limb-saving procedures. This action would take place immediately without regard for radiation hazard. During the actual Marathon event, three victims died at the scene, several others lost limbs (mostly legs because the bombs were placed at sidewalk level), and these folks could not have waited for decontamination or scene control. As quickly as possible, responders would have donned PPE and marked dangerously radioactive hot spots, cordoned them off and removed persons (including responders) to less dangerous areas. As during the actual event, officers would search for additional devices and hazards. In our fictitious RDD scenario, a team would have been designated public safety officers, and they would have begun to set up a perimeter 20 m from the center of the explosions. This area comprises the preliminary crime scene and, for RDD incidents, the initial hot zone per contamination identification. The boundary would be refined following radiation mapping wherein three strike teams, in two sequential phases, would have determined radiological contamination, located nonuniform high radiation areas, and provided survey data for mapping. The second strike team would have characterized the radiation. For example, was alpha radiation present and what size particles had been emitted? The third team would have redefined the hot zone as any area emitting 6,000 dpm/cm^2 at 0.5 cm above the ground and relocated the barriers. Ambulatory persons would have been divided into two groups. Those in the hot (radioactive) zone would have been surveyed for contamination before being released to a cold (background levels of radiation free from transported contamination) zone. They would have been examined for minor injuries at a warm (controlled contamination) zone triage center — perhaps simply an ambulance at the edge of the hot zone or something more elaborate as equipment became available — and then would have been decontaminated. Clothing may have to be removed, replaced with clean wear, and contained. Contaminated skin and hair would be washed, but the runoff must be controlled (allowable limits may be flushed into the sewer system). Personal possessions would have remained in place at the site. The bombs were placed at the finish line at 2:40 pm, so if our imaginary gamma radiation had escaped the shielding of the pressure cookers, spectators standing nearby would have been exposed to an unknown dose rate for no more than 9 minutes. Exposed persons that were not victims of the concussive explosion would likely have self-reported to hospitals if ARS symptoms presented. Those outside the hot zone — 5,700 runners continued to approach the finish line until 2:57 pm — would have been diverted by police to the Boston Common and Kenmore Square, as they were during the actual event. Ambulatory, decontaminated persons would have been provided with an instruction sheet describing possible symptoms and recommended precautions (e.g., consumption of iodized salt) and released to go home via mapped safe routes. The injured — there were 264 treated at 27 different hospitals — would have been stabilized and decontaminated (including the removal of shrapnel) to the extent possible before being transported in ambulances prepared as mobile warm zones with disposable protective liners and EMT dressed in PPE. Safe evacuation routes would have been established as part of the mapping process; this is one of the reasons why evacuation can be anticipated to begin only after an hour. The windows of apartment buildings within two city blocks of the blast were blown out. These interior spaces would need to be scanned for radioactivity and the inhabitants treated as ambulatory

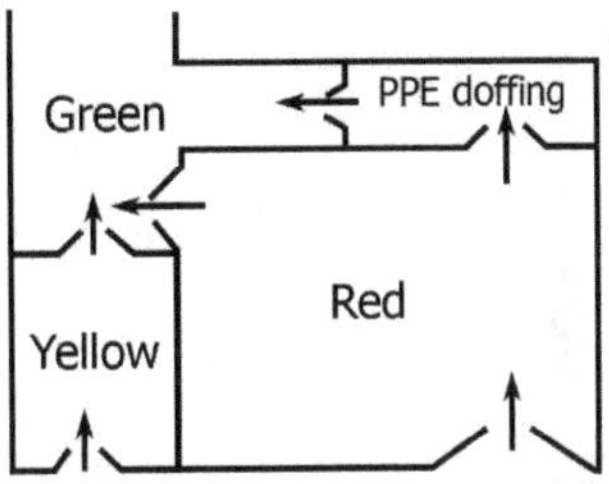

Figure 11-23 Typical hospital floorplan for radiation contamination control. Red zone (medical emergency): seriously ill or injured persons; hot zone. Bypass decontamination. Remove clothing. Emergency care. Stabilize patient. Yellow zone (medical urgency): decontaminate patient in an area designated as hot before entering zone; warm zone. Rapid radiological triage. Provide urgent care. Green zone (ambulatory patients and those received from Red and Yellow zones): decontaminate patient before entering; cold zone. Thorough radiological survey. Full service medical treatment (access to uncontrolled hospital areas). Medical personnel remove PPE when leaving the Red zone. PPE removal may take place within the Yellow zone.

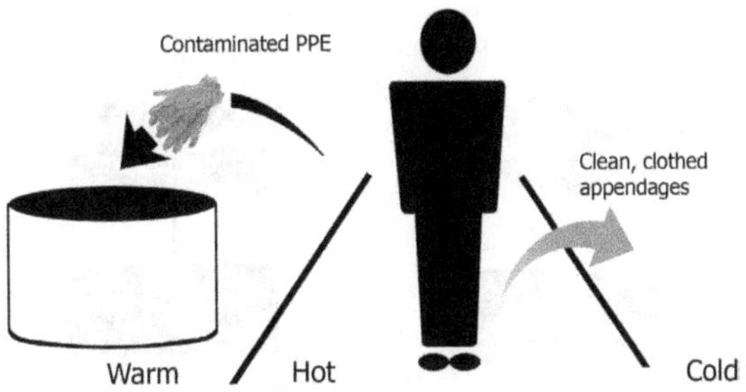

Figure 11-24 Doffing PPE. Rule of thumb: remove contaminated clothing from top to bottom. The responder stands in a designated hot zone. Head coverings are removed first, followed by external apparati such as breathing equipment, and dropped into warm area. PPE suit is opened and unrolled, peeling away like a banana to prevent contaminated outer surface from touching uncontaminated undergarments. Shoes or coverings are removed and deposited into warm area. The exposed, uncontaminated foot steps down into the cold area followed by the second foot as it becomes uncontaminated. Lastly, gloves are removed from hands still hovering over the hot zone, outer gloves removed first and deposited into warm zone, then inner gloves, deposited into warm zone. Finally, the attendant is scanned for radioactivity.

victims within the hot zone. In summary, there is a lot to be accomplished in a very short time . . . and a lot of people to be controlled in a very scary situation.

During the hour preceding the arrival of contaminated, injured victims at hospitals, administrators must transform hospital triage and care protocols to prevent exposure of additional innocents to radiation. A predetermined hospital radiation emergency floorplan for admission of contaminated patients provides guidance for hastily posted signage to direct emergency and hospital personnel to the correct areas, and to prevent the entrance of improperly attired persons. Hallways must be designated hot or cold and unidirectional traffic must be restricted to the correct route. Figure 11-23 provides an example of a receiving hospital floorplan. Three color designations indicate both medical triage and radiation control. All attending personnel must don appropriate PPE before entering the controlled zones and doff PPE according to proper protocol (Figure 11-24). Seriously injured or vomiting patients are admitted to the red zone. Prior to entrance, or in a designated area of the zone, the patient's clothing is carefully removed, and the patient is covered with a clean sheet. That simple action can eliminate 90% of the contamination. Clothing is discarded into a radiation containment bin fitted with a lid. First, the patient is medically stabilized and then surveyed for contamination. Radioactive and visible cold shrapnel gets removed first and then all wounds are cleaned, decontaminated, and sutured, if necessary. Hot spots can

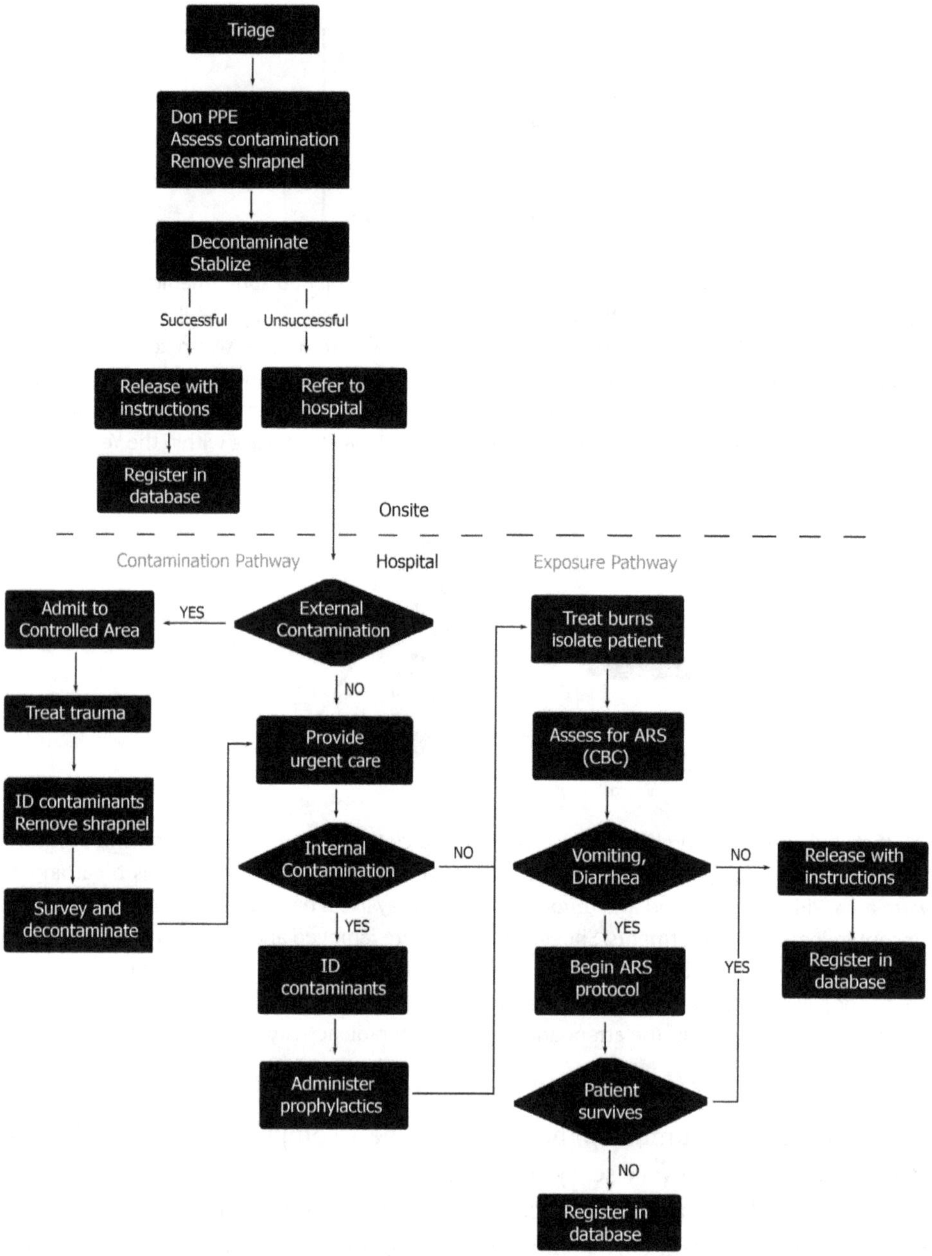

Figure 11-25 Generic triage plan for persons evacuated from a radiological incident zone. At the incident site, survivors are attended to in the order of injury severity. Apparently healthy, ambulatory survivors are self-decontaminated under supervision, scanned for radioactivity, evacuated to a controlled area, and released through a safe evacuation route. Injured and contaminated survivors are stabilized, decontaminated, and either released or transferred to a hospital for additional care under controlled conditions.

be encircled using a Sharpie pen. The patient may be decontaminated three times using mild soap and tepid water (no cold or hot water or scrubbing), including any attempts at the incident site, with a goal of no more than twice background radiation levels. Further attempts at decontamination are not productive (reductions of no more than 10%) and may be harmful. Identified residual hot spots can be covered with waterproof dressing to prevent shedding. Next, medical personnel cleanse body entrance cavities such as ears, nostrils, mouth, and eyes. All decontamination should be directed to prevent water from flushing over decontaminated skin. Also, before decontaminating the nostrils, a swab can be taken to help analyze inhaled contaminants, including alpha-emitting dust. Remember, the ^{241}Am present in our tale is an alpha emitter. The patient then can be released to the green zone for CBC analysis, x-rays (to identify embedded shrapnel), gamma camera imaging of internalized isotope sequestration, surgery, hospitalization, or release. Attending medical personnel retire to the designated buffer zone to doff PPE (Figure 11-24), enter the green zone, and don clean PPE before returning to the red zone. Ambulatory patients enter the hospital through a yellow zone where the process does not significantly differ from that performed in a red zone except that emergency treatment is unnecessary. Following decontamination and urgent care, the patient is released to the green zone for follow-up examination and procedures. Attending personnel doff PPE in a designated space within the yellow zone (with a direct cold route to the exit), enter the green zone and don clean PPE prior to reentering the yellow zone. The green zone should be scanned frequently for radiation contamination and all personnel must wear radiation monitors. All patients (including deceased) are recorded in a database and released with advisories and instructions regarding trauma — such as concussion precautions and wound care — and radiation exposure — such as ARS symptoms, risk of organ dysfunction, internalized radioisotope countermeasures, immunocompromise, and future cancer risk. Figure 11-25 provides a decision flowchart for medical/radiological triage following a radiation dispersion event.

11.9. Summary

- The RSO provides the first and last word regarding the use of ionizing radiation within his or her facility.
- Regulatory oversight falls under the jurisdiction of IAEA, ICRU, ICRP, ANSI, ASTM, DOE, NRC, NCRP, NRC, DOT, EPA, and individual state regulatory laws requiring licensing and oversight.
- Radiation protection is guided by a goal of keeping exposure ALARA.
- ALARA can be accomplished through four mechanisms: minimizing the time of exposure, maximizing the distance between source and subject, providing engineering controls, and enforcing administrative controls.
- Regulatory limits set on exposure to x-ray radiation are 50 mGy per year for adult radiation workers and 1 mGy per year for the adult public.
- During any radiation-involved situation in which you feel uncomfortable or uncertain, move away as quickly as possible — without jeopardizing the safety of other individuals.
- Radiology exposes patients through external radiation (0.01 cGy–7.0 cGy per procedure) or internalized isotopes (~1 cGy per procedure).
- Dose from nuclear medicine radiochemicals depends on the biological uptake and metabolism of the biomolecule, and physical characteristics (emission and half-life) of the radioactive tag.
- Free radioisotopes are taken up according to delivery methodology and the natural biological metabolism of periodic table neighbors (elemental characteristics).
- The deposition of inhaled particles depends on the AMAD.
- For internalized radioisotopes, radiation particle type becomes important with respect to chromosomal damage (LET) and penetration potential (target organ dose).
- Biological retention of radioactivity depends on biological half-life (T_b) and physical half-life (T_R) approximated as $\frac{1}{T_{eff}} = \frac{T_R T_b}{T_R + T_b}$.
- In practice, dose to internal target organs, derived from internalized radiation residing in source organs, is calculated as $\overline{D}(T \leftarrow S) = \widetilde{A}(nE)(\phi m^{-1})$, where values of $\overline{D}(T \leftarrow S)$, time integrated activity and specific absorbed fraction are tabulated for lookup.

- The field of radiation protection demands biological equivalency. The outcome of any radiological event is evaluated based on resultant biological effect.
- Biological effect is evaluated in sieverts (Sv).
- For a given organ, the equivalent dose, H_T, is the dose multiplied by the radiation weighting factor(s) (w_R) for each incident radiation. For an organism, the biological effect (effective dose, E) is evaluated as the sum of doses to each organ, modified by the radiation weighting factor for each radiation incident upon each organ, and the organ sensitivity factors (w_T) for each organ.
- Committed H_T and E evaluation for internalized radiation must consider diminishing activity with time. Therefore, the dose rate must be integrated over time.
- Risk estimates need not be highly precise provided the approximation overestimates the actual value, erring on the side of safety.
- Biological radiation response expresses as either deterministic or stochastic. Organ failure resulting from cell death is deterministic. Cancer occurrence is stochastic.
- To arrive at an approximation of risk to a population, the collective effective dose must be evaluated first: $S = \sum_i E \cdot N_i$.
- Risk is the likelihood of epidemiological evidence for increased occurrence over normal incidence.
- Absolute risk is additive.
- Relative risk assumes that a risk factor multiplies the nominal rate of incidence.
- The *gold standard* data for radiation risk evaluation has been collected from survivors of the WWII atomic bombing of Nagasaki and Hiroshima.
- ARS presents as physiological symptoms leading to radiation-induced illness and death.
- ARS partitions into several subsyndromes: the hematopoietic syndrome, the gastrointestinal syndrome, the neurovascular syndrome, and a multiorgan dysfunction syndrome.
- The severity of ARS is a function of dose and dose rate, exposed volume, homogeneity of dose distribution, radiation composition and it develops over time.
- The ARS time course follows distinct phases: the prodromal phase, the latent phase, manifest illness, and the final phase.
- ARS organ failure (e.g., skin burns, CBC, bone marrow collapse), and mortality are deterministic.
- Radiation-induced cancer incidence can be determined only as an increase over nominal cancer occurrence.
- Cancer risk from radiation exposure is both stochastic and linear (from ~0.2 Sv to ~10 Sv).
- The LNT model holds that cancer risk is linear at less than 0.2 Sv. Other models indicate that the risk may be nonlinearly increased or decreased at low doses.
- Heritable risk results from radiation-induced haploid genome mutation in germ cells.
- In mice, heritable mutations are dose and dose rate dependent, and stochastic.
- No heritable mutation risk has been identified in humans.
- Radiogenic risk to the unborn is gestation stage dependent.
- Zygotes and embryos irradiated preimplantation either survive unscathed or fail to implant and develop. Risk of loss occurs at >10 cGy and increases with dose.
- Abnormalities result from cell death and risk is therefore deterministic.
- As *in utero* differentiation progresses from omnipotent cells to tissue-specific stem cells, the risk of deformity progresses from low to organ-specific to a generalized reduction in growth.
- With respect to radiation protection, many agencies recommend dose limits (annual and lifetime) for various categories of persons. The limits fall into the mSv range (from about 0.5 to 50 mSv).
- All radiation accidents must be reported as soon as possible; injured persons must be stabilized immediately, radiation remediation follows.

- An historical appreciation of mass radiation exposure events promotes future radiation safety. For this reason, national and international law requires the reporting of radiation incidents.
- The EPA recommends an environmental remediation goal of one excess cancer in 10^4–10^6 persons.
- All remediation models consider the radioisotope composition, the material contaminated, the exposure scenario, the biological absorber, and the designated output.
- Remediation models for setting goals and analyzing risk are available in several forms including online calculators, downloadable programs, and Excel workbooks. Reports providing tabulated goals and input parameters for brute force calculations are also available.
- First responders are individuals responsible for preserving life as well as protection of property, evidence, and the environment during the earliest stages of an emergency.
- First receivers are healthcare workers in hospitals or other facilities where victims arrive.
- First responders and first receivers obtain protection from proper use of PPE, including the appropriate level of gear using correct donning and doffing technique.
- Radiological emergencies may remain unrecognized until radiation detection equipment arrives for use at the scene; so, first responders must be trained to anticipate undetected radiation hazard.
- Victims of radiation release suffer from one or more hazards including exposure, contamination, or internalization.
- Exposure dose can be estimated through any of three ways: the onset of vomiting, lymphocyte counts, or lymphocyte chromosomal analysis.
- For exposures of greater than 2 Gy, the treatment protocol recommends:
  - Treat only specific symptoms
  - In case of skin wounding, isolate victims to avoid infection and provide antibiotics
  - Avoid physical trauma to prevent bleeding; provide plasma infusion if hemorrhage occurs
- We strongly recommend that you bookmark the U.S. Department of HHS, Radiation Emergency Medical Management site at https://www.remm.nlm.gov/index.html.
- Potential radiation hazard should be identified within the first 5 or 6 minutes of an emergency. If radiation detection equipment is unavailable, all events should be considered a radiological event. Life- and limb-saving measures must commence at once.
- The presence of radioactivity should be confirmed or rejected within 11 minutes.
- If radiation is confirmed, the scene must be secured and managed to prevent unnecessary contamination dispersion.
- The presence of radiation hazard must be reported to the NRC.
- Following complete control of the scene, decontaminated persons may be released to return home or be transported to hospital. This should occur no more than 90 minutes after the initial event.
- Receiving hospitals set up controlled areas — red, yellow, and green — to protect patients and personnel, who don PPE as appropriate.
- Once notified, the NRC may deploy FRMAC manpower and equipment to the incident site, where FRMAC collaborates with state, local, and tribal agencies as appropriate.
- A broad collaboration of advisory agencies contributes to FRMAC including:
  - NRC
  - EPA
  - DHS
  - HHS
  - CDC
  - FDA
  - USDA

11.10. Problems

1. As the RSO working with Doctors without Borders, you need to design a vault for a medium-energy linac to be installed at Lusaka, Zambia. The machine will be used to deliver beam energies up to 6 MV for routine therapies and the dose rate will be held relatively low at 0.5 Gy per minute due to safety concerns. The hospital has a goal of delivering 50 fractions per day averaging 2.4 Gy at one meter in air per fraction. A simple floorplan resembling Figure 11-1 situates the linac head at the center of a 12m × 12m square room. Because the machine will be used to perform routine treatments, the linac head will be positioned either facing directly downward or directly upward (0^0 or 90^0) half the time and laterally (90° or 270°) another quarter of the time. The remaining quarter of the time other positions sort approximately equally. In this case, an RSO may assume a 30% average beam projection. The standard concrete mix in Europe and Africa has a density of 2.355 g/cm^3, providing a tenth value layer (TVL) of 34.3 cm for 6 MV incident x-ray radiation. How thick must the primary shielding walls (in the plane of linac head rotation — the thickest walls) be?
2. A radiology facility has purchased a new, fast scan CT to replace its older model machine. The 8 m × 10 m room was designed to provide radiation safety for a 250 keV machine and a patient load of 100 scans per week. The newer, faster machine emits up to 150 keV x-rays but will allow an increased throughput of 135 scans per week. The graph to the right enables an engineer to look up the required concrete thickness if *P* and *W* are given in Gy/week, *d* is in meters, and *W* is calculated at 1m from the source. Is the current 20 cm wall thickness sufficient to protect the radiation technologist? Show your work.

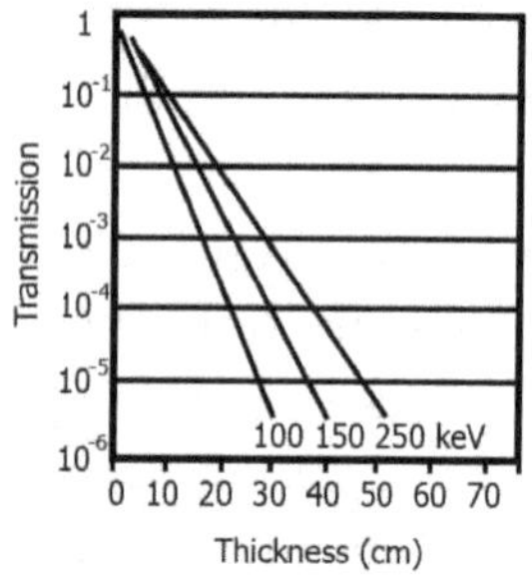

3. After administering each of the pharmaceuticals in Table 11-2, to which machine would you send the patient: PET or gamma camera? ^{11}C-choline is used specifically to visualize tumors of the brain, colon, esophagus, and lung. When would ^{11}C-choline be preferred over ^{18}FDG?
4. Isotopes incorporate into biomass in a manner similar to their stable, natural nearest neighbor on the periodic table due to chemical similarities, that is, outer electron shell configuration. The metabolic uptake difference is called the discrimination ratio. It considers the number of incorporated foreign molecules compared to the number of biologically favored molecules when the total number of each is the same. Assume 100 molecules of each. If 50 atoms of potassium (K) are metabolized to every 5 atoms of strontium (Sr), what is the discrimination ratio for Sr?
5. For contaminated ash/smoke particulates with an approximate diameter of 1 μm, what percentage of the radioactivity is anticipated to be deposited in the trachea, nasopharynx, and parenchyma?
6. According to the text, typical ^{18}FDG doses are 10^{-3} cGy per mCi, and typical administered activities are ~1,000 MBq. What is the typical patient dose from an ^{18}FDG procedure?
7. Low hematocrit may result from a decrease in RBC maturation rate or an increase in RBC loss. To determine which phenomenon is responsible for the low count, ^{51}Cr can be used to determine cell survival time. Because the isotope attaches irreversibly to RBC *in vitro*, blood can be drawn, incubated with ^{51}Cr, reinjected, and the activity can be monitored over time. If the biological half-life of bound ^{51}Cr is 30 days, what is the effective half-life?
8. Calculate the dose to a patient's thyroid if half of the injected activity, 2.0×10^7 Bq of ^{99m}Tc, is taken up by the thyroid. According to the MIRD lookup table, the ^{99m}Tc absorbed dose per disintegration within the source organ (assuming uniform distribution) can be estimated at 1.73×10^{-13} Gy(Bq s)$^{-1}$. Assume kinetics follow first-order exponential removal with a biological half-life of 4 hours.

9. Calculate the dose to a patient's bone marrow resulting from 4×10^4 Bq of ^{99m}Tc uniformly and perpetually bound in the patient's liver. The MIRD lookup table estimates the ^{99m}Tc absorbed dose per disintegration within the target organ at 1.20×10^{-16} Gy(Bq s)$^{-1}$.
10. A homeowner in Indiana requested that his basement be assessed for radon contamination prior to offering his house for sale. The report indicated that the annual effective dose from radon exposure was 2×10^{-3} Sv. What was the maximum annual equivalent dose to him and his family's lungs? What was the annual absorbed dose? What is their cancer risk if they lived in the house for 20 years?
11. To illustrate the influence of cell death on the linear-quadratic model for increased cancer incidence, suppose that $\zeta_1 = \alpha$ and $\zeta_2 = \beta$. In that case, $(\zeta_1 D + \zeta_2 D^2) = (\alpha D + \beta D^2)$. How does the increase in cancer incidence change as $(\alpha D + \beta D^2)$ goes from 1 to 10? Interpret your numerical answer.
12. A medical chest x-ray exposed an undeclared fetus to 10^{-5} Gy radiation. What is the risk of fetal detriment?
13. Using the PRG calculator's default settings for worst case scenario, find the cancer risk following our fictitious narrative of a dirty bomb incident at the Boston Marathon in 2013. Assume the brachytherapy sources included: ^{60}Co, ^{137}Cs, ^{192}Ir, ^{125}I, and ^{103}Pd. We are concerned with resident exposure and internalization of soil and tap water contaminated consumables. What is the default assumption for the quantity of soil an adult ingests per day? What is the mass loading factor for a lima bean?
14. Which RESRAD code would you use to determine human exposure following our fictitious narrative of a dirty bomb incident at the Boston Marathon in 2013? How are the contaminating isotopes selected?
15. In the RCLEA spreadsheet, how are the contaminating isotopes selected? According to the RCLEA library values, how many Brussels sprouts does the average adult eat per year? Examine the EffectiveDose and EquivDose sheets (tabs). What types of contamination are equivalent dose and effective dose used to evaluate?
16. What is the RaSoRS default parameter for human water consumption in liters per year? Transfer factors are expressed in pCi/g wet crop per pCi/g dry soil. Why?
17. List the HHS REMM instructions for all persons involved in a radiological/nuclear event.

11.11. Bibliography

Bolch, W. E., Eckerman, K. F., George, S. & Thomas, S. R., 2009. MIRD Pamphlet No. 21: A generalized schema for radiopharmaceutical dosimetry — standardization of nomenclature. *Journal of Nuclear Medicine March,* 50(3), pp. 477–484.

Chong, J. X., Buckingham, K. J., Jhangiani, S. N. & Boehm, C., 2015. The genetic basis of mendelian phenotypes: Discoveries, challenges, and opportunities. *American Journal of Human Genetics,* 97(2), pp. 199–215.

Feinendegen, L. E., Pollycove, M. & Neumann, R. D., 2012. Hormesis by low dose radiation effects: Low-dose cancer risk modeling must recognize up-regulation of protection. In: R. P. Baum, ed. *Therapeutic Nuclear Medicine.* London: Springer, pp. 789–805.

Hall, E. J. & Giaccia, A. J., 2019. *Radiobiology for the Radiologist.* 8th ed. Philadelphia: Wolters Kluwer.

International Atomic Energy Agency (IAEA), 2001. *Cytogenetic Analysis for Radiation Dose Assessment, a Manual; Technical Report Series No.405,* Vienna: IAEA.

International Commission on Radiological Protection, 1978. *Limits for Intakes of Radionuclides by Workers, ICRP Publication 30, Part 1,* Elmsford: Pergamon Press.

International Commission on Radiological Protection, 1991. *1990 Recommedations of International Commission on Radiological Protection. ICRP Publication 60,* Elmsford: Pergamon Press.

Little, M. P. *et al.*, 2009. Risks associated with low doses and low dose rates of ionizing radiation: Why linearity may be (almost) the best we can do. *Radiology,* 25(1), pp. 6–12.

National Academies of Sciences, E. A. M., 2021. *Committee on the Biological Effects of Ionizing Radiation (BEIR V).* [Online] Available at: https://www.nap.edu/initiative/committee-on-the-biological-effects-of-ionizing-radiation-beir-v [Accessed 28 April 2021].

National Council on Radiation Protection and Measurements, 1993. *Recommendations on Limits for Exposure to Ionizing Radiation. NCRP Report No. 116,* Bethesda: NCRP.

National Council on Radiation Protection and Measurements, 1996. *Screening Models for Releases of Radionuclides to Atmosphere, Surface Water and Ground,* Bethesda: NCRP.

National Urban Security Technology Laboratory, 2017. *Radiological Dispersal Device (RDD) Response Guidance: Planning for the First 100 Minutes,* Department of Homeland Security. [Online] Available at: https://www.dhs.gov/publication/st-frg-rdd-response-guidance-planning-first-100-minutes [Accessed 24 June 2022]

NIH, N. I. o. H., 2021. *National Cancer Institute.* [Online] Available at: https://www.cancer.gov/about-cancer/understanding/statistics [Accessed 24 February 2021].

RERF, 2021. *Radiation Effects Research Foundation.* [Online] Available at: https://www.rerf.or.jp/en/ [Accessed 18 February 2021].

Shubayr, N., 2017. *Overview of Radiation Risk and Dose Assessment Models for Radioactively Contaminated Sites and Selected Default Input Parameters.* [Online] Available at: https://epa-prgs.ornl.gov/radionuclides/Overview_of_Rad_R.A.Ms.pdf [Accessed 23 March 2021].

U.S. Department of Health & Human Services, 2020. *Radiation Emergency Medical Management, REMM.* [Online] Available at: https://www.remm.nlm.gov/index.html [Accessed 29 March 2021].

United Nations Scientific Committee on the Effects of Atomic Radiation, UNSCEAR, 2021. *United Nations Scientific Committee on the Effects of Atomic Radiation.* [Online] Available at: https://www.unscear.org/ [Accessed 28 April 2021].

United States Governement, 2019. *EPA, United States Environmental Protection Agency.* [Online]. Available at: https://www.epa.gov/radiation/radiation-regulations-and-laws [Accessed 24 November 2020].

United States Government, 2017. *U.S.NRC, United States Regulatory Commission.* [Online] Available at: https://www.nrc.gov/about-nrc/radiation.html [Accessed 24 November 2020].

Whicker, F. & Kirchner, T., 1987. Pathway: A dynamic food-chain model to predict radionuclide ingestion after fallout deposition. *Health Physics,* 52(6), pp. 717–737.

Appendix A

Answers to Selected Problems

A.1. Chapter 1

3. –12.419 MeV
4. 5.7011 MeV
5. $Q = 4.6787$ MeV; $T_\alpha = 4.6187$ MeV
7. $E(\beta + \nu) = 0.662$ MeV
8. $T_{(1/2)} = 8.045$ days; $T_{mean} = 11.61$ days
9. $A_d(t) = 2.60 \times 10^9\ Bq$
10. $\lambda = 0.08598\ d^{-1}$; $A(t) = 0.89mCi = 3.3 \times 10^7\ Bq$
11. 28780 ng
12. 11.52 days
13. $\beta_{max} = 1600$ keV; Energy of $\gamma_2 = 150$ keV; BR $\gamma_1 = 10\%$; BR $\gamma_3 = 15\%$
14. ~80 particles
15. ~0.3 seconds

A.2. Chapter 2

1. –12.09 eV, 6.448×10^{-27} kg.m/s, $2.92 \times 10^{15}\ s^{-1}$
4. 0.096 keV, 8.194 keV, 330.85 keV, 796.65 keV, 9,750.1 keV, 99,745 keV
5. 0.10 keV, 5.01 keV; 4.87 MeV, 0.24 MeV
6. 0.62 eV
8. 0.35 cm^{-1}
9. 5.5%
10. 10 cm, 0.69077 cm^{-1}
11. 9568.68 cpm
12. a) 3.06×10^4, b) 2.173×105 MeV, c) 2.255×105 MeV, d) 0.082 MeV
13. 1.75 MeV
14. 3.98×10^{-15} m, 6.56×10^{-4} MeV
15. 6 cm; 4615 cm
16. 46.98 MeV/cm, 0.1196 cm; 75.48 MeV/cm, 0.0764 cm; 2.21 MeV/cm, 4.839 cm; 3.785 MeV/cm, 2.905 cm

A.3. Chapter 3

1. 3
4. Same strand: T + C = 0.46; Complementary strand: T = 0.30, C = 0.24, A + G = 0.46.
5. 647 BP

A.4. Chapter 4

(NA)

A.5. Chapter 5

1. 1.2 hour, 0.785 or 0.783
6. (b)

A.6. Chapter 6

1. $D_0 = 2.2$ Gy; $D_Q = 4$ Gy; $n = 9.1$
4. $\alpha \approx 0.10$; $\beta \approx 0.02$
5. $\alpha \approx 0.10$; $\beta \approx 0.02$
7. $\eta = 138\ \text{Gy}^{-1}$
9. 2.7×10^{-8}

A.7. Chapter 7

2. $D_q = 3.0$ Gy; $D_0 = 1.2$ Gy; $n = 130$
8. 960 hours
9. 10 keV/μm
11. 75 MeV
13. $-lnS = \alpha(9.8\ \text{Gy}) + \beta(86.32\ \text{Gy}^2)$

A.8. Chapter 8

1. 0.0261, 0.0098, 3.68
2. 1.5 Gy, 1.9 Gy, 2.2
3. 5.68
4. 1.4–1.6 Gy, 1.8–3.0
7. 0.2423–0.2627, 0.01324–0.0148
9. 1.1 Gy, 2.6 Gy, (NA), 1%

A.9. Chapter 9

(NA)

A.10. Chapter 10

1. 0.00213 or 0.2%; $\cong 1$ or 100%
2. 2318 Gy/h or 38.63 Gy/m
3. 16.8 Gy
4. 18.88

6. 64.7 Gy
7. 103.3 Gy; 74.4 Gy
8. 66.084 Gy
9. 70.23 Gy; 33 Gy
10. 22.17 Gy

A.11. Chapter 11

1. >0.8m
4. 0.1
6. 0.03 cGy
7. 14.40 days
8. 2.16 cGy
9. 1.5×10^{-7} Gy
10. 16.7 mSv; 8.4×10^{-2} cGy; 0.1 per 10,000
11. 0.368 vs 0.00045; (NA)
12. 0.6 per million

Index

www.ingramcontent.com/pod-product-compliance
Lightning Source LLC
LaVergne TN
LVHW081947200926
848484LV00009B/2518
* 9 7 8 9 8 1 1 2 5 7 6 5 0 *